Nonlinear Systems and Complexity

Series Editor
Albert C. J. Luo
Southern Illinois University
Edwardsville, IL, USA

For further volumes:
http://www.springer.com/series/11433

Ricardo Carretero-González • Jesús Cuevas-Maraver
Dimitri Frantzeskakis • Nikos Karachalios
Panayotis Kevrekidis • Faustino Palmero-Acebedo
Editors

Localized Excitations in Nonlinear Complex Systems

Current State of the Art and Future Perspectives

Springer

Editors
Ricardo Carretero-González
Dept. of Mathematics and Statistics
San Diego State University
California
San Diego, USA

Jesús Cuevas-Maraver
Dept. de Física Aplicada I
Escuela Politécnica Superior
University of Sevilla
Sevilla, Spain

Dimitri Frantzeskakis
Dept. of Physics
University of Athens
Athens, Greece

Nikos Karachalios
Dept. of Mathematics
University of the Aegean
Samos, Greece

Panayotis Kevrekidis
Mathematics and Statistics
University of Massachusetts
Massachusetts
Amherst, USA

Faustino Palmero-Acebedo
Dept. de Física Aplicada I
E.T.S.I. Informática
University of Sevilla
Sevilla, Spain

ISSN 2195-9994 ISSN 2196-0003 (electronic)
ISBN 978-3-319-35031-8 ISBN 978-3-319-02057-0 (eBook)
DOI 10.1007/978-3-319-02057-0
Springer Cham New York Heidelberg Dordrecht London

Printed on acid-free paper

Springer is part of Springer Science+Business Media (www.springer.com)

Preface

Localized excitations have been at the heart of developments of nonlinear dynamics (and especially of nonlinear complex systems) during the past few decades. Their names may vary (solitons, instantons, kinks, breathers, vortices, vortex rings, quodons, etc.), yet their relevance is ubiquitous and their importance undisputed not only in a wide variety of mathematical and theoretical developments but also in a broad range of applications. The latter constitute an ever-expanding list including, but certainly not limited to, Bose-Einstein condensates in atomic physics, optical fibers, waveguide arrays and photorefractive systems in nonlinear optics, Josephson junctions and arrays thereof, molecular crystals, layered silicates, ferromagnetic and antiferromagnetic materials, granular crystals, micromechanical cantilever or simple torsion pendulum arrays, electrical transmission lines, and nematic liquid crystals. Additionally, in many other cutting edge areas, they have been conjectured to not only emerge but also play a critical role in the dynamics and the explanation of experimental observations. Such fields include graphene nanoribbons, denaturation of the DNA double strand, atomic ejection and defect migration in crystals, and protein folding. It is thus clear that this is a subject which is fundamental for both theory and applications, and the evaluation of its state-of-the-art and the identification of the challenges that lie ahead are of paramount importance. Clearly, nonlinear science matures and continues to redefine itself, playing an increasingly important and ever-expanding role within the twenty-first century.

The 2nd Conference on "Localized Excitations in Nonlinear Complex Systems" (LENCOS'12) was organized with the aim of bringing together experimental and theoretical physicists, chemists, biologists, and applied mathematicians working in the broad field of nonlinear localized modes. This workshop was the second installment in the series, following up on the successful first conference that took place 3 years earlier (LENCOS'09). The latter meeting provided the opportunity for a special volume on the themes of the workshop which appeared in the journal Discrete and Continuous Dynamical Systems – Series S (DCDS-S). Despite its short history, LENCOS has garnered considerable visibility and is already a widely appreciated venue for interdisciplinary exchanges and the promotion of research in this multifaceted and diverse thematic area. Both installments were spearheaded by the Nonlinear Physics Group of the University of Sevilla, which had also organized some previous relevant workshops, namely, "Moving Breathers in Nonlinear Lattices" (2003), "Macroscopic Effects of Nonlinear Excitations" (2004), "New Horizons in Stochastic Complexity" (2004), and "Nonlinear Excitations: Theory and Experiments" (2005). The LENCOS'09 Conference was attended by approximately 65 delegates from Europe, America, and Asia, while this number grew to about 100 (and contained also delegates from Africa and Oceania) in 2012. Both meetings have featured a vibrant mix of senior and junior researchers, and numerous productive interactions have sprung thereafter as a result of the exchanges taking place in Sevilla.

The organizing committee of the 2nd LENCOS meeting felt that given the growth in scope and number of participants and of themes represented, it would be suitable on the occasion of the 2nd LENCOS meeting to assemble a wide range of state-of-the-art (nonlinear) science contributions in the present volume. In so doing, we also conscientiously decided to target a broader set of potential contributors than merely the conference delegates. Hence, we advertised the corresponding volume more broadly to attract an as rich and diverse set of topics and authors as possible. The resulting submissions were subjected to a full refereeing process, involving at least two referees for each paper. At this point, it is appropriate for us to take the opportunity to thank all referees for their timely, careful, and valuable work and contribution to the quality of the resulting volume. We also want to thank all authors both for submitting the relevant works and for taking care of all remarks by the referees and respecting the imposed timelines. We sincerely hope that the resulting outcome will be one that will be of value to the nonlinear waves and localized modes' community, and will serve as a reference point for numerous future developments in this field.

In what follows, we give a brief overview of the topics that are represented in this volume, placing them in some rough, broad thematic categories accordingly.

Nonlinear Schrödinger Models: Continuum and Discrete. As may be natural to expect, the nonlinear Schrödinger equation (NLS) and its discrete counterpart (DNLS) have a considerable representation in the contents of this volume. The work of Achilleos, Kevrekidis, Frantzeskakis, and Carretero-González considers the context of solitary waves and vortices (and a novel type of excitation entitled a "ghost" state) in both one- and two-dimensional installments of the topic of nonlinear $\mathcal{PT}$-symmetric media; this theme has attracted considerable attention recently. On the other hand, Alfimov and Avramenko offer significant insights on the problem of complete description of nonlinear states in an NLS model with a periodic potential using encodings based on bi-infinite sequences of N-symbols. Additionally, Lepri and Casati examine the case of a nonlinear dimer embedded in an otherwise linear lattice, in a one-dimensional setting; they observe and explain the surprising feature of nonreciprocal wave propagation when wave packets scatter from the two different directions. Finally, James and Starosvetsky, motivated by a problem in granular crystals (the so-called cradle problem), examine a reduced equation, the so-called discrete-p-Schrödinger equation (which is an interesting variant of the DNLS); the authors find, through both map and quasi-continuum approaches, different analytical and numerical approximations to its solutions.

Bose-Einstein Condensates (BECs) and Nonlinear Optics. Many of the works in the volume concerned themselves with areas of application of nonlinear models, where localized excitations may arise. Among the most canonical, such areas are notably atomic physics and nonlinear optics. In the context of the former, the contribution of Balaž and Nicolin examines atomic condensates with attractive interatomic interactions under the influence of a periodic drive of the scattering length; the authors infer analytically and numerically that a resonant drive may lead the BEC into a fragmented state. On the other hand, Salasnich tackles the context of ultracold Fermi gases, examining the notion of the contact intensity, as well as using a hydrodynamic approach to explore their sound waves, response functions, and structure factors. Gerdjikov and Todorov consider the interaction of arrays of solitons (so-called N-soliton configurations) in a system of coupled NLS equations of the Manakov type; they consider, in particular, the influence of external potentials of the periodic, quadratic, and quartic type, with a view towards applications in BECs. In their work, in turn, Li, Pang, and Malomed study nonlinear localized modes trapped in a rotating ring with a modulated strength of the nonlinearity (both in a sign-definite and in a sign-changing case); the results are applicable both in BECs confined in a toroidal trap and in the propagation of light in a twisted pipe waveguide. Finally, Wang and Aceves touch upon propagation of light in two-dimensional plasmonic arrays by considering the specific setup of two-dimensional dielectric/metallic waveguides; the authors report a variety of results concerning localization, solitary wave formation, and dynamical instabilities.

Nonlinear Dynamical Lattices. Another subject that was broadly represented in the volume was that of lattice dynamical systems with a rather diverse array of associated contributions. In particular, Velarde, Ebeling, and Chetverikov consider localized excitations and anisotropic directional ordering in a two-dimensional lattice model of cuprate layers. Cisneros-Ake, Minzoni, and Velarde propose a variational approximation towards the study of electron trapping by soliton-like travelling excitations for one-dimensional lattices, emphasizing the case of an initially delocalized electron. Zolotaryuk presents a study of underdamped and strongly discrete sine-Gordon lattice equation in the presence of an oscillating force, considering the mode-locked kink states. Finally, Sánchez-Morcillo, Jiménez, Chaline, Bouakaz, and Dos Santos offer a tantalizing set of both theoretical and experimental observations suggesting deep connections between the spatiotemporal dynamics of a ring of coupled pendula (and their normal oscillation modes) with the dynamics of drops or bubbles.

Quodons and Muscovite Mica. A topic that can also be considered as a branch of nonlinear lattice dynamics, yet it received considerable attention in this volume in its own right and thus warrants a separate mention, concerns the study of quodons and related excitations in muscovite mica. In their contribution, Dubinko and Shapovalov develop a rate theory of the radiation-induced precipitation in solids as modified by the non-equilibrium fluctuations from the gas of quodons produced by irradiation. On the other hand, Russell examines the long-range propagation of multiple-kink pulses in a crystal as a result of the energy gain from the crystal, bearing in mind as an application example the nuclear scattering of relativistic muons in muscovite crystals. Finally, Archilla, Kosevich, Jiménez, Sánchez-Morcillo, and García-Raffi present a model bearing supersonic kinks in Coulomb lattices. In their case, this is also intended as a prototypical model of layered silicates such as muscovite mica where the potassium ions form a two-dimensional lattice between silicate layers.

Magnetic Metamaterials. Two contributions in this volume addressed the subject of localized excitations that can emerge in magnetic metamaterials; hence we cluster them in the final coherent thematic entity presented herein. In particular, Diblík, Fečkan, Pospíšil, Rothos, and Susanto examine the potential formation of travelling waves in nonlinear magnetic metamaterials by considering a one-dimensional discrete array of nonlinear resonators. On the other hand, Pelinovsky and Rothos explore the discrete Klein-Gordon-type model originally derived by Eleftheriou, Lazarides, and Tsironis, obtaining a general criterion for the spectral stability of the multi-site discrete breathers in that model.

Other Nonlinear Models. Naturally, there were also a number of contributions that do not fit in a definitive way in any of the above categories. In this class, we mention the following: (i) the examination of small nonlinear stochastic arrays via cumulant expansions and numerical simulations of Morillo, Gómez-Ordóñez, and Casado; (b) the study of shape-invariant solitons in nonlocal models of nematic liquid crystals and the influence of noise in their breathing (under weak noise) and even their destruction (by strong noise) of Belić, Petrović, Strinić, and Aleksić; (c) the study of conservation laws for a family of reaction-diffusion-convection equations and the presentation of the notion of weak self-adjointness by Bruzón, Gandarias, and De la Rosa; and finally (d) the study of localized excitations in nonlinear complex biological systems as induced by high-density green photons in the contribution of Comorosan, Polosan, Apostol, Popescu, Farcanasu, Paslaru, and Ionescu.

San Diego, USA	R. Carretero-González
Sevilla, Spain	J. Cuevas
Athens, Greece	D.J. Frantzeskakis
Samos, Greece	N.I. Karachalios
Amherst, USA	P.G. Kevrekidis
Sevilla, Spain	F. Palmero

Contents

Part I Nonlinear Schrödinger Models: Continuum and Discrete

List of Contributors

Alejandro B. Aceves Department of Mathematics, Southern Methodist University, Dallas, USA

V. Achilleos Department of Physics, University of Athens, Athens, Greece

Najdan B. Aleksić Institute of Physics, University of Belgrade, Belgrade, Serbia

G.L. Alfimov National Research University of Electronic Technology, Moscow, Russia

Juan F.R. Archilla Grupo de Física No Lineal, Departamento de Física Aplicada I, ETSI Informática, Universidad de Sevilla, Sevilla, Spain

Marian Apostol Department of Theoretical Physics, Institute of Nuclear Physics, Magurele, Romania

A.I. Avramenko National Research University of Electronic Technology, Moscow, Russia

Antun Balaž Scientific Computing Laboratory, Institute of Physics Belgrade, University of Belgrade, Belgrade, Serbia

Milivoj R. Belić Texas A&M University at Qatar, Doha, Qatar

A. Bouakaz Unité Mixte de Recherche "Imagerie et Cerveau", INSERM U930 – Université François Rabelais, Tours Cedex, France

M.S. Bruzón Departamento de Matemáticas, Universidad de Cádiz, Puerto Real, Cádiz, Spain

R. Carretero-González Nonlinear Dynamical Systems Group, Department of Mathematics and Statistics, Computational Science Research Center, San Diego State University, San Diego, USA

José M. Casado Facultad de Física, Área de Física Teórica, Universidad de Sevilla, Sevilla, Spain

Giulio Casati Center for Nonlinear and Complex Systems, Università degli Studi dell'Insubria, Como, Italy
Istituto Nazionale di Fisica Nucleare, Sezione di Milano, Milan, Italy

J. Chaline Unité Mixte de Recherche "Imagerie et Cerveau", INSERM U930 – Université François Rabelais, Tours Cedex, France

Alexander P. Chetverikov Department of Physics, Saratov State University, Saratov, Russia

Luis A. Cisneros-Ake Department of Mathematics, ESFM, Instituto Politécnico Nacional, Unidad Profesional Adolfo López Mateos, Mexico City, México

Sorin Comorosan Interdisciplinary Research Group, Romanian Academy, Bucharest, Romania

R. de la Rosa Universidad de Cádiz, Cádiz, Spain

J. Diblík Faculty of Electrical Engineering and Communication, Department of Mathematics, Brno University of Technology, Brno, Czech Republic

S. Dos Santos Unité Mixte de Recherche "Imagerie et Cerveau", INSERM U930 – Université François Rabelais, Tours Cedex, France

Volodymyr Dubinko NSC Kharkov Institute of Physics and Technology, Kharkov, Ukraine

Werner Ebeling Institut für Physik, Humboldt-Universität Berlin, Berlin, Germany

Ileana Farcasanu Department of Chemistry, University of Bucharest, Bucharest, Romania

M. Fečkan Department of Mathematical Analysis and Numerical Mathematics, Comenius University, Bratislava, Slovakia
Mathematical Institute of Slovak Academy of Sciences, Bratislava, Slovakia

D.J. Frantzeskakis Department of Physics, University of Athens, Athens, Greece

M.L. Gandarias Departamento de Matemáticas, Universidad de Cádiz, Puerto Real, Cádiz, Spain

Luís M. García-Raffi Instituto de Matemática Pura y Aplicada, Universidad Politécnica de Valencia, Valencia, Spain

Vladimir S. Gerdjikov Institute for Nuclear Research and Nuclear Energy, Bulgarian Academy of Sciences, Sofia, Bulgaria

José Gómez-Ordóñez Facultad de Física, Área de Física Teórica, Universidad de Sevilla, Sevilla, Spain

Elena Ionescu Department of Basic Research, Fundeni Institute, Bucharest, Romania

Guillaume James Laboratoire Jean Kuntzmann, Université de Grenoble and CNRS, Grenoble Cedex 9, France
INRIA, Bipop Team-Project, ZIRST Montbonnot, Saint Ismier, France

N. Jiménez Instituto de Investigación para la Gestión, Integrada de las Zonas Costeras, Universitat Politécnica de Valencia, Grao de Gandia, Spain

P.G. Kevrekidis Department of Mathematics and Statistics, University of Massachusetts, Amherst, USA

Yuriy A. Kosevich Semenov Institute of Chemical Physics, Russian Academy of Sciences, Moscow, Russia

Stefano Lepri Consiglio Nazionale delle Ricerche, Istituto dei Sistemi Complessi, Sesto Fiorentino, Italy

Yongyao Li Department of Applied Physics, South China Agricultural University, Guangzhou, China

Boris A. Malomed Faculty of Engineering, Department of Physical Electronics, School of Electrical Engineering, Tel Aviv University, Tel Aviv, Israel

Antonmaria A. Minzoni FENOMEC, Department of Mathematics and Mechanics, IIMAS, Universidad Nacional Autónoma de México, Mexico City, México,

Manuel Morillo Facultad de Física, Área de Física Teórica, Universidad de Sevilla, Sevilla, Spain

Alexandru I. Nicolin Horia Hulubei National Institute for Physics and Nuclear Engineering, Magurele-Bucharest, Romania
Faculty of Physics, University of Bucharest, Bucharest, Romania
Faculty of Physics, West University of Timisoara, Timisoara, Romania

Wei Pang Department of Experiment Teaching, Guangdong University of Technology, Guangzhou, China

Liliana Paslaru Department of Biochemistry, University of Medicine, Bucharest, Romania

Dmitry Pelinovsky Department of Mathematics and Statistics, McMaster University, Hamilton, Canada

Milan S. Petrović Institute of Physics, University of Belgrade, Belgrade, Serbia

Silviu Polosan National Institute of Material Physics, Magurele, Romania

Irinel Popescu Department of Basic Research, Fundeni Institute, Bucharest, Romania

M. Pospíšil Faculty of Electrical Engineering and Communication, Centre for Research and Utilization of Renewable Energy, Brno University of Technology, Brno, Czech Republic

V.M. Rothos Faculty of Engineering, Department of Mathematics, Aristotle University of Thessaloniki, Thessaloniki, Greece

F.M. Russell Heriot-Watt University, Edinburgh, UK

Luca Salasnich Dipartimento di Fisica e Astronomia, "Galileo Galilei" and CNISM, Università di Padova, Padova, Italy

V.J. Sánchez-Morcillo Instituto de Investigación para la Gestión, Integrada de las Zonas Costeras, Universitat Politécnica de Valencia, Grao de Gandia, Spain

R. Shapovalov NSC Kharkov Institute of Physics and Technology, Kharkov, Ukraine

Y. Starosvetsky Faculty of Mechanical Engineering, Technion – Israel Institute of Technology, Technion City, Haifa, Israel

Aleksandra I. Strinić Institute of Physics, University of Belgrade, Belgrade, Serbia

H. Susanto School of Mathematical Sciences, University of Nottingham, University Park, Nottingham, UK

M. D. Todorov Department of Applied Mathematics and Computer Science, Technical University of Sofia, Sofia, Bulgaria

Manuel G. Velarde Instituto Pluridisciplinar, Universidad Complutense, Madrid, Spain

Danhua Wang Department of Mathematics, Southern Methodist University, Dallas, USA

Yaroslav Zolotaryuk Bogolyubov Institute for Theoretical Physics, National Academy of Sciences of Ukraine, Kiev, Ukraine

Part I
Nonlinear Schrödinger Models: Continuum and Discrete

Solitons and Their Ghosts in $\mathcal{PT}$-Symmetric Systems with Defocusing Nonlinearities

V. Achilleos, P.G. Kevrekidis, D.J. Frantzeskakis, and R. Carretero-González

Abstract We examine a prototypical nonlinear Schrödinger model bearing a defocusing nonlinearity and Parity-Time ($\mathcal{PT}$) symmetry. For such a model, the solutions can be identified numerically and characterized in the perturbative limit of small gain/loss. There we find two fundamental phenomena. First, the dark solitons that persist in the presence of the $\mathcal{PT}$-symmetric potential are destabilized via a symmetry breaking (pitchfork) bifurcation. Second, the ground state and the dark soliton die hand-in-hand in a saddle-center bifurcation (a nonlinear analogue of the $\mathcal{PT}$-phase transition) at a second critical value of the gain/loss parameter. The daughter states arising from the pitchfork are identified as "ghost states", which are not exact solutions of the original system, yet they play a critical role in the system's dynamics. A similar phenomenology is also pairwise identified for higher excited states, with e.g. the two-soliton structure bearing similar characteristics to the zero-soliton one, and the three-soliton state having the same pitchfork destabilization mechanism and saddle-center collision (in this case with the two-soliton) as the one-dark soliton. All of the above notions are generalized in two-dimensional settings for vortices, where the topological charge enforces the destabilization of a two-vortex state and the collision of a no-vortex state with a two-vortex one, of a

V. Achilleos · D.J. Frantzeskakis
Department of Physics, University of Athens, Panepistimiopolis, Zografos, Athens 157 84, Greece
e-mail: vassosnbi@gmail.com; dfrantz@phys.uoa.gr

P.G. Kevrekidis
Department of Mathematics and Statistics, University of Massachusetts, Amherst, MA 01003-4515, USA
e-mail: kevrekid@math.umass.edu

R. Carretero-González (✉)
Nonlinear Dynamical Systems Group, Department of Mathematics and Statistics, Computational Science Research Center, San Diego State University, San Diego, CA 92182-7720, USA
e-mail: rcarretero@mail.sdsu.sdsu; http://nlds.sdsu.edu/

R. Carretero-González et al. (eds.), *Localized Excitations in Nonlinear Complex Systems*, Nonlinear Systems and Complexity 7, DOI 10.1007/978-3-319-02057-0_1,

one-vortex state with a three-vortex one, and so on. The dynamical manifestation of the instabilities mentioned above is examined through direct numerical simulations.

1 Introduction

Over the past decade, and since they were originally proposed by Bender and co-workers [1,2], systems characterized by $\mathcal{PT}$-symmetric Hamiltonians have become a subject of intense research efforts. The interest in these systems arises from their fundamental property to exhibit real spectra, while non-Hermitian, thus providing an intriguing alternative to standard Hermitian quantum mechanics. In the case of a standard Schrödinger-type Hamiltonian, with a generally complex potential U, the $\mathcal{PT}$ symmetry dictates that the potential satisfies the condition $U(x) = U^*(-x)$, where $(\cdot)^*$ stands for complex conjugation.

Optical systems appear to be ideal settings for the realization of $\mathcal{PT}$-symmetry; this was first proposed in Ref. [3], for an optical slab waveguide with a gain region and a loss region. We note that studies in optical systems with gain and loss were already performed as early as in 1992, in the setting of a twin core nonlinear coupler [4] (of course, at that time, $\mathcal{PT}$-symmetry had not been introduced). Nevertheless, an important development in the study of such models was the work of Refs. [5, 6], where both theoretical and experimental studies on $\mathcal{PT}$-symmetric optical systems were reported. In the optics context, a key element that comes into play in the physics of such systems is nonlinearity. Therefore, the considerations of Refs. [5,6] extended from bright and gap solitons to linear (Floquet-Bloch) eigenmodes in periodic potentials, examining how these coherent structures are affected by the genuinely complex, yet $\mathcal{PT}$-symmetric potentials. More recently, experimental results were reported both in nonlinear optical systems [7, 8] and electronic analogs thereof [9]. These, in turn, have triggered a wide range of theoretical studies on nonlinear lattices with either linear [10–23] or nonlinear [24–26] $\mathcal{PT}$-symmetric potentials and, more recently, on harmonic $\mathcal{PT}$-symmetric potentials [27].

In the above works, numerous features extending from bright solitons to defect modes, and from gap solitons to $\mathcal{PT}$-lattices have been examined. Nevertheless, the consideration of defocusing nonlinearities, and especially of dark solitons has received limited attention; see, e.g., Refs. [27–29]. Another theme that despite its considerable relevance has also been considered only by a few works is that of higher dimensionality. For the latter, the focusing nonlinearity case [5, 6] has been examined, especially so in the context of lattice potentials. Our aim herein is to provide a systematic analysis of $\mathcal{PT}$-symmetric Hamiltonians exhibiting defocusing nonlinearities. In particular, and building on the earlier work of Ref. [29], we give a detailed account of the existence, stability and dynamical properties of the ground state and first few excited states namely dark solitary waves and vortex two-dimensional generalizations thereof. This is done specifically in a prototypical context of the nonlinear Schrödinger type which can, in principle, be relevant both in the case of Bose-Einstein condensates [30], as well as in that of nonlinear optics [31]; see the detailed justification of the real and imaginary parts of the potential below.

Our main findings and their presentation are structured as follows.

- In Sect. 2, we put forth the general model and consider its physical relevance.
- In Sects. 3 and 4, we turn to generic potentials and the features of their ground and excited states in one-dimension. We find that odd excited states (1-soliton, 3-soliton, 5-soliton, etc.) become subject to a symmetry breaking bifurcation. The pitchfork nature of this event is rationalized through the introduction of so-called ghost states which arise from it and which are exact solutions of the steady state but remarkably not ones such of the original full problem. Nevertheless, direct numerical simulations clearly identify the ensuing symmetry breaking and manifest the dynamical role of the ghost waveforms. One more important feature identified is the nonlinear analogue of the $\mathcal{PT}$-phase transition. In particular, between the (unstable, for sufficiently strong gain/loss) saddles of the odd excited states and the centers of the even states (ground state, 2-soliton, 4-soliton, etc.), there is a pairwise collision and disappearance (blue-sky or saddle-center) event. This is strongly reminiscent of the corresponding transition of linear eigenstates of the Hamiltonian originally reported in Ref [1]. Direct numerical simulations are employed in order to identify the evolution dynamics of the unstable solitary wave states and also their dynamics past the $\mathcal{PT}$-phase transition point in Sect. 5. The ghost states are then examined separately in their own right in Sect. 6.
- In Sect. 7, we illustrate how each of the above parts of the picture is generalized in the two-dimensional variant of the relevant model. There, the charge of the vortex states imposes topological constraints enforcing that the ground state may only collide and disappear (in a nonlinear $\mathcal{PT}$-phase transition) with the two-vortex (of opposite charge, namely dipole) state. Similarly, the single vortex and triple (of alternating charge) suffer a saddle-center bifurcation and so on. Prior to these events, a destabilization of the two-vortex (and three-vortex etc.) states arises through a pitchfork event creating ghost vortex states. The latter are illustrated dynamically as well and the structural analogies of the one- and two-dimensional settings are explored both in the statics and in the dynamics. Lastly, Sect. 8 presents a brief summary of our conclusions and a number of potential directions for future study.

2 The Model and Some Analytical Insights

Our model, which can be equally applied to a variety of one-dimensional and even higher dimensional systems is a nonlinear Schrödinger (NLS) equation incorporating a complex potential. This equation, which finds applications in the contexts of nonlinear optics [31] and in the physics of atomic Bose-Einstein condensates [30], is expressed in the following dimensionless form:

$$i\partial_t u = -\frac{1}{2}\partial_x^2 u + |u|^2 u + [V(x) + iW(x)]u, \tag{1}$$

where u is a complex field, denoting the electric field envelope in the context of optics (or the macroscopic wavefunction in BECs), t denotes the propagation distance (or time in BECs), x is the transverse direction, while $V(x)$ and $W(x)$ denote, respectively, the real and imaginary parts of the external potential. For a $\mathcal{PT}$-symmetric Hamiltonian, $V(x)$ and $W(x)$ must be, respectively, an even and an odd function of x, namely:

$$\begin{cases} V(x) = V(-x), \\ W(x) = -W(-x). \end{cases} \tag{2}$$

Physically speaking, in the context of optics, $V(x)$ and $W(x)$ represent, respectively, the spatial profiles of the real and imaginary parts of the refractive index. The above requirements for the parities of $V(x)$ and $W(x)$ can be met in a case where $V(x)$ has, e.g., a parabolic profile and $W(x)$ has an anti-symmetric profile (amounting to equal and opposite gain and loss), as in the experiment of Ref. [8]. In the context of BECs, $V(x)$ represents the external trap (necessary to confine the atoms [30]) and $W(x)$ accounts for a mechanism for injecting and removing particles in equal rates. The requirement for the parity of the trap $V(x)$ can easily be met in the case of e.g., the usual parabolic potential (representing a magnetic trap) or a double-well potential (representing a combination of a magnetic trap with a suitable optical lattice) [30]. On the other hand, $W(x)$ is odd if an equal number of atoms is injected and removed from spatial regions symmetrically located around the trap center. Such a setting was originally proposed in Ref. [32] for a BEC confined in a double-well potential, but no physical mechanism for such a realization was offered (notice that although a localized particle loss can be implemented by an electron beam [33], the realization of a localized gain has not been proposed so far). Furthermore, passive parity-time symmetric analogs of the double well system (featuring only loss in one well and no gain) have also been proposed in the context of the so-called open Bose-Hubbard dimer [34]; see also Refs. [13, 14]. More recently, a very large volume of activity has focused on such double-well potentials in a balanced gain-loss (i.e., $\mathcal{PT}$-symmetric) form; see, for instance, Refs. [35–38]. In such contexts, some of the notions presented below, such as the ghost states and their emergence from suitable bifurcations and dynamical relevance, can be both numerically manifested, as well as analytically demonstrated.

We seek standing wave solutions of Eq. (1) in the form $u = \rho(x)\exp[i\phi(x) - i\mu t]$, where the real functions $\rho(x)$ and $\phi(x)$, and the real constant μ represent, respectively, the amplitude, phase and propagation constant (in optics) or chemical potential (in BECs). Substituting this ansatz into Eq. (1), and separating real and imaginary parts, we obtain the following coupled boundary-value problems (BVPs):

$$\mu\rho = -\frac{1}{2}(\rho_{xx} - \rho\phi_x^2) + \rho^3 + V(x)\rho \tag{3}$$

$$2W(x)\rho^2 = (\rho^2\phi_x)_x, \tag{4}$$

where subscripts denote partial derivatives. For generic potentials $V(x)$ and $W(x)$, the relevant $\mathcal{PT}$-symmetric problem involves solving the BVPs for ρ and ϕ. Notice the critical role of the imaginary part of the potential in dictating the phase profile $\phi(x)$. We will chiefly focus on the case of a real parabolic potential,

$$V(x) = \frac{1}{2}\Omega^2 x^2, \tag{5}$$

with strength Ω, modeling the transverse distribution of the refractive index (or the external trap in BECs) as mentioned above, while the imaginary part $W(x)$ will be considered to be an odd, localized function of space, of spatial width $\ll \Omega^{-1}$. Such a form of $W(x)$ is consistent with the experimental work of Ref. [8] in a nonlinear optics setup and could be relevant to an effective description of atom loss/gain mechanisms in trapped BECs. Our analysis will be general (independent of the particular form of $W(x)$), and we will showcase our results in the case of the following prototypical example:

$$W(x) = \varepsilon x \exp(-x^2/2), \tag{6}$$

where ε is a parameter setting the magnitude of the imaginary potential (a generalization of this model in two-dimensions will be studied in Sect. 7). We should mention that our analytical approximations (see below) can also be applied for other choices, e.g., when $W(x)$ takes the form $W(x) = \varepsilon\,\text{sech}^2(x)\tanh(x)$; in such cases, we have checked that our results remain qualitatively similar to the ones that we will present below.

We conclude this section by noting the following. Below we will study the ground state and excited states of the system (in the form of dark solitons). In that regard, it is relevant to consider the evolution of the physically relevant quantity $N = \int |u|^2 dx$, which represents power in optics or number of atoms in BECs. Employing Eq. (1), it is straightforward to find that dN/dt is governed by the equation:

$$\frac{dN}{dt} = 2\int_{-\infty}^{+\infty} |u|^2 W(x) dx. \tag{7}$$

Thus, since $W(x)$ is odd, it is obvious that the power is conserved as long as the square modulus profile of the ground state or of the excited states remains even. Below, we will show that for genuinely stationary states this is the case, indeed and we will examine the important consequences of Eq. (7) on the bifurcations and dynamics of the system at hand.

3 Ground State and Single Dark Soliton

3.1 Ground State

First, we will study the existence and stability of the most fundamental state of the system, namely the ground state. The latter is sought as a stationary solution of Eq. (1) in the form $u = u_b(x)\exp(-i\mu t)$ (where μ is the propagation constant or the chemical potential in BECs), with the background field u_b obeying the equation:

$$-\frac{1}{2}\partial_x^2 u_b + |u_b|^2 u_b + [V(x) + iW(x)]u_b - \mu u_b = 0. \tag{8}$$

For a sufficiently small imaginary potential, $W(x) = \varepsilon\tilde{W}(x)$ (with $\max\{|\tilde{W}(x)|\} = O(1)$), where $\varepsilon \ll 1$, and when the inverse width Ω^{-1} of $V(x)$ is sufficiently large so that $\Omega \sim \varepsilon$, we may find an approximate solution of Eq. (1) in the Thomas-Fermi (TF) limit [30, 39]. This solution is of the form:

$$u_b(x) = \left[\sqrt{\mu} + f(x)\right]\exp[i\phi(x)], \tag{9}$$

where the amplitude and phase $f(x)$ and $\phi(x)$ (considered to be small, of order ε^2 and ε, respectively) are given by:

$$f(x) = -\frac{1}{2\sqrt{\mu}}\left(V(x) + 2\mathcal{W}^2(x)\right), \tag{10}$$

$$\phi(x) = 2\int \mathcal{W}(x)\,dx. \tag{11}$$

where $\mathcal{W}(x) = \int W(x)dx$, and we have neglected terms of order $O(\varepsilon^3)$ (notice that the integral in Eq. (11) is an indefinite one). Contrary to the conservative case ($\varepsilon = 0$) [30], this TF background is characterized by a density dip located at the center, $x = 0$, and a nontrivial (tanh-shaped) phase distribution; similar features of the ground state were also reported in Ref. [27] (but for an extended gain/loss potential). The density and phase profiles are shown in Fig. 1, where the analytical result (dashed lines) is compared with the numerical one (solid lines); the agreement between the two is excellent. Note that we have checked that the above solution stays close to the numerically found ground state of the system up to the order $\mathcal{O}(\varepsilon^2)$ [29].

The evolution of the density of the TF background indicates that the ground state is stable (see more details in the analysis below). Importantly, a linear stability—Bogoliubov-de Gennes (BdG)—analysis (see, e.g., Ref. [39]) justifies the above argument, showing that the background $u_b(x)$ is indeed stable against small perturbations.

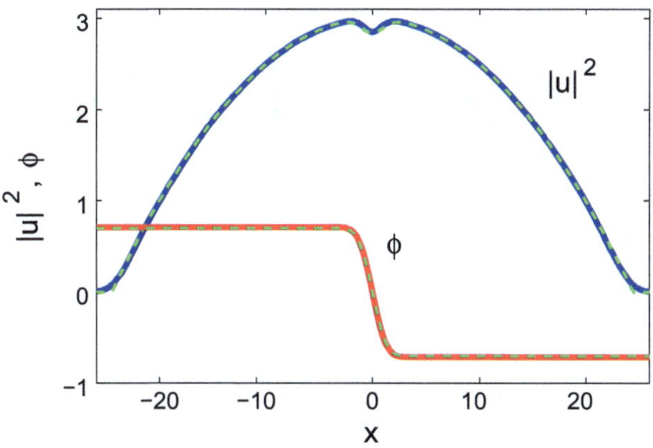

Fig. 1 The density (*solid (blue) line*) and phase (*solid (red) line*) of the numerically obtained TF background compared to the prediction of Eq. (11) (*dashed (green) lines*). The parameters values are: $\mu = 3$, $\Omega = 0.1$ and $\varepsilon = 0.3$

3.2 Single Dark Soliton

Apart from the ground state, excited states of the system—in the form of stationary dark solitons—can also be found numerically, by means of a fixed point algorithm (Newton's method). A pertinent example of the form of a single dark soliton is shown in the top panel of Fig. 2.

To analyze the dynamics of such a single dark soliton, described by function $\upsilon(x,t)$, on top of the TF background, we introduce the product ansatz: $u = u_b(x)\upsilon(x,t)$ into Eq. (1) (i.e., we decompose the solution into a soliton and its background). Then, we employ the expressions (9)–(11) for u_b, and derive the following equation for $\upsilon(x,t)$:

$$i\partial_t \upsilon + \frac{1}{2}\partial_x^2 \upsilon - |u_b|^2 \left(|\upsilon|^2 - 1\right)\upsilon = -\partial_x \ln(u_b)\partial_x \upsilon, \tag{12}$$

which is derived upon dividing with the TF background $u_b(x)$ assuming that $|x| \leq R = \sqrt{2\mu}/\Omega$ (the so-called TF radius). We then simplify the resulting equation for $\upsilon(x,t)$ by keeping only leading-order terms, up to order $\mathcal{O}(\varepsilon^2)$ in u_b (recall that the function $f(x)$ is of order $\mathcal{O}(\varepsilon^2)$). To this end, using the scale transformations $t \to \mu t$ and $x \to \sqrt{\mu}x$, we obtain the following perturbed NLS equation:

$$i\partial_t \upsilon + \frac{1}{2}\partial_x^2 \upsilon + \upsilon(1 - |\upsilon|^2) = P(\upsilon), \tag{13}$$

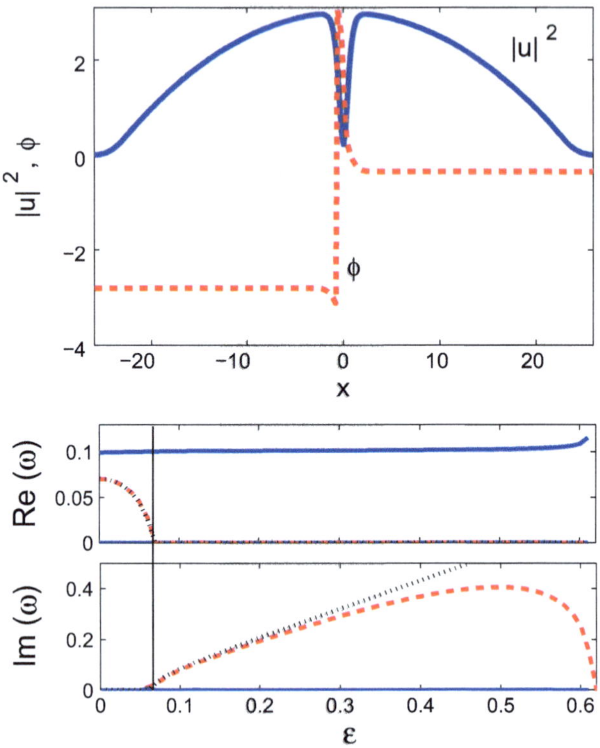

Fig. 2 *Top panel*: The density (*solid (blue) line*) and phase (*dashed (red) line*) of the first excited (single dark soliton) state for $\varepsilon = 0.3$. *Middle* and *bottom panels*: The linear spectrum of the single dark soliton branch: *middle (bottom) panel* shows the real (imaginary) part of the lowest eigenfrequencies ω, with respect to the amplitude ε. The *dashed (red) line* depicts the mode ω_α, which coincides with the anomalous mode for $\varepsilon = 0$; while the *dotted (black) line* is the analytical result of Eq. (21). The *vertical line* shows the point $\varepsilon_{\rm cr}^{(1)}$, where the bifurcation emerges. The parameters used are: $\mu = 3$ and $\Omega = 0.1$

where the functional perturbation $P(\upsilon)$ is given by

$$P(\upsilon) = \mu^{-2}\left[(1-|\upsilon|^2)\upsilon\left(V+2\mathcal{W}^2\right)+\upsilon_x\left(\frac{1}{2}V_x-2(W-i)\mathcal{W}\right)\right]. \quad (14)$$

Applying the perturbation theory for dark solitons devised in Ref. [40] (see also Ref. [39]), we seek a solution of Eq. (13) in the form of the dark soliton of the unperturbed system ($P(\upsilon) = 0$):

$$\upsilon(x,t) = \cos\varphi(t)\tanh\xi + i\sin\varphi(t). \quad (15)$$

Here, $\xi \equiv \cos\varphi(t)\,[x - x_0(t)]$, while the unknown, slowly-varying functions $\varphi(t)$ and $x_0(t)$ represent, respectively, the phase ($|\varphi| \le \pi/2$) and center of the soliton, while $dx_0/dt = \sin\varphi$ accounts for the soliton velocity. In the framework of the adiabatic approximation, the perturbation-induced evolution equations for $x_0(t)$ and $\varphi(t)$ read:

$$\frac{dx_0}{dt} \approx \varphi(t), \tag{16}$$

$$\frac{d\varphi}{dt} \approx -\frac{1}{2}\partial_x V(x) - \int \operatorname{sech}^4(\xi)\left[\tanh(\xi)\mathcal{W}^2(x) + W(x)\mathcal{W}(x)\right]dx, \tag{17}$$

where we have assumed almost black solitons with sufficiently small phase angles $\phi(t)$, such that $\sin\phi \approx \phi$; such an assumption is relevant for solitons moving close to the trap center ($x \approx 0$), an assumption already used in the derivation of Eq. (17)—see more details and discussion in Ref. [39]. Thus, generally, for a given imaginary potential $W(x)$, and by calculating the above integral, we can derive an equation of motion for the dark soliton center x_0 in the form:

$$\frac{d^2x_0}{dt^2} = -\frac{\partial V_{\text{eff}}}{\partial x_0}, \tag{18}$$

where V_{eff}, is the effective potential felt by the single soliton.

In the case where the imaginary part of the potential takes the form of Eq. (6), we numerically calculate the integral in Eq. (17) and find that the effective potential can be approximated as:

$$V_{\text{eff}}(x_0) = \frac{1}{2}\left(\frac{\Omega}{\sqrt{2}}\right)^2 x_0^2 + \frac{\varepsilon^2 c_1}{4c_2}\operatorname{sech}^4(c_2 x_0). \tag{19}$$

where $c_1 \approx 1.9$ and $c_2 \approx 0.6$. Notice that the effective potential contains essentially two contributions, the first from the external parabolic trap (as is the case in the context of BECs [30, 39, 41]), and the second from the presence of the imaginary potential $W(x)$.

To examine the stability of the equilibrium at $x_0 = 0$ (i.e., the stability of a quiescent black soliton located at the trap center), we Taylor expand the effective potential of Eq. (19) around $x_0 = 0$, and find—to leading order—the following equation of motion for the soliton center [29]:

$$\frac{d^2x_0}{dt^2} = -\omega_{\text{osc}}^2\, x_0, \tag{20}$$

$$\omega_{\text{osc}}^2 \approx \left(\frac{\Omega}{\sqrt{2}}\right)^2 - 1.14\varepsilon^2. \tag{21}$$

Equation (21) implies that if the amplitude ε of $W(x)$ is less than a critical value $\varepsilon_{\text{cr}}^{(1)} \approx 0.66\,\Omega$, the soliton performs oscillations in the complex potential with frequency ω_{osc}; on the other hand, if $\varepsilon > \varepsilon_{\text{cr}}^{(1)}$ the soliton becomes unstable. In fact, the prediction has the quintessential characteristics of a pitchfork bifurcation. For $\varepsilon < \varepsilon_{\rm cr}^{(1)}$, the relevant potential is monostable at $x_0 = 0$, while beyond the critical point, two symmetric minima arise (in this effective potential picture) and the center position becomes the saddle point that separates them.

The above prediction about the change of stability of the critical point at $x_0 = 0$ has been confirmed numerically, both by means of direct simulations and by employing a BdG analysis. The latter reveals that the considered stationary dark soliton is characterized by a mode ω_α (initially coinciding with the anomalous mode [39, 41] for $\varepsilon = 0$), which is real for $\varepsilon < \varepsilon_{\rm cr}^{(1)}$ (in this case, $\omega_\alpha = \omega_{\rm osc}$), and it becomes imaginary for $\varepsilon > \varepsilon_{\rm cr}^{(1)}$, thus signaling the onset of the spontaneous symmetry breaking (SSB) instability of the dark soliton (which should be expected to displace the soliton from the trap center).

The dependence of ω_α on the amplitude ε of the imaginary potential $W(x)$, as found by the BdG analysis, is illustrated in the middle and bottom panels of Fig. 2. The mode, corresponding to the oscillatory motion of the dark soliton, is the first non-zero mode in the real part of the spectrum, indicated with a dashed (red) line—cf. middle panel of Fig. 2. We observe that it initially moves towards the spectral plane origin, and past the critical point, $\varepsilon_{\rm cr}^{(1)}$ (cf. vertical line), ω_α becomes imaginary, and thus the soliton becomes unstable. The corresponding pair of imaginary eigenfrequencies is shown in the bottom panel of Fig. 2. The solid (black) line in both panels, shows the analytical result of Eq. (21). An excellent agreement between the analytical prediction and the BdG numerical result is observed even beyond the SSB bifurcation point, while for larger values of the parameter ε, perturbation theory fails and, as expected, the agreement becomes worse.

Importantly, for large values of the parameter ε, the imaginary eigenfrequencies start moving towards the spectral plane origin and collide with it at a second critical point, $\varepsilon_{\rm cr}^{(2)}$, after which the branch is terminated. To better understand how the branch ceases to exist, we first observe (bottom panel of Fig. 1) that the density profile of the soliton becomes increasingly shallower (i.e., more "grey") as ε grows and the second critical point is approached. This is due to the development of an increasingly strong *even* imaginary part of the solution. Furthermore, the stable background solution $u_b(x)$ (cf. Eqs. (11) and top panel of Fig. 1) develops an *odd* imaginary part resembling a (progressively darker) grey soliton. Finally, at $\varepsilon = \varepsilon_{\rm cr}^{(2)}$, the profiles of these modes become identical and disappear in a blue-sky bifurcation through their collision. In this saddle-center bifurcation, the dark soliton plays the role of the saddle, while the ground state of the system is the center. This observation leads to the conclusion that there is a "phase transition" where the system loses its $\mathcal{PT}$-symmetry, since the nonlinear problem eigenvalues become complex; this is in direct analogy with the so-called linear $\mathcal{PT}$ "phase transition" as per the terminology of Ref. [1]. This will be discussed in more detail below.

However, we should highlight here that one more nontrivial question persists in the context of the single dark soliton. Past the critical point $\varepsilon_{\rm cr}^{(1)}$, it is expected that the instability of the soliton at the center will provide us with "daughter states" which are spontaneously symmetry broken dark solitons centered at a positive or at a negative value of x_0. For real chemical potentials/ propagation constants, such states can, nevertheless, not be identified. This appears to be inconsistent with the pitchfork nature of the bifurcation and constitutes a point to which we will return after we visit the dark soliton evolution to appreciate the dynamical implications of this instability.

4 Multiple Dark Soliton States and Nonlinear $\mathcal{PT}$ Phase Transitions

In this section, we will study higher excited states, in the form of multiple dark solitons. Such states are known to exist in the presence of $V(x)$, and have been studied extensively in the context of BECs [39, 41, 42]. Here, we will consider them in the presence of the imaginary part of the potential $W(x)$. Particularly, we will focus on the two- and three-dark-soliton states; qualitatively similar results can also be obtained for higher excited states.

4.1 Two-Dark-Soliton State

The density and phase of a two-dark-soliton-state are shown in the top panel of Fig. 3. Similarly to the ground state branch (see Fig. 1), we observe that a density dip is formed at the center, which becomes deeper as ϵ is increased. Additionally, we observe that the two solitons are located away from the center, at some finite equilibrium distance $\pm X$. This equilibrium corresponds to the situation where the repulsive force between the solitons and the "expulsive" (with respect to the trap center) effect, for sufficiently large ε, of the imaginary part of the potential $W(x)$ analyzed in the last section is counter balanced by the confining nature of the real part of the potential $V(x)$ [39, 41].

Below we will study analytically how the presence of the imaginary potential $W(x)$ modifies the equilibrium distance X characterizing the stationary two-soliton-state, and also study small-amplitude oscillations of the solitons around their fixed points. To obtain an approximate equation of motion (analogous to Eq. (21)) for the two soliton centers, we assume that each of the two solitons feels the effective external potential of Eq. (19) and, at the same time, the solitons interact with each other. We assume that the pertinent interaction potential, $V_{\rm int}$, for two (almost black) solitons, located at position x_1 and x_2, has the form obtained from the exact two-soliton solution of the underlying unperturbed NLS system (see details in Ref. [41]), namely:

$$V_{\rm int}(x_1, x_2) = \frac{\mu}{2\sinh^2[\sqrt{\mu}(x_2 - x_1)]}. \tag{22}$$

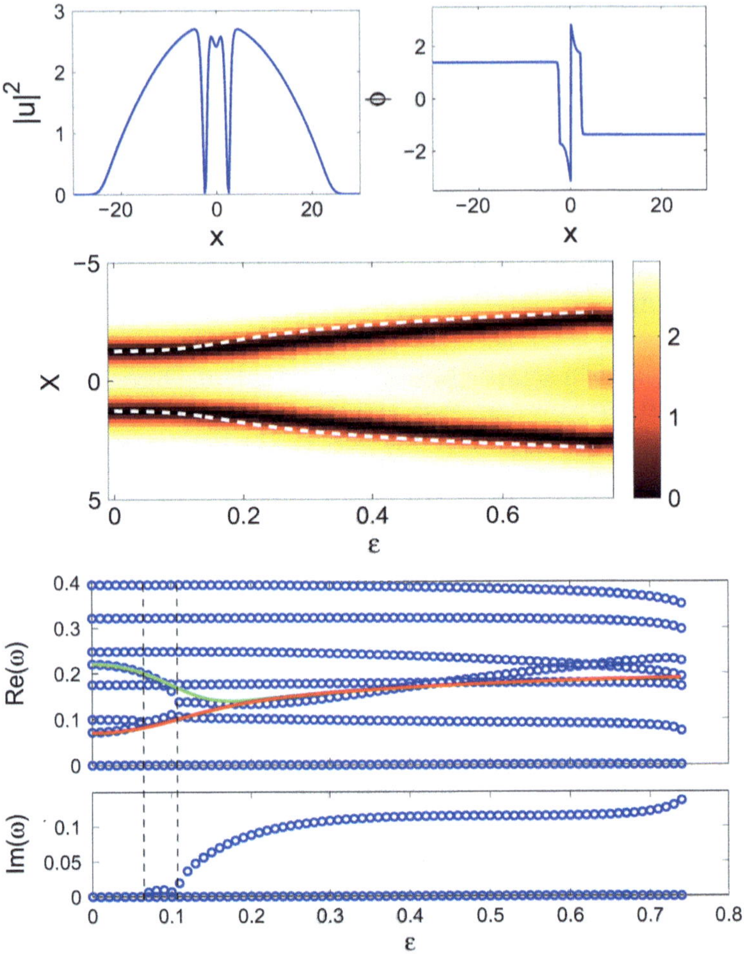

Fig. 3 *Top panels*: the density (*left*) and phase (*right*) profiles of a stationary two-dark-soliton state. *Middle panel*: the dependence of the equilibrium configuration of the two-soliton solution as a function of ε; *dashed (white) lines* show the theoretical predictions for the position of the centers of the waves, cf. fixed points of Eqs. (24) and (25). *Bottom panels*: the linear spectrum of the two soliton branch, namely the real (*top*) and imaginary (*bottom*) parts of the eigenfrequencies as functions of ε. *Solid red* and *green lines*, correspond to the analytical result of Eq. (30). The value of these modes at $\varepsilon = 0$, corresponds to the first and second anomalous modes of the Hamiltonian case

The equation of motion for the two soliton centers (x_1 and x_2) can now be obtained from the Lagrangian:

$$L = \frac{1}{2}\dot{x_1}^2 + \frac{1}{2}\dot{x_2}^2 - V_{\text{eff}}(x_1) - V_{\text{eff}}(x_2) - 2\, V_{\text{int}}(x_1, x_2), \tag{23}$$

where the external effective potential V_{eff} is given in Eq. (19). Using the Euler-Lagrange equations, and assuming that the two solitons are far away from each other, i.e., $|x_2 - x_1| \gg 0$, we approximate their mutual interaction term in Eq. (23) by its exponential asymptote and obtain the following equations of motion:

$$\ddot{x}_1 = -8\mu^{3/2}e^{-2\sqrt{\mu}(x_2-x_1)} - \tilde{\omega}^2 x_1 + \varepsilon^2 c_1 \text{sech}^4(c_2x_1)\tanh(c_2x_1), \tag{24}$$

$$\ddot{x}_2 = 8\mu^{3/2}e^{-2\sqrt{\mu}(x_2-x_1)} - \tilde{\omega}^2 x_2 + \varepsilon^2 c_1 \text{sech}^4(c_2x_2)\tanh(c_2x_2). \tag{25}$$

where $\tilde{\omega} = \Omega/\sqrt{2}$ is the oscillation frequency of the single dark soliton in the absence of $W(x)$ [39]. The fixed points $X_2 = -X_1 = X$ of the above system can be found numerically, by setting the left-hand side equal to zero, and employing a fixed-point algorithm.

Next, we can study the stability of these equilibrium positions X_1 and X_2, by considering small deviations, η_1 and η_2, around them. By Taylor expanding the nonlinear terms, and keeping only linear terms in η_1 and η_2, we derive the following set of linearized equations of motion:

$$\ddot{\eta}_1 = 16\mu^2 e^{-4\sqrt{\mu}X}(\eta_2-\eta_1) - \tilde{\omega}^2\eta_1 + \varepsilon^2 c_1c_2\left[5\,\text{sech}^6(c_2X) - 4\,\text{sech}^4(c_2X)\right]\eta_1, \tag{26}$$

$$\ddot{\eta}_2 = -16\mu^2 e^{-4\sqrt{\mu}X}(\eta_2-\eta_1) - \tilde{\omega}^2\eta_2 + \varepsilon^2 c_1c_2\left[5\,\text{sech}^6(c_2X) - 4\,\text{sech}^4(c_2X)\right]\eta_2. \tag{27}$$

Let us now consider the normal modes of the linearized system and seek solutions of the form $\eta_i = \eta_i^0 e^{i\omega t}$, where ω is the common oscillation frequency of both dark solitons. Then, substituting this ansatz into Eqs. (26) and (27), we rewrite the equations of motion as a matrix eigenvalue equation, namely:

$$-\omega^2\boldsymbol{\eta} = \begin{pmatrix} -\tilde{\omega}^2 + f(X) & 16\mu^2 e^{-4\sqrt{\mu}X} \\ 16\mu^2 e^{-4\sqrt{\mu}X} & -\tilde{\omega}^2 + f(X) \end{pmatrix}\boldsymbol{\eta}. \tag{28}$$

where $\boldsymbol{\eta} = (\eta_1, \eta_2)^T$ and the function $f(X)$ is given by:

$$f(X) = -16\mu^2 e^{-4\sqrt{\mu}X} + \varepsilon^2 c_1c_2\left[5\,\text{sech}^6(c_2X) - 4\,\text{sech}^4(c_2X)\right]. \tag{29}$$

To this end, it is possible to obtain from Eq. (28) the two characteristic frequencies

$$\omega_{1,2} = \sqrt{\tilde{\omega}^2 + f(X) \pm 16\mu^2 e^{-4\sqrt{\mu}X}}, \tag{30}$$

where ω_1 (ω_2) correspond to the in-phase (out-of-phase) oscillations of the two solitons [39,41].

The validity of the above analytical result can now be compared with the corresponding numerical result obtained by the numerical existence theory and the BdG analysis. The latter provides the equilibrium configuration, as well as the real/imaginary parts of the eigenfrequencies as functions of the imaginary potential strength ε, as shown in the middle and bottom panels of Fig. 3, respectively. In the middle panel, we can observe a very good agreement between the positions of the theoretically predicted solitary wave centers (cf. the fixed points of Eqs. (24) and (25)) and their numerically exact counterparts.

For the linearization analysis of the bottom panels, we observe the following. In the Hamiltonian case, there exist two anomalous modes (and more generally, n anomalous modes for a n-soliton state—see, e.g., Ref. [39]). The first, which is the lowest nonzero mode in the spectrum, corresponds to the in-phase oscillation frequency of the two dark solitons, while the second being the fourth finite eigenfrequency for our parameters, corresponds to out-of-phase oscillations. The two-soliton branch is initially stable up to a critical point (indicated by the left vertical line), where the first finite mode collides with the dipole (alias Kohn) mode [39]; this mode is located, for $\varepsilon = 0$, at $\mathrm{Re}(\omega) = \Omega = 0.1$. This collision results in the emergence of an instability, which is identified by the appearance of an imaginary part of the eigenfrequency (see the bottom panel of Fig. 3). As ε is increased, this mode is eventually *detached* from the dipole mode and this instability band ceases to exist for a very narrow parametric interval. However, the quadrupole mode (located, for $\varepsilon = 0$, at $\mathrm{Re}(\omega) \approx \sqrt{3}\Omega = 0.17$), which has been scattered at about $\varepsilon = 0.08$, by the branch just above, now collides with the detached mode (right vertical line). Upon the latter collision, a new (and persistent hereafter) second instability emerges (see the bottom panel of Fig. 3). This way, another quartet of complex eigenfrequencies is created, indicating that the two-soliton state becomes unstable with a growth rate that continues to increase as the gain/loss parameter ε increases.

The analytical result for the two eigenfrequencies $\omega_{1,2}$ (cf. Eq. (30) and solid red and green lines in Fig. 3) essentially coincides with the two branches initialized at the two anomalous modes of the Hamiltonian case, for sufficiently small ε; however, for larger ε, the analytical result is somewhat less accurate. The observed discrepancy is, at least in part, due to the fact that our approximate result is based on a particular ansatz of well-separated (thus weakly-interacting), almost black solitons. However, observing the density profile of the two soliton state (top panel of Fig. 3), one can see that the small density dip in the center (which is not included in our ansatz) is certainly affecting the interaction between the two solitons; importantly, this dip becomes larger as ε is increased. Thus, for relatively large values of ε, our perturbative approach may be expected to be of lesser value.

Additionally we note that, as in the case of the single-soliton-state, for values $\epsilon > 0.72$ no stationary two-soliton-state could be identified. This is again a by-product of the nonlinear analogue of the $\mathcal{PT}$-phase transition. More specifically, the two-soliton branch collides with the three-soliton one and they pairwise annihilate, as is occurring to higher (linear) eigenvalue pairs in the linear $\mathcal{PT}$-phase transition of Ref. [1]. We illustrate the generality of this effect (and the relevant cascade of nonlinear eigenvalue collisions) in more detail below.

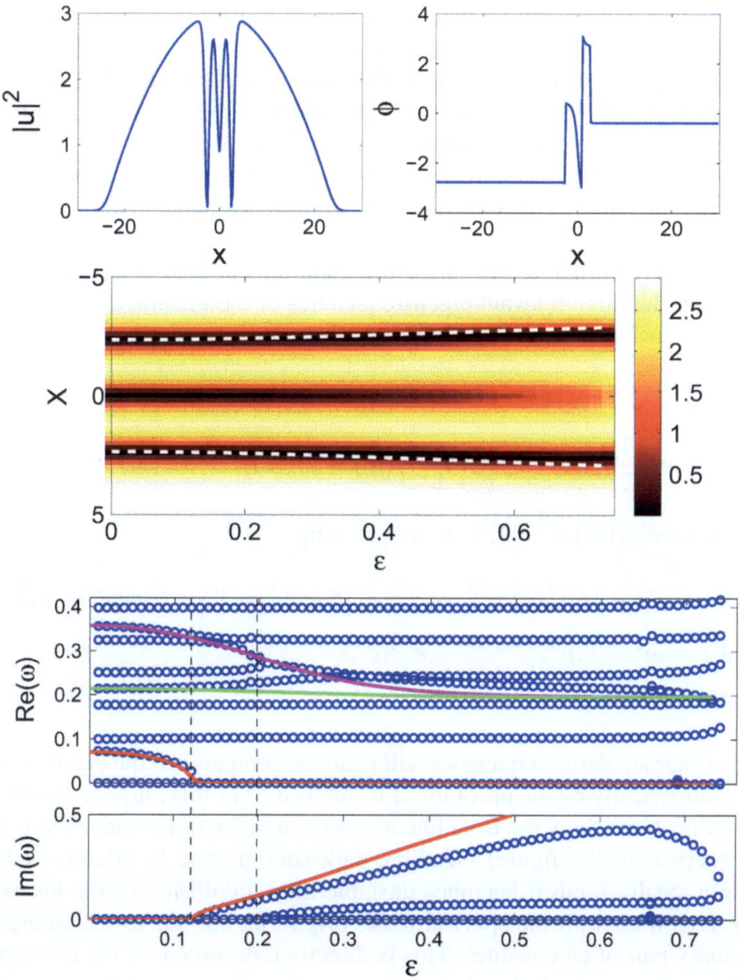

Fig. 4 Same as Fig. 3, but for the three soliton state. In the *bottom panels*, *solid red*, *green* and *magenta lines*, correspond to the analytical result of Eqs. (31) and (32). The value of these modes at $\varepsilon = 0$ correspond to the first, second and third anomalous modes of the Hamiltonian case

4.2 Three-Dark-Soliton State

We now proceed with the investigation of the three-soliton branch; examples of the density and phase of such a state are shown in the top panel of Fig. 4. This state shows similar behavior with the single soliton state, in the sense that the soliton located at the center becomes shallower as ε is increased.

The statics and dynamics around equilibria of the three-dark-soliton state can be analyzed by the methodology adopted in the case of the two-dark-soliton state.

Particularly, we will determine the equilibrium positions of the three solitons and study their small-amplitude oscillations around their fixed points. First we note that, from symmetry arguments, we expect that the fixed points are $X_2 = 0$ (for the central soliton), and $X_3 = -X_1 = X$ (for each of the two outer solitons). For this configuration, we can again calculate the fixed points for the outer solitons numerically and compare them to the particle theory analytical predictions, as illustrated in the middle panel of Fig. 4, obtaining good agreement for small values of ε.

Subsequently, we derive the linearized equations around the fixed points, and finally obtain the three relevant eigenfrequencies of the system. This way, we find that the frequencies of the normal modes of the system are:

$$\omega_1 = \sqrt{f_1(X)} \tag{31}$$

$$\omega_{2,3} = \left\{\frac{1}{2} f_1(X) \pm f_2(X)\left[\left(f_1(X) - f_2(X)\right)^2 + 8 f_3(X)^2\right]^{1/2}\right\}^{1/2}, \tag{32}$$

where functions $f_i(X)$ ($i = 1, 2, 3$) are given by:

$$f_1(X) = \tilde{\omega}^2 - 16\mu^2 e^{-2\sqrt{\mu}X} + \varepsilon^2 c_1 c_2 \left[5\,\mathrm{sech}^6(c_2 X) - 4\,\mathrm{sech}^4(c_2 X)\right], \tag{33}$$

$$f_2(X) = \tilde{\omega}^2 - 32\mu^2 e^{-2\sqrt{\mu}X} + \varepsilon^2 c_1 c_2, \tag{34}$$

$$f_3(X) = 16\mu^2 e^{-2\sqrt{\mu}X}. \tag{35}$$

The above characteristic frequencies will again be compared to the eigenfrequencies obtained numerically by means of the BdG analysis. For this purpose, in the bottom panels of Fig. 4, we show the BdG linear spectrum of the three-dark-soliton branch.

As observed in the figure, the three-dark-soliton state is initially stable (for sufficiently small ε), but it becomes unstable after a collision of the lowest mode (dashed (red) line) with the spectral plane origin, giving rise to the emergence of an imaginary pair of eigenvalues. This is directly reminiscent of the corresponding instability of the single dark soliton state and the mechanism of the instability is expected to persist for *any* configuration with an odd number of dark solitons in this system (we have also checked that it arises in the case of a five-soliton state). For larger values of ε, another collision between the fifth and sixth mode, results in the emergence of an imaginary eigenfrequency pair (right vertical dashed line), and thus the three soliton state remains unstable.

At the same time, and as in the single-soliton case, the imaginary eigenfrequency corresponding to the first nonzero mode, eventually returns to the origin and collides with it, at $\varepsilon = 0.72$. This is the critical point of the collision of the two-soliton and the three-soliton branch and of their pairwise annihilation. As the relevant critical value of ε is approached, the center grey soliton of the three-soliton state becomes grayer and eventually becomes identical to the central density dip of the two-soliton state with which it collides. The above characteristic frequencies will again be compared to the eigenfrequencies obtained numerically by means of the

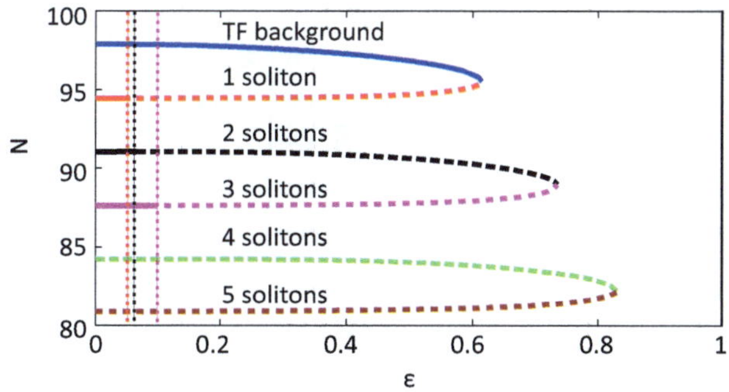

Fig. 5 The power N as a function of ε, for the six lowest states. *Solid lines* correspond to linearly stable states and *dashed lines* to unstable states, with $\mathrm{Im}(\omega) \neq 0$. The *vertical lines* denote the critical values of ε found for the single-, two- and three-dark-soliton states. This bifurcation diagram illustrates the nonlinear $\mathcal{PT}$-phase transition points, at which stationary states disappear in pairs through blue-sky bifurcations

BdG analysis. For this purpose, in the bottom panels of Fig. 4, we show the BdG linear spectrum of the three-dark-soliton branch.

The analytical results for the three eigenfrequencies (cf. Eqs. (31) and (32) and solid red, green and magenta lines in Fig. 4) are in a good agreement with the respective modes obtained numerically via the BdG analysis, as long as ε is sufficiently small. In particular, the critical value of ε where the first finite mode collides with the origin (thus giving rise to the instability) is very well predicted; furthermore, the analytically found "trajectory" of the other two branches follow, in a fairly good accuracy, the respective numerical result, even for moderate and large values of ε, with an expected progressive discrepancy as ε becomes large (when the problem is outside the realm of a perturbative treatment).

4.3 Nonlinear $\mathcal{PT}$ Phase Transitions

In order to further elaborate on the emergence of the nonlinear $\mathcal{PT}$-phase transition, in Fig. 5 we plot the power N as function of the parameter ε, for the lowest six (ground state and the first five excited) states of the system. The top solid (blue) branch depicts the stable ground state (i.e., the TF background u_b—cf. Eqs. (9)–(11)), which ultimately collides with the single-dark-soliton branch (dashed (red) line), i.e., the first excited state at $\varepsilon \approx 0.62$ (for $\mu = 3$). Furthermore, the next pair of branches, corresponding to the two- and three-dark-soliton states (depicted by the dashed black and magenta lines, respectively), also collide at $\varepsilon \approx 0.72$. This picture of colliding and annihilating pairs of excited states remains the same even for higher excited states—cf. the lowest dashed (green and brown) curves in Fig. 5, showing the annihilation of the five- and six-dark-soliton states as well.

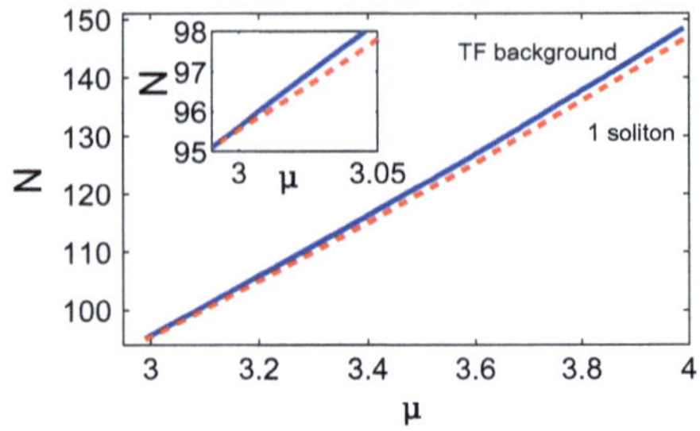

Fig. 6 The power N as a function of μ, for the TF branch (*solid (blue) line*) and the single-soliton branch (*dashed (red) line*, indicating instability); the two branches collide approximately at $\mu = 3$. The gain/loss parameter is $\varepsilon \approx 0.613$ and the trap strength is $\Omega = 0.1$

An important general remark, regarding the structure of the bifurcations shown in Fig. 5, is that higher excited states sustain these collisions, saddle-center bifurcations and corresponding blue-sky annihilations for progressively larger values of ε. This feature motivates us to refer to this process as the *nonlinear analogue* of the "traditional" $\mathcal{PT}$ transition, in direct correspondence with the pairwise collisions in Ref. [1] (see, e.g., Fig. 1 of that reference) for the linear setting. Although the two figures are similar, there are some interesting differences. One such is that in the linear picture the lowest eigenvalues collide the last, while in the 1D nonlinear case the lowest eigenstates disappear first.[1] This feature has important consequences for the supercritical evolution of the system which we will revisit in the dark soliton dynamics section below.

To conclude this section we also add the following comments and compare our results with those reported in Ref. [27]. In the latter work, a bifurcation diagram $N(\mu)$ was obtained, where pairs of branches starting from the *linear* limit collided and disappeared in the nonlinear regime with the defocusing nonlinearity. Such a form of the bifurcation diagram was due to the assumed form of the $\mathcal{PT}$-symmetric potential, which allowed for the existence of linear eigenstates with real eigenvalues for all values of the gain/loss coefficient. However, in our case, progressively more linear eigenstates acquire complex—rather than real—eigenvalues past corresponding critical (maximum) values of the gain/loss coefficient ε. As a result, a $N(\mu)$ bifurcation diagram in our case would typically have a considerably different form than the one of Ref. [27]: starting from the highly nonlinear Thomas-Fermi regime (where $\mu/\Omega \gg 1$ and the length scale of the solitary waves is much shorter than that of the condensate) and decreasing μ, pairs of branches would collide at a critical value of μ and disappear via a saddle-center bifurcation. This behavior is shown in Fig. 6 for a value of $\varepsilon \approx 0.613$ where the first two branches, i.e. the TF branch and the one-soliton branch, coming from the far nonlinear limit with decreasing μ, collide at $\mu = 3$. For smaller μ, only states with complex μ exist. For this value of ε, we have numerically obtained the first two linear eigenvalues: $\lambda_{1l} \approx 0.149 + 0.093i$

[1]Another important difference that we do not touch upon here is that in the model of Ref. [1], the ground state does not collide with some other linear mode of the system, contrary to what is the case herein.

and $\lambda_{2l} \approx 0.283 + 0.034i$, which are complex, and thus one does not expect to find nonlinear states bifurcating from these two linear states.

4.4 The "Free Space" Case

In our previous considerations, we have studied the case where the real part of the external potential was parabolic. Nevertheless, and as a complementary setting, it is also interesting to consider the "free space" case, where the real trapping potential is absent, i.e., $V(x) = 0$. Such a situation may occur, e.g., in the context of optics where—instead of a graded index medium—one may consider a medium with a constant linear refractive index, i.e., $V(x) = V_0 =$ const.; in this case, the pertinent term $V_0 u$ that would appear in Eq. (1) can straightforwardly be removed by a trivial gauge transformation.

An interesting feature characterizing the free space case is that stationary multiple dark soliton states can no longer exist: in this case, the repulsion between dark solitons cannot be counter balanced by any restoring force as the $V(x)$ producing the latter is absent. As a result, the most fundamental states pertinent to this setting are the ground state and the single-dark-soliton state. An approximate solution for the free-space ground state u_0, can be found employing Eq. (11) and setting $V(x) = 0$, namely:

$$u_0 = \mu - \frac{1}{\sqrt{\mu}} \mathcal{W}^2(x) \exp\left(2i \int \mathcal{W}(x)\, dx\right). \tag{36}$$

Using the above result, and following the analysis of the previous sections, we find that the evolution of the dark soliton center is described by (cf. Eq. (21)):

$$\frac{d^2 x_0}{dt^2} = \frac{6}{5}\varepsilon^2 x_0. \tag{37}$$

The above equation dictates that a quasi-stationary (almost black) dark soliton is always unstable for *any finite* ε, since it experiences an effective expulsive force, which tends to displace it from the origin and set it into motion with an increasing (magnitude of) acceleration, at least for small values of x_0. This is in contrast to the case of a parabolic $V(x)$, where the respective force exerted on the soliton and the effective expulsive force induced by the imaginary potential $W(x)$ could balance each other. Thus, the presence of the real parabolic trap (or, more generally, of a confining potential) is crucial for the existence of stable stationary soliton solutions.

The top panel of Fig. 7 shows the relevant eigenfrequency for the linear stability spectrum of the dark soliton, with respect to the parameter ε. As predicted by Eq. (37), as soon as ε becomes non-vanishing, the dark soliton becomes unstable. Its spectrum shows a pair of imaginary eigenfrequencies emerging from the spectral plane origin in good agreement (for small/intermediate values of ε) with the

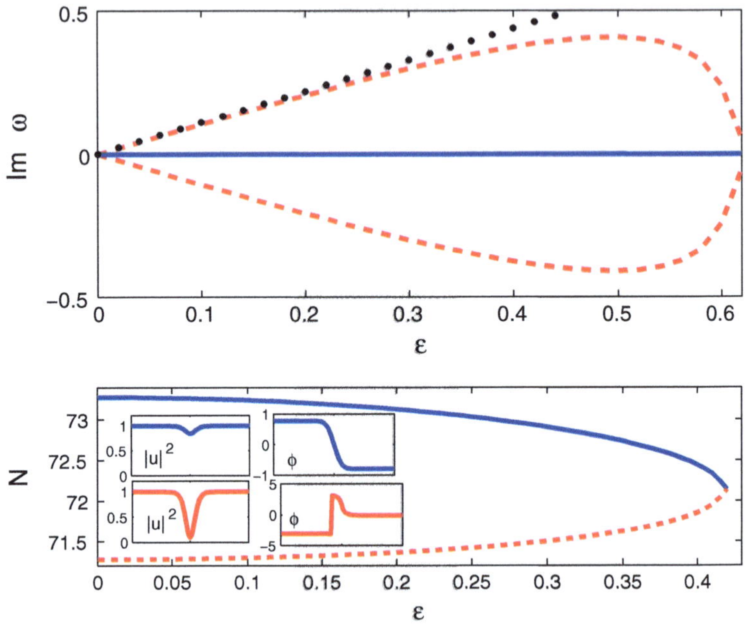

Fig. 7 *Top panel*: pertinent eigenfrequency of the linear spectrum for the free-space dark soliton with respect to ε. *Dashed (red) line* depicts the imaginary eigenfrequency pair arising due to the presence of $W(x)$ and the *dotted line* represents the corresponding theoretical prediction. *Bottom panel*: the $\mathcal{PT}$-phase transition diagram in the (N, ϵ) plane. *Solid (blue) line* shows the background branch, while *dashed (red) line* the single-dark-soliton state. The *upper* and *lower insets* show characteristic densities and phases for the ground state and the dark soliton, respectively. In this case, $\mu = 1$

corresponding theoretical prediction. Similarly to the case with the real parabolic potential (see Fig. 2), this pair of eigenfrequencies eventually returns to—and collides with—the spectral plane origin, signaling the termination of the branch and the $\mathcal{PT}$-transition. In the bottom panel of Fig. 7, the bifurcation diagram for the ground state and the single-dark-soliton in the free-space case is plotted; we observe a behavior similar to the one found in the case with the parabolic $V(x)$, but with the major difference that beyond the critical point ($\varepsilon \approx 0.43$ in this case) *no stationary states* exist, as the only ones such have collided and disappeared in the saddle-center bifurcation discussed above.

5 Dark Soliton Dynamics

We continue by considering the dynamics of the nonlinear excited states, by numerically integrating Eq. (1) in various different regimes in ε. To be more specific, our aim is to illustrate the dynamics of the dark soliton upon its destabilization after

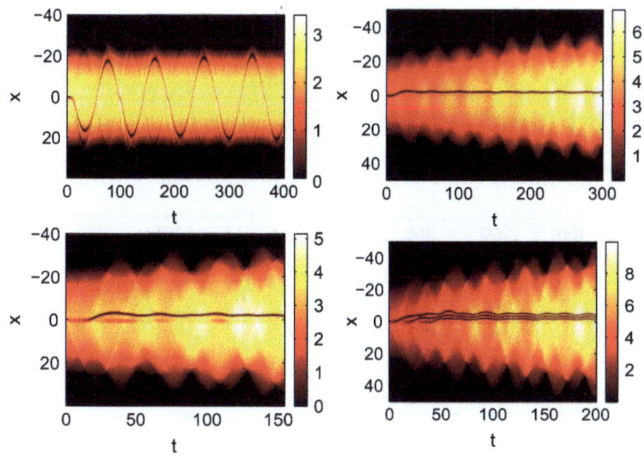

Fig. 8 Bifurcation-induced dynamics. *Top panels*: two potential manifestations of the SSB destabilization scenarios for a single unstable dark soliton past $\varepsilon = \varepsilon_{\rm cr}^{(1)} \approx 0.066$. *Bottom panels*: soliton "sprinklers" spontaneously leading to a single dark soliton (*left*) and the formation of three dark solitons (*right*), using as an initial condition the ground state for $\varepsilon > \varepsilon_{\rm cr}^{(2)} \approx 0.62$. The parameters are $\mu = 3$ and $\Omega = 0.1$ and $\varepsilon = 0.4$ (*top row*), $\varepsilon = 0.63$ (*bottom left*), and $\varepsilon = 0.66$ (*bottom right*)

the SSB bifurcation at $\varepsilon_{\rm cr}^{(1)}$, and also beyond the nonlinear $\mathcal{PT}$-phase transition. Generally, to study the dynamics of unstable solitons, we prepare a stationary dark soliton solution in the unstable regime (dashed parts of the soliton branches in Fig. 5), we add a small random perturbation, and then allow the resulting configuration to evolve in time. We have found that the dynamics of the system strongly depends on whether the soliton will be displaced—as a result of the instability—towards $x > 0$ (the "gain side" of $W(x)$) or $x < 0$ (the "lossy side" of $W(x)$).

First, we consider a case where a dark soliton is spontaneously ejected to the gain side of the imaginary potential; a pertinent example is shown in the top left panel of Fig. 8. It is observed that, in this case, the solitary wave starts to perform oscillations of large amplitude, a behavior which is generic for all solitons that are initially located at $x > 0$. On the other hand, if the soliton is initially kicked towards the "lossy" side of $W(x)$, upon an initial displacement of a finite ε-dependent size, it stops moving and remains quiescent, while the background on which it "lives" begins to grow in amplitude and width. A relevant example is illustrated in the top right panel of Fig. 8. We once again highlight the apparent incompatibility of the above scenarios with the expectation of a double well effective potential for the soliton and the SSB bifurcation of the soliton at $x_0 = 0$. This paves the way for the examination of the ghost states in the next section. For completeness, we also show the evolution of the power $N(t) = \int |u(x,t)|^2 dx$ in the top panels of Fig. 9, for both the above scenarios. In the left panel, which corresponds to the oscillating dark soliton, the power (after a short time interval during which it decreases) oscillates

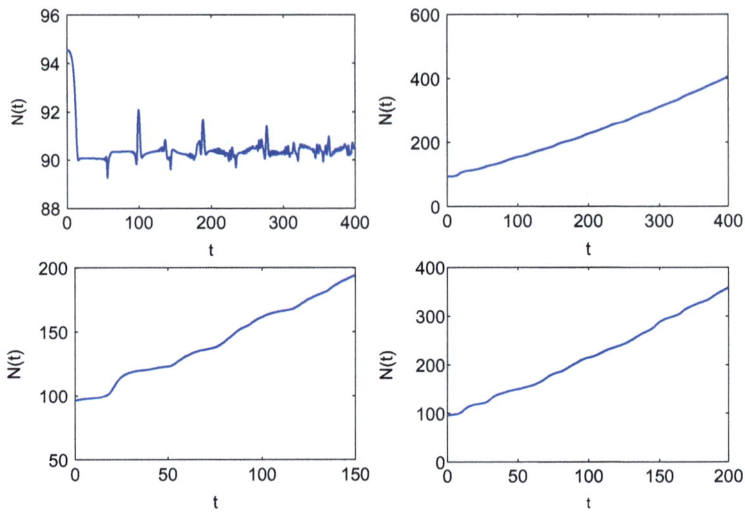

Fig. 9 Evolution of the power $N(t)$; each panel corresponds to the respective panels of Fig. 8. The *top left panel*, pertaining to the case of a oscillating dark soliton, exhibits oscillatory behavior. In all other cases the power is monotonically increasing

around a fixed value of $N \approx 90$. On the other hand, in the case of an unstable soliton localizing on the lossy side of $W(x)$, the power is monotonically increasing with time, as depicted in the top right panel of Fig. 9. The above two cases are the prototypical possibilities (independently of ε) for the fate of the dark soliton upon its destabilization, when the dark soliton state still exists.

We now turn to the examination of what happens beyond $\varepsilon = \varepsilon_{\rm cr}^{(2)}$ (where the TF background and the single soliton are no longer stationary solutions of Eq. (1)). There, we use, e.g., an initial condition in the form of a TF background $u(x,0) = \sqrt{\mu - V(x)}$, pertinent to the conservative system [30] as a means of exploring how the system responds when its fundamental states are no longer present. We have found that a dark soliton train is spontaneously formed, with an increasingly larger number of solitons as larger values of ε are employed. The bottom two rows of panels of Fig. 8 depict two examples of this phenomenon. This constitutes a form of what can be dubbed a "soliton sprinkler". This can, at least in part, also be intuitively connected to the observation of Fig. 2 that higher excited multi-soliton states persist for larger ε than lower ones. Nevertheless, it should also be highlighted that the observations typically suggest that the solitons are nucleated and stay in the vicinity of the global minimum of $W(x)$, and more generally tend to become stationary residing on the "lossy" side of the imaginary potential, similarly to the dynamical state observed above and accompanied by the same kind of background growth. The time evolution of the power for these multi-soliton states is shown in the bottom panels of Fig. 9; in both cases, the power increases monotonously with time, in line

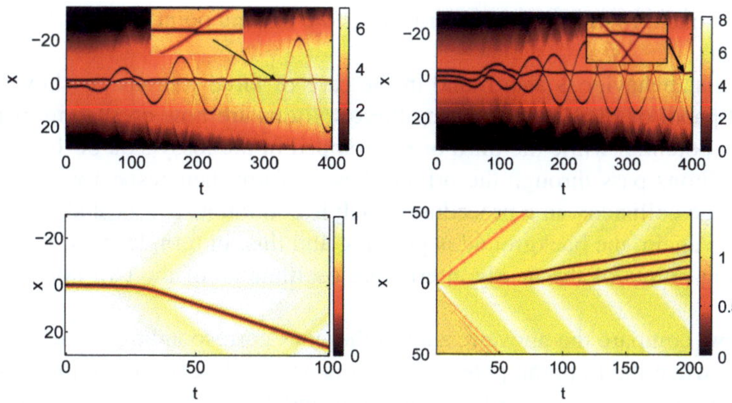

Fig. 10 Bifurcation-induced dynamics. *Top panels*: Manifestation of the dynamical instability of the two soliton state (oscillatory instability of the *top left panel*) and of the three soliton state (exponential and oscillatory instability of the *top right panel*). Both end up with one solitary wave sitting on the lossy side and the rest impinging on it through collisions between turning points within a continuously growing background. The *bottom panel* shows the case without $V(x)$ where the single soliton manifests its instability by being set into motion (*bottom left panel*), while a uniform initial condition past the $\varepsilon_{cr}^{(2)}$ starts emitting dark solitons in the form of a sprinkler. The parameters are $\mu = 3$ and $\Omega = 0.1$ and $\varepsilon = 0.4$ (*top left*), $\varepsilon = 0.4$ (*top right*), while $\mu = 1$ and $\varepsilon = 0.1$ (*bottom left*), and $\varepsilon = 0.45$ (*bottom right*)

with the localization of such solitary waves on the lossy side of the potential (cf. top right panel of Figs. 8 and 9).

As an additional aspect of the investigation of the dynamics of higher-excited (multi-soliton) states, and how their instabilities manifest themselves when they are subject to small perturbations, we consider the runs of the top panel of Fig. 10. In the left panel, the dynamics of a two-soliton state is shown. At $t = 0$ the two solitons are in equilibrium, but the soliton on the gain side starts to execute oscillations within the trap, as was the case before in Fig. 8 for the soliton displaced towards the gain side (top left panel of the figure). Thus, inevitably, this soliton, upon hitting a turning point in its oscillation, returns and collides with the other one. The result of the multiple collisions that ensue is that one of the solitons continues to execute oscillations of increasing amplitude while the other one remains effectively stationary on the lossy side. While the above sequence of oscillations and collisions is continuously repeated, at the same time, the background grows both in amplitude and width. It is interesting to note that the two solitons behave in a way that resembles a combination of the two different scenarios observed in the single soliton case, with the relevant dynamics resembling a "superposition" of the ones shown in the top two panels of Fig. 8. Yet there is one non-trivial difference. While the amplitude of oscillation in the top left panel of Fig. 8 remains roughly unaltered, the one of the top left panel of Fig. 10 keeps increasing due to the continuously expanding background (which, in turn, shifts the turning point of the corresponding oscillation).

A similar behavior is also found in the case of three solitons, as shown in the top right panel of Fig. 10. Two of the three solitons are now oscillating inside the harmonic potential while the third one is trapped in the lossy side of the potential. As the solitons pass through the origin, they perform their respective oscillations and undergo collisions, in a procedure which is continuously repeated; at the same time, once again, the background is growing and this, in turn, leads to a growth of the turning points of the nearly alternating oscillations of the two non-stationary solitons.

We now compare the above results to the ones corresponding to the free-space case, i.e., when the trapping potential $V(x)$ is absent. Recall that in this case only two stationary solutions exist (the ground state and the single soliton) and the soliton is always unstable. Since the background wave is now of constant density, when the soliton is perturbed, it will start to move away from the origin and will accelerate according to Eq. (37) up to the point where it is practically no longer under the influence of the (exponentially localized) imaginary part of the potential. Thereafter, it will acquire a nearly constant velocity and will keep moving in that direction, as shown in the bottom left panel of Fig. 10 (the absence of the real part of the potential leads to a lack of a turning point in this case).

Finally, we consider the parameter regime $\varepsilon > \varepsilon_{\text{cr}}$, i.e., after the occurrence of the $\mathcal{PT}$-phase transition where no stationary solutions exist. We numerically integrate Eq. (1), with an initial condition of the form of a plane wave $|u(x,0)| = \sqrt{\mu}$, and the result is shown in the right bottom panel of Fig. 10. The initial plane wave begins to emit dark solitons, which are all starting to propagate along the "lossy" direction. Contrary to the confined case where the trapping potential was keeping the solitons near the origin, the solitons are continuously created and are now free to propagate along the x axis. Hence, once the "sprinkling" process begins, it continues indefinitely.

6 Ghost States and Their Dynamical Role

It is now relevant to try to address some of the remaining questions, concerning the nature of the daughter states within the pitchfork bifurcation (of Fig. 2 and by extension e.g. in Fig. 4) and the unexpected asymmetric behavior of the dynamical evolution of unstable dark solitary waves in Fig. 8. A related issue concerns the feature that solitary waves become stationary on the lossy side of the potential (while the corresponding background starts growing); see Fig. 8 and by extension also Fig. 10.

A first way to try to address this problem is to consider the possibility (as suggested by the direct numerical simulations) of a state involving a stationary dark solitary wave off of $x_0 = 0$ (and, in particular, for $x_0 < 0$). Then the steady state problem would read:

$$\mu u = \mathcal{L}u + iW\,u + |u|^2 u, \tag{38}$$

where $\mathcal{L}$ is used to denote the linear, Hermitian part of the right hand side operator. The corresponding conjugate equation then reads:

$$\mu^* u^* = \mathcal{L}u^* - iW\,u^* + |u|^2 u^*. \tag{39}$$

Now, by multiplying Eq. (38) by u^*, and Eq. (39) by u, integrating and subtracting the second equation from the first, we obtain the self-consistency condition for the imaginary part of the chemical potential:

$$\mathrm{Im}(\mu) = \frac{1}{N}\int W(x)|u|^2 dx. \tag{40}$$

Interestingly, the right hand side is directly related to the one arising in Eq. (7) and, thus, also controls the time dependence of the atom number/optical power (or mathematically the L^2 norm).

The key realization that emerges from Eq. (40) is that since $|u|^2$ is an even function for solitary structures centered at $x_0 = 0$ (and $W(x)$ is anti-symmetric), the only way for our states to be centered at $x_0 \neq 0$ is if $\mu_I \equiv \mathrm{Im}(\mu) \neq 0$. However, this is an atypical feature for states arising in the context of the nonlinear Schrödinger (NLS) equation. More specifically, a fundamental underlying assumption behind the identification of stationary states, critically employing the U(1) invariance of the model, is the use of the product ansatz for stationary states of the form $u(x,t) = e^{-i\mu t} f(x)$. Now, we find ourselves at a junction where solutions may exist with a complex μ. This, more specifically, implies that they will be indeed exact solutions of the *stationary* NLS problem, yet their incompatibility with the above ansatz *precludes* them from being exact solutions of the original dynamical problem of Eq. (1).

The natural next questions then involve whether these states can, in fact, be identified as such (exact solutions of the steady state problem) and, perhaps more importantly, what is their role in the observed dynamics. The first of these questions is answered in Fig. 11, where we have performed a continuation of these states as a function of ε, starting from the Hamiltonian limit in the absence of gain/loss. The top left panel of the figure presents the center of the solution as a function of the parameter, while the top right one evaluates the imaginary part of the chemical potential, by self-consistently enforcing the condition (40). It is key to mention here that for the solutions with negative x_0, it is straightforward to see that they correspond to a setting with $\mu_I > 0$ (from Eq. (40)). On the other hand, the term $e^{-i\mu t}$ will then involve a part associated with $e^{-i(i\mu_I)t}$ and hence this state will be associated with growth over time. On the other hand, the state with $\mu_I < 0$ will be connected to decay over time and does not appear to be relevant for physical observations. It is for that reason that one of the two states (the robust one) is denoted with a bold solid line in the top panels of Fig. 11, while the other one is denoted by

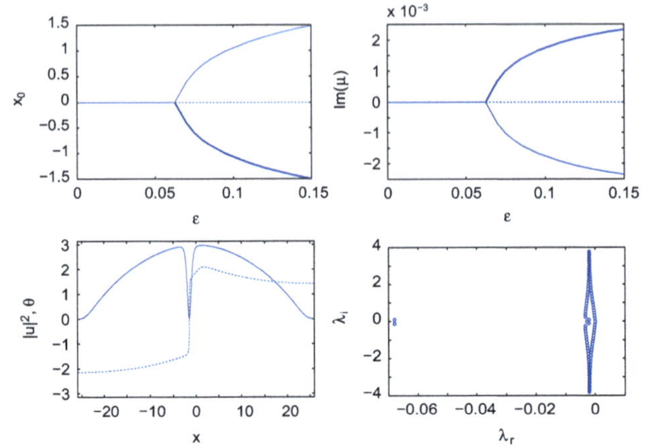

Fig. 11 Properties of the ghost states. The *top left panel* illustrates their (off-center) location with the *bold line* illustrating the dynamically robust one. The *center line* corresponds to the dark soliton at $x_0 = 0$ and the *solid* to *dashed* transition corresponds to the SSB bifurcation. Another diagram illustrating the pitchfork character of the transition is given in the *top right panel* through the imaginary part of the chemical potential μ_I as a function of the gain/loss parameter. Again, the persisting state is given in *bold*. A typical example of the squared density and phase (*solid* and *dashed*, respectively) of a ghost state for $\varepsilon = 0.15$ is given in the *bottom left panel*. Finally, a suggestive linear stability analysis (as if this were a true steady state of the system) is performed on the *bottom right* suggesting also the dynamical robustness of the relevant state due to the absence of eigenvalues with a positive real part (i.e., of ones corresponding to growth)

a thin line (and unstable states centered at $x_0 = 0$ in the diagram are denoted by a dashed line).

First of all, this observation restores the canonical (and expected) pitchfork nature of the observed bifurcation (see top panels of Fig. 11 and the daughter state of the bottom left panel). But, at the same time it reconciles that nature with the absence of any "regular" daughter states. A stability analysis of these states is not particularly meaningful per se (as these are not stationary states of the original problem), yet it is natural to expect that it will be suggestive regarding the evolution of perturbations in the vicinity of such states. This, as shown in the bottom right panel for the displaced state of the bottom left one in Fig. 11, illustrates the effective stability of the resulting ghost state.

We now turn to the dynamical implications of the existence of such ghost modes. The asymmetry between the growth of the mode with the dark soliton at $x_0 < 0$ and the decay of the one with $x_0 > 0$ justifies the observed asymmetry of the dynamical evolution of the perturbations in Fig. 8. To illustrate the relevance of the ghost states in the case where the random perturbation manifests the instability of the $x_0 = 0$ soliton by kicking it to the left, we performed some relevant numerical experiments in Fig. 12, for $\varepsilon = 0.15$. Specifically, by initiating the soliton at $x_0 = 0$, we confirm that it goes and sits precisely at the location (for that particular ε) where the ghost state has the soliton centered (cf. with the out of scale profile of the shifted soliton

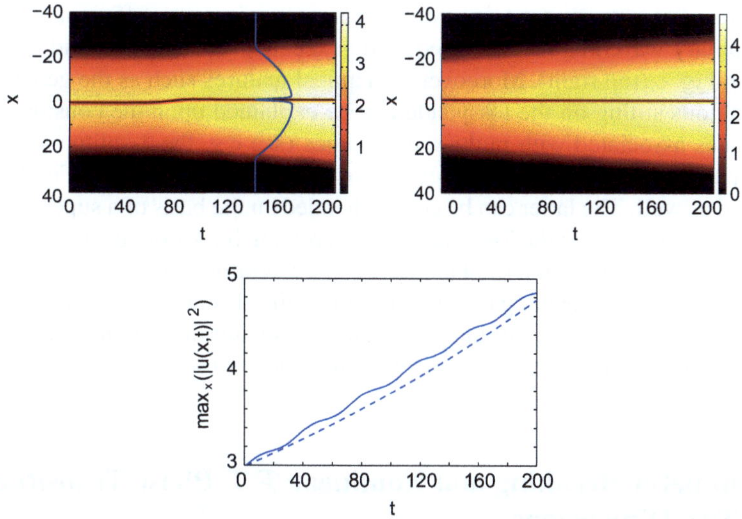

Fig. 12 The *top left panel* shows the space-time evolution of the contour plot of the unstable dark soliton at $x_0 = 0$ for $\varepsilon = 0.15$. For comparison (and out of scale), we plot (using a *solid (blue) line*) the exact obtained ghost state for the same gain/loss parameter. Clearly the dynamics gets attracted to this state, as is also corroborated by its evolution as initial condition in the top right panel (the dark soliton stays put, but the background width and amplitude grows in a way directly analogous to the case on the *left*). Finally, the comparison of the amplitude background growth $A(t)$ (*solid line*) with the law implied by the growth rate of the ghost state $A(t) = A(0)\exp(\mu_I t)$ (*dashed*) is shown on the *bottom panel*, suggesting good agreement between the two

within the graph). On the other hand, for a clear comparison, in the top right panel of Fig. 12, we have evolved the dynamical equations with exactly that proper ghost state for the particular ε as initial condition (wherein the soliton is located at -1.5 for $\varepsilon = 0.15$; see the right panel of the figure). The dynamics clearly illustrates that the soliton stays immobile (as expected by the nature of the ghost state), while the background increases in amplitude and also the width of the solution increases. These latter growth features are strongly reminiscent of the evolution of the left panel of the figure past the point of destabilization and convergence to a solitary wave centered at $x_0 = -1.5$. To further cement this proximity, we have computed the growth of the background amplitude i.e., the maximum absolute value of the field $A(t)$ and have directly compared it to the simplistic law (implied by the ghost state growth rate μ_I) $A(t) = A(0)\exp(\mu_I t)$, yielding the kind of agreement shown in the bottom panel of the figure. The natural conclusion is that while the ghost states are not genuine (steady) attractors of the original dynamics, nevertheless the evolution of the system closely shadows the relevant states and utilizes their asymmetry.

As for the case where the dark soliton moves to $x_0 > 0$ due to the perturbation, then the ghost state is simply untenable as a dynamical state (due to its decay in time), hence the dark soliton proceeds to escape from the region of influence of the gain and moves towards a turning point. This motion imparts sufficient momentum to the solitary wave that it can subsequently execute oscillations around $x_0 = 0$, as shown in Fig. 8 (top right). Moreover, dynamical features such as the generation of multi-solitons sitting on the lossy side can be explained upon the consideration of ghost states associated with multi-soliton states (see e.g. the 3-soliton state of the bottom panel of Fig. 8), while features such as those of the top panel of Fig. 10 can be explained as well. The latter can be comprehended on the basis of a superposition of a ghost state centered on the lossy side with a growing background, and of one (as in top left, or more as in the top right panel of Fig. 12) soliton(s) moving to the opposite direction and hence executing oscillations due to the absence of a ghost state to trap it. The above characteristics of the ghost states thus naturally lend themselves to the explanation of the static and dynamic phenomenology observed herein.

7 Symmetry Breaking and Nonlinear $\mathcal{PT}$ Phase Transitions in Two Dimensions

In order to consider the generalization of the ideas presented herein to higher dimensions, we now consider the case of a 2D $\mathcal{PT}$-symmetric potential ($W(-x,-y) = -W(x,y)$) with

$$W(x,y) = \varepsilon(x+y)e^{-(x^2+y^2)/4}. \tag{41}$$

The configurations sustained in the 2D setting follow a similar symmetry breaking bifurcation scenario as in their 1D counterparts. However, the bifurcation structure is naturally expected to be more complex for the higher dimensional case. Figure 13 displays the bifurcation scenario for 2D solutions with and without topological charge for $\Omega = 0.2$. The bifurcation diagram includes the chargeless ground state (the TF background cloud), solutions bearing from one to six vortices, and the dark soliton stripe. Similar to the 1D case, the TF background is stable in all its domain of existence and collides, as ε increases, with an excited state (the two-vortex solution or vortex-dipole) in a blue-sky bifurcation at $\varepsilon = \varepsilon_{\rm cr}^{(2)}$. However, in contrast to the 1D case where this collision occurs with the first excited state, in the 2D case the collision happens with the *second* excited state consisting of a vortex-dipole (a vortex pair with opposite charges). This is due to the fact that the TF background has no topological charge and, in turn, the vortex-dipole has also no net topological charge. This allows for the emergence of the vortex-dipole from a central dip on the TF background (see top panel of Fig. 14a) that becomes deeper as ε increases. This blue-sky bifurcation is depicted in Fig. 13 where the stable (center) TF branch (solid blue curve denoted by `TF`) collides with the unstable (saddle) vortex-dipole branch (red dashed curve denoted by `2-vort`) at the 1D equivalent of the critical point $\varepsilon = \varepsilon_{\rm cr}^{(2)}$.

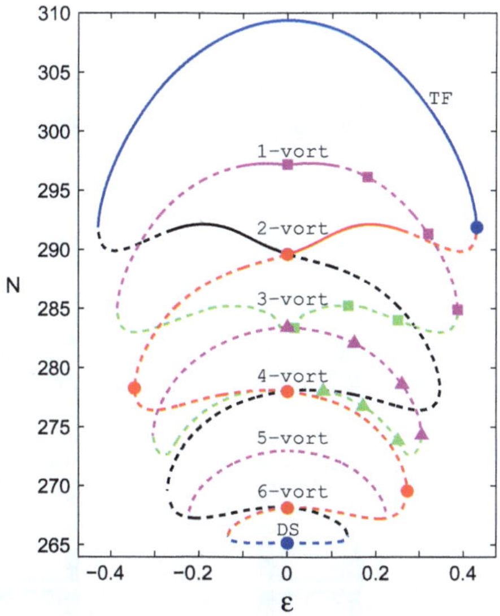

Fig. 13 Bifurcation diagram of stationary profiles in the 2D case. Stable and unstable branches, as determined by the corresponding BdG analysis, are depicted, respectively, with *solid* and *dashed lines*. The large symbols (*circles*, *squares* and *triangles*) denote the location in parameter space for the series of profiles depicted in Fig. 14. The parameter values are: $\mu = 2$ and $\Omega = 0.2$

It is possible to further follow the vortex-dipole branch by decreasing ε where we note that it regains stability for sufficiently smaller values of ε (see transition between the dashed and solid portions of the red line for the branch denoted by `2-vort`). As this branch is followed further (from top to bottom in Fig. 13) a series of bifurcations arise where the vortices are drawn towards the edge of the cloud, a central dip deepens leading eventually to the emergence of a new vortex dipole in the middle of the cloud (namely, the emergence of a higher excited state). Therefore, the branches with *even* number of vortices are all inter-connected in this bifurcation scenario and only the TF branch and part of the vortex-dipole branch are stable. However, as more and more vortex pairs emerge, the extent of the cloud "saturates" as it can no longer support new vortex pairs and therefore is replaced by a dark soliton stripe (see dashed blue line denoted by `DS` in Fig. 13). This overall bifurcating scenario for even number of vortices is symmetric such that if $\varepsilon \to -\varepsilon$ the solutions are just flipped by $(x, y) \to (-x, -y)$. Figure 14a depicts the density and phase profiles for different representatives along this branch.

Let us now describe the bifurcation scenario for *odd* number of vortices. The first excited state, corresponding to a single vortex, starts with a single vortex at the origin for $\varepsilon = 0$ and it is stable for small values of ε. As in the 1D case, this first excited state suffers a SSB bifurcation at $\varepsilon = \varepsilon_{\rm cr}^{(1)}$ where it loses its stability. As ε is increased, the single vortex moves towards the periphery of the cloud and a dip at the center of the cloud deepens until a vortex dipole emerges at the center in the same way as new vortex-dipoles emerged for the branches with even number of vortices described above. This scenario connects, again through a blue-sky bifurcation, the one-vortex branch (magenta line denoted by `1-vort` in Fig. 13) with the *asymmetric* three-vortex ($+ - +$ vortex tripole) branch (green dashed

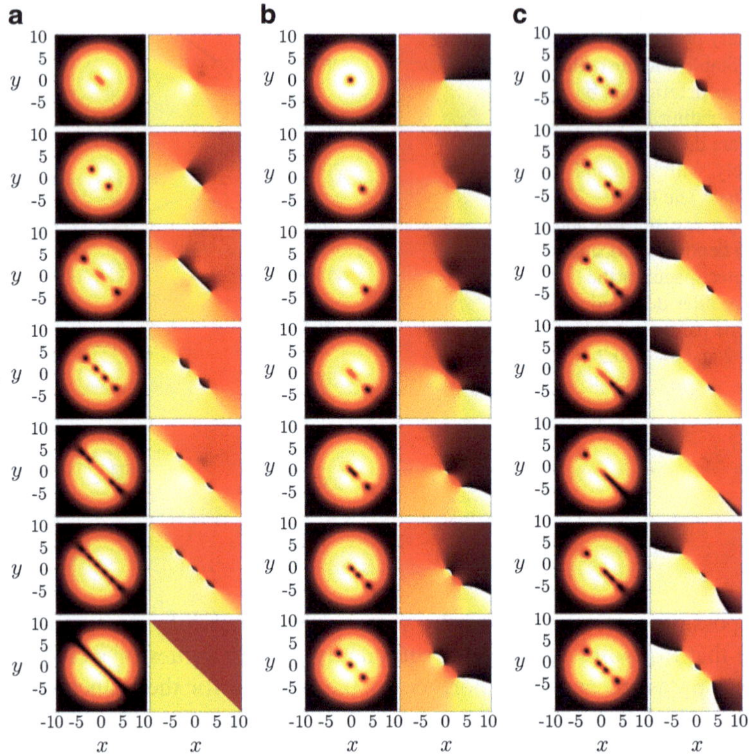

Fig. 14 Stationary states in the 2D case corresponding to the main bifurcation branches in Fig. 13. The respective *left* (*right*) columns depict the density (phase) profiles of the solutions. (**a**) Profiles for the branch containing only even number of vortices corresponding to the circles in Fig. 13 (from top to bottom). (**b**) Profiles for the branch starting with a single vortex and connecting with three vortex branch corresponding to the squares in Fig. 13 (from top to bottom). (**c**) Profiles for the branch starting with three symmetric vortices and ending with four vortices corresponding to the triangles in Fig. 13 (from top to bottom). The parameter values are: $\mu = 2$ and $\Omega = 0.2$

line denoted by `3-vort` in Fig. 13). The series of snapshots at the parameters depicted by the squares in Fig. 13 is depicted in Fig. 14b. It is nevertheless relevant to note that the blue-sky bifurcation for the single-triple vortex state (and more generally for any higher order pair of vortices) happens for values of ε lower than that of the bifurcation involving the TF ground state and the vortex-dipole state. Interestingly, although the structure of the relevant bifurcations is reminiscent of its one-dimensional analogue, there are also non-trivial differences, including the topological charge issue mentioned above, as well as the order in which the bifurcations occur (between more fundamental and more highly excited states) which is reversed. As it is evident from the figure, the asymmetric three-vortex branch eventually connects with the symmetric one for values of $\varepsilon \to 0$. A similar bifurcation occurs with the symmetric three-vortex branch, which becomes asymmetric with a deepening dip at the center where a vortex pair emerges (at the same time that a vortex is lost at the periphery), connecting in this way with the four-vortex branch (see series of snapshots in panels Fig. 14c). As a relevant aside, let us also mention that all of the above true stationary states of the system (with real chemical potential or propagation constant) clearly form along the diagonal $x+y=0$ on which there is no gain or loss according to the prescription of Eq. (41).

It is important to mention that the precise structure of the bifurcation diagram depends on the values of the trap strength Ω and the propagation constant μ. In general terms, for weaker Ω and/or larger μ the spatial extent of the TF background will be larger and thus allowing for a longer bifurcating chain of higher-order vortex states (before all the vortices finally merge into a dark soliton stripe). Nonetheless, the displayed SSB instabilities and the nonlinear $\mathcal{PT}$ phase transition involving the cascade of blue-sky bifurcations appear to be universal in confining $\mathcal{PT}$-symmetric potentials as considered herein.

Let us now describe the dynamics of the unstable steady states described in the above bifurcation scenario. We first start by describing the unstable dynamics of single vortex states. In this case, in analogy with the 1D case, single vortex states tend to migrate towards the lossy side of the potential. This tendency is depicted by the series of density snapshots in Fig. 15a, b where *unstable* single vortex states for $\varepsilon = 0.2$ and $\varepsilon = 0.35$, respectively, migrate to different positions within the lossy side. This phenomenology can be very clearly understood on the basis of our ghost state interpretation. Once again, ghost states are present here corresponding to a unique shift of the vortex center to a location such that $x_0 + y_0 < 0$. This state is associated with $\mu_I > 0$, due to the fact that $\iint W(x,y)\,|u(x,y)|^2\,dx\,dy > 0$ and hence leads to growth of the amplitude and width of the condensate, a trait which can be clearly observed in the relevant panels. It is also important to mention that the stable single vortex, close to where it loses stability, is only weakly stable and can be 'kicked out' by a relatively small perturbation towards the lossy side. This scenario is depicted in Fig. 15c where the *stable* single vortex configuration for $\varepsilon = 0.1$ is perturbed by applying a small rotation of 5°. Although the rotation is very small and that the unperturbed configuration is stable, we note that the vortex eventually migrates (after a long transient, see that time in the series of snapshots runs to t = 3,000) towards the lossy side. Further numerics shows that angles larger than 4° display the same dynamics while smaller angles correspond to the single vortex returning periodically towards its original position (see for example the periodic orbit depicted in Fig. 15d where the stable vortex is initially rotated by 4°). To complement the snapshots for the evolution of the different configurations, we depict in Fig. 16 the evolution of the corresponding power $N(t) = \iint |u(x,y,t)|^2\,dx\,dy$ and the total angular momentum $L_z(t) = -i \iint u^*\,\partial_\theta u\,dx\,dy$. As it may be observed from panel (a), the migration of the unstable vortices towards the lossy side does not initially (see flat portion in the figures for relatively short times) affect significantly the total power, nor the angular momentum. However, once the vortex settles in the lossy side and the background starts growing the power and angular momentum increase rapidly. In contrast, as depicted in Fig. 16b, for the case of a stable vortex slightly perturbed

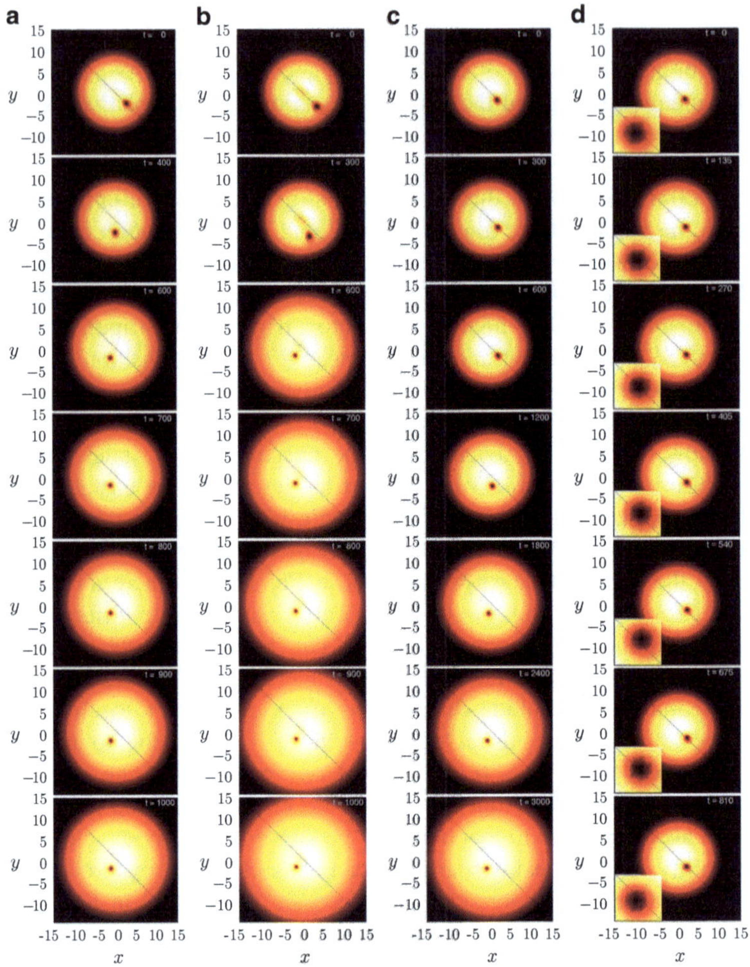

Fig. 15 Dynamics of unstable steady states in the 2D case. All panels depict the density profiles at the times indicated for $\mu = 2$ and $\Omega = 0.2$. The *thin diagonal line* in all panels depicts the border between the loss and gain sides of $W(x, y)$. (**a** and **b**) Unstable single vortex state for $\varepsilon = 0.2$ and $\varepsilon = 0.35$, respectively. (**c**) Stable single vortex configuration for $\varepsilon = 0.1$ perturbed by a small rotation of 5°. (**d**) Stable single vortex configuration for $\varepsilon = 0.1$ perturbed by a small rotation of 4° performs a periodic oscillation around its original position (a single period is shown, see zoom in the *inset*)

from its original location the power and angular momentum perform oscillations corresponding to the oscillations of the perturbed vortex observed in Fig. 15d. Another interesting example is that of panel (e), corresponding to Figs. 18c, d and 19a, b below. Notice that a fundamental difference in this case (in analogy to what we observed in the 1D setting) is the fact that the vortices do not settle on the lossy side (in the form of ghost states) but rather oscillate between the gain and the loss

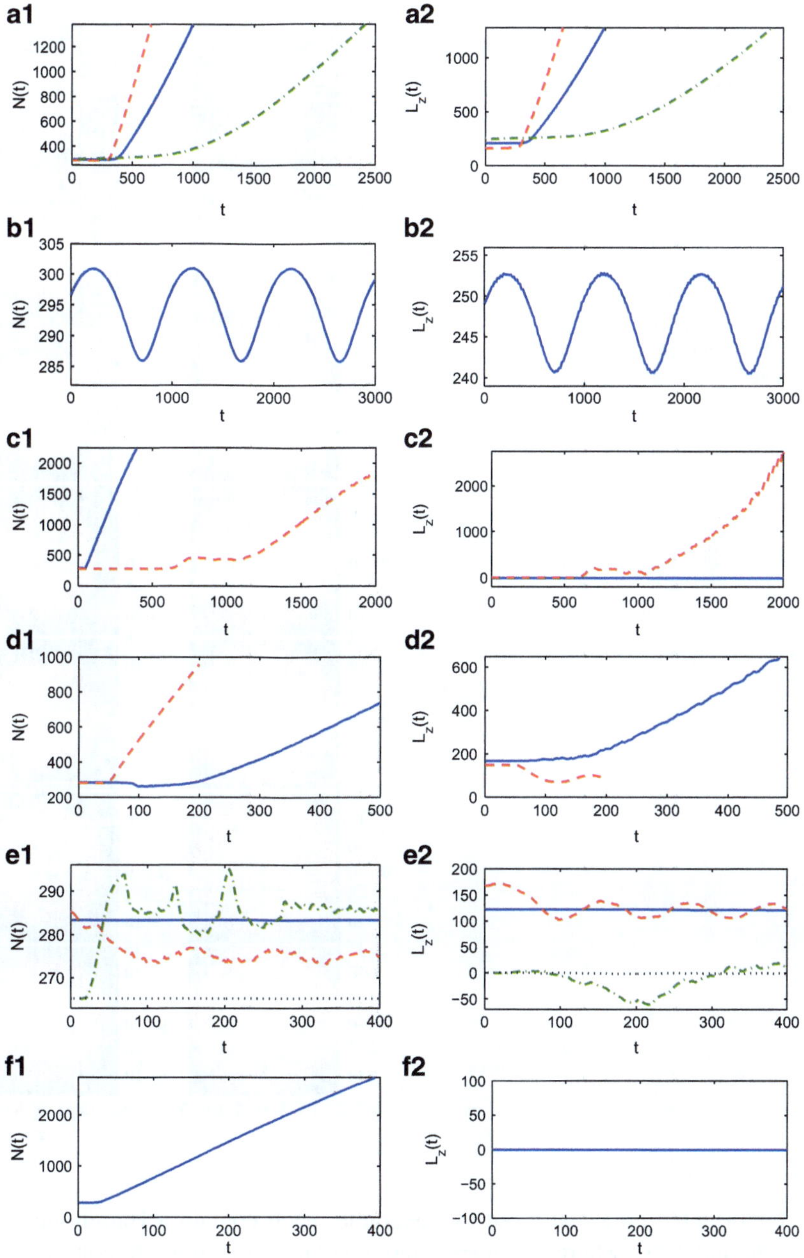

Fig. 16 Evolution of the power $N(t)$ *(left column)* and the angular momentum $L_z(t)$ *(right column)* corresponding to the dynamics for the following configurations: (**a**) Fig. 15a–c depicted with *solid (blue)*, *dashed (red)*, and *dash-dotted (green) lines*, respectively. (**b**) Fig. 15d. (**c**) Fig. 17a, b depicted with *solid (blue)* and *dashed (red) lines*, respectively. (**d**) Fig. 18a, b depicted with *solid (blue)* and *dashed (red) lines*, respectively. (**e**) Figs. 18c, d and 19a, b depicted with *solid (blue)*, *dashed (red)*, *dash-dotted (green)*, and *dotted (black) lines*, respectively. (**f**) Fig. 18c

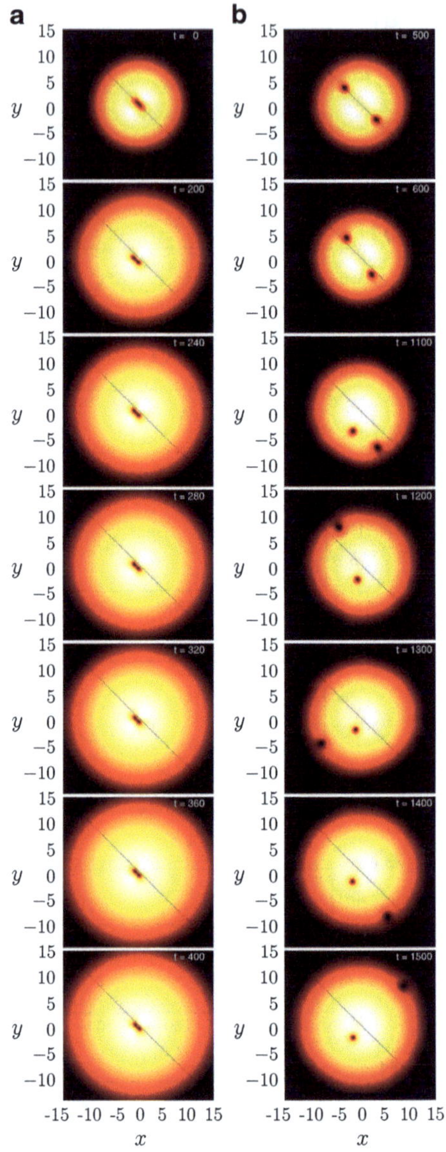

Fig. 17 Same as in Fig. 15 for the following two-vortex scenarios. (**a**) Unstable tight vortex-dipole state belonging to the branch directly connecting with the TF cloud (depicted by the *red dashed line* denoted by `2-vort` in Fig. 13) for $\varepsilon = 0.4$. (**b**) Unstable, well separated, two-vortex state belonging to the branch connecting with the four-vortex branch (depicted by the *black dashed line* denoted by `2-vort` in Fig. 13) for $\varepsilon = 0.2$

side. Importantly, note that these two examples each contain a Hamiltonian case (Figs. 18c and 19b), where the energy and angular momentum are conserved and a gain-loss one where the evolution is oscillatory, yet does not lead to indefinite growth as in other examples of Fig. 16.

Let us now describe the evolution of unstable vortex-dipoles. In Fig. 17a we depict the evolution of a *tight* vortex-dipole close to the parameter values where it is created (i.e., in the dip at the center of the cloud). As the figure shows, the

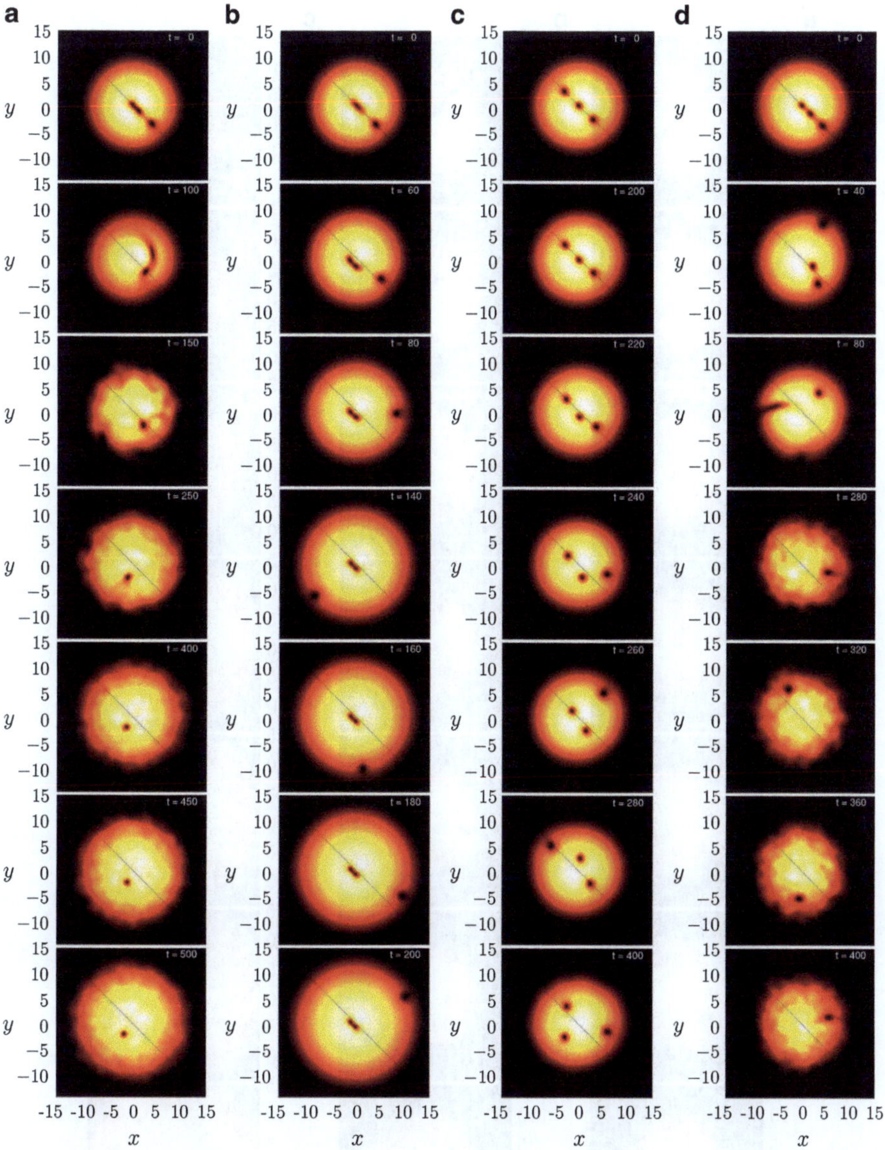

Fig. 18 Same as in Fig. 15 for the following three-vortex scenarios. (**a** and **b**) Unstable asymmetric three-vortex states (depicted by the *green dashed line* denoted by `3-vort` in Fig. 13) for $\varepsilon = 0.2$ and $\varepsilon = 0.35$, respectively. (**c**) Unstable symmetric three-vortex state for $\varepsilon = 0$. (**d**) Unstable asymmetric three-vortex state (depicted by the *green dashed line* denoted by `3-vort` in Fig. 13) for $\varepsilon = 0.1$

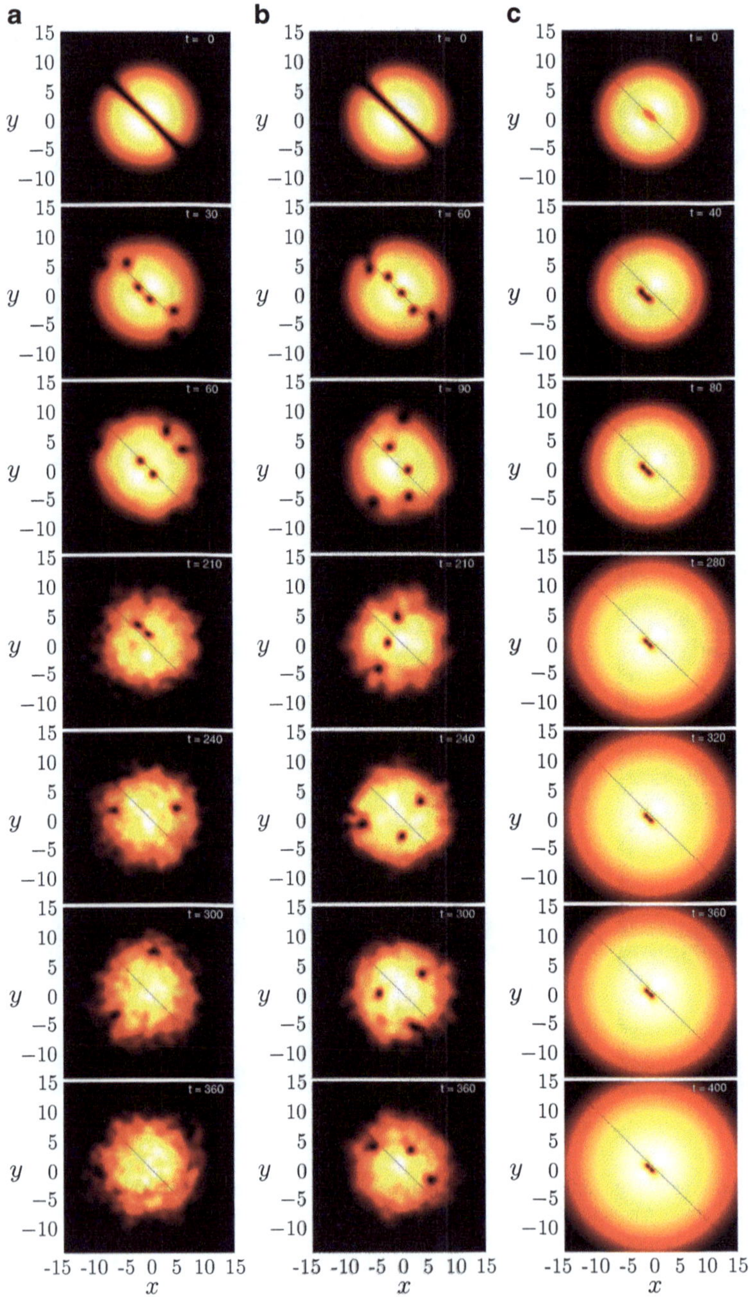

Fig. 19 Same as in Fig. 15 for the following scenarios. (**a** and **b**) Unstable dark soliton stripes (depicted by the lowest *blue dashed line* denoted by DS in Fig. 13) for $\varepsilon = 0.1$ and $\varepsilon = 0$, respectively. (**c**) Thomas-Fermi (ground) state obtained for $\varepsilon = 0.4$ and let to evolve for $\varepsilon = 0.43$

tight vortex-dipole undergoes a small excursion into the lossy side and, as time progresses, it re-approaches to the center of the cloud at the same time that the cloud grows (as in the 1D case). Clearly the relevant interpretation here involves the existence of a ghost state which attracts the dynamics and contains both vortex members of the dipole on the lossy side of the system. In this case, as it is depicted by the solid (blue) line in Fig. 16c, the power increases but the angular momentum remains equal to zero since the vortices are of opposite charge. In Fig. 17b we depict the dynamics arising from a *well separated* unstable vortex-dipole state. We typically observe that, for a well separated unstable vortex dipole, one vortex is trapped at a suitable location within the lossy side while the other vortex circles around slowly approaching the periphery of the cloud where it is finally absorbed. This is in a number of ways tantamount to the two-soliton dynamics that we explored in Fig. 10. Generally, it is clear that the three possible dynamical scenario for the vortices constituting the dipole and involving ghost states here are that either both vortices transition to the lossy side (Fig. 17a), or one (Fig. 17b) or none (not shown here).

Let us now briefly describe some typical evolution examples for higher order unstable states. In particular, let us start by discussing the three-vortex states. Figure 18a depicts the evolution of an unstable asymmetric 3-vortex profile. In this case, the outer vortex migrates towards the attracting basin within the lossy side while the remaining vortex-dipole gets violently ejected and the two vortices annihilate each other. This again confirms the dynamical relevance of the single-vortex corresponding ghost state. Another possible scenario for the 3-vortex configuration is depicted in Fig. 18b where this time the central vortex-dipole slowly migrates towards the lossy side and slowly re-approaches the center of the trap as time progresses while the outer vortex performs large excursions around the periphery of the cloud where it is eventually absorbed. In this setting, we observe once again the dynamics of a vortex-dipole ghost state (with the background in this, as well as in the previous example featuring the analyzed growth and concurrent spreading). For contrasting purposes, we depict in Fig. 18c the dynamics of an unstable *symmetric* three-vortex state for $\varepsilon = 0$. In this case, the external potential is conservative and the motion of the three vortices is Hamiltonian and thus, typically, the three vortices will remain orbiting inside the cloud for very long times. Nevertheless, notice in the latter case the well established (and generic) instability of the Hamiltonian vortex tripole which leads to its break-up into a vortex dipole and a single vortex [43, 44]. Finally, in Fig. 18d we depict the evolution of another unstable asymmetric state. In this case it can be observed that two vortices annihilate each other and the third does not get absorbed by the lossy side as in Fig. 18a, but it performs large oscillations close to the periphery of the cloud where it is finally absorbed. The precession of this vortex due to its initial placement within the gain side suggests a direct analogy of this case (and after the dipole annihilation) with the top left panel of Fig. 8.

Finally, we briefly explore the dynamics of unstable dark soliton stripes. Dark soliton stripes are known to be unstable in the Hamiltonian case ($\varepsilon = 0$) due to the so-called snaking instability [45, 46] (although they can be rendered stable by a very tightly confining trap [47] or a wall-like external potential [48]). Figure 19a depicts the typical evolution of an unstable dark soliton stripe solution in the non-Hamiltonian ($\mathcal{PT}$-symmetric) case. In this case, the dark soliton stripe decays into a chain of alternating charge vortices induced by the snaking instability and their long term behavior includes annihilation and large oscillations where they get eventually absorbed at the periphery of the cloud leaving behind a highly perturbed (chargeless) Thomas-Fermi ground state. In contrast, as depicted in Fig. 19b, for the Hamiltonian case of $\varepsilon = 0$, most of the vortices that formed from the snaking instability remain interacting for long times (i.e., the latter are more likely to survive here, analogously to what we also saw for the tripole case above). Finally, in Fig. 19c we depict the evolution when starting with a TF cloud for ε past the critical value of existence of the stationary TF cloud. In this example, we compute the stationary (stable), chargeless, steady TF profile for $\varepsilon = 0.4$ and run it with $\varepsilon = 0.43$ (i.e., past the blue-sky bifurcation between the TF cloud and the vortex dipole states). We observe in this simulation that the central dip of the TF (which contains no charge) rapidly gets converted into a tight vortex-dipole state which does a small excursion towards the lossy side (very similar to the dynamics of the unstable tight vortex-dipole profile of Fig. 17a), clearly suggesting the persistence and dynamical selection of the corresponding vortex-dipole ghost state. This behavior is akin to what we observed in 1D for values of ε past $\varepsilon_{\text{cr}}^{(2)}$ (the dark soliton 'sprinkler' case) where excited states with topological charge are dynamically created from chargeless configurations. Hence, the two-dimensional setting may be used as a "vortex sprinkler".

8 Conclusions

In the present work, we offered a basic, yet reasonably self-contained view on the fundamental modifications that the phenomenology of nonlinear entities sustains in the presence of $\mathcal{PT}$-symmetric confining potentials (although the case where there is no real potential was also touched upon as well). The prototypical among the observed features of such systems were found to be the emergence of symmetry breaking bifurcations that destabilize single, triple, five, etc. soliton states in one dimension and similarly one- or two- or higher vortex states in two-dimensions. The second important and unexpected feature was the emergence of a nonlinear analogue of the $\mathcal{PT}$-phase-transition, whereby e.g. the ground and first excited, the second and third excited, the fourth and fifth excited state and so on (in 1D) are led to pairwise collisions and blue-sky bifurcations. Generalizations of these effects which, however, properly respect the topological nature of the states (e.g. the zero-vortex with the vortex dipole, the single vortex with the vortex tripole etc.) are found to arise in the two-dimensional setting. In addition, the pitchfork nature of the symmetry-breaking bifurcation was illustrated through the examination of the so-called ghost states and the proof, not only of their genuinely complex eigenvalue parameter, but also of their direct relevance for the dynamics of the full problem, despite the fact that they do not constitute exact solutions thereof (but rather only solutions of the static problem). The dynamical evolution of unstable soliton and vortex states was examined for all the identified instabilities, revealing a wealth of possibilities involving ghost states, soliton/vortex sprinklers, and oscillations between the loss and gain sides.

It would be interesting to offer a detailed map of the ghost states of the system. Here, we have illustrated their existence and observed many of their dynamical implications, yet a systematic characterization of their emergence (and perhaps the more challenging understanding of their basins of attraction) certainly merits further examination. On the other hand, a feature that has been examined in some detail in the context of $\mathcal{PT}$-symmetric linear problems which is, arguably, worthwhile to consider here concerns the analytic continuation of the states past the point of the $\mathcal{PT}$-phase-transition. This was already touched upon in the nonlinear realm for double well potentials in the work of Refs. [35–38], but it would be relevant to offer a definitive view of such analytically continued states for general potentials. Finally, a canonical set of investigations which is still missing concerns the effects of such potentials in three-dimensional continuum or one- and higher-dimensional (infinite) lattice contexts. These topics will be pursued in future publications.

Acknowledgements V.A. gratefully acknowledges support from the A.G. Leventis Foundation. P.G.K. gratefully acknowledges support from the National Science Foundation under grants DMS-0806762 and CMMI-1000337, as well as from the Alexander von Humboldt Foundation, the Alexander S. Onassis Public Benefit Foundation and the Binational Science Foundation. The work of D.J.F. was partially supported by the Special Account for Research Grants of the University of Athens. R.C.G. gratefully acknowledges support from the National Science Foundation under grant DMS-0806762.

References

1. C.M. Bender, S. Boettcher, Phys. Rev. Lett. **80**, 5243 (1998)
2. C.M. Bender, S. Boettcher, P.N. Meisinger, J. Math. Phys. **40**, 2201 (1999); C.M. Bender, Rep. Prog. Phys. **70**, 947 (2007)
3. A. Ruschhaupt, F. Delgado, J.G. Muga, J. Phys. A: Math. Gen. **38**, L171 (2005)
4. Y. Chen, A.W. Snyder, D.N. Payne, IEEE J. Quantum Electron. **28** 239 (1992)
5. Z.H. Musslimani, K.G. Makris, R. El-Ganainy, D.N. Christodoulides, Phys. Rev. Lett. **100**, 030402 (2008)
6. K.G. Makris, R. El-Ganainy, D.N. Christodoulides, Z.H. Musslimani, Phys. Rev. A **81**, 063807 (2010)
7. A. Guo, G.J. Salamo, D. Duchesne, R. Morandotti, M. Volatier-Ravat, V. Aimez, G.A. Siviloglou, D.N. Christodoulides, Phys. Rev. Lett. **103**, 093902 (2009)
8. C.E. Rüter, K.G. Makris, R. El-Ganainy, D.N. Christodoulides, M. Segev, and D. Kip, Nat. Phys. **6**, 192 (2010)
9. J. Schindler, A. Li, M.C. Zheng, F.M. Ellis, T. Kottos, Phys. Rev. A **84**, 040101 (2011)
10. H. Ramezani, T. Kottos, R. El-Ganainy, and D.N. Christodoulides, Phys. Rev. A **82**, 043803 (2010)
11. A.A. Sukhorukov, Z. Xu, Yu.S. Kivshar, Phys. Rev. A **82**, 043818 (2010)
12. M.C. Zheng, D.N. Christodoulides, R. Fleischmann, T. Kottos, Phys. Rev. A **82**, 010103(R) (2010)

13. E.M. Graefe, H.J. Korsch, A.E. Niederle, Phys. Rev. Lett. **101**, 150408 (2008)
14. E.M. Graefe, H.J. Korsch, A.E. Niederle, Phys. Rev. A **82**, 013629 (2010)
15. Z. Lin, H. Ramezani, T. Eichelkraut, T. Kottos, H. Cao, D.N. Christodoulides, Phys. Rev. Lett. **106**, 213901 (2011)
16. K. Li, P.G. Kevrekidis, Phys. Rev. E **83**, 066608 (2011)
17. S.V. Dmitriev, S.V. Suchkov, A.A. Sukhorukov, Yu.S. Kivshar, Phys. Rev. A **84**, 013833 (2011)
18. S.V. Suchkov, B.A. Malomed, S.V. Dmitriev, Yu.S. Kivshar, Phys. Rev. E **84**, 046609 (2011)
19. R. Driben, B.A. Malomed, Opt. Lett. **36**, 4323 (2011)
20. R. Driben, B.A. Malomed, Europhys. Lett. **96**, 51001 (2011)
21. F.Kh. Abdullaev, V.V. Konotop, M. Ögren, M.P. Sørensen, Opt. Lett. **36**, 4566 (2011)
22. N.V. Alexeeva, I.V. Barashenkov, A.A. Sukhorukov, Yu.S. Kivshar, Phys. Rev. A **85**, 063837 (2012)
23. A.A. Sukhorukov, S.V. Dmitriev, Yu.S. Kivshar, Opt. Lett. **37**, 2148 (2012)
24. A.E. Miroshnichenko, B.A. Malomed, Yu.S. Kivshar, Phys. Rev. A **84**, 012123 (2011)
25. F. Kh. Abdullaev, Y.V. Kartashov, V.V. Konotop, D.A. Zezyulin, Phys. Rev. A **83**, 041805 (2011)
26. D.A. Zezyulin, Y.V. Kartashov, V.V. Konotop, Europhys. Lett. **96**, 64003 (2011)
27. D.A. Zezyulin, V.V. Konotop, Phys. Rev. A **85**, 043840 (2012)
28. H.G. Li, Z.W. Shi, X.J. Jiang, X. Zhu, Opt. Lett. **36**, 3290 (2011)
29. V. Achilleos, P.G. Kevrekidis, D.J. Frantzeskakis, R. Carretero-González, Phys. Rev. A **86**, 013808 (2012)
30. P.G. Kevrekidis, D.J. Frantzeskakis, R. Carretero-González, *Emergent Nonlinear Phenomena in Bose-Einstein Condensates: Theory and Experiment* (Springer, Heidelberg, 2008)
31. Yu.S. Kivshar, G.P. Agrawal, *Optical solitons: from fibers to photonic crystals* (Academic, San Diego, 2003)
32. S. Klaiman, U. Günther, N. Moiseyev, Phys. Rev. Lett. **101**, 080402 (2008)
33. P.Würtz, T. Langen, T. Gericke, A. Koglbauer, H. Ott, Phys. Rev. Lett. **103**, 080404 (2009)
34. M. Hiller, T. Kottos, A. Ossipov, Phys. Rev. A **73**, 063625 (2006)
35. H. Cartarius, G. Wunner, Phys. Rev. A **86**, 013612 (2012); H. Cartarius, D. Haag, D. Dast, G. Wunner, J. Phys. A: Math. Theor. **45**, 444008 (2012)
36. E.-M. Graefe, J. Phys. A: Math. Theor. **45**, 444015 (2012)
37. D. Dast, D. Haag, H. Cartarius, G. Wunner, R. Eichler, J. Main, Fortschr. Phys. (in press). doi:10.1002/prop.201200080
38. A.S. Rodrigues, K. Li, V. Achilleos, P.G. Kevrekidis, D.J. Frantzeskakis, C.M. Bender, Romanian Reports in Physics **65**, 5 (2013). arXiv:1207.1066
39. D.J. Frantzeskakis, J. Phys. A: Math. Theor. **43**, 213001 (2010)
40. Yu.S. Kivshar, B. Luther-Davies, Phys. Rep. **298**, 81 (1998)
41. G. Theocharis, P.G. Kevrekidis, M.K. Oberthaler, D.J. Frantzeskakis, Phys. Rev. A 76, 045601 (2007); A. Weller, J.P. Ronzheimer, C. Gross, J. Esteve, M.K. Oberthaler, D.J. Frantzeskakis, G. Theocharis, P.G. Kevrekidis, Phys. Rev. Lett. **101**, 130401 (2008); G. Theocharis, A. Weller, J.P. Ronzheimer, C. Gross, M.K. Oberthaler, P.G. Kevrekidis, D.J. Frantzeskakis, Phys. Rev. A **81**, 063604 (2010)
42. P.G. Kevrekidis, D.J. Frantzeskakis, Discr. Cont. Dyn. Sys. S **4**, 1199 (2011)
43. S. Middelkamp, P.G. Kevrekidis, D.J. Frantzeskakis, R. Carretero-González, P. Schmelcher, Phys. Rev. A **82**, 013646 (2010)
44. J. Stockhofe, S. Middelkamp, P.G. Kevrekidis, P. Schmelcher, EPL **93**, 20008 (2011)
45. E.A. Kuznetsov, J.J. Rasmussen, Phys. Rev. E **51**, 4479 (1995)
46. D.E. Pelinovsky, Yu.A. Stepanyants, Yu.S. Kivshar, Phys. Rev. E **51**, 5016 (1995)
47. P.G. Kevrekidis, G. Theocharis, D.J. Frantzeskakis, A. Trombettoni, Phys. Rev. A **70**, 023602 (2004)
48. Manjun Ma, R. Carretero-González, P.G. Kevrekidis, D.J. Frantzeskakis, B.A. Malomed, Phys. Rev. A **82**, 023621 (2010)

Coding of Nonlinear States for NLS-Type Equations with Periodic Potential

G.L. Alfimov and A.I. Avramenko

Abstract The problem of complete description of nonlinear states for NLS-type equations with periodic potential is considered. We show that in some cases all nonlinear states for equations of such kind can be coded by bi-infinite sequences of symbols of N-symbol alphabet (*words*). Sufficient conditions for one-to-one correspondence between the set of nonlinear states and the set of these bi-infinite words are given in the form convenient for numerical verification (Hypotheses 1–3). We report on numerical check of these hypotheses for the case of Gross-Pitaevskii equation with cosine potential and indicate regions in the space of governing parameters where this coding is possible.

1 Introduction

The model equation

$$u_{xx} + (\mu - V(x))u - F(u) = 0, \quad \mu \in \mathbb{R}, \quad V(x), u(x) \in \mathbb{R} \tag{1}$$

where $V(x)$ is a periodic function, has been widely used to study steady states in nonlinear periodic media. In particular, Eq. (1) describes nonlinear modes

$$\psi(x,t) = \exp(-i\mu t)u(x) \tag{2}$$

for NLS-type equation

$$i\psi_t = -\psi_{xx} + f(|\psi|^2)\psi + V(x)\psi. \tag{3}$$

G.L. Alfimov (✉) · A.I. Avramenko
National Research University of Electronic Technology, Zelenograd, Moscow, 124498, Russia
e-mail: galfimov@yahoo.com; avramenkoaleksey@gmail.com

R. Carretero-González et al. (eds.), *Localized Excitations in Nonlinear Complex Systems*, Nonlinear Systems and Complexity 7, DOI 10.1007/978-3-319-02057-0_2,

Equation (3) arises in various physical applications such as optics, plasma physics and theory of Bose-Einstein condensation (BEC). In BEC context μ has the sense of chemical potential and $\psi(x,t)$ describes the macroscopic wave function. The periodic potential $V(x)$ corresponds to an optical trap to confine the condensate, see the surveys [1–3]. Let us give some examples.

(i) The classical version of Eq. (3) corresponds to the case $f(\xi) = g\xi$ where $g = \pm 1$ (both the signs "+" and "−" are admissible from physical viewpoint). In this case Eq. (3) becomes the Gross-Pitaevskii equation [4]

$$i\psi_t = -\psi_{xx} + g|\psi|^2\psi + V(x)\psi. \tag{4}$$

The case $g = 1$ corresponds to repulsive interparticle interactions, $g = -1$ – to attractive interactions. Using ansatz (2) and assuming that $u(x)$ is real we arrive at the following version of Eq. (1)

$$u_{xx} + (\mu - V(x))u - gu^3 = 0. \tag{5}$$

(ii) In order to describe complex nonlinear modes for Eq. (4) one can separate amplitude and phase of the solution and use the substitution

$$\psi(x,t) = u(x)\exp\{-i(\mu t + \phi(x))\}$$

instead of (2). Then one arrives at the equation for the amplitude $u(x)$

$$u_{xx} + (\mu - V(x))u - gu^3 - \frac{M^2}{u^3} = 0, \tag{6}$$

where M is an arbitrary real constant [5]. The phase $\phi(x)$ can be found from the relation $u^2\phi_x = M$. Equation (6) is a particular case of Eq. (1) with $F(u) = gu^3 + M^2/u^3$.

(iii) In some situations the cubic nonlinearity in (4) must be replaced by a nonlinearity of more general type, for instance, quintic or cubic-quintic one [6,7]. Taking $f(\xi) = g_1\xi + g_2\xi^2$ and assuming that $u(x)$ is real one arrives at the equation

$$u_{xx} + (\mu - V(x))u - g_1u^3 - g_2u^5 = 0. \tag{7}$$

It is known that even in the simplest case of cubic nonlinearity and model cosine potential

$$V(x) = A\cos 2x \tag{8}$$

Eq. (1) describes a great variety of nonlinear states. Among them are: bright and dark gap solitons [5,8–10], nonlinear periodic structures (nonlinear Bloch waves) [8,11], domain walls [12], gap waves [13] and so on. However, to best of our knowledge, up to the moment a classification of all the possible types of nonlinear modes for Eq. (1) has not been constructed.

In the present contribution we address the problem of *complete description of nonlinear states* covered by Eq. (1). Our approach is based on the following observation: for some class of nonlinearities the "most part" of the solutions for Eq. (1) are *singular* i.e. they collapse (tend to infinity) at some finite point of real axis. In particular, such a situation takes place in the case of GPE with repulsive nonlinearity, $g = 1$, Eqs. (5), (6), and for cubic-quintic generalization of GPE if $g_2 > 0$, Eq. (7). In these cases the set of bounded in $\mathbb{R}$ solutions for Eq. (1) is quite scanty, it can be found numerically by properly organized scanning of initial data plane for the Cauchy problem. Then the transformations of this initial data under the action of Poincaré map T can be interpreted in terms of symbolic dynamics.

This contribution is aimed to present *a general scheme* of the coding technique. The proofs of theorems are omitted (see the paper [14] for details). The main results are presented for Eq. (5) with $g = 1$, i.e.

$$u_{xx} + (\mu - V(x))u - u^3 = 0, \tag{9}$$

where $\mu, V(x), u(x) \in \mathbb{R}$ and the potential $V(x)$ is π-periodic. In Sect. 2 we introduce some notations and definitions which will be used throughout the rest of the text and formulate some statements about them. In Sect. 3 we formulate some statements from symbolic dynamics theory and apply them to describe the dynamics of iterated Poincaré map. The main outcome of Sect. 3 is formulated in a form of three Hypotheses. These Hypotheses give sufficient conditions for one-to-one correspondence between the bounded solutions of Eq. (9) and coding sequences. These conditions should be verified numerically. Section 4 contains application of this approach to the case of the cosine potential (8). Section 5 includes short summary and discussion.

2 Bounded and Singular Solutions of Stationary States

2.1 Some Definitions

In what follows we refer to a solution $u(x)$ of Eq. (9) as *a singular solution* if for some $x = x_0$

$$\lim_{x\to x_0} u(x) = +\infty \quad \text{or} \quad \lim_{x\to x_0} u(x) = -\infty.$$

In this case we say that the solution $u(x)$ *collapses* at x_0. Also let us introduce the following definitions:

Collapsing and non-collapsing points: A point (u_0, u_0') of the plane $\mathbb{R}^2 = (u, u')$ is

- *L-collapsing forward point,* $L > 0$, if the solution of Cauchy problem for Eq. (9) with initial data $u(0) = u_0$, $u_x(0) = u_0'$ collapses at value $x = x_0$ and $0 < x_0 < L$;
- *L-non-collapsing forward point,* $L > 0$, if the solution of Cauchy problem for Eq. (9) with initial data $u(0) = u_0$, $u_x(0) = u_0'$ does not collapse at any value $x = x_0, 0 < x_0 \le L$.
- *L-collapsing backward point* if the corresponding solution of Cauchy problem for Eq. (9) collapses at some value $x = -x_0$ and $0 < x_0 < L$;
- *L-non-collapsing backward point* if the corresponding solution of Cauchy problem for Eq. (9) does not collapse at any value $x = -x_0, 0 < x_0 \le L$;
- *A collapsing point* if it is either L-collapsing forward point or L-collapsing backward point for some L.

Functions $h^{\pm}(\tilde{u}, \tilde{u}')$. The functions $h^+(\tilde{u}, \tilde{u}')$ and $h^-(\tilde{u}, \tilde{u}')$ are defined in $\mathbb{R}^2$ as follows: $h^+(\tilde{u}, \tilde{u}') = x_0$ if the solution of Cauchy problem for Eq. (9) with initial data $u(0) = \tilde{u}$, $u_x(0) = \tilde{u}'$ collapses at value $x = x_0$, $x_0 > 0$. Similarly, $h^-(\tilde{u}, \tilde{u}') = -x_0$ if the solution of Cauchy problem for Eq. (9) with initial data $u(0) = \tilde{u}$, $u_x(0) = \tilde{u}'$ collapses at value $x = x_0$, $x_0 < 0$.

The sets $\mathcal{U}_L^{\pm}$ and $\mathcal{U}_L$. We denote the set of all L-non-collapsing forward points by $\mathcal{U}_L^+$ and the set of all L-non-collapsing backward points by $\mathcal{U}_L^-$. In terms of the functions $h^{\pm}(u, u')$ these sets are

$$\mathcal{U}_L^+ = \{(u, u') \in \mathbb{R}^2 | \, h^+(u, u') > L\}, \quad \mathcal{U}_L^- = \{(u, u') \in \mathbb{R}^2 | \, h^-(u, u') > L\}.$$

The intersection of $\mathcal{U}_L^+$ and $\mathcal{U}_L^-$ will be denoted by $\mathcal{U}_L$. Evidently, if $L_1 < L_2$ then $\mathcal{U}_{L_2}^+ \subset \mathcal{U}_{L_1}^+, \mathcal{U}_{L_2}^- \subset \mathcal{U}_{L_1}^-$ and $\mathcal{U}_{L_2} \subset \mathcal{U}_{L_1}$.

2.2 Some Statements About Collapsing Points

It turns out that the "most part" of the plane (u, u') of initial data are either L-collapsing forward points or L-collapsing backward points for some L. This follows from Theorem 1 below:

Theorem 1. *Let the potential $V(x)$ be continuous and bounded on $\mathbb{R}$. Then for each L there exist $\tilde{u}_L$ and $\tilde{u}_L'$ such that the set $\mathcal{U}_L$ is situated in the rectangle $-\tilde{u}_L < u < \tilde{u}_L$, $-\tilde{u}_L' < u' < \tilde{u}_L'$.*

The proof of Theorem 1 is rather technical and can be found in [14]. Another important statement is as follows:

Theorem 2. *Let the potential $V(x)$ be continuous and bounded on $\mathbb{R}$ and $h^+(u_0, u_0') = L < \infty$. Then $h^+(u, u')$ is a continuous function in some vicinity of the point (u_0, u_0').*

The proof of Theorem 2 also is presented in [14]. It worth commenting Theorem 2 as follows:

(i) Analogous statement is valid for the function $h^-(u, u')$;
(ii) It follows from Theorem 2 that the boundary of the set $\mathcal{U}_L^+$ *is continuous* and corresponds to the level line $h^+(u, u') = L$ of the function $h^+(u, u')$. This boundary consists of the points $(\tilde{u}, \tilde{u}')$ such that the solution of Eq. (9) with initial data $u(0) = \tilde{u}$, $u_x(0) = \tilde{u}'$ satisfies one of the conditions

$$\lim_{x \to L} u(x) = +\infty \quad \text{or} \quad \lim_{x \to L} u(x) = -\infty$$

Correspondingly, the boundary of the set $\mathcal{U}_L^-$ is also continuous and consists of the points $(\tilde{u}, \tilde{u}')$ such that the solution of Eq. (9) with initial data $u(0) = \tilde{u}$, $u_x(0) = \tilde{u}'$ satisfy one of the conditions

$$\lim_{x \to -L} u(x) = +\infty \quad \text{or} \quad \lim_{x \to -L} u(x) = -\infty$$

The set of solutions for Eq. (9) that collapse at given point $x = x_0$ can be described more precisely in terms of asymptotic expansions.

Theorem 3. *Let $x = x_0$ be an arbitrary fixed real. Assume that $U(x) = \omega - V(x)$ in a vicinity of $x = x_0$ can be represented as follows*

$$U(x) = U_0 + U_1\delta + U_2\delta^2 + U_3\delta^3 + o\left(\delta^3\right)$$

where $\delta \equiv x - x_0$. Then the solutions of Eq. (9) which satisfy the condition

$$\lim_{x \to x_0} u(x) = +\infty \tag{10}$$

obey the asymptotic expansion

$$u(\delta) = \frac{\sqrt{2}}{\delta} + A_1\delta + A_2\delta^2 + A_3\delta^3 \ln|\delta| + \mathbf{C}\delta^3 + A_4\delta^4 + o\left(\delta^4\right) \tag{11}$$

Here $\mathbf{C} \in \mathbb{R}$ is a free parameter and

$$A_1 = \frac{\sqrt{2}U_0}{6}; \quad A_2 = \frac{\sqrt{2}U_1}{4}; \quad A_3 = -\frac{\sqrt{2}U_2}{5}, \quad A_4 = \frac{\sqrt{2}}{6}\left(\frac{U_0 U_1}{12} - U_3\right)$$

The statement above follows from straightforward substitution of series (11) into Eq. (9). Theorem 3 should be commented as follows:

(i) The free parameter **C** is "internal" parameter of *continuous one-parameter set* of solutions which tend to $+\infty$ at the point $x = x_0$. This agrees well with the fact that the "most part" of solutions Eq. (9) are singular.
(ii) Since Eq. (9) is invariant with respect to the symmetry $u \to -u$, the solutions of Eq. (9) which satisfy the condition

$$\lim_{x \to x_0} u(x) = -\infty$$

obey the same, up to sign, asymptotic expansion (11).

3 Periodic Potential: Poincaré Map

3.1 Poincaré Map

Assume now that the potential $V(x)$ is *continuous* and π*-periodic*

$$V(x + \pi) = V(x).$$

The prototypical example is the cosine potential (8) which appears as a basic model in numerous studies. Define the *Poincaré map* $T : \mathbb{R}^2 \to \mathbb{R}^2$ associated with Eq. (9) as follows: if $p = (\tilde{u}, \tilde{u}') \in \mathbb{R}^2$ then $Tp = (u(\pi), u_x(\pi))$ where $u(x)$ is a solution of Eq. (9) with initial data $u(0) = \tilde{u}$, $u_x(0) = \tilde{u}'$. The map T is an *area-preserving diffeomorphism*. It is important that T is defined *not* in the whole $\mathbb{R}^2$, but only on the set of π-non-collapsing forward points for Eq. (9), i.e. $\mathcal{U}_\pi^+$. Inverse map T^{-1} is defined on the set $\mathcal{U}_\pi^-$. Evidently, $T\mathcal{U}_\pi^+ = \mathcal{U}_\pi^-$.

If, in addition, the potential $V(x)$ is *even*, $V(x) = V(-x)$, Eq. (9) is reversible. Denote I the reflection with respect to u axis in the plane (u, u'). Due to reversibility of Eq. (9), if $p \in \mathcal{U}_\pi^+$ then $Ip \in \mathcal{U}_\pi^-$ and

$$T^{-1}Ip = ITp. \tag{12}$$

In general, the sets $\mathcal{U}_\pi^+$ and $\mathcal{U}_\pi^-$ are connected by the relations $I\mathcal{U}_\pi^+ = \mathcal{U}_\pi^-$, $I\mathcal{U}_\pi^- = \mathcal{U}_\pi^+$. The set $\mathcal{U}_\pi = \mathcal{U}_\pi^+ \cap \mathcal{U}_\pi^-$ consists of the points which has both T-image and T-pre-image. It follows from Sect. 2 that $\mathcal{U}_\pi$ is bounded.

We call *an orbit* a sequence of points (finite, infinite or bi-infinite) $\{p_n\}$, such that $Tp_n = p_{n+1}$. The fixed points of T correspond to π-periodic solutions of Eq. (9). These solutions do exist for quite general periodic potential $V(x)$, see [15]. For a fixed point p let us denote DT_p the operator of linearization of T at p. Denote $\lambda_{1,2}$ the eigenvalues of DT_p. Since the map is area-preserving, $\lambda_1\lambda_2 = 1$. Depending on the behavior of T in a vicinity of a fixed point, it may be of elliptic or hyperbolic type [16]. In the case of hyperbolic fixed point both $\lambda_{1,2}$ are real and in the case of the elliptic point they are complex conjugated, $|\lambda_{1,2}| = 1$. Also we call a *k-cycle* of T an orbit which consists of points $p_1, \dots, p_k \in \mathbb{R}^2$ such that

$$Tp_1 = p_2, \quad Tp_2 = p_3, \dots, \quad Tp_k = p_1.$$

Evidently p_1 is a fixed point for T^k. The k-cycles correspond to $k\pi$-periodic solutions of Eq. (9). A k-cycle is elliptic if p_1 is an elliptic fixed point for T^k and it is hyperbolic if p_1 is a hyperbolic fixed point for T^k.

The orbits of interest for this study are those which are (i) bi-infinite and (ii) remain within the set $\mathcal{U}_\pi$ for any $n \in \mathbb{Z}$.

3.2 Symbolic Dynamics: Theory

Let (u, u') be Cartesian coordinates in $\mathbb{R}^2$ and $\nu(S)$ be a measure of set S in $\mathbb{R}^2$. Let us remind that the function $f(u)$ is *γ-Lipschitz function* on some interval $J \in \mathbb{R}$ if for any $u_1, u_2 \in J$ the relation holds

$$|f(u_2) - f(u_1)| \leq \gamma |u_2 - u_1|.$$

Introduce the following definitions.

Definition. We call *an island* a curvilinear quadrangle D formed by nonintersecting curve segments α^+, β^+, α^-, β^- (α^+ and α^- are opposite sides of the quadrangle and have no common points as well as β^+ and β^-) such that

- The segments α^+ and α^- are graphs of monotone non-decreasing/non-increasing γ-Lipschitz functions $u' = v_\pm(u)$ for some γ;
- The segments β^+ and β^- are graphs of monotone non-increasing/non-decreasing γ-Lipschitz functions $u = h_\pm(u')$ for the same γ;
- If $v_\pm(u)$ are monotone non-decreasing functions then $h_\pm(u)$ are monotone non-increasing functions and vice versa.

Definition. Let γ be a fixed real and D be an island bounded by curve segments $\alpha^+, \beta^+, \alpha^-, \beta^-$. We call *v-curve* a curve segment β with endpoints on α^- and α^+ which is

- A graph of monotone non-decreasing γ-Lipschitz function $u' = v(u)$ if $\beta^\pm$ are graphs of monotone non-decreasing functions;
- A graph of monotone non-increasing γ-Lipschitz function $u' = v(u)$ if $\beta^\pm$ are graphs of monotone non-increasing functions.

Similarly, we call *h-curve* a curve segment α with endpoints on β^- and β^+ which

- Is a graph of monotone non-decreasing γ-Lipschitz function $u = h(u')$ if $\alpha^\pm$ are graphs of monotone non-decreasing functions;

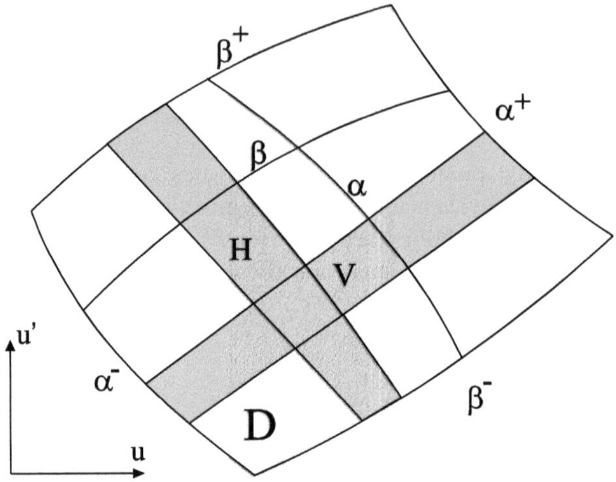

Fig. 1 An island D with v-curve β, v-strip V, h-curve α and h-strip H

- Is a graph of monotone non-increasing γ-Lipschitz function $u = h(u')$ if $\alpha^{\pm}$ are graphs of monotone non-increasing functions.

Definition. Let D be an island. We call *v-strip* an area concluded between two nonintersecting v-curves. Similarly, we call *h-strip* an area concluded between two nonintersecting h-curves.

Comment. According to the given definition of island any island is both h-strip and v-strip simultaneously.

Figure 1 illustrates schematically the definitions introduced above.

Let T be a diffeomorphism defined on a set $D = \bigcup_{i=1}^{N} D_i$ where each D_i, $i = 1, \ldots, N$ is an island and all the islands are disjoined. Introduce the set $\mathcal{P}$ of bi-infinite orbits $\{\ldots, p_{-1}, p_0, p_1, \ldots\}$, $Tp_i = p_{i+1}$, where all p_i, $i = 0, \pm 1, \pm 2, \ldots$ belong to D. Let us denote Ω^N the set of bi-infinite sequences $\{\ldots, i_{-1}, i_0, i_1, \ldots\}$ where $i_k \in \{1, 2, \ldots, N\}$. Define a map $\Sigma : \mathcal{P} \to \Omega^N$ as follows: i_k is the number i of the component D_i where the point p_k lies. Evidently, each orbit from $\mathcal{P}$ defines uniquely a sequence from Ω^N. However, under some conditions each element from Ω^N also defines uniquely an orbit from $\mathcal{P}$, therefore Σ establishes one-to-one correspondence between $\mathcal{P}$ and Ω^N. The following statement is valid:

Theorem 4. *Assume that a diffeomorphism T is defined on a set of N disjoined islands D_i, $i = 1, \ldots, N$, $D = \bigcup_{i=1}^{N} D_i$. Assume that:*

(i) For any i, $i = 1, \ldots, N$, and for each v-strip $V \in D_i$ the intersection $TV \cap D_j$, $j = 1, \ldots, N$ is non-empty and is also a v-strip. Similarly, for any i, $i = 1, \ldots, N$, and for each h-strip $H \in D_i$ the intersection $T^{-1}H \cap D_j$, $j = 1, \ldots, N$ is non-empty and is also an h-strip.

(ii) Denote

$$\Delta_0^+ = D, \quad \Delta_n^+ = T\Delta_{n-1}^+ \cap D,$$
$$\Delta_0^- = D, \quad \Delta_n^- = T^{-1}\Delta_{n-1}^- \cap D.$$

Then

$$\lim_{n\to\infty} \nu(\Delta_n^+) = 0, \quad \lim_{n\to\infty} \nu(\Delta_n^-) = 0.$$

Then each element $\mathbf{s} \in \Omega^N$ *defines uniquely an orbit* $\mathbf{p} \in \mathcal{P}$ *such that* $\Sigma\mathbf{p} = \mathbf{s}$.

Theorems of such kind are well-known in dynamical system theory. The conditions (i)–(ii) can be viewed as some version of Conley-Moser conditions [16]. In particular, Theorem 4 allows to describe coding of orbits for Smale horseshoe [16] (in this case $N = 2$). The proof of Theorem 4 is presented in [14].

3.3 Symbolic Dynamics: Application

Basing on Theorem 4 a strategy for coding of bounded solutions for Eq. (9) would consist in verifying the following hypotheses.

Hypothesis 1. The set $\mathcal{U}_\pi$ consists of N disjoined islands D_i, $i = 1, \dots, N$, i.e. of N curvilinear quadrangles bounded by curves which possess some monotonic properties (see Sect. 3.2).

Hypothesis 2. The Poincaré map T associated with Eq. (9) is such that

(a) T maps v-strips of any D_i, $i = 1, \dots, N$, in such a way that for any v-strip V, $V \in D_i$, all the intersections $TV \cap D_j$, $j = 1, \dots, N$, are nonempty and are v-strips.
(b) the inverse map T^{-1} maps h-strips of any D_i, $i = 1, \dots, N$, in such a way that for any h-strip H, $H \in D_i$, the intersections $T^{-1}H \cap D_j$, $j = 1, \dots, N$, are nonempty and are h-strips.

Comment. If the periodic potential $V(x)$ is *even* the point (b) of Hypothesis 2 follows from the point (a). In fact, if H is an h-strip then IH is a v-strip where I is a reflection with respect to ψ axis. Then the statement (b) follows from the relation (12).

Hypothesis 3. The sequences of sets $\Delta_n^\pm$ defined as follows

$$\Delta_0^+ = \mathcal{U}_\pi, \quad \Delta_n^+ = T\Delta_{n-1}^+ \cap \mathcal{U}_\pi,$$
$$\Delta_0^- = \mathcal{U}_\pi, \quad \Delta_n^- = T^{-1}\Delta_{n-1}^- \cap \mathcal{U}_\pi,$$

is such that $\lim_{n\to\infty} \nu(\Delta_n^\pm) = 0$.

Comment. In the case of even periodic potential $V(x)$ the relation (12) implies that $\nu(\Delta_n^+) = \nu(\Delta_n^-)$ for any n. Therefore in order to verify Hypothesis 3 in this case it is enough to check the condition $\lim_{n\to\infty} \nu(\Delta_n^+) = 0$ only.

It follows from Theorem 4 that if Hypotheses 1–3 hold then there exists one-to-one correspondence between all bounded in $\mathbb{R}$ solutions of Eq. (9) and the sequences from Ω^N which can be regarded as *codes* for these solutions. The "coding" sequence from Ω^N consists of numbers i of D_i in the order the orbit "visit" them. In the next section we describe the results of numerical study of Eq. (9) with cosine potential (8) i.e. for the equation

$$u_{xx} + (\mu - A\cos 2x)u - u^3 = 0. \tag{13}$$

We present numerical arguments that Hypotheses 1–3 hold for vast areas in the plane of parameters (μ, A).

4 Steady States with Cosine Potential

First, let us remind some facts from the theory of linearized equation

$$u_{xx} + (\mu - A\cos 2x)u = 0 \tag{14}$$

which is the Mathieu equation [17]. In the plane of parameters (μ, A) there exist zones where all the solutions of Eq. (14) are bounded (*bands*) and where all of them are unbounded (*gaps*). Figure 2 reproduces classical picture of gap/band structure for Eq. (14) in the plane (μ, A). It is known that this structure also play an important role in the theory of nonlinear equation (13) (see e.g. [1]). In terms of Poincaré map T associated with Eq. (13), if the point (μ, A) is situated in a band then the origin $O(0,0)$ is an elliptic fixed point for T and if (μ, A) lies in a gap then $O(0,0)$ is a hyperbolic fixed point for T.

The numerical study of Eq. (13) was fulfilled using interactive software elaborated specifically for this equation. It allows to visualize orbits of map T, to seek for fixed points of T^n, $n = 1, 2, \ldots$, to scan the plane of initial data (u, u') in order to construct the sets $\mathcal{U}_L^{\pm}$ and $\mathcal{U}_L$, to measure the areas of these sets and possesses some other useful features.

Let us now summarize the results of the numerical study.

Orbits for *T*. Figure 3 represents successive iterations of the map T for various initial points. Vast empty areas outskirts of the central region in panels a-d correspond to collapsing solutions Eq. (13). For these points the map T is not defined.

The sets $\mathcal{U}_\pi^{\pm}$ and $\mathcal{U}_\pi$. In order to visualize the sets $\mathcal{U}_L^{+}$ numerical scanning of the initial data plane (u, u') was fulfilled. Numerical study shows that for any

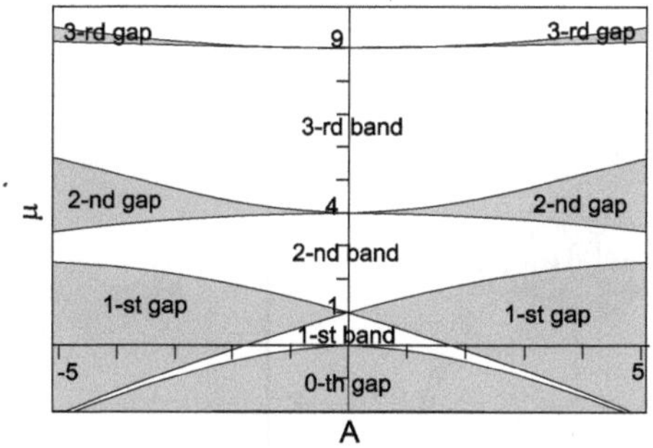

Fig. 2 Band/gap structure for the Mathieu equation

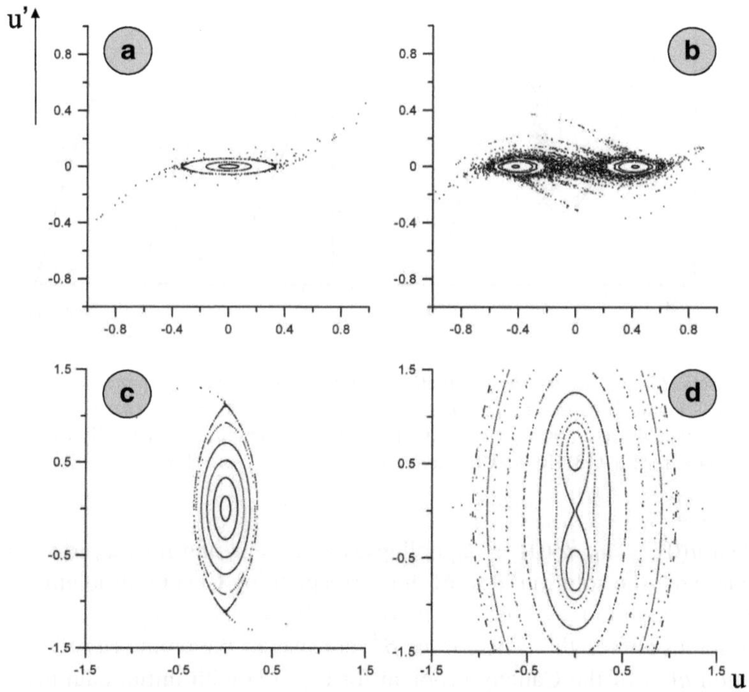

Fig. 3 Orbits for various initial points in the plane of initial data (iterated Poincaré maps), $A = -1$. (**a**) $\mu = 0$, 1st band; (**b**) $\mu = 0.6$, 1st gap; (**c**) $\mu = 2$, 2nd band; (**d**) $\mu = 4.05$, 2nd gap. Vast empty areas outskirts the central region in panels **a–d** correspond to collapsing solutions of Eq. (13). For these data the map T is not defined

values of parameters μ and A the sets $\mathcal{U}_\pi^\pm$ are *infinite curvilinear strips*. The typical shapes of the sets $\mathcal{U}_\pi^\pm$ for Eq. (13) are shown in Fig. 4. The boundary of $\mathcal{U}_\pi^+$ is represented by two continuous curves $\alpha^\pm$. The curve α^+ consists of the such points (u_0, u_0') that the solution $u(x)$ of the Cauchy problem for Eq. (13) with

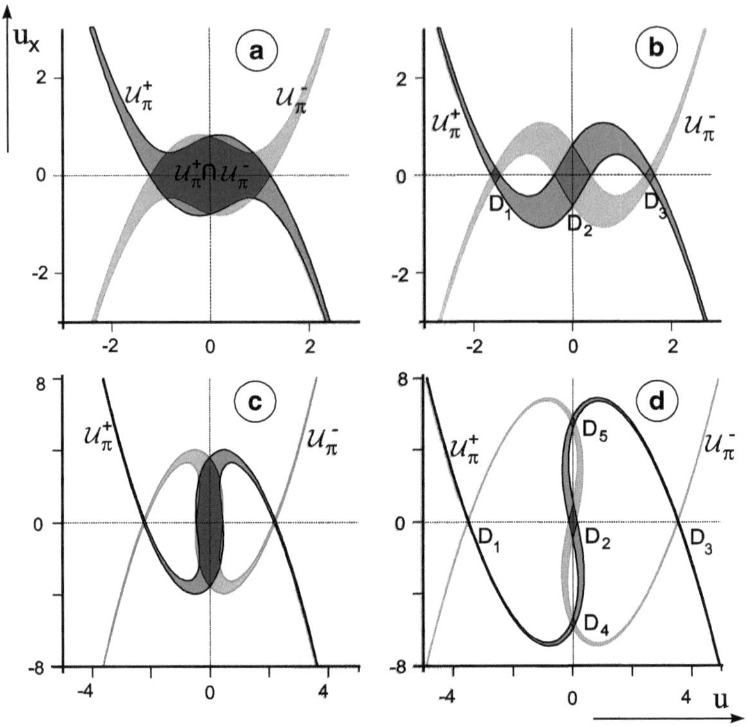

Fig. 4 The sets $\mathcal{U}_\pi^+$ and $\mathcal{U}_\pi^-$ for the parameters μ and A lying in the first (panels **a** and **b**) and the second (panels **c** and **d**) gaps. The parameters are: (**a**) $\mu = 1$, $A = -1$; (**b**) $\mu = 1$, $A = -3$; (**c**) $\mu = 4$, $A = -4$; (**d**) $\mu = 4$, $A = -10$. The sets were obtained numerically by scanning of initial data plane for Eq. (13). In *dark* the areas $\mathcal{U}_\pi = \mathcal{U}_\pi^+ \cap \mathcal{U}_\pi^-$ are shown

initial data $u(0) = u_0$, $u'(0) = u_0'$ collapses at $x = \pi$ and $\lim_{x\to\pi} u(x) = +\infty$. At the curve α^- the solution $u(x)$ of the corresponding Cauchy problem obeys the condition $\lim_{x\to\pi} u(x) = -\infty$. Similarly, the boundary of $\mathcal{U}_\pi^-$ is represented by two continuous curves $\beta^\pm$. The curves $\beta^\pm$ consists of the points (u_0, u_0') such that the solution $u(x)$ of the Cauchy problem for Eq. (13) with initial data $u(0) = u_0$, $u'(0) = u_0'$ collapses at $x = -\pi$ and $\lim_{x\to-\pi} u(x) = \pm\infty$. It is important that the set $\mathcal{U}_\pi = \mathcal{U}_\pi^+ \cap \mathcal{U}_\pi^-$ may consist of *several disjoined components*, as it is in panel b and d of Fig. 4.

The sets $\mathcal{U}_{n\pi}^\pm$. Figure 5 exhibits the sets $\mathcal{U}_{6\pi}^+$ for $\mu = 1$ and various values of A. The sets $\mathcal{U}_{6\pi}^-$ are the reflection of the sets $\mathcal{U}_{6\pi}^+$ with respect to u axes. In Fig. 5 the sets $\mathcal{U}_{6\pi}^\pm$ have quite a complex layered structure. The structure of $\mathcal{U}_{n\pi}^\pm$ becomes more sophisticated as n grows resembling *fractals*. The situation is similar to one described in [18] for Eq. (9) in the case of delta-comb potential (see Fig. 7 in Ref. [18]).

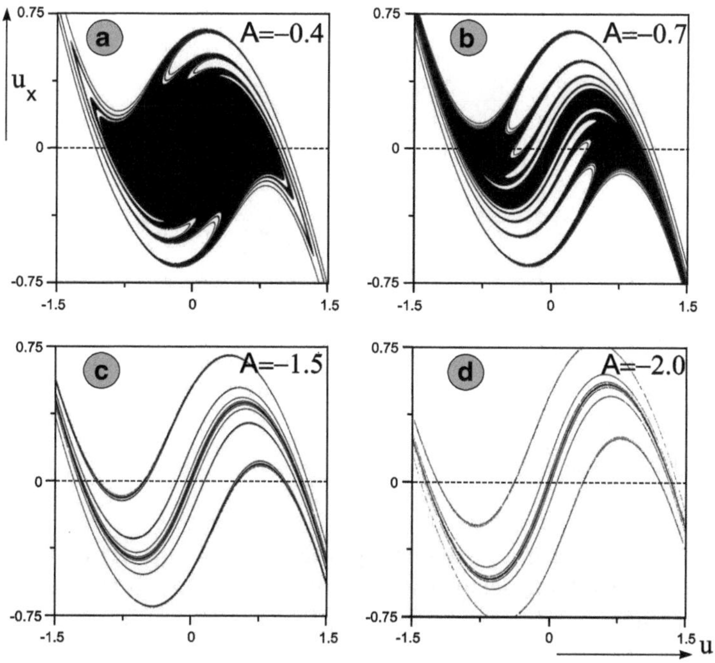

Fig. 5 The sets $\mathcal{U}_{6\pi}^{+}$ and $\mathcal{U}_{6\pi}^{-}$. In all the cases $\mu = 1$ and (**a**) $A = -0.4$; (**b**) $A = -0.7$; (**c**) $A = -1.5$; (**d**) $A = -2.0$. The steps of scanning with respect to u and u' are 0.0005 and 0.0002 correspondingly

4.1 *Hypothesis 1*

In all the cases the sets $\mathcal{U}_{\pi}^{+}$ and $\mathcal{U}_{\pi}^{-}$ shown in Fig. 4 are *curvilinear strips*. We found that this is *a general feature* of Eq. (13) for all parameters μ and A which we considered. The shapes of the strips $\mathcal{U}_{\pi}^{\pm}$ may be quite complex and their intersection $\mathcal{U}_{\pi}$ may consist of different number of disjoined sets. Since the strips $\mathcal{U}_{\pi}^{+}$ and $\mathcal{U}_{\pi}^{-}$ and related to each other by symmetry with respect to u axis, the typical situation is that $\mathcal{U}_{\pi}$ consists of *some number of curvilinear deltoids* (see Fig. 4, panels b and d) which are symmetrical with respect to u or u' axes.

Figure 6 shows the regions in the parameter plane (μ, A) where such decomposition of $\mathcal{U}_{\pi}$ takes place (the half-plane $A < 0$ is depicted). In the area above the curve $N = 3$ and below the curve $N = 5$ the set $\mathcal{U}_{\pi}$ consists of three connected components, in the area above the curve $N = 5$ it consists of five connected components etc. It is important that *these components are islands* (in the sense of Sect. 3.2) in the areas between the marked curves and upper boundaries of the gaps, shown in Fig. 6 in dark.

The possible numbers of islands are related (indirectly) to numbers of fixed points of the Poincaré map T. In its turn the number of fixed points of T is

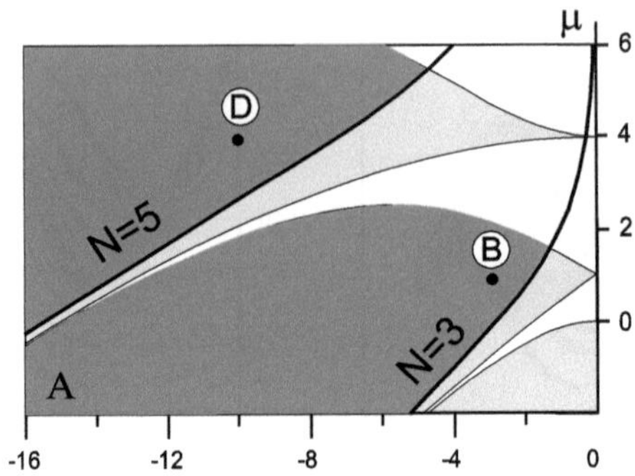

Fig. 6 The plane of parameters (μ, A) with band and gap zones. The boundaries of the regions where $\mathcal{U}_\pi$ is disjoined in three (the curve $N = 3$) or five (the curve $N = 5$) connected components are shown. These components are islands (in the sense of Sect. 3.2) in *dark areas* between the *marked curves* and *upper boundaries* of the corresponding gaps. The *points* corresponding to panels b and d of Fig. 4 are marked

determined by the number of band or gap where the point (μ, A) is situated in the plane of parameters. More detailed analysis of these relations is an interesting issue for a further study.

4.2 Hypothesis 2

Let DT_p be the operator of linearization of T at point p. Let

$$\mathbf{e}_1 = \begin{pmatrix} 1 \\ 0 \end{pmatrix}; \quad \mathbf{e}_2 = \begin{pmatrix} 0 \\ 1 \end{pmatrix}. \tag{15}$$

Define the functions

$$g_1(p) = (DT_p\mathbf{e}_1, \mathbf{e}_1) \cdot (DT_p\mathbf{e}_2, \mathbf{e}_1),$$
$$g_2(p) = (DT_p\mathbf{e}_1, \mathbf{e}_2) \cdot (DT_p\mathbf{e}_2, \mathbf{e}_2).$$

The way to verify the conditions of Hypothesis 2 is based on the following statement:

Theorem 5. *Assume that the potential $V(x)$ is even and the following conditions hold:*

- $\mathcal{U}_\pi^+$ *is an infinite curvilinear strips;*
- $\mathcal{U}_\pi^+ \cap \mathcal{U}_\pi^- = \mathcal{U}_\pi = \bigcup_{i=1}^N D_i$ *where D_i are nonintersecting islands;*
- *For each pair (i, j), if*

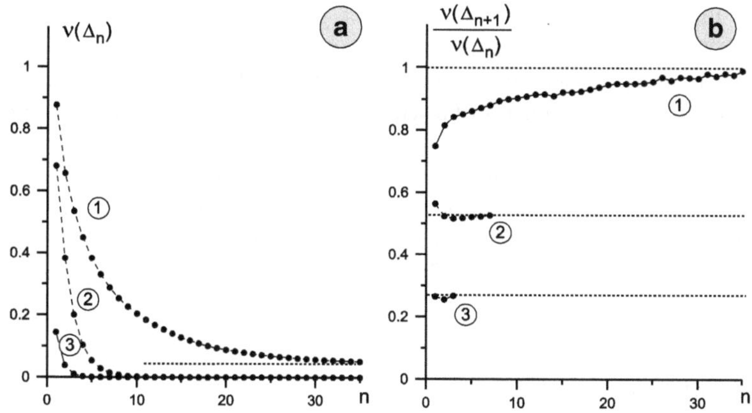

Fig. 7 Panel **a**: The area of the set Δ_n^+ versus n for $\mu = 1$ and *(1)* $A = -0.7$, *(2)* $A = -1.2$ and *(3)* $A = -2$. Panel **b**: The ratio $R_n = \nu(\Delta_{n+1}^+)/\nu(\Delta_n^+)$ for the same values of μ and A. Only few points are shown in panel **b** for the cases *(2)* and *(3)* since accuracy of calculations drastically falls due to division to small denominator

– $\beta_i^\pm$ *are graphs of monotone increasing functions then for any* $p \in T^{-1}D_j \cap D_i$ *the relations* $g_1(p) > 0$, $g_2(p) > 0$ *hold;*
– $\beta_i^\pm$ *are graphs of monotone decreasing functions then for any* $p \in T^{-1}D_j \cap D_i$ *the relations* $g_1(p) < 0$, $g_2(p) < 0$ *hold.*

Then the conditions of Hypothesis 2 take place.

The proof of Theorem 5 is presented in [14]. It is convenient to incorporate the calculation of the values $g_1(p)$ and $g_2(p)$ in general cycle of numerical scanning in the plane (u, u'). Our numerical study indicates that *Hypotheses 2 holds for the parameters* μ *and* A *lying in the dark areas in Fig. 6.*

4.3 *Hypothesis 3*

The computation of $\nu(\Delta_n^+)$ was incorporated in the general cycle of numerical scanning of the plane (u, u'). The numerical results for some μ and A are depicted in Fig. 7. It follows from Fig. 7 that Hypothesis 3 is valid not in all considered cases.

A natural obstruction for the conditions of Hypothesis 3 to hold is presence of elliptic fixed points of T or elliptic periodic orbits (cycles) of T. Due to KAM theory in vicinity of an elliptic fixed point (or cycle) generically there exists a set of positive measure that consists of points which do not leave this vicinity after any number of iterations of T. This means that Hypothesis 3 is not valid when the parameters μ and A are in band zones, because of presence of elliptic fixed point $O(0, 0)$ of T. This situation takes place in the case (1) in Fig. 7.

When the point (μ, A) is situated in a gap, the point $O(0,0)$ becomes a hyperbolic fixed point of T. However there exists some region in the gap adjacent to its lower boundary where elliptic periodic orbits of T persist. They are related to the fixed point $O(0,0)$ by means of a sequence of period doubling bifurcations. The cascade of bifurcations is similar to the period doubling cascade described in [18] for Eq. (9) with delta-comb potential. At the same time *if μ and A are situated in dark zones in Fig. 6 the computations give a numerical evidence that Hypothesis 3 holds.* Moreover, our results allow to assume *exponential* convergence of $\nu(\Delta_n^+)$ to zero. This takes place for the cases (2) and (3), see panel b of Fig. 7.

4.4 Examples

Let us give some examples how the coding technique can be applied.

Example 1. Let the parameters (μ, A) belong to dark zone in the first gap, see Fig. 6. Then all the bounded in $\mathbb{R}$ solutions of Eq. (13) can be coded by bi-infinite sequences of three symbols. These symbols mark entering of orbit into D_1, D_2 and D_3 correspondingly. They may be chosen as "+", "0" and "−". Then

- There exists well-known solution $u(x)$ in the form of bright gap soliton, localized in one well of the potential. This solution corresponds to the code $(\ldots 00+00\ldots)$. Also there exist the gap soliton solution $-u(x)$ with the code $(\ldots 00-00\ldots)$;
- There exists two π-periodic solutions of Eq. (13) with codes $(\ldots+++\ldots)$ and $(\ldots---\ldots)$;
- There exists infinitely many dark soliton type solution corresponding to the codes $(\ldots---+++\ldots)$, $(\ldots---0+++\ldots)$, etc;
- There exists "domain wall"-type solutions corresponding to the codes $(\ldots 000+++\ldots)$, $(\ldots---000\ldots)$. These objects were found to exist in the case of GPE with attractive nonlinearity [12]. So, they exist in the case of repulsive nonlinearity also.

Example 2. Assume again that the parameters (μ, A) belong to dark zone in the first gap. Consider boundary value problem for Eq. (13) on the interval $[-4\pi, 0]$ with Neumann boundary conditions at $x=-4\pi$ and $x=0$. These solutions can be regarded as reductions to the interval of length 4π of periodic solutions with period 8π which satisfy additional symmetry conditions

$$u(x) = u(-x); \quad u(-4\pi + x) = u(-4\pi - x).$$

The codes for these solutions are of the form

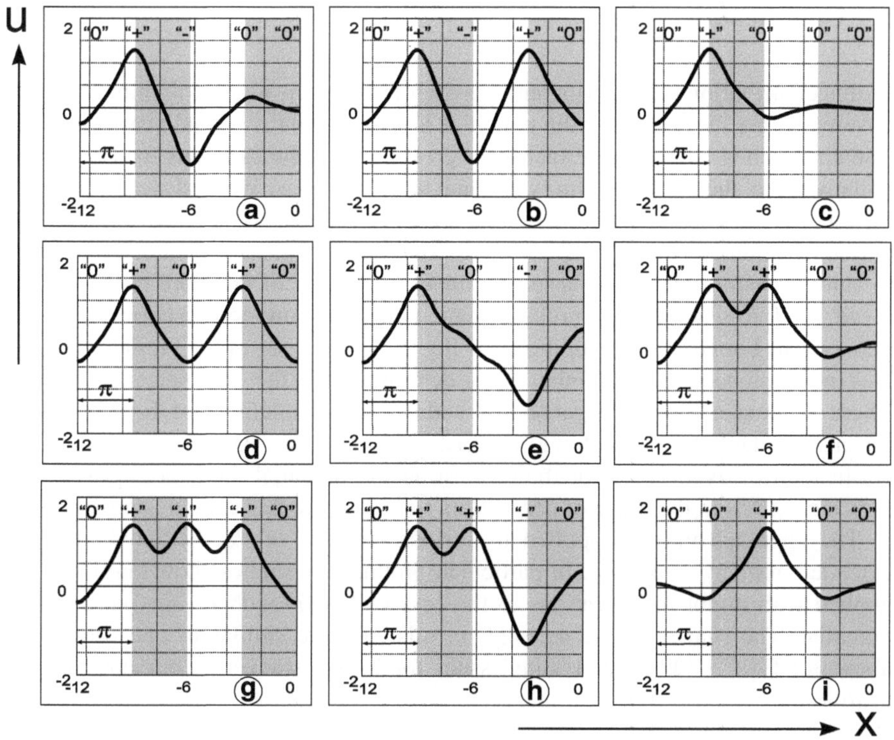

Fig. 8 The solutions of Eq. (13) ($A = -2$, $\mu = 1$) on the interval $[-4\pi, 0]$ with Neumann boundary conditions at $x = -4\pi$ and $x = 0$. The codes (the blocks $(\theta_1\theta_2\theta_3\theta_4\theta_5)$) are: **a**: $(0+-00)$; **b**: $(0+-+0)$; **c**: $(0+000)$; **d**: $(0+0+0)$; **e**: $(0+0-0)$; **f**: $(0++00)$; **g**: $(0+++0)$; **h**: $(0++-0)$; **i**: $(00+00)$

$$(\ldots\theta_5\theta_4\theta_3\theta_2\,\underbrace{\theta_1\theta_2\theta_3\theta_4\theta_5\theta_4\theta_3\theta_2}_{\text{the period}}\,\theta_1\theta_2\theta_3\theta_4\theta_5\ldots),$$

where θ_i, $i = 1 \div 5$, is one of the symbols "+", "0" or "−". Therefore there are $3^5 = 243$ solutions of this type. In Fig. 8 nine of these solutions are depicted for $A = -2$, $\mu = 1$. All the solutions shown in Fig. 8 have the codes with "0" as the first and last symbols on the semi-period. Therefore they can be regarded as approximations for localized modes which have the domain of localization of length 4π. These localized modes correspond to the codes

$$(\ldots 00\;\theta_2\theta_3\theta_4\;00\ldots).$$

There are 27 sequences of this type but only 10 of them are different in the sense that they are not related to each other by symmetry reductions. All of them, except zero solution with the code $(\ldots 00000\ldots)$ are presented in Fig. 8.

5 Conclusion

In this contribution we describe an approach for classification of nonlinear modes covered by Eq. (5) with periodic potential $V(x)$ and repulsive nonlinearity $g = 1$. The approach is based on the fact that "great part" of the solutions for Eq. (5) are collapsing i.e. they tend to infinity at some point of x axis. We analyze the location of initial data for solutions which does not exhibit collapsing behavior within the period of potential (the sets $\mathcal{U}_\pi^\pm$). Then we applied methods of symbolic dynamics to study transformations of these sets under the action of Poincaré map T. As a result, we formulate three hypotheses which guarantee one-to-one correspondence between all bounded in $\mathbb{R}$ solutions of Eq. (5) and the set of bi-infinite words of N-symbol alphabet. The number N is determined by the parameters of Eq. (5). These sequences can be regarded as codes for the solutions of Eq. (5). For a given $V(x)$ the hypotheses should be verified numerically. We report on numerical verification of hypotheses for the case of cosine potential, i.e., for Eq. (13), and indicate regions in the plane of parameters (μ, A) where this coding is possible.

The approach may be applied to various versions of Eq. (1) and it may be extended in many directions. In particular, preliminary studies show that it can be applied with minor modifications to the case of Eq. (7) with $g_2 < 0$ and cosine potential. The shapes of the sets $\mathcal{U}_\pi^\pm$ in this case are similar to ones described above for Eq. (5). Other interesting issues for further studies, in our opinion, are as follows:

- Is it possible to modify the method described above to study the dynamics in the case of partially overlapping islands?
- What peculiarities can appear in the situation when the potential $V(x)$ has several wells of different depth on each period?
- What type of coding can be used to describe the nonlinear modes for Eq. (6)?

Acknowledgements The work of authors was supported by Russian federal program "Scientific and educational personnel of the innovative Russia", grant 14.B37.21.1273.

References

1. V.A. Brazhnyi, V.V. Konotop, Mod. Phys. Lett. B, **18**, 627 (2004)
2. L.P. Pitaevskii, Phys. Usp. **49**, 333 (2006)
3. O. Morsch, M. Oberthaler, Rev. Mod. Phys. **78**, 179 (2006)
4. L. Pitaevskii, S. Stringari, *Bose-Einstein condensation* (Clarendon Press, Oxford, 2003)
5. G.L. Alfimov, V.V. Konotop, M. Salerno, Europhys. Lett. **58**, 7 (2002)
6. F.K. Abdullaev, M. Salerno, Phys. Rev. A **72**, 033617 (2005)
7. G.L. Alfimov, V.V. Konotop, P. Pacciani, Phys. Rev. A **75**, 023624 (2007)
8. P.J.Y. Louis, E.A. Ostrovskaya, C.M. Savage, Yu.S. Kivshar, Phys. Rev. A **67**, 013602 (2003)
9. V.V. Konotop, M. Salerno, Phys. Rev. A **65**, 021602 (2002)
10. D.E. Pelinovsky, A.A. Sukhorukov, Yu.S. Kivshar, Phys. Rev. E **70**, 036618 (2004)
11. B. Wu, Q. Niu, Phys. Rev. A **64**, 061603(R) (2001)

12. P.G. Kevrekidis, B.A. Malomed, D.J. Frantzeskakis, A.R. Bishop, H. Nistazakis, R. Carretero-González, Math. Comput. Simul. **69**, 334 (2005)
13. T.J. Alexander, E.A. Ostrovskaya, Yu.S. Kivshar, Phys. Rev. Lett. **96**, 040401 (2006)
14. G.L. Alfimov, A.I. Avramenko, Physica D **254**, 29 (2013)
15. P.J. Torres, Nonlinear Anal. **65**, 841 (2006)
16. S. Wiggins, *Introduction to applied dynamical systems and chaos* (Sprigner, New York, 1990)
17. M. Abramowitz, I. Stegun (eds.), *Handbook of mathematical functions* (Dover, New York, 1970)
18. D. Witthaut, K. Rapedius, H.J. Korsch, J. Nonlinear Math. Phys. **16**, 207 (2009)

Nonreciprocal Wave Propagation Through Open, Discrete Nonlinear Schrödinger Dimers

Stefano Lepri and Giulio Casati

Abstract We consider asymmetric (nonreciprocal) wave transmission through a layered nonlinear, non mirror-symmetric system described by the one-dimensional Discrete Nonlinear Schrödinger equation with spatially varying coefficients embedded in an otherwise linear lattice. Focusing on the simplest case of two nonlinear sites (the dimer), we compute exact scattering solutions such that waves with the same frequency and incident amplitude impinging from left and right directions have different transmission coefficients. The stability of some particular solutions is addressed. We show that oscillatory instability may lead to the formation of stable extended states coexisting with a nonlinear defect mode oscillating at a different frequency. Numerical simulations of wave packet scattering are presented. Gaussian wave packets with the same amplitude arriving from opposite directions on the dimer are indeed trasmitted very differently. Moreover, asymmetric transmission is sensitively dependent on the input parameters, akin to the case of chaotic scattering.

1 Introduction

The possibility to control energy and/or mass flows using nonlinear features of physical systems is a fascinating issue both from the point of view of basic science as well as from the applied one. In the context of wave propagation trough nonlinear

---

S. Lepri (✉)
Consiglio Nazionale delle Ricerche, Istituto dei Sistemi Complessi, via Madonna del piano 10, I-50019 Sesto Fiorentino, Italy
e-mail: stefano.lepri@isc.cnr.it

G. Casati
Center for Nonlinear and Complex Systems, Università degli Studi dell'Insubria, Como, Italy

Istituto Nazionale di Fisica Nucleare, Sezione di Milano, Milan, Italy
e-mail: giulio.casati@uninsubria.it

R. Carretero-González et al. (eds.), *Localized Excitations in Nonlinear Complex Systems*, Nonlinear Systems and Complexity 7, DOI 10.1007/978-3-319-02057-0_3,

media, the simplest form of control would be to devise a "wave diode" in which electromagnetic or elastic waves are transmitted differently along two opposite propagation directions.

In a linear, time-reversal symmetric system this possibility is forbidden by the reciprocity theorem. As stated by Lord Rayleigh in his treatise *The theory of sound* [1]:

> Let A and B be two points [...] between which are situated obstacles of any kind. Than a sound originating at A is perceived at B with the same intensity as that with which an equal sound originating at B would be perceived at A. In acoustics [...] in consequence of the not insignificant value of the wavelength in comparison with the dimension of ordinary obstacles the reciprocal relation is of considerable interest.

To achieve the desired effect one must thus violate the hypothesis of the theorem. In linear systems, a standard way is to break the time-reversal symmetry by applying a magnetic field as done, for instance, in the case of optical isolators. An entirely alternative possibility is instead to consider *nonlinear* media. At least in principle, this option would offer a whole new range of possibilities of propagation control based on intrinsic material properties rather than an external field.

Asymmetric wave propagation induced by nonlinearity arises in several different domains. Among the first examples discussed in the literature is the asymmetric phonon transmission through a nonlinear interface layer between two very dissimilar crystals [2]. In the field of nonlinear optics a relatively vast number of approaches exist. A so-called all-optical diode has been proposed first in Refs. [3,4] and later on realized experimentally [5]. There are also proposals to employ left-handed metamaterials [6], quasiperiodic systems [7], coupled nonlinear cavities [8] or $\mathcal{PT}$-symmetric waveguides [9, 10]. Extension to the quantum regime in which few-photon states display a diode effect has been proposed [12].

In the realm of acoustics the possibility of realizing a diode has been demonstrated for nonlinear phononic media [13, 14]. Another promising context is the propagation of acoustic pulses through granular systems. Indeed, experimental studies demonstrated a change of solitary wave reflectivity from the interface of two granular media [15]. More recently, demonstration of rectification of mechanical energy at sonic frequencies in a one-dimensional array of particles has been also reported [16].

Despite the variety of physical contexts, the basic underlying rectification mechanisms rely on nonlinear phenomena as, for instance, second-harmonic generation in photonic [17] or phononic crystals [13], or bifurcations [16]. In those examples the rectification depends on whether some harmonic (or subharmonic) of the fundamental wave is transmitted or not.

A related question is the possibility that the transmitted power at the *same frequency and incident amplitude* would be sensibly different in the two opposite propagation directions. In this chapter we address the above problem with the Discrete Nonlinear Schrödinger (DNLS) equation [18, 19] with site-dependent coefficients. It has been demonstrated [20] that DNLS equation can be a sensible approximation for the evolution of longitudinal Bloch waves in layered photonic or phononic crystals. Variable coefficients describe different nonlinear properties of each layer and the presence of defects. In the realm of the physics of cold atomic gases, the equation is an approximate semiclassical description of bosons trapped in periodic optical lattices (see e.g. Ref. [21] and references therein for a recent survey). Beyond its relevance in many different physical contexts, the DNLS equation has the big advantage of being among the simplest dynamical systems amenable to a complete theoretical analysis. For our purpose, it is particularly convenient as it allows to solve the scattering problem exactly without the complications of having to deal with wave harmonics [22].

In Sect. 2 we outline the model and show some examples of asymmetric plane-wave solutions. The issue of their stability is briefly addressed in Sect. 3. It is also shown that oscillatory instability may lead to the formation of stable extended states coexisting with a nonlinear defect mode oscillating at a different frequency. In Sect. 4 we report some numerical simulation of wave packet's scattering and illustrate its dependence on initial intensity. Finally, a brief summary of the results is given in Sect. 5.

2 The Model

The DNLS equation with spatially varying coefficients, defined on an infinite one-dimensional lattice is given by

$$i\dot{\phi}_n = V_n\phi_n - \phi_{n+1} - \phi_{n-1} + \alpha_n|\phi_n|^2\phi_n, \tag{1}$$

where units have been chosen such as the coupling $C = 1$.

We will assume the usual scattering setup where V_n and α_n are non vanishing only for $1 \leq n \leq N$. The two semi infinite portions ($n < 1, n > N$) of the lattice, model two leads where the wave can propagate freely [23]. Let us look for solutions of the associated stationary transmission problem $\phi_n(t) = u_n \exp(-i\mu t)$

$$\mu u_n = V_n u_n - u_{n+1} - u_{n-1} + \alpha_n|u_n|^2 u_n \qquad 1 \leq n \leq N \tag{2}$$

of the form

$$u_n = \begin{cases} R_0 e^{ikn} + Re^{-ikn} & n \leq 1 \\ Te^{ikn} & n \geq N \end{cases} \tag{3}$$

where $\mu = -2\cos k$ and $0 \leq k \leq \pi$ for the wave coming from the left direction; R_0, R and T are the incident, reflected and transmitted amplitudes respectively. The solution sought must be complex in order to carry a non vanishing current $J = 2|T|^2 \sin k$.

To break the mirror symmetry with respect to the center of the nonlinear portion, one must choose at least one of the two sets of coefficients V_n, α_n such that

$V_n \neq V_{N-n+1}$, $\alpha_n \neq \alpha_{N-n+1}$. Note that the transmission of the right-incoming wave with the same R_0 and μ is computed by solving the problem with $(V_n, \alpha_n) \longrightarrow (V_{N-n+1}, \alpha_{N-n+1})$ (i.e. "flipping the sample"). In the following, we will adopt the convention to label with $-k$ the right-incoming solutions with wave number k. Nonlinearity is essential as for $\alpha_n = 0$ the transmission coefficient is the same for waves coming from the left or right side, independently on V_n due to time-reversal invariance of the underlying equations of motion [24].

The standard way to solve the problem is to introduce the (backward) transfer map [23, 25–27]

$$u_{n-1} = -v_n + (V_n - \mu + \alpha_n |u_n|^2)u_n, \quad v_{n-1} = u_n. \tag{4}$$

Note that these are complex quantities therefore the map is nominally four-dimensional. However, due to conservation of energy and norm, it can be reduced to a two-dimensional area-preserving map [23, 25–27] with an additional control parameter (the conserved current J). The solutions are straightforwardly found by iterating (4) from the initial point $u_N = T\exp(ikN)$, $v_N = T\exp(ik(N+1))$ dictated by the boundary conditions of Eq. (3). For fixed T and k, the incident and reflected amplitudes are determined as

$$R_0 = \frac{\exp(-ik)u_0 - v_0}{\exp(-ik) - \exp(ik)}, \qquad R = \frac{\exp(ik)u_0 - v_0}{\exp(ik) - \exp(-ik)}$$

and the transmission coefficient is $t(k, |T|^2) = |T|^2/|R_0|^2$. Note that if $(u_0, v_0) = (u_N, v_N)$ (periodic point of the map) then $t = 1$.

For very short chains (oligomers), t can be computed analytically. For instance for the dimer $N = 2$:

$$t = \left| \frac{e^{ik} - e^{-ik}}{1 + (\delta_2 - e^{ik})(e^{ik} - \delta_1)} \right|^2 \tag{5}$$

where

$$\begin{aligned} \delta_1 &= V_2 - \mu + \alpha_2 T^2, \\ \delta_2 &= V_1 - \mu + \alpha_1 T^2 [1 - 2\delta \cos k + \delta_1^2]. \end{aligned}$$

For the trimer $N = 3$

$$t = \left| \frac{e^{ik} - e^{-ik}}{e^{ik} - \delta_1 + (e^{ik} - \delta_3)(1 - \delta_2(\delta_1 - e^{ik}))} \right|^2 \tag{6}$$

where

$$\delta_3 = V_3 - \mu + \alpha_3 |T|^2$$

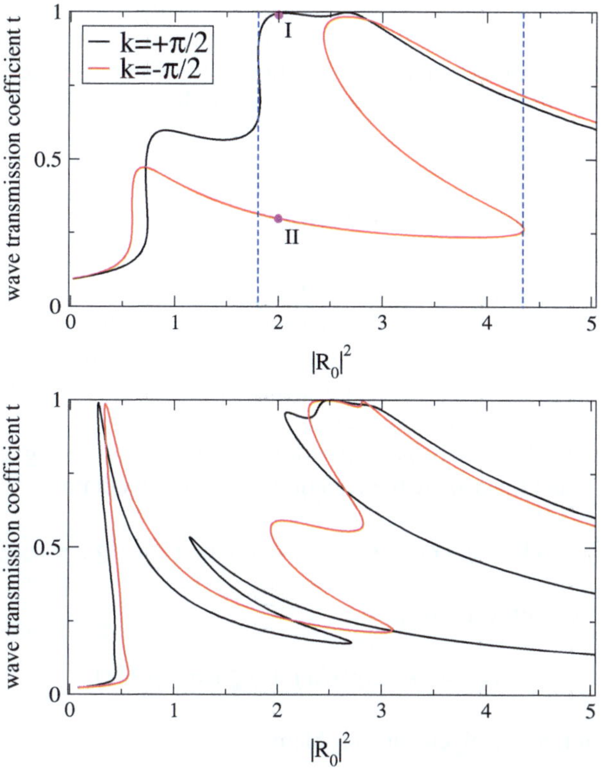

Fig. 1 Transmission coefficients as a function of the input intensity $|R_0|^2$ for $k = \pi/2$, $\alpha_n = 1$. *Upper panel*: dimer $V_{1,2} = V_0(1 \pm \varepsilon)$. *Lower panel*: trimer $V_{1,3} = V_0(1 \pm \varepsilon)$, $V_2 = 0$; $V_0 = -2.5$ $\varepsilon = 0.05$

$$\delta_2 = V_2 - \mu + \alpha_2 |T|^2 |\delta_3 - e^{ik}|^2$$
$$\delta_1 = V_1 - \mu + \alpha_1 |T|^2 |1 + \delta_2(e^{ik} - \delta_3)|^2.$$

The formulas apply for $k > 0$ (left-incoming waves); the transmission for right-incoming waves is obtained by a suitable exchange of subscripts in the above equations.

Two examples of the dependence of t on the input power are shown in Fig. 1. The curves display a multistable behavior and, for strong enough intensities, are sizeably different indicating nonreciprocity. The effect is maximal in the vicinity of the nonlinear resonances that are detuned differently for the $k > 0$ and $k < 0$ cases yielding intervals of input values were multiple solutions exist only for one propagation direction [22]. The lower panel of Fig. 1 shows that even a moderate increase of the number of sites ($N = 3$) dramatically increase the complexity of the curves as expected due to the mixed phase-space of the underlying transfer map [25].

To conclude this section, we mention that an alternative approach would be to use the Green's function formalism previously used to compute the stationary states for an electron moving on a chain with nonlinear impurities [28]. Indeed, for the case of a symmetric nonlinear dimer, resonance phenomena are demonstrated that lead to complete transmission through the dimer [28]. Of course, we expect such an approach to yield the same results when applied to the present case.

3 Stability of Scattering Solutions

An important issue is the dynamical stability of the solutions. To the best of our knowledge no systematic study of scattering solutions of the type described above has been presented in the literature (see [29] for the case of NLS equation with concentrated nonlinearities in the continuum case and [30] for an analysis of a related problem, the nonlinear Fano effect).

The linear stability analysis is performed [18, 19] by letting $\phi_n = (u_n + \chi_n)\exp(-i\mu t)$, and linearizing the equation of motion to obtain

$$i\dot{\chi}_n = (V_n - \mu)\chi_n - \chi_{n+1} - \chi_{n-1} + \alpha_n\left(2|u_n|^2\chi_n + u_n^2\chi_n^*\right) \tag{7}$$

Note that χ_n is complex. Letting

$$\chi_n = A_n \exp(i\lambda t) + B_n^* \exp(-i\lambda^* t)$$

then λ is solution of the eigenvalue problem

$$\begin{aligned}\lambda A_n &= -(\varepsilon_n - \mu)A_n + A_{n+1} + A_{n-1} - \theta_n B_n \\ \lambda B_n &= +(\varepsilon_n - \mu)B_n - B_{n+1} - B_{n-1} + \theta_n^* A_n\end{aligned} \tag{8}$$

with $\varepsilon_n \equiv V_n + 2\alpha_n|u_n|^2$, $\theta_n \equiv \alpha_n u_n^2$. Note that, at variance with the case of e.g. breather solutions, the solutions are complex, and also the coefficients θ_n in (8) are complex as well. As it is known, the eigenvalues come in quadruplets of the form $\pm\lambda$, $\pm\lambda^*$. If eigenvalues have a nonzero imaginary part then the solution is unstable. Generally speaking, equilibria of an Hamiltonian system can lose spectral (and therefore linear) stability in two ways: a pair of real eigenvalues can either (i) merge at the origin and split onto the imaginary axis (saddle-node bifurcation) or (ii) collide at a nonzero point and split off into the complex plane, forming a complex quadruplet (Krein bifurcation). The latter case correspond to an oscillatory instability.

To solve the linear problem (8) exactly one should impose to the solution a definite plane wave form (with complex wave numbers) in the two seminiifinite linear parts of the chain. The matching of such waves through the nonlinear portion reduces the infinite-dimensional problem (8) to an homogeneous linear system of $2N$ equations, whose solvability condition, along with the dispersion relations, yields a set of nonlinear equations for the unknowns. The details of this method

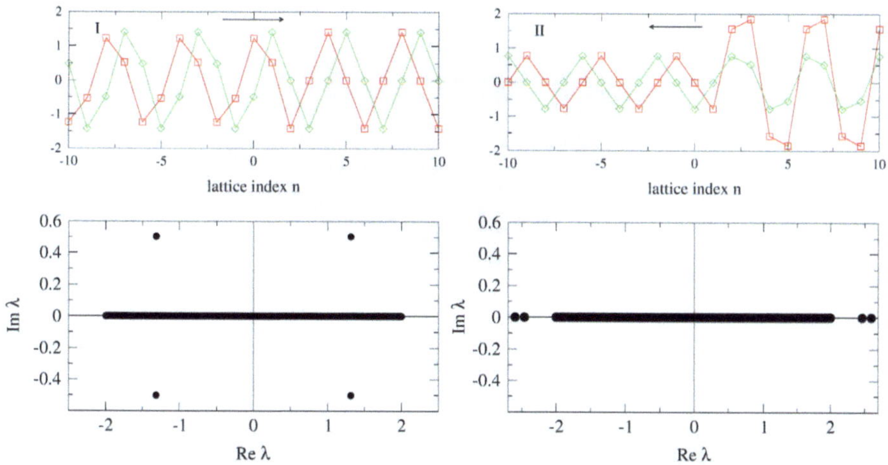

Fig. 2 *Upper panels*: real (*squares*) and imaginary (*diamonds*) parts of the two solutions marked with I and II in Fig. 1: they correspond to incident waves with the same input $|R_0|^2 = 2$ having transmission coefficients $t = 0.99$ and 0.30 respectively. *Lower panels*: the spectrum in the complex plane $N_p = 200$. Isolated eigenvalues for solution I are ± 1.316 and $\pm 0.502i$ for II, ± 2.60 and ± 2.46 respectively

(which is technically more involved than a straightforward diagonalization) are presented elsewhere [11]. Here, we limit to illustrate the stability properties of some representative solutions by a more direct approach, i.e. by solving numerically the eigenvalue problem for a finite truncation $-N_p \leq n \leq N + N_p$ of the chain (M sites with $M = 2N_p + N + 1$), checking that the relevant eigenvalues of the resulting $2M \times 2M$ matrix are not affected by the truncation error.

Of course, for small enough nonlinearities/amplitudes the solutions should be stable. Since our main object of interest here is in the large asymmetry effects, we concentrate on the cases of strongly nonlinear waves. Figure 2 shows two examples of two such solutions corresponding to the same input (marked by dots in the upper panel of Fig. 1) along with their eigenvalue spectra. As expected, in both cases there is a continuum component filling densely the interval $[-2, 2]$ on the real axis corresponding to propagation of linear waves.[1] In addition, isolated eigenvalues indicate that the solution I undergoes an oscillatory instability while II is stable. The components of the corresponding eigenvectors are exponentially localized around the nonlinear portion of the chain. This is intuitively clear as only perturbations located there can destabilize the solutions.

[1]Examining the eigenvalue spectra for different sizes M of the matrix, reveals that this continuum part of the spectra is mostly affected by truncation error. Typically the numerical eigenvalues have a spurious imaginary part which is of order $1/M$. This is not surprising since the corresponding eigenvectors are extended waves and thus more sensitive to boundary effects.

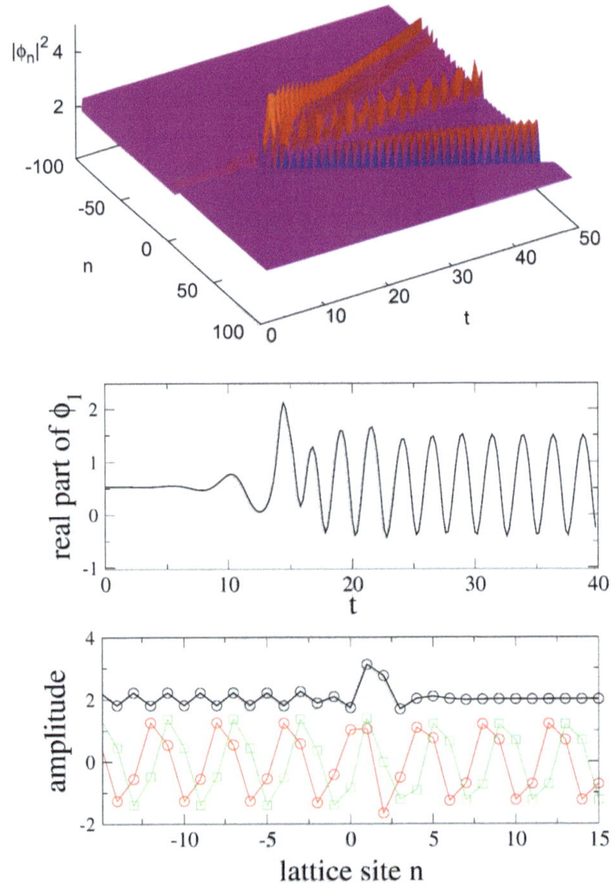

Fig. 3 *Upper panel*: unstable evolution of the solution *I* depicted in Fig. 2; numerical integration of DNLS with initial condition corresponding to the stationary orbit with a small perturbation of size 10^{-5} to the site $n = 1$; *Middle panel*: evolution of $Re\phi_1$: initial oscillations are at an angular frequency close to the imaginary part of the unstable eigenvalue, $Re\lambda = 1.316$. At the later stage, a stable periodic oscillation sets in with frequency increases to 2.5 which is outside of the band of linear waves. *Lower panel*: snapshots of real (*squares*) and imaginary (*diamonds*) parts of central part the chain at $t = 400$. The *upper curve* is $|\phi_n|^2$ showing the appearance of a localized excitation residing on the dimer

For comparison we also integrate the time-dependent DNLS equation setting as initial condition $\phi_n(0) = u_n$ and imposing the boundary conditions $\phi_n(t) = u_n e^{-i\mu t}$ at the two edge sites $n = -N_p$ and $n = N + N_p$ to simulate the infinite system. As seen in Fig. 3 the destabilization of solution *I* occurs by an exponential growth of ϕ_n on the central sites accompanied by an oscillatory behavior with the frequency prescribed by the stability spectrum (see the middle panel of Fig. 3). Since this frequency is in the band of linear waves, this process is accompanied by emission of some radiation (traveling peaks in the upper panel of Fig. 3) until the amplitude and frequency become large enough leading to a stable localized object (lower panel in Fig. 3). This state is reminiscent of a nonlinear defect mode [19]. There is however an important difference as the localized mode is superimposed to a plane wave and that the overall evolution is quasi-periodic.

4 Scattering of Wave Packets

In this section we illustrate the consequences of the above results on the transmission of wave packets. In a nonlinear system where the superposition principle no longer holds, the connection between the two problems is not trivial. We solved numerically the time-dependent DNLS on a finite lattice $|n| \leq M$ with open boundary conditions, for the case of the dimer discussed in [22]. We take as initial condition a Gaussian wave packet (for $n_0 < 0$)

$$\phi_n(0) = I \exp\left[-\frac{(n-n_0)^2}{w^2} + ik_0 n\right] \qquad (9)$$

where $k_0 > 0$ for $n_0 < 0$ (left-incoming packet) and where $k_0 < 0$ for $n_0 > N$ (right-incoming packet). The upper panels of Fig. 4 display the evolution of two packets with the same I and opposite wave number k_0 impinging on the nonlinear dimer. The asymmetry of their propagation is manifest. In both cases, the packets are significantly distorted after scattering, and the emerging envelope may vary wildly. However, the Fourier analysis shows that they remain almost monochromatic at the incident wave number k_0 (lower panels of Fig. 4), with some small background amplitude radiation leaking throughout the lattice.

To quantify the asymmetry of the scattering, and to compare with the above analysis, we measured the wave packet transmission coefficient as the ratio between the transmitted norm at the end t_{fin} of the run divided by the initial one, namely (for $n_0 < 0$)

$$t_p = \frac{\sum_{n>N} |\phi_n(t_{fin})|^2}{\sum_{n<1} |\phi_n(0)|^2}.$$

In Fig. 5 we show the transmission coefficient as a function of the packet intensity $|I|^2$ for two widths of the initial packet. Comparison with Fig. 1 show that the region of wave amplitudes where the non-reciprocal behavior is maximal (that between the vertical dotted lines) corresponds qualitatively well to the range where packets with comparable intensities are transmitted more asymmetrically.

Besides this, the wider packets show a highly irregular behavior in the sense that there is *sensitive dependence* on initial packets' parameters. This is somehow reminiscent of chaotic scattering (see e.g. [31]). Empirically, we found that this behavior depends very much on the width w: for small w the curves are much

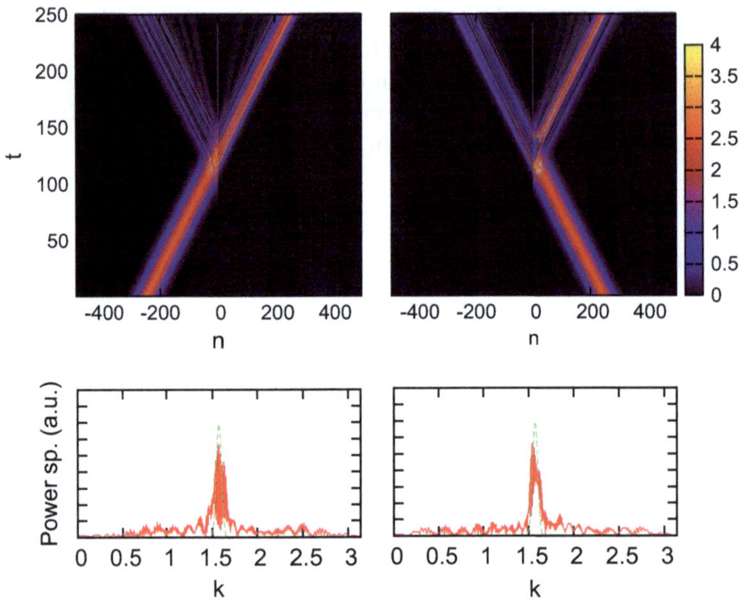

Fig. 4 Numerical simulations of the propagation of Gaussian wave packets, Eq. (9) impinging on a DNLS dimer. Here $V_0 = -2.5$, $|k_0| = 1.57$, $\varepsilon = 0.05$, $M = 500$, $|I|^2 = 2.5$, $w = 50$ and $n_0 = \mp 250$, respectively. *Lower panels*: Power spectra of the real part of ϕ_n at times $t = 0$ (*green dashed line*) and $t = 250$ (*red solid line*)

smoother. Qualitatively speaking, this can be understood as follows. The dimer can be seen as an integrable, two-degrees of freedom system perturbed by a time-dependent force (the incoming wave packet) and subject to dissipation (the radiation towards the linear leads) [32]. If the packet is very narrow, the duration of the perturbation is short and one may argue that this just determines the initial condition for the dynamical system. The subsequent evolution will be almost regular and a slight change of the forcing will have a little effect. On the contrary, a wider packet will result in a more complex (possibly chaotic) dynamics yielding large changes in the transmission.

5 Summary and Conclusions

We discussed the scattering problem for a linear Schrödinger chain with an embedded nonlinear, non-mirror symmetric dimer or trimer. The simplicity of the model allows to compute analytically a whole family of non reciprocal plane wave solutions, such as left- and right-incoming waves, with the same incident amplitude R_0 and frequency μ, have sizeably different transmission coefficients t.

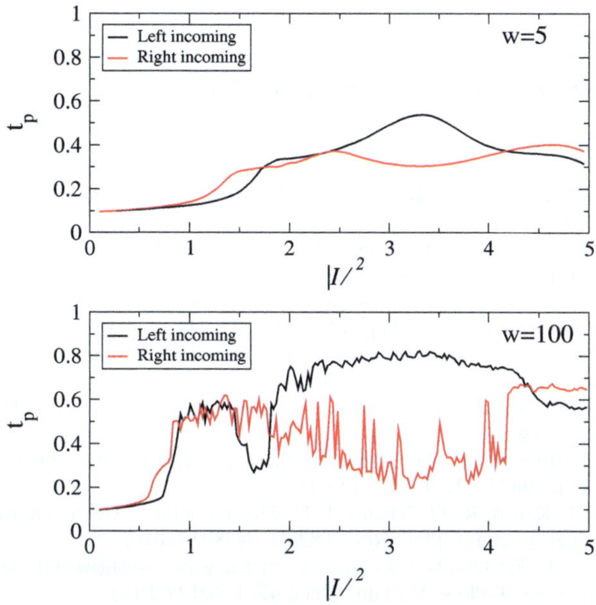

Fig. 5 Dependence of wave packet transmission coefficients on intensity for two different wave packet widths w. Other parameters as in the previous figure

We than addressed the question of dynamical stability for some specific values of R_0 and μ. Solutions with a large enough t generically undergo oscillatory instabilities (see again Fig. 3). At the initial stage, the development of the instability occurs through growth and oscillation localized around the dimer sites, accompanied by emission of radiation towards infinity. Eventually, a stable extended state emerges coexisting with a nonlinear defect mode oscillating at a different frequency. This numerical finding suggests that the lattice may support exact stable quasiperiodic, nonreciprocal solutions.

Numerical simulations demonstrated that equal Gaussian wave packets impinging on the dimer from the two opposite directions have indeed very different transmission coefficients t_p. The packet amplitudes I for which such non-reciprocal behavior is maximal qualitatively correspond to the range in which extended waves with the same wavenumber are transmitted more asymmetrically. Moreover, t_p is sensitively dependent on the input parameters, like for instance the initial width w and amplitude I (see Fig. 5), a feature which is reminiscent of chaotic scattering.

Acknowledgements SL thanks P.G. Kevrekidis and J. D'Ambroise for a useful correspondence regarding the stability problem. This work is part of the Miur PRIN 2008 project *Efficienza delle macchine termoelettriche: un approccio microscopico*.

References

1. J.W.S. Rayleigh, *The theory of sound* (Dover, New York, 1945)
2. Yu.A. Kosevich, Fluctuation subharmonic and multiharmonic phonon transmission and Kapitza conductance between crystals with very different vibrational spectra. Phys. Rev. B **52**, 1017 (1995)
3. M. Scalora, J.P. Dowling, C.M. Bowden, M.J. Bloemer, The photonic band edge optical diode. J. Appl. Phys. **76**(4), 2023–2026 (1994)
4. M.D. Tocci, M.J. Bloemer, M. Scalora, J.P. Dowling, C.M. Bowden, Thin-film nonlinear optical diode. Appl. Phys. Lett. **66**(18), 2324–2326 (1995)
5. K. Gallo, G. Assanto, K.R. Parameswaran, M.M. Fejer, All-optical diode in a periodically poled lithium niobate waveguide. Appl. Phys. Lett. **79**, 314 (2001)
6. M.W. Feise, I.V. Shadrivov, Y.S. Kivshar, Bistable diode action in left-handed periodic structures. Phys. Rev. E **71**(3), 037602 (2005)
7. F. Biancalana, All-optical diode action with quasiperiodic photonic crystals. J. Appl. Phys. **104**(9), 093113 (2008)
8. V. Grigoriev, F. Biancalana, Nonreciprocal switching thresholds in coupled nonlinear microcavities. Opt. Lett. **36**(11), 2131–2133 (2011)
9. H. Ramezani, T. Kottos, R. El-Ganainy, D.N. Christodoulides, Unidirectional nonlinear PT-symmetric optical structures. Phys. Rev. A **82**(4), 043803 (2010)
10. J. D'Ambroise, P.G. Kevrekidis, S. Lepri, Asymmetric wave propagation through nonlinear PT symmetric oligomers. J. Phys. A: Math. Theor. **45**, 444012 (2012)
11. J. D'Ambroise, P.G. Kevrekidis, S. Lepri, Eigenstates and instabilities of chains with embedded defects. Chaos **23**, 023109 (2013)
12. D. Roy, Few-photon optical diode. Phys. Rev. B **81**(15), 155117 (2010)
13. B. Liang, B. Yuan, J.C. Cheng, Acoustic diode: rectification of acoustic energy flux in one-dimensional systems. Phys. Rev. Lett. **103**(10), 104301 (2009)
14. B. Liang, X.S. Guo, J. Tu, D. Zhang, J.C. Cheng, An acoustic rectifier. Nat. Mater. **9**(12), 989–992 (2010)
15. V.F. Nesterenko, C. Daraio, E.B. Herbold, S. Jin, Anomalous wave reflection at the interface of two strongly nonlinear granular media. Phys. Rev. Lett. **95**(15), 1–4 (2005)
16. N. Boechler, G. Theocharis, C. Daraio, Bifurcation-based acoustic switching and rectification. Nat. Mater. **10**(9), 665–668 (2011)
17. V.V. Konotop, V. Kuzmiak, Nonreciprocal frequency doubler of electromagnetic waves based on a photonic crystal. Phys. Rev. B **66**(23), 235208 (2002)
18. J.C. Eilbeck, P.S. Lomdahl, A.C. Scott, The discrete self-trapping equation. Physica D **16**, 318–338 (1985)
19. P.G. Kevrekidis, *The discrete nonlinear Schrödinger equation* (Springer, Berlin, 2009)
20. A.M. Kosevich, M.A. Mamalui, Linear and nonlinear vibrations and waves in optical or acoustic superlattices (photonic or phonon crystals). J. Exp. Theor. Phys. **95**(4), 777 (2002)
21. R. Franzosi, R. Livi, G.L. Oppo, A. Politi, Discrete breathers in Bose–Einstein condensates. Nonlinearity **24**, R89 (2011)
22. S. Lepri, G. Casati, Asymmetric wave propagation in nonlinear systems. Phys. Rev. Lett. **106**(16), 164101 (2011)
23. D. Hennig, G.P. Tsironis, Wave transmission in nonlinear lattices. Phys. Rep. **307**(5–6), 333–432 (1999)
24. B. Lindquist, Design of filters in a one-dimensional tight-binding system. Phys. Rev. E **63**(5), 56605 (2001)
25. F. Delyon, Y.E. Lévy, B. Souillard, Nonperturbative bistability in periodic nonlinear media. Phys. Rev. Lett. **57**(16), 2010–2013 (1986)
26. Y. Wan, C.M. Soukoulis, One-dimensional nonlinear Schrödinger equation: a nonlinear dynamical approach. Phys. Rev. A **41**(2), 800–809 (1990)

27. Q. Li, C.T. Chan, K.M. Ho, C.M. Soukoulis, Wave propagation in nonlinear photonic band-gap materials. Phys. Rev. B **53**(23), 15577–15585 (1996)
28. M.I. Molina, G.P. Tsironis, Nonlinear impurities in a linear chain. Phys. Rev. B **47**(22), 15330 (1993)
29. B.A. Malomed, M.Ya. Azbel, Modulational instability of a wave scattered by a nonlinear center. Phys. Rev. B **47**(16), 10402–10406 (1993)
30. A.E. Miroshnichenko, Instabilities and quasi-localized states in nonlinear Fano-like systems. Phys. Lett. A **373**(39), 3586–3590 (2009)
31. A.S. Pikovsky, The simplest case of chaotic wave scattering. Chaos **3**(4), 505–506 (1993)
32. S. Tietsche, A. Pikovsky, Chaotic destruction of Anderson localization in a nonlinear lattice. Europhys. Lett. **84**, 10006 (2008)

Breather Solutions of the Discrete p-Schrödinger Equation

Guillaume James and Yuli Starosvetsky

Abstract We consider the discrete p-Schrödinger (DpS) equation, which approximates small amplitude oscillations in chains of oscillators with fully-nonlinear nearest-neighbors interactions of order $\alpha = p - 1 > 1$. Using a mapping approach, we prove the existence of breather solutions of the DpS equation with even- or odd-parity reflectional symmetries. We derive in addition analytical approximations for the breather profiles and the corresponding intersecting stable and unstable manifolds, valid on a whole range of nonlinearity orders α. In the limit of weak nonlinearity ($\alpha \to 1^+$), we introduce a continuum limit connecting the stationary DpS and logarithmic nonlinear Schrödinger equations. In this limit, breathers correspond asymptotically to Gaussian homoclinic solutions. We numerically analyze the stability properties of breather solutions depending on their even- or odd-parity symmetry. A perturbation of an unstable breather generally results in a translational motion (traveling breather) when α is close to unity, whereas pinning becomes predominant for larger values of α.

G. James (✉)
Laboratoire Jean Kuntzmann, Université de Grenoble and CNRS, BP 53, 38041 Grenoble Cedex 9, France

INRIA, Bipop Team-Project, ZIRST Montbonnot, 655 Avenue de l'Europe, 38334 Saint Ismier, France
e-mail: Guillaume.James@imag.fr

Y. Starosvetsky
Faculty of Mechanical Engineering, Technion – Israel Institute of Technology, Technion City, Haifa 32000, Israel
e-mail: staryuli@techunix.technion.ac.il

R. Carretero-González et al. (eds.), *Localized Excitations in Nonlinear Complex Systems*, Nonlinear Systems and Complexity 7, DOI 10.1007/978-3-319-02057-0_4,

1 Introduction and Main Results

In this paper we consider the discrete p-Schrödinger (DpS) equation defined by

$$i\frac{d}{dt}\phi_n = (\Delta_p\phi)_n, \quad n \in \mathbb{Z}, \tag{1}$$

where $\phi(t) = (\phi_n(t))_{n\in\mathbb{Z}}$ denotes a time-dependent complex sequence and

$$(\Delta_p\phi)_n = (\phi_{n+1} - \phi_n)\,|\phi_{n+1} - \phi_n|^{p-2} - (\phi_n - \phi_{n-1})\,|\phi_n - \phi_{n-1}|^{p-2}$$

the discrete p-Laplacian with $p > 2$. This model has a Hamiltonian structure associated with the energy

$$H(\phi) = \frac{2}{p}\sum_{n\in\mathbb{Z}} |\phi_{n+1} - \phi_n|^p. \tag{2}$$

It is reminiscent of the so-called discrete nonlinear Schrödinger (DNLS) equation [6, 15], with the difference that it contains a fully nonlinear inter-site coupling term. The DpS equation was recently introduced to describe small amplitude oscillations in a class of mechanical systems consisting of a chain of touching beads confined in smooth local potentials [11, 14, 26], the most well known example of such systems being Newton's cradle [10]. In this context, the p-Laplacian involved in (1) accounts for the fully-nonlinear character of Hertzian interactions between beads (with $p = 5/2$ in the case of contacting spheres). More generally, the DpS equation arises in the study of chains of oscillators involving fully nonlinear nearest-neighbors interactions. It can be derived as a modulation equation when nonlinear interactions dominate on-site anharmonicity, and allows to describe the slow evolution of the envelope of small amplitude oscillations over long (but finite) times [3]. A variant of the DpS equation combining an on-site cubic nonlinearity and a cubic intersite nonlinearity (case $p = 4$) was also introduced in [23] in a case when nonlinear interactions and on-site anharmonicity have equal orders.

Numerical simulations of the DpS equation with $p = 5/2$ have revealed the existence of breather solutions (i.e. spatially localized oscillations) of different types, either static (i.e. pinned to some lattice sites and time-periodic) or traveling along the lattice. Such solutions can be generated from localized initial conditions or may arise from modulational instabilities of periodic waves [11, 14, 26]. Although such properties are classical in the context of anharmonic Hamiltonian lattices [9], energy localization is particularly strong in the DpS equation. Indeed, it is proved in [3] that the solution of (1) does not disperse for any nonzero initial condition in $\ell_2(\mathbb{Z})$ (i.e. with finite power). Moreover, static breather solutions decay doubly-exponentially in space, as numerically illustrated in [11] (see also [13] for an analytical decay estimate).

In this paper, we prove an existence theorem for static breather solutions of the DpS equation valid for all $p > 2$ and introduce different methods to approach their profiles. The breather solutions are searched in the form

$$\phi_n(t) = |\Omega|^{\frac{1}{p-2}} a_n \, e^{i \, \Omega \, t}, \tag{3}$$

with $a = (a_n)_{n\in\mathbb{Z}}$ a real sequence and Ω an arbitrary nonvanishing constant. Equation (1) admits time-periodic solutions of the form (3) if and only if a satisfies the stationary DpS equation

$$-\, s \, a_n = (\Delta_p a)_n, \quad n \in \mathbb{Z}, \tag{4}$$

where $s = \mathrm{Sign}(\Omega) = 1$ or -1. We shall prove the following theorem, which ensures the existence of nontrivial solutions of (4) homoclinic to 0 (i.e. satisfying $\lim_{n\to\pm\infty} a_n = 0$) for $s = 1$, and the nonexistence of (nonzero) bounded solutions for $s = -1$.

Theorem 1. *The stationary DpS equation (4) with $s = 1$ admits solutions a_n^i ($i = 1, 2$) satisfying*

$$\lim_{n\to\pm\infty} a_n^i = 0,$$

$(-1)^n \, a_n^i > 0$, $|a_n^i| > |a_{n-1}^i|$ *for all $n \leq 0$, and*

$$a_n^1 = a_{-n}^1, \quad a_n^2 = -a_{-n+1}^2, \quad \textit{for all } n \in \mathbb{Z}.$$

Moreover, for $s = -1$ the only bounded solution of (4) is $a_n = 0$.

For $s = 1$, the solutions a_n^i ($i = 1, 2$) of (4) homoclinic to 0 correspond to breathers solutions of (1) taking the form (3) with arbitrary $\Omega > 0$. The case $i = 1$ yields the so-called even-parity modes (or site-centered solutions), and the case $i = 2$ corresponds to odd-parity modes (bond-centered solutions). These solutions were numerically computed in [11, 14, 26] for $p = 5/2$ (see also [9, 23] and references therein in the case $p = 4$ with an additional on-site cubic nonlinearity).

Theorem 1 is proved using a map approach reminiscent of previous works of Flach [8] and Qin and Xiao [22] (see also [23]). More precisely, a proof of existence of homoclinic solutions has been proposed in [8] (when $p \geq 4$ is an even integer), relying on a reformulation of (4) as a two-dimensional reversible mapping. However, the analysis was not complete because the existence of stable and unstable manifolds of the "hyperbolic" fixed point at the origin was not analytically justified. This point requires a particular attention because the corresponding map is not differentiable at the origin. This problem was addressed in [8] by adding a small linear Laplacian $\epsilon(\Delta_2 a)_n$ in (4), and studying the limit $\epsilon \to 0$ numerically. In the same spirit, Eq. (4) can be recast into the form of a generalized stationary discrete nonlinear Schrödinger (DNLS) equation reminiscent of a model studied in [22], except the resulting nonlinearity is not differentiable at the origin. The work [22] establishes the existence of homoclinic solutions for this class of models when C^1 stable and unstable manifolds exist, a point requiring special justification in our context. In this paper, we give equivalent reformulations of (4) as two-dimensional mappings (Sect. 2), and prove the existence of smooth stable and unstable manifolds of the origin (Sect. 3). This allows us to show the existence of homoclinic solutions of (4) for $s = 1$ using Qin and Xiao's result (Sect. 4). The nonexistence of nontrivial bounded solutions for $s = -1$ follows from elementary arguments (see the appendix).

Apart from existence results, several works have introduced analytical approximations of homoclinic orbits of the stationary DpS equation having compact supports, relying on two different approaches. The first one is reminiscent of the method of successive approximations [20, 25], but presents some limitations since a convergence analysis is not available. The second one relies on formal continuum approximations which provide quantitatively correct results at least for $p = 5/2$ and $p = 4$ [14, 17, 26]. However, in these works the continuum approximation is performed with a finite mesh size, so that the continuum problem is not properly justified.

In this paper, we proceed differently and provide analytical approximations of the stable and unstable manifolds of the origin (Sect. 5). This allow us to approximate the intersections of the invariant manifolds and thus the corresponding homoclinic orbits. A similar approach was previously introduced in [4] to approximate homoclinic solutions of the DNLS equation, but the technique employed to approximate the invariant manifolds was different. In our case, two different methods are used in this context. The first one employs a leading order approximation of the local stable manifold in conjunction with backward iterations, a method especially efficient when p is far from 2. The second method is based on a continuum limit obtained when p is close to 2, where one recovers a logarithmic stationary nonlinear Schrödinger equation with a Gaussian homoclinic solution. This limiting procedure is a first step towards the rigorous justification of previous formal continuum approximations [14, 17, 26]. Moreover, it is interesting to note that the case $p \approx 2$ is physically sound in the context of granular chain models [16, 19, 27].

We complete the above existence and approximation results by a numerical study of the stability of site- and bond-centered breathers (Sect. 6). Similarly to the case $p = 5/2$ considered in [14], we find that bond-centered breathers are spectrally stable and site-centered breathers are unstable, at least for all $p \in [2.2, 4.4]$. When p is close enough to 2 (e.g. for $p \leq 5/2$), site-centered breather instability is very weak, and perturbations along a marginal mode [1] lead to traveling breathers propagating at an almost constant velocity. When p becomes larger (e.g. for $p \geq 4$), suitable perturbations of a site-centered breather can induce a translational motion over a few sites, but trapping of the localized solution occurs subsequently. We relate this pinning effect to a sharp increase of the Peierls-Nabarro energy barrier [18] separating site- and bond-centered breathers.

Another interest of the analytical approximations of stable and unstable manifolds obtained in Sect. 5 concerns the analysis of breather bifurcations induced by a localized defect. Indeed, as shown in [12] (see also [21]), such bifurcations possess a geometrical interpretation for DNLS-type systems. In the presence of an isolated defect, an homoclinic orbit exists when the image of the unstable manifold by some linear transformation (which depends on the defect strength) intersects the stable manifold. Consequently, homoclinic bifurcations occur when the set of these intersections changes topologically. This typically corresponds to tangent or pitchfork bifurcations of some intersections for critical defect values. Obviously, using this approach to analyze defect-induced homoclinic bifurcations requires a good knowledge of the geometrical structure of the stable and unstable manifolds, whereas their geometry is often hard to establish rigorously. In that case, a good strategy consists in approximating the stable and unstable manifolds and perform the above analysis on the approximate invariant manifolds, which provides approximations of critical defect values [12]. This approach is briefly discussed in Sect. 7, and will be used in a forthcoming paper to analyze defect-induced breather bifurcations in the DpS equation.

2 Two-Dimensional Mappings Equivalent to the Stationary DpS Equation

In this section, we introduce some reformulations of the stationary DpS equation as two-dimensional reversible mappings. The corresponding variables will be denoted by force variables and mixed (amplitude-force) variables, in analogy with the Newton's cradle system [11]. For spatially homogeneous systems, the force variables described in Sect. 2.1 allow to convert the stationary DpS equation into a generalized DNLS equation with an on-site (fractional power) nonlinearity, which allows to prove the existence of homoclinic orbits by classical arguments (see Sect. 4). However, this formulation would lead to some complications when considering DpS equations with a local inhomogeneity (see Sect. 7). The mixed variables described in Sect. 2.2 will turn out to be more convenient in that case. They will be used in Sect. 5 when approximating the stable and unstable manifolds of the origin.

In this section we restrict to the case $s = 1$ of (4). The even- and odd-parity localized solutions of (4) described in Theorem 1 consist of spatial modulations of binary oscillations, i.e. a_n and a_{n+1} have opposite signs. This leads us to perform the so-called staggering transformation $a_n = (-1)^n\, u_n$, which yields the equivalent form of (4) given by

$$s\, u_n = P_\alpha(u_{n+1} + u_n) + P_\alpha(u_n + u_{n-1}), \quad n \in \mathbb{Z}, \tag{5}$$

where

$$P_\alpha(x) = x\,|x|^{\alpha-1}$$

and $\alpha = p-1 > 1$. With this change of variables, the even- and odd-parity localized solutions of (4) correspond to one-sign localized solutions u_n of (5).

The nonlinear system (5) with $s = 1$ can be rewritten as a two-dimensional mapping in different ways. The most direct one is obtained by introducing $w_n = u_{n+1}$ and using the variable (u_n, w_n). Since the map P_α defines an homeomorphism of $\mathbb{R}$ and $P_\alpha^{-1} = P_{1/\alpha}$, (5) can be rewritten

$$\begin{pmatrix} u_{n+1} \\ w_{n+1} \end{pmatrix} = M \begin{pmatrix} u_n \\ w_n \end{pmatrix}, \tag{6}$$

where M is the nonlinear map defined by

$$M \begin{pmatrix} u \\ w \end{pmatrix} = \begin{pmatrix} w \\ P_{1/\alpha}[w - P_\alpha(u+w)] - w \end{pmatrix}. \tag{7}$$

Notice that $P_{1/\alpha}$ and M are not differentiable at the origin since $\alpha > 1$. Because the expression of the map M is quite cumbersome, we shall replace (6) by two different formulations, which involve the force and mixed variables described below.

2.1 Force Variables

Defining

$$x_n = P_\alpha(u_{n+1} + u_n) \tag{8}$$

and summing (5) (with $s = 1$) at ranks n and $n+1$ yields the following equivalent problem,

$$(\Delta x)_n = q(x_n), \quad x = (x_n)_{n \in \mathbb{Z}}, \tag{9}$$

where Δ is the usual discrete Laplacian

$$(\Delta x)_n = x_{n+1} - 2x_n + x_{n-1}, \quad x = (x_n)_{n \in \mathbb{Z}},$$

and for all $x \in \mathbb{R}$

$$q(x) = P_{1/\alpha}(x) - 4x \tag{10}$$

(we omit the α-dependency of q in notations for simplicity). System (9) falls within the class of generalized stationary DNLS equations studied by Qin and Xiao in [22], except q is not differentiable at the origin. Introducing $z_n = x_{n+1}$, one can reformulate (9) as a two-dimensional mapping as in [22]

$$\begin{pmatrix} x_{n+1} \\ z_{n+1} \end{pmatrix} = T \begin{pmatrix} x_n \\ z_n \end{pmatrix}, \tag{11}$$

where

$$T \begin{pmatrix} x \\ z \end{pmatrix} = \begin{pmatrix} z \\ q(z) + 2z - x \end{pmatrix}. \tag{12}$$

The map T is an homeomorphism of $\mathbb{R}^2$ with inverse

$$T^{-1} \begin{pmatrix} x \\ z \end{pmatrix} = \begin{pmatrix} q(x) + 2x - z \\ x \end{pmatrix}.$$

Moreover, T is reversible with respect to the symmetry

$$R_1 = \begin{pmatrix} 0 & 1 \\ 1 & 0 \end{pmatrix}$$

and the nonlinear symmetry $R_2 = R_1 \circ T$, i.e. we have $R_i \circ T = T^{-1} \circ R_i$ for $i = 1, 2$. In other words, any orbit $X_n = (x_n, z_n)^T$ of (11) yields by symmetry other orbits $\tilde{X}_n^i$ $(i = 1, 2)$ taking the form $\tilde{X}_n^i = R_i(X_{-n})$.

In order to stress the equivalence between formulations (6) and (11), let us consider the homeomorphism of $\mathbb{R}^2$

$$h \begin{pmatrix} u \\ w \end{pmatrix} = \begin{pmatrix} P_\alpha(u + w) \\ w - P_\alpha(u + w) \end{pmatrix}, \tag{13}$$

whose inverse reads

$$h^{-1} \begin{pmatrix} x \\ z \end{pmatrix} = \begin{pmatrix} P_{1/\alpha}(x) - x - z \\ x + z \end{pmatrix}. \tag{14}$$

Note that h defines a diffeomorphism on $\mathbb{R}^2 \setminus \{0\}$. The following result can be checked by a simple computation.

Lemma 1. *The maps T and M are topologically conjugate by h, i.e. they satisfy $M = h^{-1} \circ T \circ h$.*

2.2 Mixed Variables

Let us consider $v_n = u_{n-1}$ and

$$y_n = x_{n-1} = P_\alpha(v_{n+1} + v_n). \tag{15}$$

Equation (5) reads for $s = 1$

$$v_{n+1} = y_{n+1} + y_n. \tag{16}$$

This yields the mapping

$$\begin{pmatrix} v_{n+1} \\ y_{n+1} \end{pmatrix} = F \begin{pmatrix} v_n \\ y_n \end{pmatrix}, \tag{17}$$

where

$$F \begin{pmatrix} v \\ y \end{pmatrix} = \begin{pmatrix} P_{1/\alpha}(y) - v \\ P_{1/\alpha}(y) - y - v \end{pmatrix}. \tag{18}$$

The first component of (17) corresponds to Eq. (15) and the second to (16).

The maps T and F are topologically conjugate by the linear transformation

$$J = \begin{pmatrix} 1 & 1 \\ 0 & 1 \end{pmatrix},$$

i.e. one has $F = J \circ T \circ J^{-1}$. It follows in particular that F is reversible with respect to the symmetry

$$\mathcal{R}_1 = J\, R_1 J^{-1} = \begin{pmatrix} 1 & 0 \\ 1 & -1 \end{pmatrix} \tag{19}$$

and the nonlinear symmetry $\mathcal{R}_2 = \mathcal{R}_1 \circ F$ taking the form

$$\mathcal{R}_2 \begin{pmatrix} v \\ y \end{pmatrix} = \begin{pmatrix} P_{1/\alpha}(y) - v \\ y \end{pmatrix}.$$

3 Stable and Unstable Manifolds

In this section we construct the stable and unstable manifolds of the origin for the different maps introduced in Sect. 2. These results will be used in Sect. 4 to obtain homoclinic solutions as intersections of these manifolds, and in Sect. 5 to approximate the stable and unstable manifolds and homoclinic orbits.

3.1 Heuristics

Let us first consider the generalized DNLS equation (9) in the usual case when

$$q \in C^1(\mathbb{R}), \quad q'(0) > 0 \tag{20}$$

and the corresponding mapping (11). Condition (20) implies that the origin is a hyperbolic fixed point of T, since the Jacobian matrix $DT(0)$ has a pair of real eigenvalues Λ, Λ^{-1} given by

$$\Lambda = 1 + \frac{q'(0)}{2} + \left(\frac{q'(0)^2}{4} + q'(0) \right)^{1/2} > 1. \tag{21}$$

This yields the existence of C^1 stable and unstable manifolds of the origin, denoted respectively by $W^{\mathrm{s}}(0)$ and $W^{\mathrm{u}}(0)$. At the origin, $W^{\mathrm{u}}(0)$ is tangent to the unstable subspace corresponding to the line $x = \Lambda^{-1} z$, and $W^{\mathrm{s}}(0)$ is tangent to the stable subspace $z = \Lambda^{-1} x$.

For system (9)–(10), q is not differentiable at the origin and $\lim_{x \to 0^\pm} q'(x) = +\infty$. This situation corresponds formally to the existence of an hyperbolic fixed point of T at the origin, the eigenvalues of the singular "Jacobian matrix" being $\Lambda = +\infty$ and $\Lambda^{-1} = 0$, the unstable subspace corresponding to the axis $x = 0$ and the stable subspace to the axis $z = 0$. In what follows, we put these arguments onto a rigorous footing, showing the existence of C^1 stable and unstable manifolds of the origin. As usual, we shall first prove the existence of local stable and unstable manifolds which can be written as graphs near the origin, and then construct the global stable and unstable manifolds.

3.2 Stable Manifold Theorem

In order to obtain local stable and unstable manifolds for (11), we first prove a similar result for system (5), which has the advantage of being C^1. For completeness we shall treat the two cases $s = 1$ and $s = -1$ simultaneously, but the applications considered subsequently will only concern the case $s = 1$.

Let us consider the usual Banach space $\ell_\infty(\mathbb{N})$ consisting of real bounded sequences, equipped with the supremum norm. For all $u_0 \in \mathbb{R}$ and $u = (u_n)_{n \geq 1} \in \ell_\infty(\mathbb{N})$, we define $N(u_0, u) \in \ell_\infty(\mathbb{N})$ by

$$(N(u_0, u))_n = P_\alpha(u_{n+1} + u_n) + P_\alpha(u_n + u_{n-1}), \quad n \geq 1.$$

System (5) restricted to $n \geq 1$ takes the form

$$g(u_0, u) = 0 \tag{22}$$

where $g(u_0, u) = s\,u - N(u_0, u)$ satisfies $g \in C^1(\mathbb{R}\times \ell_\infty(\mathbb{N}), \ell_\infty(\mathbb{N}))$, $g(0,0) = 0$, $D_u g(0,0) = s\,\mathrm{Id}$. The following result is one of the main steps to construct the local stable manifold of the origin. Below, $B_\infty(R)$ denotes the open ball of radius $R > 0$ centered at the origin in $\ell_\infty(\mathbb{N})$ and we denote $B(\varepsilon) = (-\varepsilon, \varepsilon)$.

Theorem 2. *There exist $\varepsilon, R > 0$ and an odd function $\psi \in C^1(B(\varepsilon), B_\infty(R))$, such that $g(u_0, \psi(u_0)) = 0$ for all $u_0 \in B(\varepsilon)$. Moreover, for all solution of (22) in $B(\varepsilon) \times B_\infty(R)$ one has $u = \psi(u_0)$. In addition, $\psi = (\psi_k)_{k\geq 1}$ possesses the following properties.*

(i) The function $u_0 \mapsto \psi_1(u_0)$ is increasing for $s = 1$, decreasing for $s = -1$, and satisfies

$$|\psi_1(u_0)| < |u_0| \text{ for all } u_0 \in B(\varepsilon), \tag{23}$$

$$\psi_1(u_0) = s\,P_\alpha(u_0) + O(|u_0|^{2\alpha-1}) \text{ when } u_0 \to 0. \tag{24}$$

(ii) The functions $u_0 \mapsto \psi_k(u_0)$ $(k \geq 1)$ satisfy

$$\psi_k = \overbrace{\psi_1 \circ \psi_1 \cdots \circ \psi_1}^{k \text{ times}}, \tag{25}$$

$$s\,\psi_1 = P_\alpha \circ (\mathrm{Id} + \psi_1) \circ \psi_1 + P_\alpha \circ (\mathrm{Id} + \psi_1). \tag{26}$$

(iii) For all $u_0 \in B(\varepsilon)$ we have $\lim_{k\to+\infty} \psi_k(u_0) = 0$. Moreover, for all $C > 1$, there exists $\varepsilon_C \leq \varepsilon$ such that for all $u_0 \in B(\varepsilon_C)$ we have

$$|\psi_k(u_0)| \leq C^{\frac{1}{1-\alpha}} \left(C^{\frac{1}{\alpha-1}} |u_0|\right)^{\alpha^k}. \tag{27}$$

Proof. By a direct application of the implicit function theorem, the solutions of (22) sufficiently close to 0 in $\mathbb{R} \times \ell_\infty(\mathbb{N})$ have the form $u = \psi(u_0)$, where ψ is odd due to the invariance $u_n \to -u_n$ of (5). Since we have

$$s\,\psi(u_0) = N(u_0, \psi(u_0)), \tag{28}$$

it follows that

$$\psi(u_0) = \mathrm{O}(|u_0|^\alpha) \text{ in } \ell_\infty(\mathbb{N}) \text{ when } u_0 \to 0. \tag{29}$$

In particular, by choosing ε small enough we have

$$\|\psi(u_0)\|_\infty < |u_0| \text{ for all } u_0 \in B(\varepsilon), \tag{30}$$

hence (23) holds true. Moreover, by identifying the first terms of the sequences at both sides of (28), we get

$$s\,\psi_1(u_0) = P_\alpha(\psi_2(u_0) + \psi_1(u_0)) + P_\alpha(\psi_1(u_0) + u_0), \tag{31}$$

which yields (24) after elementary computations based on estimate (29).

System (22) is invariant by index shift, i.e. if $(u_0, (u_n)_{n\geq 1})$ is a solution then, for all $k \geq 1$, $(u_k, (u_{k+n})_{n\geq 1})$ is also a solution. Combining this invariance with the above reduction implies $\psi_{n+k}(u_0) = \psi_n(\psi_k(u_0))$ for all $k, n \geq 1$. In particular, $\psi_{1+k}(u_0) = \psi_1(\psi_k(u_0))$ yields (25), and (26) follows by combining (31) and (25). In addition, differentiating (26) with respect to u_0 yields after lengthy but straightforward computations

$$\psi_1'(u_0) = s\,P_\alpha'(u_0) + \mathrm{O}(|u_0|^{2\alpha-2}),$$

hence ψ_1 is increasing on $B(\varepsilon)$ for $s = 1$ and decreasing on $B(\varepsilon)$ for $s = -1$ provided ε is chosen small enough.

Properties (25) and (23) imply $\lim_{k\to+\infty} \psi_k(u_0) = 0$ for all $u_0 \in B(\varepsilon)$. Moreover, according to (24), for all $C > 1$ there exists $\varepsilon_C > 0$ such that for $|u_0| < \varepsilon_C$ we have

$$|\psi_1(u_0)| \leq C\,|u_0|^\alpha, \tag{32}$$

which gives in conjunction with (25)

$$|\psi_{k+1}(u_0)| \leq C\,|\psi_k(u_0)|^\alpha, \text{ for all } k \geq 1. \tag{33}$$

Then using both (32) and (33) yields (27) by induction. □

3.3 Stable and Unstable Manifolds for the Maps *M*, *T*, *F*

In this section we consider the case $s = 1$ of (5), and use Theorem 2 to construct the stable and unstable manifolds of the origin for the maps M, T, F.

Introducing $w_n = u_{n+1}$, one can reformulate (5) as a two-dimensional mapping defined by (6). Let us consider ε, R, ψ_1 as in Theorem 2. Since the statement of Theorem 2 remains true replacing ε by $\varepsilon' < \varepsilon$, one can assume $\varepsilon < R$ without loss of generality. Let us define $\Omega = (-\varepsilon, \varepsilon)^2$ and consider the C^1 one-dimensional submanifold of $\mathbb{R}^2$

$$W^s_{\mathrm{loc}}(0) = \{\,(u_0, \psi_1(u_0)) \in \mathbb{R}^2, |u_0| < \varepsilon\,\} \subset \Omega. \tag{34}$$

The following result establishes that $W^s_{\text{loc}}(0)$ is the local stable manifold of the fixed point 0 of M. In the sequel we call a curve Γ *negatively invariant* by an invertible map M if $M^{-1}(\Gamma) \subset \Gamma$, and *positively invariant* by M if $M(\Gamma) \subset \Gamma$.

Theorem 3. *The manifold $W^s_{loc}(0)$ possesses the following properties.*

(i) $W^s_{loc}(0)$ *is positively invariant by M.*
(ii) *If $(u_0, w_0) \in W^s_{loc}(0)$, then the corresponding solution of (6) satisfies*

$$\lim_{n\to+\infty} (u_n, w_n) = 0.$$

(iii) *All solution of (6) such that $(u_n, w_n) \in \Omega$ for all $n \geq 0$ satisfies $(u_n, w_n) \in W^s_{loc}(0)$ for all $n \geq 0$.*

Proof. Let $(u_0, w_0) \in W^s_{\text{loc}}(0)$ and $(u_n, w_n) = (u_n, u_{n+1})$ the corresponding solution of (6). Since $w_0 = \psi_1(u_0)$, we have $(u_n, w_n) = (\psi_n(u_0), \psi_{n+1}(u_0))$ for all $n \geq 1$, by uniqueness of the solution of (5) given an initial condition. Then properties *(i)* and *(ii)* follow from properties *(ii)* and *(iii)* of Theorem 2.

Let $(u_n, w_n) = (u_n, u_{n+1})$ denote a solution of (6) staying in Ω for all $n \geq 0$. Since $|u_0| < \varepsilon$ and $\sup_{n\geq 1} |u_n| \leq \varepsilon < R$, Theorem 2 ensures that $u_n = \psi_n(u_0)$ for all $n \geq 1$, hence $(u_n, w_n) \in W^s_{\text{loc}}(0)$ for all $n \geq 0$. □

Now, to obtain a local stable manifold of the origin for the map T defined by (12), we use the fact that T and M are topologically conjugate by the homeomorphism h defined in (13). Let us define

$$\tilde{\Omega} = h(\Omega), \quad \tilde{W}^s_{\text{loc}}(0) = h(W^s_{\text{loc}}(0)). \tag{35}$$

The following result establishes that $\tilde{W}^s_{\text{loc}}(0)$ is a C^1 local stable manifold for the fixed point 0 of T. The smoothness of the stable manifold requires a particular treatment, given the fact that T and h^{-1} are not differentiable at the origin.

Theorem 4. *The manifold $\tilde{W}^s_{loc}(0)$ possesses the following properties.*

(i) $\tilde{W}^s_{loc}(0)$ *is positively invariant by T.*
(ii) *If $(x_0, z_0) \in \tilde{W}^s_{loc}(0)$, then the corresponding solution of (11) satisfies*

$$\lim_{n\to+\infty} (x_n, z_n) = 0.$$

(iii) *All solution of (11) such that $(x_n, z_n) \in \tilde{\Omega}$ for all $n \geq 0$ satisfies $(x_n, z_n) \in \tilde{W}^s_{loc}(0)$ for all $n \geq 0$.*
(iv) *There exist $\varepsilon_1 > 0$ and a function $\gamma \in C^1((-\varepsilon_1, \varepsilon_1), \mathbb{R})$ such that*

$$\tilde{W}^s_{loc}(0) = \{\, (x, \gamma(x)) \in \mathbb{R}^2, |x| < \varepsilon_1 \,\} \subset \tilde{\Omega}. \tag{36}$$

The function γ is odd, increasing and satisfies $\gamma(x) = P_\alpha(x) + o(|x|^\alpha)$ when $x \to 0$.

Proof. Properties *(i)–(iii)* are direct consequences of Theorem 3 and Lemma 1. Let us prove property *(iv)*. Using (35) and (26), one obtains after elementary computations

$$\tilde{W}^{\mathrm{s}}_{\mathrm{loc}}(0) = \{ ((P_\alpha \circ \varphi)(u_0), (P_\alpha \circ \varphi)(\psi_1(u_0))) \in \mathbb{R}^2, |u_0| < \varepsilon \}$$

where $\varphi = \mathrm{Id} + \psi_1$ is odd and defines a C^1-diffeomorphism from $(-\varepsilon, \varepsilon)$ onto its image since $\varphi' > 0$. This yields the parametrization (36), where $x = (P_\alpha \circ \varphi)(u_0)$, $\varepsilon_1 = \varphi(\varepsilon)^\alpha$ and $\gamma = P_\alpha \circ \varphi \circ \psi_1 \circ \varphi^{-1} \circ P_{1/\alpha}$. The function γ is odd, increasing and belongs to $C^1((0, \varepsilon_1), \mathbb{R})$. Using (24), one obtains

$$(\psi_1 \circ \varphi^{-1} \circ P_{1/\alpha})(x) = x + o(x) \text{ when } x \to 0^+,$$

so that the non-differentiability of $P_{1/\alpha}$ at $x = 0$ is compensated by ψ_1. Consequently, one finds

$$\gamma(x) = x^\alpha + o(x^\alpha) \text{ when } x \to 0^+,$$

which implies $\gamma \in C^1((-\varepsilon_1, \varepsilon_1), \mathbb{R})$. □

Due to the reversibility of T under R_1, the fixed point of T at the origin also admits an unstable manifold, which is simply

$$\tilde{W}^{\mathrm{u}}_{\mathrm{loc}}(0) = R_1 \tilde{W}^{\mathrm{s}}_{\mathrm{loc}}(0) = \{ (\gamma(z), z) \in \mathbb{R}^2, |z| < \varepsilon_1 \}, \tag{37}$$

as stated in the following theorem.

Theorem 5. *The manifold $\tilde{W}^{u}_{loc}(0)$ possesses the following properties.*

(i) $\tilde{W}^{u}_{loc}(0)$ is negatively invariant by T.
(ii) If $(x_0, z_0) \in \tilde{W}^{u}_{loc}(0)$, then the corresponding solution of (11) satisfies

$$\lim_{n \to -\infty} (x_n, z_n) = 0.$$

(iii) All solution of (11) such that $(x_n, z_n) \in R_1 \tilde{\Omega}$ for all $n \leq 0$ satisfies $(x_n, z_n) \in \tilde{W}^{u}_{loc}(0)$ for all $n \leq 0$.

As a consequence of Theorem 4,

$$\tilde{W}^{\mathrm{s}}(0) = \cup_{n \geq 0} T^{-n}(\tilde{W}^{\mathrm{s}}_{\mathrm{loc}}(0)) \tag{38}$$

defines the (global) stable manifold of the origin, i.e. the set of initial conditions (x_0, z_0) such that $\lim_{n \to +\infty} (x_n, z_n) = 0$. In the same way,

$$\tilde{W}^{\mathrm{u}}(0) = \cup_{n \geq 0} T^{n}(\tilde{W}^{\mathrm{u}}_{\mathrm{loc}}(0)) \tag{39}$$

defines the (global) unstable manifold of the origin, i.e. the set of initial conditions (x_0, z_0) such that $\lim_{n\to-\infty}(x_n, z_n) = 0$.

In the same way, the stable and unstable manifolds of the origin for the map F are respectively

$$\mathcal{W}^{\rm s}(0) = J\,\tilde{W}^{\rm s}(0), \quad \mathcal{W}^{\rm u}(0) = J\,\tilde{W}^{\rm u}(0) = \mathcal{R}_1\mathcal{W}^{\rm s}(0). \tag{40}$$

In addition, we shall later refer to the local stable and unstable manifolds defined by

$$\mathcal{W}^{\rm s}_{\rm loc}(0) = J\,\tilde{W}^{\rm s}_{\rm loc}(0), \quad \mathcal{W}^{\rm u}_{\rm loc}(0) = J\,\tilde{W}^{\rm u}_{\rm loc}(0) = \mathcal{R}_1\mathcal{W}^{\rm s}_{\rm loc}(0). \tag{41}$$

We have more explicitly

$$\mathcal{W}^{\rm s}_{\rm loc}(0) = \{\, (\gamma(x) + x, \gamma(x)) \in \mathbb{R}^2, |x| < \varepsilon_1 \,\}, \tag{42}$$

$$\mathcal{W}^{\rm u}_{\rm loc}(0) = \{\, (\gamma(z) + z, z) \in \mathbb{R}^2, |z| < \varepsilon_1 \,\}. \tag{43}$$

4 Homoclinic Solutions of the Stationary DpS Equation

This section is devoted to the proof of Theorem 1. For this purpose, we use the results of Sect. 3 combined with a reformulation of the results of [22] adapted to our context.

4.1 Homoclinic Solutions of Generalized DNLS Equations

We consider system (9) where $q : \mathbb{R} \to \mathbb{R}$ is an odd function. In the sequel we assume $q \in W^{1,1}_{\rm loc}(\mathbb{R})$, where $W^{1,1}_{\rm loc}$ refers to a classical Sobolev space [7]. Introducing $z_n = x_{n+1}$, one can reformulate (9) as a two-dimensional reversible mapping given by (11). The following existence theorem for homoclinic solutions is essentially proved in [22], but we shall provide the whole proof for completeness.

Theorem 6. *Assume the function $q \in W^{1,1}_{loc}(\mathbb{R})$ is odd, and there exists $x^* > 0$ such that $q(x^*) = 0$ and*

$$q > 0 \text{ on } (0, x^*), \quad q' < 0 \text{ on } (x^*, +\infty). \tag{44}$$

Moreover, assume the existence of a curve Γ_0^u negatively invariant by T and taking the form

$$\Gamma_0^u = \{\, (x, z) \in \mathbb{R}^2, x = \gamma_u(z),\ 0 < z < \epsilon_1 \,\}, \tag{45}$$

where $\epsilon_1 \in (0, x^*)$, $\gamma_u \in C^1([0, \epsilon_1], \mathbb{R})$ and $0 < \gamma_u(z) < z$ for all $z \in (0, \epsilon_1]$.

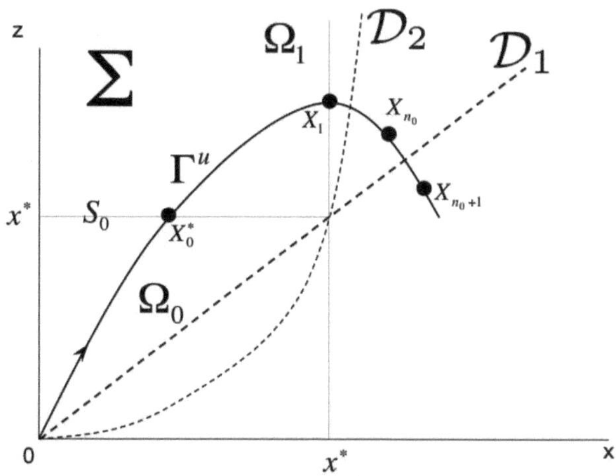

Fig. 1 Sketch of the different subsets of the plane used in the proof of Theorem 6

Under the above conditions, there exist symmetric homoclinic solutions of (11) taking the form $X_n = \pm(x_n^i, x_{n+1}^i)^T$ *(*$i = 1, 2$*), with* $\lim_{n\to\pm\infty} x_n^i = 0$, $x_n^i > x_{n-1}^i > 0$ *for all* $n \leq 0$, $x_0^i > x^*$ *and*

$$x_n^1 = x_{-n+1}^1, \quad x_n^2 = x_{-n}^2, \quad \textit{for all } n \in \mathbb{Z}.$$

Remark 1. Under the above assumptions, q has exactly three real zeros, at $x = \pm x^*$ and $x = 0$, and T has three fixed points $X = \pm(x^*, x^*)^T$ and $X = 0$.

Remark 2. Due to the reversibility of (12), the assumption made on Γ_0^{u} in Theorem 6 is equivalent to the existence of a curve

$$R_1\Gamma_0^{\mathrm{u}} = \{\,(x, z) \in \mathbb{R}^2, z = \gamma_{\mathrm{u}}(x),\ 0 < x < \epsilon_1\,\}$$

positively invariant by T.

Remark 3. In the work [22], the authors have considered the smooth situation when condition (20) is satisfied. In that case, the assumptions made on the curve Γ_0^{u} are automatically satisfied. Indeed, condition (20) implies that the origin is a hyperbolic fixed point of T, since $DT(0)$ has a pair of real eigenvalues Λ, Λ^{-1} given by (21). This yields the existence of a C^1 local unstable manifold of the origin, $W_{\mathrm{loc}}^{\mathrm{u}}(0) = -\Gamma_0^{\mathrm{u}} \cup \Gamma_0^{\mathrm{u}}$, with Γ_0^{u} taking the form (45) (at the origin, $W_{\mathrm{loc}}^{\mathrm{u}}(0)$ is tangent to the unstable subspace corresponding to the line $x = \Lambda^{-1} z$).

Proof. The proof given below can be more easily followed using Fig. 1. Since $q \in W_{\mathrm{loc}}^{1,1}(\mathbb{R})$ and Γ_0^{u} is negatively invariant by T, $\Gamma_n^{\mathrm{u}} = T^n(\Gamma_0^{\mathrm{u}}) \subset \Gamma_{n+1}^{\mathrm{u}}$ defines an increasing sequence of continuous rectifiable curves. Let us consider $\Gamma^{\mathrm{u}} = \cup_{n\geq 0}\Gamma_n^{\mathrm{u}}$. Since Γ_0^{u} lies in the sector

$$\Sigma = \{ (x,z) \in (0,+\infty)^2,\, z > x \},$$

and T^{-1} maps the z-axis outside Σ (onto the x-axis), the curve Γ^{u} cannot exit Σ by crossing the z-axis. In what follows we show that Γ^{u} exits Σ crossing the line $\mathcal{D}_1 = \mathrm{Fix}(R_1) = \mathrm{Span}((1,1)^T)$.

For this purpose, we first observe that for all $X = (x,z)^T$, one has

$$\mathrm{dist}(X, \mathcal{D}_1) = (z - x)/\sqrt{2}$$

and thus

$$\mathrm{dist}(T(X), \mathcal{D}_1) = \frac{q(z)}{\sqrt{2}} + \mathrm{dist}(X, \mathcal{D}_1). \tag{46}$$

Let us denote by Ω_0 the triangular domain

$$\Omega_0 = \{ (x,z) \in (0,+\infty) \times (0,x^*),\, z > x \}$$

containing Γ^{u}_0. For all $X = (x,z)^T \in \Omega_0$, $X' = T(X) = (x',z')^T$ satisfies

$$x' = z, \quad x^* > x' > x, \tag{47}$$

$$\mathrm{dist}(X', \mathcal{D}_1) > \mathrm{dist}(X, \mathcal{D}_1), \tag{48}$$

where we used (46) and the fact that $q > 0$ on $(0,x^*)$. By using inequalities (47)–(48), one can rule out the two following situations (for details see [22], Theorem 2.3),

(i) $\Gamma^{\mathrm{u}}_n \subset \Omega_0$ for all $n \geq 0$,
(ii) Γ^{u} exits Ω_0 by intersecting $\mathcal{D}_1$.

Consequently, Γ^{u} exits Ω_0 by intersecting the segment $\mathcal{S}_0 = (0,x^*) \times \{x^*\}$ at some point $X^*_0 = (x_0, x^*)^T$, thereby entering the domain

$$\Omega_1 = \{ (x,z) \in (0,+\infty) \times (x^*,+\infty),\, z > x \}.$$

Defining $X_n = T^n(X^*_0) = (x_n, z_n)^T = (x_n, x_{n+1})^T$, one can check that $X_1 = (x^*, z_1) \in \Omega_1$. If $X_n \in \Omega_1$ for $1 \leq n \leq n_0$, we find by induction for all $n = 2, \ldots, n_0$,

$$z_n > z_1 > x^*, \quad \mathrm{dist}(X_{n+1}, \mathcal{D}_1) < \frac{q(z_1)}{\sqrt{2}} + \mathrm{dist}(X_n, \mathcal{D}_1), \tag{49}$$

thanks to equality (46) with $q' < 0$ on $(x^*,+\infty)$. Since $q(z_1) < 0$, (49) provides a nonzero minimal decrease of $\mathrm{dist}(X_n, \mathcal{D}_1)$ at each step, hence X_n must leave Ω_1 for large enough n. Consequently, Γ^{u} connects two iterates $X_{n_0} \in ([x^*,+\infty) \times$

$\mathbb{R}) \cap \Omega_1$ and $X_{n_0+1} \in ([x^*, +\infty) \times \mathbb{R}) \setminus \Omega_1$, hence Γ^{u} intersects the boundary $\mathcal{D}_1 \cap (x^*, +\infty)^2$ of Ω_1. Consequently, we have shown that Γ^{u} exits Σ by crossing the line $\mathcal{D}_1 = \mathrm{Fix}(R_1)$.

Now let us show that for all initial condition $X_k \in \Gamma^{\mathrm{u}}$, the corresponding solution $X_n = (x_n, z_n)^T = (x_n, x_{n+1})^T$ of (11) satisfies $\lim_{n \to -\infty} X_n = 0$. Let $X_k \in \Gamma_k^{\mathrm{u}}$ and $X_0 = T^{-k}(X_k) \in \Gamma_0^{\mathrm{u}}$. We have $x_1 = z_0 \in (0, \epsilon_1]$ and $x_{-n-1} = \gamma_{\mathrm{u}}(x_{-n}) \in (0, x_{-n})$ for all $n \geq -1$, since Γ_0^{u} is negatively invariant by T. Consequently, there exists $\ell \in [0, \epsilon_1)$ such that $\lim_{n \to -\infty} x_n = \ell$. It follows that $\ell = 0$, since 0 is the only fixed point of γ_{u} in $[0, \epsilon_1)$.

Let us consider an intersection X_0 of Γ^{u} and $\mathcal{D}_1$ and denote by $X_n = (x_n^1, x_{n+1}^1)^T$ the corresponding solution of (11). The above result implies that $\lim_{n \to -\infty} X_n = 0$. Moreover, since $(R_1 X_{-n})_{n \in \mathbb{Z}}$ and $(X_n)_{n \in \mathbb{Z}}$ define two solutions of (11) equal at $n = 0$, both solutions coincide for all $n \in \mathbb{Z}$. We have then $R_1 X_{-n} = X_n$, i.e. $x_n^1 = x_{-n+1}^1$ for all $n \in \mathbb{Z}$. This implies $\lim_{n \to +\infty} X_n = 0$, i.e. $(X_n)_{n \in \mathbb{Z}}$ is a solution homoclinic to 0. Now we consider the intersection X_0^1 of Γ^{u} and $\mathcal{D}_1 \cap (x^*, +\infty)^2$ which is the closest to the origin along Γ^{u} (i.e. the part of Γ^{u} joining 0 and X_0^1 has minimal arclength). Since the arc joining 0 and X_0^1 does not exit Σ, we have $x_n^1 > 0$ and $x_n^1 > x_{n-1}^1$ for all $n \leq 0$.

There remains to check that Γ^{u} intersects

$$\mathcal{D}_2 = \mathrm{Fix}(R_2) = \{ (x, z) \in \mathbb{R}^2,\ x = z + \frac{q(z)}{2} \}.$$

Since $q < 0$ on $(x^*, +\infty)$, the curve $\mathcal{D}_2$ divides Ω_1 into two connected components. The boundary of the first component contains the segment $\mathcal{S}_0$, and the boundary of the second the half line $\mathcal{D}_1 \cap (x^*, +\infty)^2$. Since Γ^{u} intersects $\mathcal{S}_0$ at X_0^* and $\mathcal{D}_1 \cap (x^*, +\infty)^2$ at X_0^1, there exists an intersection $X_{-1}^2 = (x_{-1}^2, x_0^2)^T$ between Γ^{u} and $\mathcal{D}_2$ (with $x_0^2 > x^*$). As above, since $R_2 X_{-1}^2 = X_{-1}^2$, the solution of (11) given by $X_n = T^{n+1}(X_{-1}^2) = (x_n^2, x_{n+1}^2)^T$ satisfies $R_2 X_{-n-2} = X_n$, i.e. $x_n^2 = x_{-n}^2$ for all $n \in \mathbb{Z}$. Consequently, $(X_n)_{n \in \mathbb{Z}}$ is a solution homoclinic to 0. Moreover, since the arc joining 0 and X_{-1}^2 does not exit Σ, we have $x_n^2 > 0$ and $x_n^2 > x_{n-1}^2$ for all $n \leq 0$. □

4.2 *Application to the DpS Equation*

Theorem 6 can be applied to system (9) with nonlinearity q defined by (10), which is equivalent to the stationary DpS equation. Indeed, q' is locally integrable and the assumptions made on q in Theorem 6 are satisfied with $x^* = 4^{\frac{\alpha}{1-\alpha}}$. Moreover, the existence of the invariant curve Γ_0^{u} is established in Theorem 5, where $\tilde{W}_{\mathrm{loc}}^{\mathrm{u}}(0) = -\Gamma_0^{\mathrm{u}} \cup \Gamma_0^{\mathrm{u}}$. This leads to the following result.

Theorem 7. *There exist solutions x_n^i $(i = 1, 2)$ of (9)–(10) satisfying*

$$\lim_{n \to \pm\infty} x_n^i = 0,$$

$x_n^i > x_{n-1}^i > 0$ *for all* $n \leq 0$, $x_0^i > 4^{\frac{\alpha}{1-\alpha}}$ *and*

$$x_n^1 = x_{-n+1}^1, \quad x_n^2 = x_{-n}^2, \quad \textit{for all } n \in \mathbb{Z}.$$

Returning to the DpS equation in its original form (4), we get the following result, which establishes Theorem 1 in conjunction with Lemma 3 proved in the appendix.

Theorem 8. *The stationary DpS equation (4) with* $s = 1$ *admits solutions* $\tilde{a}_n^1$, a_n^2 *satisfying the properties*

$$\lim_{n \to \pm\infty} a_n = 0, \quad (-1)^n a_n > 0, \quad |a_n| > |a_{n-1}| \; \forall n \leq 0,$$

and

$$\tilde{a}_n^1 = \tilde{a}_{-n}^1, \quad a_n^2 = -a_{-n+1}^2, \quad \textit{for all } n \in \mathbb{Z}. \tag{50}$$

Proof. From the localized solutions x_n^i of Theorem 7, let us define

$$\tilde{u}_n^1 = x_n^1 + x_{n+1}^1, \quad \tilde{a}_n^1 = (-1)^n \tilde{u}_n^1, \tag{51}$$

$$u_n^2 = x_n^2 + x_{n-1}^2, \quad a_n^2 = (-1)^n u_n^2. \tag{52}$$

Since x_n^1 and x_n^2 are solution of (9)–(10), we get

$$\tilde{u}_n^1 + \tilde{u}_{n-1}^1 = P_{1/\alpha}(x_n^1), \quad u_n^2 + u_{n+1}^2 = P_{1/\alpha}(x_n^2),$$

hence

$$x_n^1 = P_\alpha(\tilde{u}_n^1 + \tilde{u}_{n-1}^1), \quad x_n^2 = P_\alpha(u_n^2 + u_{n+1}^2).$$

Reporting the above identities in definitions (51) and (52), one finds that $\tilde{u}_n^1, u_n^2$ define solutions of (5) for $s = 1$ (note that $u_n^1 := \tilde{u}_{n-1}^1$ and x_n^1 are linked by equality (8)), and thus $\tilde{a}_n^1, a_n^2$ are solutions of (4). The remaining properties of $\tilde{a}_n^1, a_n^2$ directly follow from those of x_n^1 and x_n^2. □

Figures 2 and 3 illustrate the profiles of the homoclinic solutions $\tilde{a}_n^1$, a_n^2 for different values of α. The solutions of the stationary DpS equation (4) with $s = 1$ are computed with a Newton-type method (we use the MATLAB function fsolve), for a finite lattice of 21 particles with zero boundary conditions. When α converges towards unity, the homoclinic solutions become more extended and their amplitude goes to 0. This phenomenon will be explained in Sect. 5.2 by introducing a suitable continuum limit of the stationary DpS equation.

One can notice that the localized solutions provided by Theorem 8 are "staggered", i.e. a_{n+1} and a_n have opposite signs. As shown in the following lemma, this remains true for all localized solutions a_n of (4) provided $|n|$ is sufficiently large.

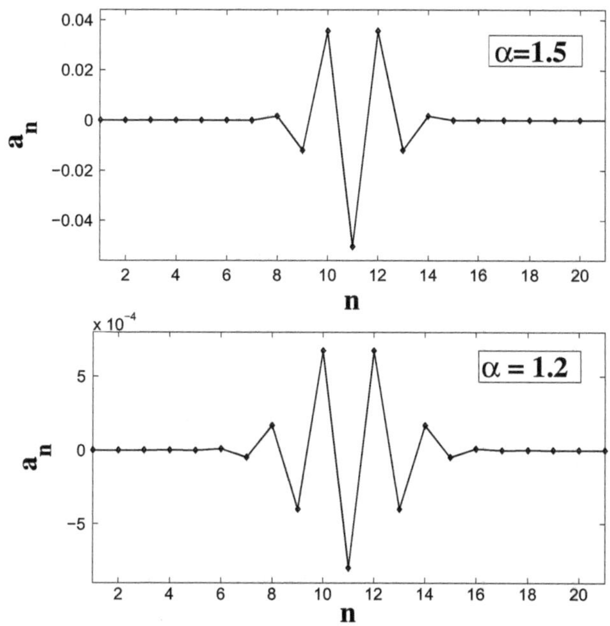

Fig. 2 Homoclinic solution $\tilde{a}_n^1$ of the stationary DpS equation (4) with $s = 1$. The solution is computed numerically for $\alpha = 3/2$ (*top panel*) and $\alpha = 1.2$ (*bottom panel*)

Moreover, we prove below that all localized solutions of (4) (in particular $\tilde{a}_n^1$, a_n^2) decay at least doubly exponentially at infinity.

Lemma 2. *Let* $(a_n)_{n\in\mathbb{Z}}$ *denote a solution of (4) such that* $\lim_{n\to\pm\infty} a_n = 0$ *and fix* $q \in (0, 1)$. *There exists* $n_0 \geq 0$ *such that for all* $n \in \mathbb{Z}$ *with* $|n| \geq n_0$, a_{n+1} *and* a_n *have opposite signs and*

$$|a_n| \leq q^{1+\alpha^{|n|-n_0}}. \tag{53}$$

Proof. According to Lemma 3 proved in the appendix, nonzero localized solutions of (4) exist only for $s = 1$. To prove that localized solutions are staggered for sufficiently large n, let us consider a solution a_n of (4) such that $\lim_{n\to+\infty} a_n = 0$. One can define a solution of (5) vanishing as $n \to +\infty$ through $u_n = (-1)^n a_{n+n_0}$, with $\sup_{n\geq 0} |u_n|$ arbitrarily small provided n_0 is sufficiently large. By Theorem 2, it follows that $u_n = \psi_n(u_0)$ for all $n \geq 1$ provided n_0 is large enough. For $s = 1$, $\psi_n(u_0)$ and u_0 have the same sign for all $n \geq 1$ when u_0 is small enough (due to properties *(i)* and *(ii)* of Theorem 2), hence $(u_n)_{n\geq 0}$ has a sign provided n_0 is large enough. Consequently, a_{n+1} and a_n have opposite signs for all $n \geq n_0$ if n_0 is sufficiently large. In the same way (using the reflectional symmetry of (4)), if $\lim_{n\to-\infty} a_n = 0$ then a_{n+1} and a_n have opposite signs for all $n \leq -n_0$ if $n_0 \geq 0$ is sufficiently large. Inequality (53) follows similarly from Theorem 2,

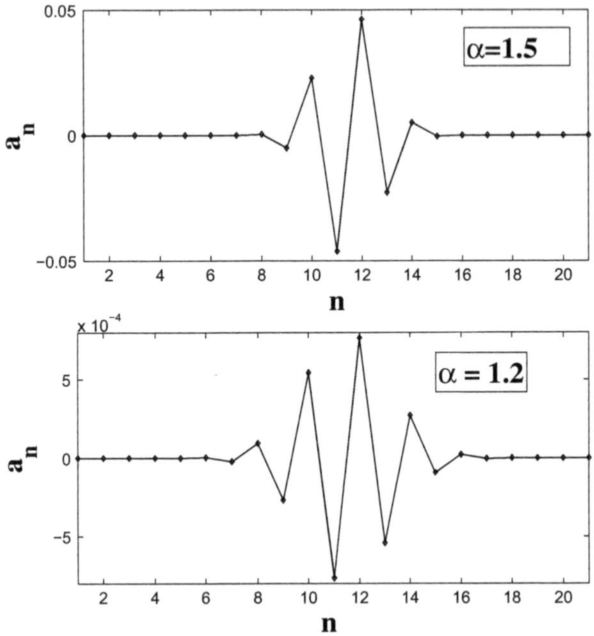

Fig. 3 Homoclinic solution a_n^2 of the stationary DpS equation (4) with $s = 1$. The solution is computed numerically for $\alpha = 3/2$ (*top panel*) and $\alpha = 1.2$ (*bottom panel*)

after elementary computations based on the decay estimate (27). One can fix $C = q^{1-\alpha}$ in (27) and observe that $q^{-1}\,|u_0| \leq q$ provided $|n_0|$ is large enough, which yields (53). □

5 Analytical Approximations of Stable and Unstable Manifolds

In previous sections, we have proved the existence of stable and unstable manifolds of the origin for maps equivalent to the stationary DpS equation (4), and the existence of homoclinic orbits for $s = 1$. In what follows, we illustrate the shape of these manifolds and orbits and provide analytical approximations thereof, depending on the nonlinearity exponent $\alpha = p - 1 > 1$. We shall work with the map F defined in (18) because it will be suitable for future extensions to this work taking into account lattice defects, as briefly discussed in Sect. 7.

Numerical computation of the exact stable manifold is performed in two stages. In the first stage we construct a local stable manifold. To this end we choose N equally distant points $0 < u_{0,1} < u_{0,2} < \cdots < u_{0,N}$ with $u_{0,N}$ small enough. From each point $u_{0,i}$ we compute an orbit on the local stable manifold from the corresponding solution $(u_n)_{n\geq 0}$ of Eq. (22) with $u_0 = u_{0,i}$. This problem is solved with a Newton-type method, for a finite lattice of size L sufficiently large and the fixed boundary condition $u_L = 0$. This yields a set of orbits $(v_{n,i}, y_{n,i})_{n\geq 1}$ on the stable manifold, where $v_{n,i} = u_{n-1}$ and $y_{n,i} = P_\alpha(u_n + u_{n-1})$. In a second stage, to construct the global stable manifold, we recursively apply the inverse mapping F^{-1} to the set of points $(v_{1,i}, y_{1,i})$ computed previously on the local stable manifold. The construction of the unstable manifold is performed in a similar manner.

In addition, we shall resort to two different methods to obtain analytical approximations of stable and unstable manifolds. The first one is based on a leading order approximation of the local stable manifold and the computation of some backward iterates (Sect. 5.1), which turns out to be efficient when α is far from unity. The second method is based on a continuum limit obtained when α is close to unity, where one recovers a logarithmic stationary nonlinear Schrödinger equation (Sect. 5.2). These approximations will be computed for the stable manifold, and their analogues for the unstable manifold can be obtained using the reversibility symmetry $\mathcal{R}_1$. For the approximation of homoclinic orbits we shall restrict to the site-centered solution $\tilde{a}_n^1$ described in Theorem 8, but the bond-centered solution a_n^2 could be approximated similarly.

5.1 Method of Local Approximation and Backward Iterates

Let us consider the map F defined in (18) and the stable and unstable manifolds of the origin $\mathcal{W}^{\mathrm{u}}(0)$ and $\mathcal{W}^{\mathrm{s}}(0)$ defined in Sect. 3.3. Using parametrization (42) and the fact that $y(x) = P_\alpha(x) + o(|x|^\alpha)$ when $x \to 0$, we derive the following approximations of the local stable and unstable manifolds

$$\mathcal{W}^{\mathrm{s}}_{\mathrm{app}} = \{\, (v, y) = (P_\alpha(x) + x, P_\alpha(x)) \in \mathbb{R}^2, x \in \mathbb{R}\,\}, \tag{54}$$

$$\mathcal{W}^{\mathrm{u}}_{\mathrm{app}} = \{\, (v, y) = (P_\alpha(z) + z, z) \in \mathbb{R}^2, z \in \mathbb{R}\,\}, \tag{55}$$

which are valid close enough to the origin. To improve their validity domain, we shall consider the backward iterates of the approximate stable manifold

$$\mathcal{W}^{\mathrm{s},(k)}_{\mathrm{app}} = F^{-k}(\mathcal{W}^{\mathrm{s}}_{\mathrm{app}}), \tag{56}$$

where

$$F^{-1}\begin{pmatrix} v \\ y \end{pmatrix} = \begin{pmatrix} P_{1/\alpha}(v-y) - v \\ v - y \end{pmatrix}.$$

We define by symmetry

$$\mathcal{W}^{\mathrm{u},(k)}_{\mathrm{app}} = \mathcal{R}_1 \mathcal{W}^{\mathrm{s},(k)}_{\mathrm{app}}.$$

One can use the parametrization

$$\mathcal{W}_{\rm app}^{s,(k)} = \{ (v,y) = (v^{(k)}(x), y^{(k)}(x)) \in \mathbb{R}^2, x \in \mathbb{R} \}, \tag{57}$$

where $v^{(0)} = P_\alpha + \mathrm{Id}$, $y^{(0)} = P_\alpha$ and the functions $(v^{(k)}, y^{(k)})$ are defined by induction for $k \geq 1$, with

$$\begin{pmatrix} v^{(k+1)} \\ y^{(k+1)} \end{pmatrix} = \begin{pmatrix} P_{1/\alpha}\left(v^{(k)} - y^{(k)}\right) - v^{(k)} \\ v^{(k)} - y^{(k)} \end{pmatrix}. \tag{58}$$

We have in particular

$$\begin{pmatrix} v^{(1)}(x) \\ y^{(1)}(x) \end{pmatrix} = \begin{pmatrix} P_{1/\alpha}(x) - P_\alpha(x) - x \\ x \end{pmatrix}. \tag{59}$$

When α is sufficiently far from unity, one observes numerically that $\mathcal{W}_{\rm app}^{s,(1)}$ and $\mathcal{W}_{\rm app}^{u,(1)}$ provide good approximations of $\mathcal{W}^{s}(0)$ and $\mathcal{W}^{u}(0)$ close to the reversible homoclinics. This result is illustrated by Fig. 4 for $\alpha = 3$. The quality of the approximation decreases when α becomes close to unity, as shown in Fig. 5 for $\alpha = 3/2$. However, the precision can be improved by considering $\mathcal{W}_{\rm app}^{s,(k)}$ and $\mathcal{W}_{\rm app}^{u,(k)}$ with larger values of k (at the expense of working with more complex parametrizations of the manifolds). It seems that the precision can be improved arbitrarily (in some fixed neighborhood of $(v,y) = 0$) by increasing k, but a convergence proof is not yet available.

As an application, let us use the analytic approximations $\mathcal{W}_{\rm app}^{s,(1)}$ and $\mathcal{W}_{\rm app}^{u,(1)}$ to approximate the site-centered homoclinic solution of the stationary DpS equation. The fixed point $(v_1, y_1)^T$ of $\mathcal{R}_1$ lying on $\mathcal{W}_{\rm app}^{s,(1)}$ satisfies

$$v_1 = 2\, y_1, \quad v_1 = P_{1/\alpha}(y_1) - P_\alpha(y_1) - y_1,$$

hence $y_1 > 0$ satisfies

$$3 = y_1^{(1/\alpha)-1} - y_1^{\alpha-1}, \tag{60}$$

which admits a unique solution $y_1 \in (0,1)$ since the right side of (60) is monotone decreasing. This solution corresponds to an approximate solution of the stationary DpS equation (4) determined by the initial condition

$$a_0 = v_1 = 2\, y_1, \quad a_1 = v_1 - y_1^{(1/\alpha)}. \tag{61}$$

Obviously, due to the sensitivity of the map F to initial conditions, the above approximation is only meaningful for a finite number of sites away from $n = 0$.

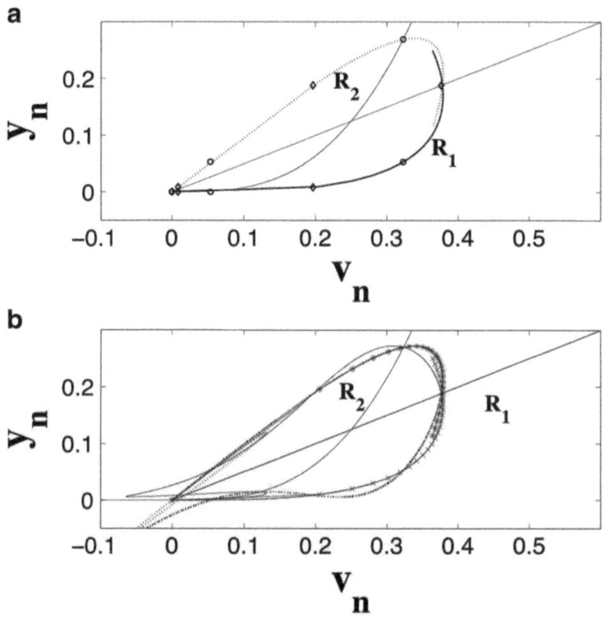

Fig. 4 (**a**): Approximate stable manifold $\mathcal{W}_{\text{app}}^{s,(1)}$ (*bold solid line*) and approximate unstable manifold $\mathcal{W}_{\text{app}}^{u,(1)}$ (*dashed thin curve*) in the case $\alpha = 3$ (only a limited part of the curves has been represented). *Dots* correspond to numerically computed homoclinic orbits. The intersections of $\text{Fix}(\mathcal{R}_1)$ and $\text{Fix}(\mathcal{R}_2)$ with $\mathcal{W}_{\text{app}}^{s,(1)}$ and $\mathcal{W}_{\text{app}}^{u,(1)}$ are very close to exact homoclinic intersections. (**b**): Same as above, except the exact stable (*solid curve*) and unstable (*dashed curve*) manifolds $\mathcal{W}^s(0)$ and $\mathcal{W}^u(0)$ are now represented. The approximate stable manifold $\mathcal{W}_{\text{app}}^{s,(1)}$ is denoted with a solid curve with (x) marks and $\mathcal{W}_{\text{app}}^{u,(1)}$ is denoted with a solid curve with ($*$) marks. The exact and approximate stable and unstable manifolds almost coincide before crossing the line $\text{Fix}(\mathcal{R}_1)$. $\text{Fix}(\mathcal{R}_1)$ and $\text{Fix}(\mathcal{R}_2)$ are denoted on both (**a**) and (**b**) with *solid lines* marked with the appropriate labels R_1 and R_2

In order to obtain an approximation of the solution of (60), we introduce a new variable

$$\eta = y_1^{\frac{\alpha-1}{\alpha}}. \tag{62}$$

Introducing (62) into (60) yields

$$\eta^{\alpha+1} = 1 - 3\eta. \tag{63}$$

Using the method of successive iterations to approximate η, we fix $\eta_0 = 0$ and consider the following recurrence relation

$$\eta_i = \left(1 - \eta_{i-1}^{\alpha+1}\right)/3. \tag{64}$$

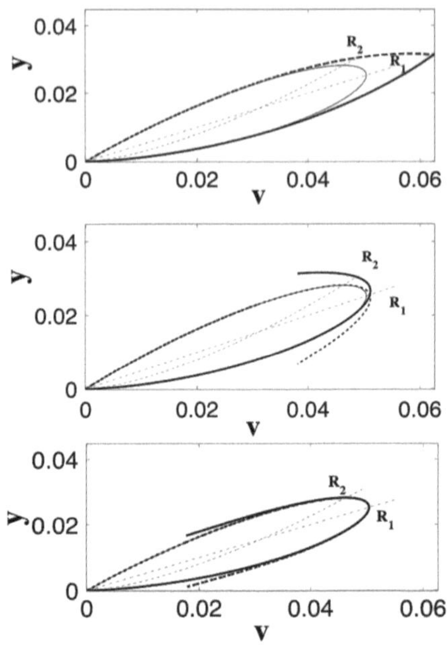

Fig. 5 Comparison of the exact invariant manifolds $\mathcal{W}^{s}(0)$ and $\mathcal{W}^{u}(0)$ (*solid lines*) with the approximate ones $\mathcal{W}^{s,(k)}_{app}$ (*bold solid curve*) and $\mathcal{W}^{u,(k)}_{app}$ (*bold dashed curve*) in the case $\alpha = 3/2$. The *thin, dotted lines* correspond to $\mathrm{Fix}(\mathcal{R}_1)$ and $\mathrm{Fix}(\mathcal{R}_2)$. The *top panel* corresponds to $k = 1$. The analytic approximation corresponding to $\mathcal{W}^{s,(1)}_{app}$ and $\mathcal{W}^{u,(1)}_{app}$ doesn't provide a good match with the exact invariant manifolds. To make the approximation more precise we perform additional iterations, yielding the improved approximations $\mathcal{W}^{s,(2)}_{app}$, $\mathcal{W}^{u,(2)}_{app}$ (*middle panel*), and $\mathcal{W}^{s,(3)}_{app}$, $\mathcal{W}^{u,(3)}_{app}$ (*bottom panel*)

Using η_2 as an approximation of the solution of (63) leads to

$$y_1 \approx \eta_2^{\frac{\alpha}{\alpha-1}} = 3^{\frac{\alpha}{1-\alpha}} \left(1 - 3^{-\alpha-1}\right)^{\frac{\alpha}{\alpha-1}} . \tag{65}$$

According to the results presented in Fig. 6, using η_2 gives a good approximation of the branch of solutions we are looking for, provided α is chosen within the range of values where approximation (60) is valid.

In Fig. 7, we compare for different values of α the approximate site-centered homoclinic orbit determined by (65) and (61) with the exact site-centered homoclinic orbit (computed by a Newton-type method from the stationary DpS equation). We plot the amplitude $u_n = (-1)^n a_n$ obtained after the staggering transformation. An approximation error is computed as

$$\mathrm{Err} = \frac{\|\mathbf{u}_{\mathrm{exact}} - \mathbf{u}_{\mathrm{approx}}\|_2}{\|\mathbf{u}_{\mathrm{exact}}\|_2} \tag{66}$$

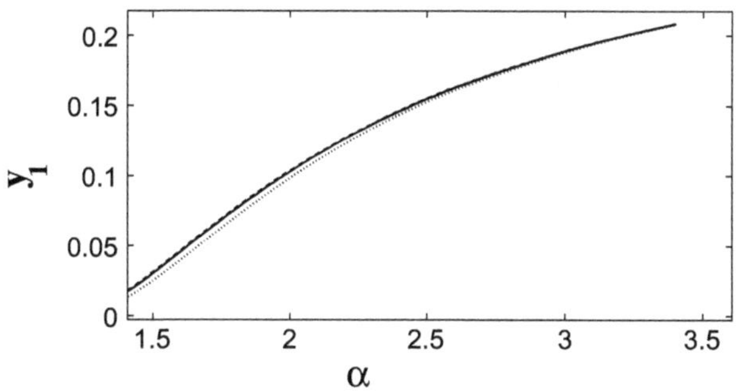

Fig. 6 Comparison of the exact value of y_1 for the site-centered homoclinic orbit (*dotted line*), its approximation given by the exact solution of (60) (*dashed line*) and the analytical approximation (65) (*solid line*), for different values of α. The exact site-centered homoclinic orbit is obtained by solving the stationary DpS equation with a Newton-type method, and the same iterative procedure is employed to solve (60). It appears that the solution of (60) yields in fact a slightly less precise approximation compared to analytical approximation (65)

where $\mathbf{u} = [u_{-2}, \ldots, u_2]$. As it comes out from Fig. 7, analytical approximation (65) is in fairly good agreement with the results of direct numerical computations for the higher values of α, but it provides unsatisfactory results for α close to unity (e.g. $\alpha = 3/2,\ 2$).

To improve the above approximation we resort to the improved approximate stable manifold $\mathcal{W}^{s,(2)}_{\text{app}}$, parametrized in the following way using (58),

$$\begin{pmatrix} v^{(2)}(x) \\ y^{(2)}(x) \end{pmatrix} = \begin{pmatrix} P_{1/\alpha}\left(P_{1/\alpha}(x) - P_{\alpha}(x) - 2x\right) - P_{1/\alpha}(x) + x + P_{\alpha}(x) \\ P_{1/\alpha}(x) - P_{\alpha}(x) - 2x \end{pmatrix}. \tag{67}$$

Thus, setting $(v_1, y_1) = (v^{(2)}(x), y^{(2)}(x))$ and $v_1 = 2\,y_1$, we obtain an approximation of the homoclinic intersection determined by

$$\begin{aligned} &P_{1/\alpha}\left(P_{1/\alpha}(x) - P_{\alpha}(x) - 2x\right) - 3P_{1/\alpha}(x) + 3P_{\alpha}(x) + 5x = 0, \\ &y_1 = P_{1/\alpha}(x) - P_{\alpha}(x) - 2x. \end{aligned} \tag{68}$$

Here it is important to emphasize that in contrast to the case of the first order approximation (59) where the intersection point y_1 could be analytically approximated and is given explicitly in (65), the solution of (68) is calculated numerically by a Newton-type method.

In Table 1, we compare the accuracies of the approximate site-centered homoclinic solutions computed from the approximate stable manifolds $\mathcal{W}^{s,(k)}_{\text{app}}$, for $k = 1, 2, 3$ and $\alpha = 1.5,\ 2,\ 3$. The agreement obtained with the improved approximation (68) is excellent (with a strong improvement compared to approximation (60)), but at the expense of working with the more complex formula (68).

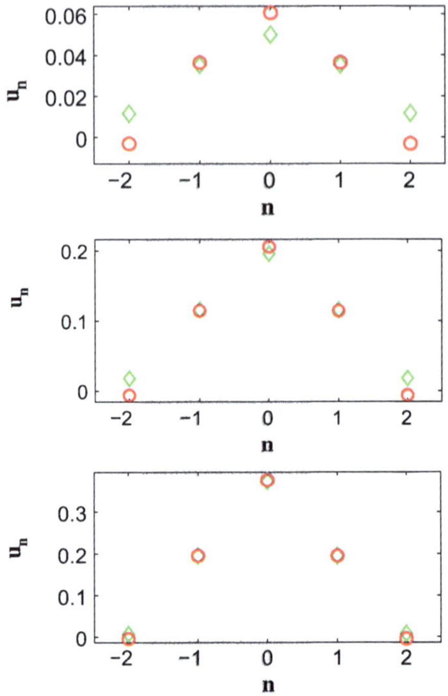

Fig. 7 Comparison of the approximate site-centered homoclinic orbit determined by (65) and (61) with the exact site-centered homoclinic solution $\tilde{a}_n^1$ described in Theorem 8 (this solution is computed numerically). The exact homoclinic orbit $\tilde{u}_n = (-1)^n \tilde{a}_n^1$ corresponds to *green diamonds* and the approximate homoclinic orbit is represented with *red circles*. The site-centered homoclinic orbit is spanned over approximately five particles of interest (i.e. $n = -2, \ldots, 2$). *Top panel*: $\alpha = 3/2$ (relative error: 31.64 %), *middle panel*: $\alpha = 2$ (relative error: 13.98 %), *bottom panel*: $\alpha = 3$ (relative error: 3.83 %). The relative error drops to 1.19 % for $\alpha = 4$ and 0.39 % for $\alpha = 5$ (profiles not shown)

Table 1 Approximations $y_1^{(1)}, y_1^{(2)}, y_1^{(3)}$ of the exact value of y_1 for the site-centered homoclinic orbit. The approximations correspond to the intersection between Fix(R_1) and the approximate stable manifolds $\mathcal{W}_{\text{app}}^{s,(1)}, \mathcal{W}_{\text{app}}^{s,(2)}, \mathcal{W}_{\text{app}}^{s,(3)}$. The relative errors have been calculated between the exact site-centered homoclinic solution $\tilde{a}_n^1$ described in Theorem 8 and the approximate ones, according to formula (66) (five lattice sites have been used to evaluate the error). The approximate site-centered homoclinic solutions are computed from the iteration of the map starting from the initial conditions $(v_1, y_1) = (2\,y_1^{(i)}, y_1^{(i)})$ with $i = 1, 2, 3$

α	Approx. $y_1^{(1)}$	Relative error on a_n	Approx. $y_1^{(2)}$	Relative error on a_n	Approx. $y_1^{(3)}$	Relative error on a_n
3/2	0.0312	37.54 %	0.0255	1.78 %	0.0252	0.0079 %
2	0.1038	16.72 %	0.0984761	0.02 %	0.0984678	$4.37{\cdot}10^{-7}$ %
3	0.1890611	4.69 %	0.1880980	$8.69{\cdot}10^{-5}$ %	0.1880981	$3.32{\cdot}10^{-7}$ %

As a conclusion, the approximation method of invariant manifolds and homoclinic orbits described in this section is efficient when α is far enough from unity. For α close to unity, it becomes unpractical because approximate stable manifolds

$\mathcal{W}_{\text{app}}^{s,(k)}$ involving increasingly complex parametrizations must be used to reach a good precision.

5.2 *Continuum Limit for Weak Nonlinearities*

In this section we develop another approximation method for invariant manifolds and homoclinic orbits based on a continuum limit, which is obtained when α is close to unity. This method completes the one described in Sect. 5.1, which was efficient for α far enough from unity.

5.2.1 Continuum Limit Model with Logarithmic Nonlinearity

We consider Eq. (5) for $s = 1$. As seen in Sect. 2.1, $y_n = x_{n-1}$ satisfies

$$y_{n+1} - 2y_n + y_{n-1} = P_{1/\alpha}(y_n) - 4y_n, \quad n \in \mathbb{Z}. \tag{69}$$

We renormalize the problem by setting

$$y_n = 4^{\frac{\alpha}{1-\alpha}} z_n \tag{70}$$

(the prefactor corresponds to a fixed point of (69)) and define $f_\alpha(z) = P_{1/\alpha}(z) - z$. Dividing (69) by $\alpha - 1$ and using (70) yields

$$\frac{z_{n+1} - 2z_n + z_{n-1}}{\alpha - 1} = \frac{4}{\alpha - 1} f_\alpha(z_n), \quad n \in \mathbb{Z}. \tag{71}$$

This system has a well-defined formal limit when $\alpha \to 1$. Indeed, we observe that for all $z > 0$,

$$\lim_{\alpha \to 1} \frac{f_\alpha(z)}{\alpha - 1} = -z \ln(z).$$

Let us set

$$z_n = z\left(\sqrt{\alpha - 1}\,(n - n_0)\right), \tag{72}$$

where z denotes a sufficiently smooth function and n_0 will be specified later. Letting $\alpha \to 1$ in (71) yields the ODE

$$\frac{d^2 z}{dx^2} + 4z \ln(z) = 0. \tag{73}$$

This equation can be seen as a (one-dimensional) stationary logarithmic nonlinear Schrödinger equation, a system originally introduced in the context of nonlinear wave mechanics [2]. It admits the first integral of motion

$$\left(\frac{dz}{dx}\right)^2 - 4z^2\left(\frac{1}{2} - \ln z\right) = C \tag{74}$$

and the Gaussian homoclinic solution

$$z(x) = \sqrt{e}\, e^{-x^2}. \tag{75}$$

In what follows we deduce an approximate solution of the stationary DpS equation from the above computations, and we compare it to the site-centered homoclinic solution of the DpS equation denoted in Theorem 8 by $\tilde{a}_n^1 = (-1)^n \tilde{u}_n^1$. The site-centered symmetry of $\tilde{a}_n^1$ and $\tilde{u}_n^1$ (i.e. $\tilde{u}_n^1 = \tilde{u}_{-n}^1$) corresponds to a bond-centered symmetry of $y_n = x_{n-1}^1 = P_\alpha(\tilde{u}_{n-1}^1 + \tilde{u}_{n-2}^1)$, i.e. we have $y_{1-n} = y_{2+n}$. In analogy with Eq. (51), the corresponding approximate solution of (5) is thus defined by $u_n^{\text{app}} = y_{n+1} + y_{n+2}$, with

$$y_n = 4^{\left(\frac{\alpha}{1-\alpha}\right)} z_n, \quad z_n = z\left(\sqrt{\alpha - 1}\,(n - 3/2)\right) = \exp\left(\frac{1}{2} - (\alpha - 1)\,(n - 3/2)^2\right), \tag{76}$$

where we have fixed $n_0 = 3/2$ in (72). After some elementary computations, we get the expression

$$u_n^{\text{app}} = 2^{\frac{1+\alpha}{1-\alpha}}\, \sqrt{e}\, e^{-(\alpha-1)\,(n^2+\frac{1}{4})} \cosh((\alpha - 1)\, n). \tag{77}$$

Figure 8 compares the exact site-centered homoclinic solution $\tilde{u}_n^1$ of (5) (computed numerically by a Newton-type method) and the continuum approximation (77). The left panel displays the relative error, which decreases to 0 (almost linearly) when α approaches unity. The right panel shows the very good agreement between the approximate and numerical homoclinic solutions for $\alpha = 1.05$. According to Fig. 8, the relative error drops below 10 % approximately when $\alpha \leq 1.4$, but a refined approximation is required to approximate the homoclinic solution precisely for higher values of α. We shall address this problem in the sequel.

5.2.2 Improved Continuum Limit

In this section, we show that the continuum limit approximation derived in the previous section can be substantially improved. Instead of letting $\alpha \to 1$ at both sides of (71), we retain the right side as it is, yielding for $z \geq 0$

$$z'' = \frac{4}{\alpha - 1}\left[z^{1/\alpha} - z\right]. \tag{78}$$

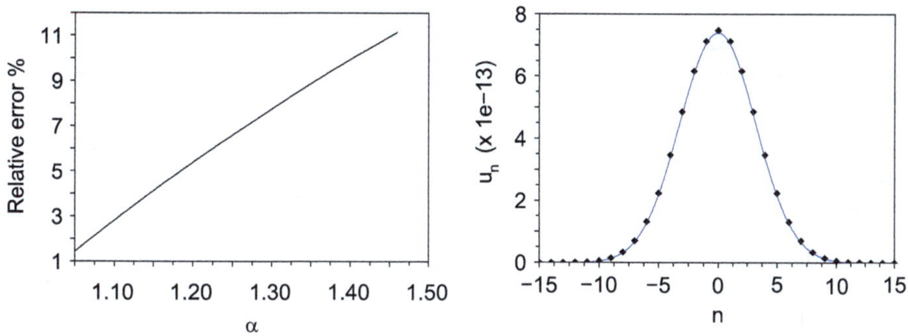

Fig. 8 Comparison of the continuum approximation defined by (77) with the exact site-centered homoclinic solution of (5) computed numerically. The *left graph* provides the relative error as a function of α (31 lattice sites are used to compute the error). The *right graph* displays both profiles for the particular value $\alpha = 1.05$. The *dots* correspond to the numerically computed homoclinic orbit, and the *blue curve* to the continuum approximation

Equation (78) possesses the first integral of motion,

$$\frac{z'^2}{2} = \frac{4}{\alpha - 1}\left[\frac{\alpha}{1+\alpha} z^{(1/\alpha)+1} - \frac{z^2}{2}\right] + C. \tag{79}$$

Below we give a method for approximating the stable manifold of the origin $\mathcal{W}^s(0)$ for the map F and the site-centered homoclinic orbit using the above continuum limit. Using (70)–(72) and the centered finite difference approximation

$$z'\left(\sqrt{\alpha-1}\,(n-n_0)\right) \approx \frac{z_{n+1} - z_{n-1}}{2\sqrt{\alpha-1}}$$

in the left side of (79), one obtains for $y_n \geq 0$

$$(y_{n+1} - y_{n-1})^2 = 8\left[\frac{\alpha}{1+\alpha} y_n^{(1/\alpha)+1} - 2y_n^2\right]. \tag{80}$$

Let us assume this equality holds true for all orbit of F along $\mathcal{W}^s(0)$, lying near the origin for all $n \geq 0$. Using (69) and (16) at rank $n-1$, we obtain

$$\left(y_n^{1/\alpha} - 2v_n\right)^2 = 8\left[\frac{\alpha}{1+\alpha} y_n^{(1/\alpha)+1} - 2y_n^2\right].$$

Consequently, we obtain an approximation of (a part of) the stable manifold of the origin, given for $y \geq 0$ by the parametrization

$$v = \frac{1}{2} y^{1/\alpha} + \sqrt{2}\left[\frac{\alpha}{1+\alpha} y^{(1/\alpha)+1} - 2y^2\right]^{1/2}. \tag{81}$$

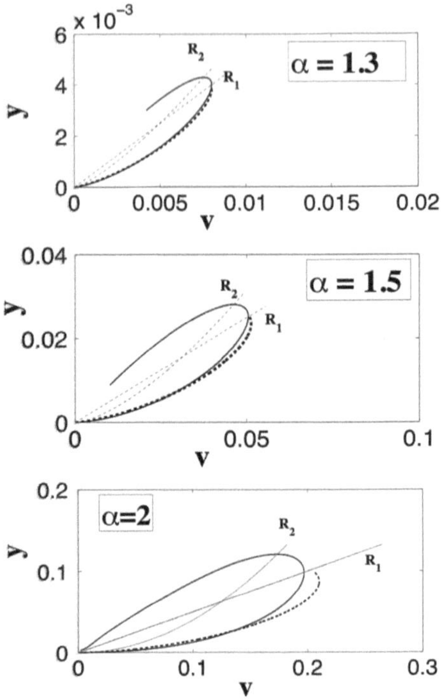

Fig. 9 Comparison of the numerically computed stable manifold of the origin for the map F (*bold solid curve*) with continuum approximation (81) (*bold dashed curve*). System parameter: $\alpha = 1.3$ (*top*), $\alpha = 1.5$ (*middle*), $\alpha = 2$ (*bottom*). The fixed sets of reversibility symmetries $\mathcal{R}_1$, $\mathcal{R}_2$ are also plotted (*thin dashed lines*)

In Fig. 9 we compare the analytic approximation (81) of the stable manifold with the numerically computed stable manifold, and find a good agreement provided α is below 1.5.

To further illustrate the improvement of the continuum limit model, we compare the approximate site-centered homoclinic orbit deduced from the above computations with the exact solution computed numerically. The approximate site-centered homoclinics can be obtained from the point of intersection between the approximate stable manifold and $\text{Fix}(\mathcal{R}_1)$. Thus setting $v = 2y$ in (81), we derive a scalar nonlinear equation for the point of intersection $(v, y) = (2y_{\text{i}}, y_{\text{i}})$ which belongs to the homoclinic orbit, namely

$$2y_{\text{i}} = \frac{1}{2}y_{\text{i}}^{1/\alpha} + \sqrt{2}\left[\frac{\alpha}{1+\alpha}y_{\text{i}}^{(1/\alpha)+1} - 2y_{\text{i}}^2\right]^{1/2}. \qquad (82)$$

The point of intersection can be obtained by solving (82) iteratively as described above. With a good approximation of y_{i} at hand, we then iterate the map F to obtain the approximate homoclinic orbit on a few lattice sites. We iterate the map until y_n becomes negative or starts to increase (such errors are due to the sensitivity of iterates of F to initial conditions), after what the values of the approximate solution are set to 0.

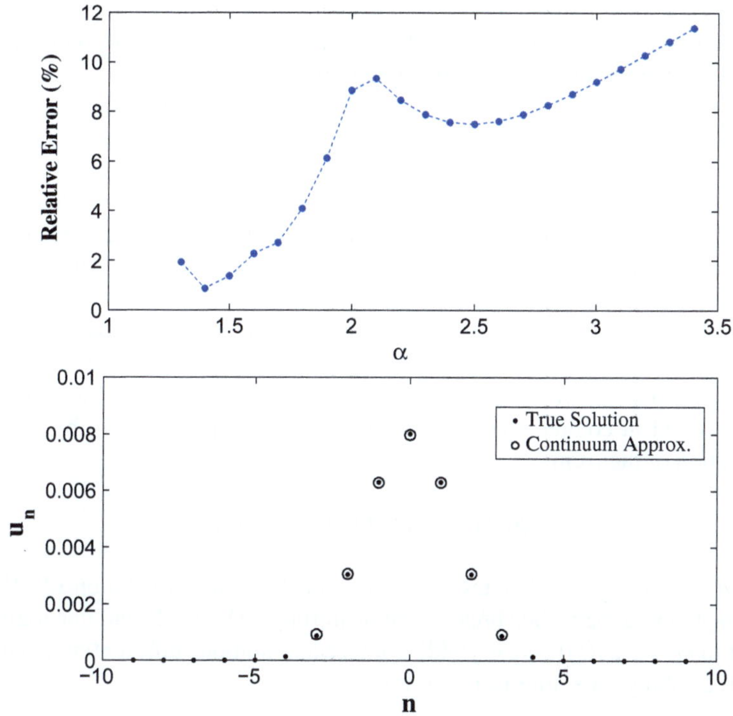

Fig. 10 *Top plot*: relative error between approximate and exact site-centered homoclinic orbits vs. exponent α. The approximate solution is deduced from (82) (see text) and the exact one is computed numerically from the stationary DpS equation. The relative error is evaluated over the support of the approximate solution (the support being computed as indicated in the text). *Bottom plot*: comparison of the exact (*bold solid dots*) and approximate (*circles*) homoclinic orbits for $\alpha = 1.3$

In Fig. 10 we compare the approximate and exact homoclinic orbits. We observe an excellent agreement even for much higher values of α than in the previous case. Indeed, for $\alpha \leq 3$ the relative error doesn't exceed 10 %. It is interesting to notice that the error does not vary monotonically with respect to α.

6 Breather Stability and Mobility

In this section we numerically analyze the spectral stability of site-centered and bond-centered breather solutions of (1) given by (3) (with $\Omega > 0$), in connection with their possible mobility (translational motion) under suitable perturbations. These solutions correspond respectively to the site-centered homoclinic orbit a_n^1 and bond-centered homoclinic orbit a_n^2 described in Theorem 1. Their profiles are illustrated on Figs. 2 and 3 for $\alpha = 3/2$ and $\alpha = 1.2$. This stability analysis was performed in the work [14] in the particular case $\alpha = 3/2$, and we shall extend this study to some interval of physically meaningful values of α.

The stability analysis can be simplified by taking advantage of the scale invariance of (1). More precisely, if $\phi(t) = (\phi_n(t))_{n\in\mathbb{Z}}$ is a solution of (1) then so is $\Omega^{\frac{1}{\alpha-1}}\phi(\Omega\, t)$ for all $\Omega > 0$. This scaling transformation allows one to generate the full family of breather solutions (3) from the particular case $\Omega = 1$. Consequently, the stable or unstable character of these periodic solutions is independent of Ω, both for nonlinear, linear and spectral stability. For this reason we shall restrict to the case $\Omega = 1$ of (3) without loss of generality.

The bond- and site-centered homoclinic solutions of the stationary DpS equation (4) are computed numerically with a Newton-type method, for a finite lattice of $N = 21$ particles with zero boundary conditions. Their spectral stability can be determined in complete analogy with the DNLS equation [15], by adopting a perturbation of the form

$$\phi_n(t) = \exp(i\, t)\,(a_n + \varphi_n(t)), \tag{83}$$

where $a = (a_n)_{1\leq n\leq N}$ denotes the (real-valued) localized solution of (1) corresponding to the unperturbed breather. Substituting (83) into (1) and linearizing with respect to $\varphi = (\varphi_n(t))_{1\leq n\leq N}$ yields a linear autonomous differential equation for $(\varphi, \varphi^\star)$. Looking for solutions of the form

$$\varphi_n(t) = \alpha_n \exp(\lambda t) + \beta_n^\star \exp(\lambda^\star t) \tag{84}$$

yields a linear eigenvalue problem $MV = i\,\lambda\, V$, where $M \in M_{2N}(\mathbb{R})$ and $V = (\alpha_n, \beta_n)^T_{1\leq n\leq N}$. Due to the Hamiltonian character of (1), whenever λ is an eigenvalue, so are $\lambda^\star$, $-\lambda$ and $-\lambda^\star$.

Figure 11 displays the resulting eigenvalues in the complex plane, for the site- and bond-centered homoclinic solutions and $\alpha = 3/2$ (the eigenvalue problem is solved by standard numerical linear algebra solvers). For all values of α in the interval $[1.2, 3.4]$, we find that the bond-centered breather is spectrally stable (i.e. the eigenvalues λ are purely imaginary), whereas the site-centered breather is unstable through a simple real positive eigenvalue. In Fig. 12 we plot $M^* = \mathrm{Max}\{\mathrm{Re}(\lambda)\}$, i.e. the maximal real part reached by the eigenvalues for the site-centered breather, versus $\alpha \in [1.2, 3.4]$. These results show that the strength of the instability increases with α, and indicate that the unstable eigenvalue goes to 0 in the limit $\alpha \to 1^+$.

The above property can be interpreted intuitively in connection with the continuum limit derived in Sect. 5.2, where one obtains asymptotically a translationally invariant family of Gaussian breather solutions. Such families of solutions usually correspond to an extra double (non-semi-simple) eigenvalue 0 associated with a translation mode [5], and perturbations along the associated marginal mode (generalized eigenvector) generate a translational motion [1]. However, this analogy is purely heuristic since breather solutions delocalize when $\alpha \to 1^+$.

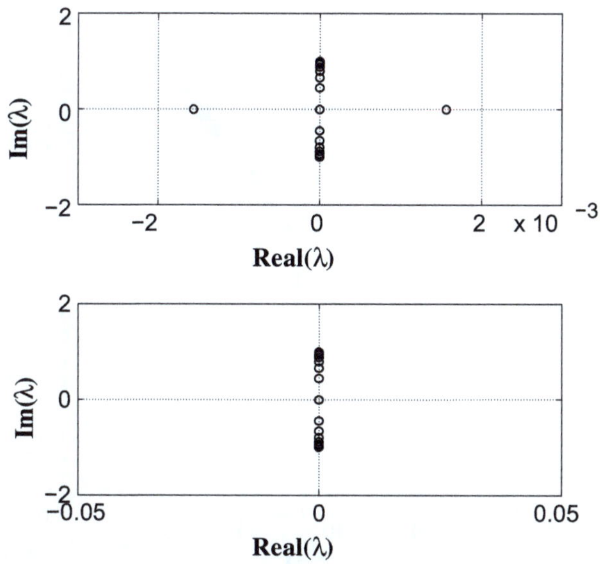

Fig. 11 Eigenvalues of the linearization of the DpS equation around the site-centered breather solution (*top plot*) and the bond-centered breather solution (*bottom plot*), computed for $\alpha = 1.5$ and $\Omega = 1$

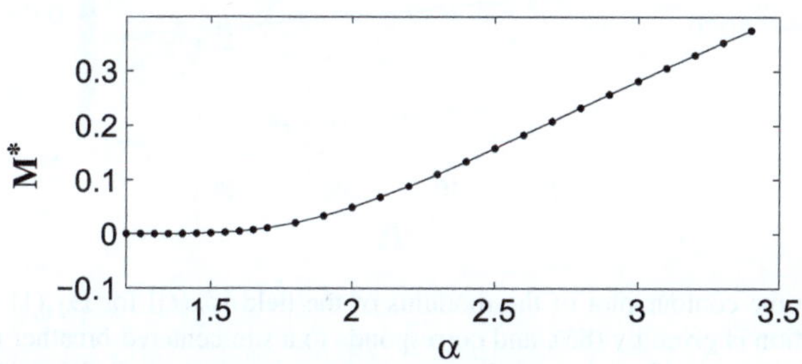

Fig. 12 Maximal real part of the eigenvalues of the linearization of the DpS equation around the site-centered breather solution, plotted for different values of α and $\Omega = 1$

To illustrate the dynamical effects of breather instability we attempt the perturbation of the site-centered breather with $\Omega = 1$ for different values of α. We consider an eigenvector $V^{\mathrm{u}} = (\alpha_n, \alpha_n^*)_{1 \le n \le N}$ associated with the real unstable eigenvalue $\lambda > 0$, and the eigenvector $V^{\mathrm{s}} = (\alpha_n^*, \alpha_n)_{1 \le n \le N}$ associated with the stable eigenvalue $-\lambda$. In formal analogy with the work [1], we perturb the unstable breathers along the *approximate* marginal mode $W = V^{\mathrm{u}} - V^{\mathrm{s}}$, which corresponds to fixing

$$\phi_n(0) = a_n + c\, i\, \mathrm{Im}(\alpha_n), \tag{85}$$

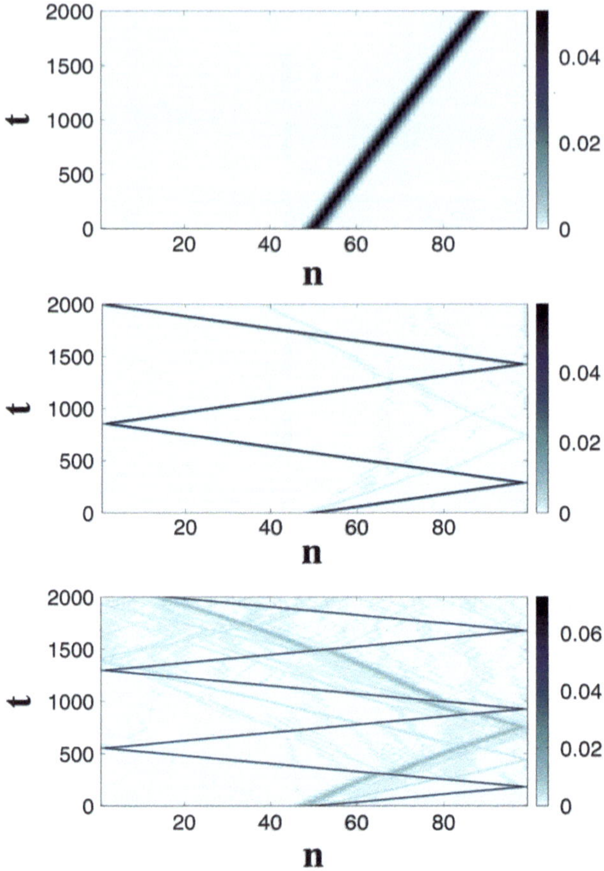

Fig. 13 Space-time contour plot of the modulus of the field $|\phi_n(t)|$ for Eq. (1) with $\alpha = 3/2$. The initial condition is given by (85), and corresponds to a site-centered breather (with frequency $\Omega = 1$) perturbed in the direction of the approximate marginal mode W. Different levels of relative energy perturbation are considered: $\delta H_{\text{rel}} \approx 0.001$ (*top plot*), $\delta H_{\text{rel}} \approx 0.1$ (*middle plot*), $\delta H_{\text{rel}} \approx 0.5$ (*bottom plot*). The perturbation leads to the manifestation of the instability of the site-centered breather which, in turn, leads to its translational motion

with $c \in \mathbb{R}$. The parameter c is tuned in order to achieve a desired level of relative energy perturbation $\delta H_{\text{rel}} = (H(\phi) - H(a))/H(a)$, where the energy H is defined through (2).

Figure 13 illustrates the dynamics of a perturbed site-centered breather for $\alpha = 3/2$. For different strengths of the perturbation, the instability of the site-centered breather leads to its translational motion. One can notice that additional breathers with smaller amplitudes are also emitted when the initial perturbation is sufficiently large. Note that random perturbations (which generally possess a nonzero component along the marginal mode) also typically generate a translational

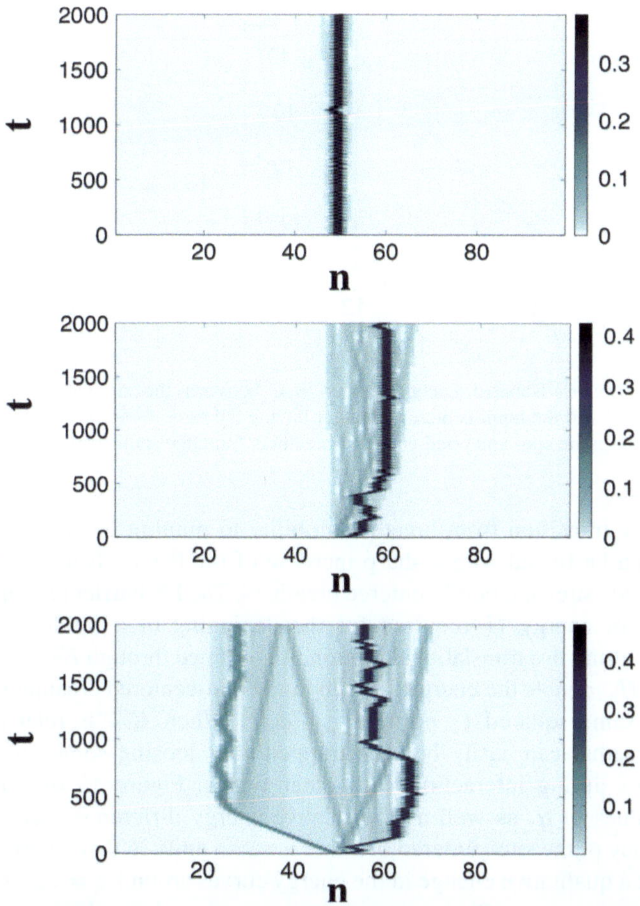

Fig. 14 Same as in Fig. 13, but in the case $\alpha = 3$. For $\delta H_{\rm rel} \approx 0.001$ (*top plot*), the perturbed breather is wandering between two neighboring sites. For $\delta H_{\rm rel} \approx 0.1$ (*middle plot*) and $\delta H_{\rm rel} \approx 0.5$ (*bottom plot*), the breather is able to propagate over a few sites but get pinned subsequently. Breather trapping can be interrupted by phases of irregular motion which may originate from the interaction of the localized solution with other waves. For the largest initial perturbation, one can notice the emission of additional breathers with smaller amplitudes

motion [14], albeit the propagation velocity is generally smaller compared to a "pure" perturbation (of similar strength) along the marginal mode.

In Fig. 14, the same type of computation is performed for $\alpha = 3$. The situation is strikingly different compared to Fig. 13. The resulting dynamics is now mainly characterized by a pinning of the localized excitation alternating with phases of irregular motion when the initial perturbation is large enough. Such behavior may be induced by the interaction of the localized solution with other localized or extended waves generated by the initial perturbation, a situation reminiscent of numerical observations made in [24].

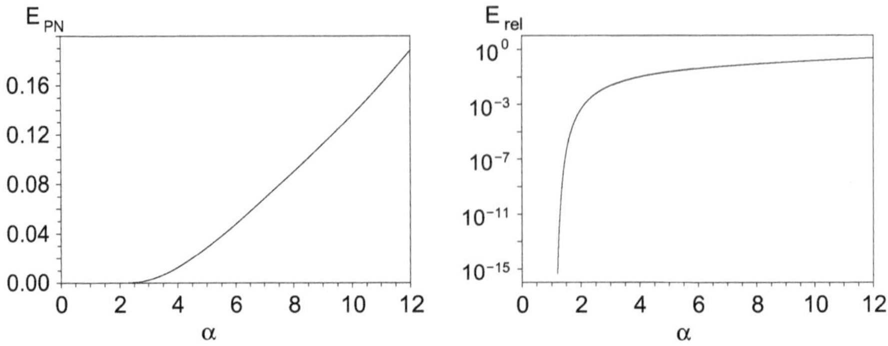

Fig. 15 *Left*: Peierls-Nabarro energy barrier E_{PN} between the site-centered breather with frequency $\Omega = 1$ and the bond-centered breather having the same ℓ_2 norm. *Right*: relative energy difference between the site- and bond-centered breathers (note the semi-logarithmic scale)

The above transition from breather mobility to pinning phenomena when α is increased can be linked with a sharp increase of the Peierls-Nabarro (PN) energy barrier between site- and bond-centered breathers. The PN barrier [18] approximates the amount of energy H required for the depinning of a stable bond-centered breather, resulting in a translational motion. It is defined through $E_{\mathrm{PN}} = |H_{\mathrm{sc}} - H_{\mathrm{bc}}|$, where H_{sc}, H_{bc} denote the energies of site- and bond-centered breather solutions (3) having the same squared ℓ_2 norm $\sum_{n\in\mathbb{Z}} |\phi_n|^2$. When E_{PN} is relatively high, a traveling breather can easily become trapped after loosing some energy through dispersion or during interaction with other waves. Figure 15 displays E_{PN} for different values of α, as well as the relative energy difference $E_{\mathrm{rel}} = E_{\mathrm{PN}}/H_{\mathrm{sc}}$ (the frequency of the site-centered breather is set to unity in this computation). One can observe a qualitative change in the energy curves around $\alpha = 2.5$. For $\alpha \leq 2.5$, the absolute and relative PN barriers are very small, and they become much larger for $\alpha \in [3, 12]$ (e.g. E_{rel} increases by 5 orders of magnitude between $\alpha = 3/2$ and $\alpha = 3$). In this interval the growth of E_{PN} becomes almost linear in α. The increase of E_{PN} explains why traveling breathers are easily generated from static breathers when α is close to unity, whereas pinning dominates for sufficiently large values of α.

7 Discussion

In this paper, we have proved an existence theorem for homoclinic orbits of the stationary DpS equation (4), in agreement with previous numerical results [11, 14, 26]. These homoclinics correspond to breather solutions of the time-dependent DpS equation (1). This result implies the existence of long-lived breather solutions in infinite chains of oscillators with fully nonlinear nearest-neighbors interactions [3], in particular granular chains with Hertzian contact interactions [11, 14, 26]. In addition, we have obtained analytical approximations of breather profiles and of the corresponding intersecting stable and unstable manifolds. For $p \approx 2$, this was achieved by deriving a suitable continuum limit of the stationary DpS equation, consisting of a logarithmic nonlinear Schrödinger equation. We have also numerically determined the spectral stability of breather solutions depending on their odd- or even-parity symmetry, and we have studied their mobility properties.

An interesting extension to this work would be to consider a DpS equation including a localized defect

$$i \frac{d}{dt} \phi_n - \delta_{0,n}\, \nu\, \phi_n = (\Delta_p \phi)_n, \quad n \in \mathbb{Z}, \tag{86}$$

where ν is a real parameter defining the defect strength and $\delta_{m,n}$ denotes the Kronecker delta. For example, system (86) can be derived in a similar way as in [11], by considering a Newton's cradle including a small mass defect or a small stiffness variation of one oscillator in the chain. If $\nu \neq 0$, the Ansatz (3) defines a solution of (86) if

$$-s\, a_n - \delta_{0,n}\, \lambda\, a_n = (\Delta_p a)_n, \quad n \in \mathbb{Z}, \tag{87}$$

where we have set $\nu = |\Omega|\, \lambda$ and $s = \mathrm{Sign}(\Omega)$. An interesting problem is to analyze bifurcations of homoclinic solutions of (87) when λ is varied, or equivalently when the defect strength and breather frequency are tuned. For example, the associated solutions of (86) may correspond to localized modes excited by a traveling breather reaching a defect [14]. In addition, it is also interesting to understand how the breathers existing in a spatially homogeneous system are affected by spatial inhomogeneities, since the latter are always present in real systems, or could be introduced dynamically to manipulate localized excitations.

Similarly to what we have seen, one can reformulate (87) as a two-dimensional (nonautonomous) mapping. The map possesses a particularly simple structure at the defect site when the mixed variables introduced in Sect. 2.2 are used. Indeed, the mapping (17) becomes

$$\begin{pmatrix} v_{n+1} \\ y_{n+1} \end{pmatrix} = \left(\mathrm{Id} + \delta_{0,n}\, \lambda \begin{pmatrix} 0 & 0 \\ 1 & 0 \end{pmatrix} \right) F \begin{pmatrix} v_n \\ y_n \end{pmatrix}. \tag{88}$$

This yields an interpretation of the defect as the composition of F with the linear shear

$$A(\lambda) = \begin{pmatrix} 1 & 0 \\ \lambda & 1 \end{pmatrix}, \tag{89}$$

which is quite convenient to analyze bifurcations of homoclinic solutions (note that using the maps M and T introduced in Sect. 2 would lead to more complex perturbations induced by the defect). Indeed, the mapping (88) admits an orbit homoclinic to 0 if and only if $A(\lambda)\,\mathcal{W}^{\mathrm{u}}(0)$ and $\mathcal{W}^{\mathrm{s}}(0)$ intersect (see Sect. 3.3 for the definition of the stable and unstable manifolds). In that case, each intersection (v_1, y_1) determines an homoclinic orbit of (88), hence an homoclinic solution of (87) with $s = 1$. Consequently, the approximations of $\mathcal{W}^{\mathrm{u}}(0)$ and $\mathcal{W}^{\mathrm{s}}(0)$ introduced in Sect. 5 provide a theoretical tool to predict defect-induced breather bifurcations, and approximate the shape and energy of bifurcating solutions. This problem will be treated in detail in a forthcoming paper.

Acknowledgements G.J. acknowledges financial support from the Rhône-Alpes Complex Systems Institute (IXXI). Y.S. is grateful to Israel Science Foundation (Grant 484 /12) for financial support.

Appendix

Analysis of the Stationary DpS Equation for $s = -1$

In this appendix, we consider the stationary DpS equation with $s = -1$ and prove the non-existence of nontrivial bounded solutions stated in Theorem 1.

Lemma 3. *For $s = -1$, the only bounded solution of (4) is $a_n = 0$.*

Proof. We restrict to the case $a_0 \geq 0$ due to the invariance $a_n \to -a_n$ of (4). Let us first assume

$$a_1 > 0, \quad a_0 \geq 0, \quad a_1 \geq a_0. \tag{90}$$

Using (4), one can show by induction that $(a_n)_{n\geq 0}$ is a positive non-decreasing sequence. Then (4) implies $(a_{n+1} - a_n)^\alpha \geq a_n$ for all $n \geq 0$. It follows that $a_{n+1} - a_n \geq a_1^{1/\alpha}$ for all $n \geq 1$, hence $\lim_{n\to+\infty} a_n = +\infty$.

If $0 \leq a_1 < a_0$, then $\tilde{a}_n = a_{-n+1}$ defines a solution of (4) satisfying (90), and thus $\lim_{n\to-\infty} a_n = +\infty$.

If $a_1 \leq 0 \leq a_0$ and $(a_0, a_1) \neq (0, 0)$, the case $n = 1$ of (4) yields $a_2 < a_1 \leq 0$. Then $\tilde{a}_n = -a_{n+1}$ defines a solution of (4) satisfying (90), and thus $\lim_{n\to+\infty} a_n = +\infty$. □

References

1. S. Aubry, T. Cretegny, Physica D **119**, 34 (1998)
2. I. Bialynicki-Birula, J. Mycielski, Ann. Phys. **100**, 62 (1976)
3. B. Bidégaray-Fesquet, E. Dumas, G. James, From Newton's cradle to the discrete p-Schrödinger equation. SIAM J. Math. Anal. (2013, to appear). arXiv: 1306.2105 [math.DS]

4. J. Cuevas, G. James, P.G. Kevrekidis, B. Malomed, B. Sánchez-Rey, J. Nonlinear Math. Phys. **15**(supplement 3), 124 (2008)
5. S.V. Dmitriev, P.G. Kevrekidis, N. Yoshikawa, D.J. Frantzeskakis, J. Phys. A Math. Theor. **40**, 1727 (2007)
6. J.C. Eilbeck, M. Johansson, The discrete nonlinear Schrödinger equation – 20 years on, in *Conference on Localization and Energy Transfer in Nonlinear Systems*, San Lorenzo de El Escorial, ed. by L. Vazquez, R.S. MacKay, M.-P. Zorzano (World Scientific, River Edge, 2003), p. 44
7. L.C. Evans, R.F. Gariepy, *Measure Theory and Fine Properties of Functions*. Studies in Advanced Mathematics (CRC, Boca Raton, 1992)
8. S. Flach, Phys. Rev. E **51**, 1503 (1995)
9. S. Flach, A. Gorbach, Phys. Rep. **467**, 1 (2008)
10. S. Hutzler, G. Delaney, D. Weaire, F. MacLeod, Am. J. Phys. **72**, 1508 (2004)
11. G. James, Math. Models Methods Appl. Sci. **21**, 2335 (2011)
12. G. James, B. Sánchez-Rey, J. Cuevas, Rev. Math. Phys. **21**, 1 (2009)
13. G. James, J. Cuevas, P.G. Kevrekidis, Breathers and surface modes in oscillator chains with Hertzian interactions, in *Proceedings of the 2012 International Symposium on Nonlinear Theory and Its Applications (NOLTA 2012)*, Palma, Majorca, 22–26 Oct 2012, pp. 470–473
14. G. James, P.G. Kevrekidis, J. Cuevas, Physica D **251**, 39 (2013)
15. P.G. Kevrekidis, *The Discrete Nonlinear Schrödinger Equation: Mathematical Analysis, Numerical Computations and Physical Perspectives* (Springer, Heidelberg, 2009)
16. D. Khatri, D. Ngo, C. Daraio, Granul. Matter **14**, 63 (2012)
17. Yu. S. Kivshar, Phys. Rev. E **48**, R43 (1993)
18. Yu. S. Kivshar, D.K. Campbell, Phys. Rev. E **48**, 3077 (1993)
19. D. Ngo, S. Griffiths, D. Khatri, C. Daraio, Granul. Matter (2013). doi:10.1007/s10035-012-0377-5
20. J.B. Page, Phys. Rev. B **41**, 7835 (1990)
21. F. Palmero, R. Carretero-González, J. Cuevas, P.G. Kevrekidis, W. Królikowski, Phys. Rev. E **77**, 036614 (2008)
22. W.-X. Qin, X. Xiao, Nonlinearity **20**, 2305 (2007)
23. P. Rosenau, S. Schochet, Chaos **15**, 015111 (2005)
24. S. Sen, T.R. Krishna Mohan, Phys. Rev. E **79**, 036603 (2009)
25. A.J. Sievers, J.B. Page, Unusual anharmonic local mode systems, in *Dynamical Properties of Solids*, vol. 7, ed. by G.K. Norton, A.A. Maradudin (North-Holland, Amsterdam, 1995), p. 137
26. Y. Starosvetsky, M. Arif Hasan, A.F. Vakakis, L.I. Manevitch, SIAM J. Appl. Math. **72**, 337 (2012)
27. D. Sun, C. Daraio, S. Sen, Phys. Rev. E **83**, 066605 (2011)

Part II
Bose-Einstein Condensates (BECs) and Nonlinear Optics

Fragmentation of a Bose-Einstein Condensate Through Periodic Modulation of the Scattering Length

Antun Balaž and Alexandru I. Nicolin

Abstract Motivated by recent experimental results on the dynamics of a cigar-shaped ^{7}Li Bose-Einstein condensate subject to periodic modulation of the scattering length, we investigate here the dynamics of the condensate by numerical and analytical means and show that for resonant drives of large amplitude the condensate reaches a fragmented state. The fragmented state is a hybrid state of the condensate which consists of a quadrupole mode on which a longitudinal resonant density wave is grafted.

1 Introduction

Investigations of the collective dynamics of Bose-Einstein condensates (BECs) have become one of the recurrent themes in the study of ultra-cold quantum gases. Part of the appeal comes from the high precision with which the collective modes are determined experimentally (e.g., the determination of characteristic frequencies), and fresh impetus is also received from the experimental side, as there is a growing number of methods which can be used to excite the collective modes. Following

A. Balaž
Scientific Computing Laboratory, Institute of Physics Belgrade, University of Belgrade, Pregrevica 118, 11080 Belgrade, Serbia
e-mail: antun@ipb.ac.rs

A.I. Nicolin (✉)
Horia Hulubei National Institute for Physics and Nuclear Engineering, 30 Reactorului, Magurele-Bucharest, 077125, Romania

Faculty of Physics, University of Bucharest, 405 Atomistilor, Magurele-Bucharest, 077125, Romania

Faculty of Physics, West University of Timisoara, 4 Parvan, Timisoara 300223, Romania
e-mail: alexandru.nicolin@nipne.ro

R. Carretero-González et al. (eds.), *Localized Excitations in Nonlinear Complex Systems*, Nonlinear Systems and Complexity 7, DOI 10.1007/978-3-319-02057-0_5,

the pioneering JILA experiments on the collective excitations of a ^{87}Rb condensate [1] induced by an inhomogeneous oscillatory force of adjustable frequency and symmetry, a couple of theoretical approaches have been proposed. The most important ones are the hydrodynamic treatment of high-density BECs [2] and the variational treatment of low-density condensates [3] which yield simplified versions of the equation which describes the $T = 0$ K dynamics of a condensate, the so-called Gross-Pitaevskii equation (GPE), and predict frequencies of the collective modes which are in good agreement with the experimental data. For more precise descriptions of the condensate there are now very efficient recipes for the numerical solution of the GPE based on spectral and finite-differences methods in one-, two- and three-dimensional settings (see Refs. [4–9] for an overview of the main results).

The subsequent experimental progress in the parametric excitation of BECs through periodic modulations of the strength of the confining potential has catalyzed theoretical investigations in distinct research veins such as the control, management and collision of solitons [10], nonlinear excitations [11], parametric and geometric resonances [12–14], the generation of density waves [15–17] and many others (see Refs. [18, 19] for a comprehensive review). Particularly noteworthy are the non-polynomial Schrödinger equations able to present in an explicit form such transverse (longitudinal) modulations in an effectively longitudinal (transverse) equation which describes the dynamics of the condensate wave function [20–23]. Faraday and resonant waves in cigar-shaped parametrically-excited condensates have been described in detail using the aforementioned equations and the experimental results of Engels et al. [24] on density waves in ^{87}Rb condensate are now well-understood [25,26]. Similar density waves have been predicted in binary BECs [27] and dipolar condensates [28].

Finally, Pollack et al. [29] excited the lowest-lying quadrupole mode of a cigar-shaped ^{7}Li condensate through periodic modulation of the scattering length and identified the resonances located at the natural quadrupole frequency and at the first harmonic. The observed excitation spectra are in good qualitative agreement with the existing linear theory [30], but it was shown in Ref. [11] through a detailed quantitative analysis that the excitation spectra also exhibit prominent nonlinear features such as mode coupling, higher harmonics generation, and significant shifts in the frequencies of collective modes. In this paper we revisit the experiments of Pollack et al. [29] and show that for resonant drives of large amplitude the condensate reaches a fragmented state, which is a hybrid state that consists of a quadrupole mode on which a longitudinal resonant density wave is grafted.

The rest of the paper is structured as follows: in Sect. 2 we briefly review the classification of density waves through a variational treatment, which allows us to separate between Faraday and density waves and show how the collective modes couple to the longitudinal density waves, while in Sect. 3 we present our detailed numerical results on the fragmentation of the condensate. In this section we show that our results on the dynamics of the longitudinal extent of the condensate are in good quantitative agreement with the experimental data and that the excited longitudinal wave is resonant in nature, namely it has an intrinsic oscillation frequency equal to that of the radial frequency of the trap. Finally, in Sect. 4 we present our concluding remarks.

2 Classification of Density Waves

In this section we briefly review a recent variational treatment (see Ref. [25] for further details) which is able to describe both the collective modes and the density waves of cigar-shaped condensates subject to parametric excitations. This variational treatment generalizes the one used in Ref. [29] to determine the collective properties of the condensate and allows us to classify the longitudinal density waves based on their intrinsic frequency. As customary in variational treatments, we compute the Lagrangian of the condensate

$$L(t) = \int d\mathbf{r}\mathcal{L}(\mathbf{r},t), \tag{1}$$

using the Lagrangian density

$$\mathcal{L}(\mathbf{r},t) = \frac{i}{2}\left(\psi\psi_t^* - \psi^*\psi_t\right) + \frac{1}{2}\left|\nabla\psi\right|^2 + \frac{g(t)N}{2}\left|\psi\right|^4 + V\left|\psi\right|^2, \tag{2}$$

for convenience written here by taking $\hbar = m = 1$, $g(t) = 4\pi a(t)$, $\psi = \psi(\mathbf{r},t)$, $V = V(\mathbf{r},t) = 1/2 \cdot \left(\Omega_\rho^2(t)\rho^2 + \Omega_z^2(t)z^2\right)$ and $\rho^2 = x^2 + y^2$, and using the cylindrically-symmetric hybrid trial wave-function

$$\psi(\mathbf{r},t) = f(k, w_z(t), w_\rho(t), u(t), v(t)) \exp\left(-\frac{\rho^2}{2w_\rho^2(t)} + i\rho^2\beta(t)\right)$$
$$\times \exp\left(-\frac{z^2}{2w_z^2(t)} + iz^2\alpha(t)\right)(1 + (u(t) + iv(t))\cos kz)\,. \tag{3}$$

This trial wave function accounts both for the bulk part of the condensate (through the Gaussian envelopes) and the density wave (through the periodic function), and function f is such that the wave function is normalized to 1. It contains three pairs of conjugate variables, namely $\{w_z(t), \alpha(t)\}$ – the longitudinal width and its associated phase, $\{w_\rho(t), \beta(t)\}$ – the radial width and its associated phase and $\{u(t), v(t)\}$ – the real and imaginary amplitude of the density wave, which (after integrating the Lagrangian density, see Ref. [26] for details) yield three pairs of Euler-Lagrange equations. The Euler-Lagrange equations

$$\frac{\partial L}{\partial q} = \frac{d}{dt}\left(\frac{\partial L}{\partial \dot{q}}\right) \tag{4}$$

for $q \in \{\alpha(t), \beta(t), w_z(t), w_\rho(t)\}$ are given by

$$\dot{w}_z(t) = 2w_z(t)\alpha(t), \tag{5}$$

$$\dot{w}_\rho(t) = 2w_\rho(t)\beta(t), \tag{6}$$

$$\dot{\alpha}(t) = -\frac{\Omega_z^2(t)}{2} + \frac{1}{2w_z^4(t)} + \frac{Ng(t)}{4\sqrt{2}\pi^{3/2}w_\rho^2(t)w_z^3(t)} - 2\alpha^2(t), \tag{7}$$

$$\dot{\beta}(t) = -\frac{\Omega_\rho^2(t)}{2} + \frac{1}{2w_\rho^4(t)} + \frac{Ng(t)}{4\sqrt{2}\pi^{3/2}w_\rho^4(t)w_z(t)} - 2\beta^2(t), \tag{8}$$

while those for $q \in \{u(t), v(t)\}$ are given by

$$\dot{v}(t) = -u(t)\left(\frac{k^2}{2} + \frac{Ng(t)}{\sqrt{2}\pi^{3/2}w_\rho^2(t)w_z(t)}\right), \tag{9}$$

$$\dot{u}(t) = \frac{k^2}{2}v(t). \tag{10}$$

Equations (5)–(8) describe the dynamics of the bulk of the condensate and have become textbook material, while Eqs. (9) and (10) describe the dynamics of the density waves. The last two equations can usually be written in the form of the standard Mathieu equation

$$\ddot{u}(\tau) + [a(k,\omega) + b(k,\omega)\sin 2\tau]\,u(\tau) = 0, \tag{11}$$

where $\tau = \omega t/2$, while $a(k,\omega)$ and $b(k,\omega)$ depend also on the widths of the condensate, the number of atoms in the condensate, the scattering length, etc., and are cumbersome to write down [26]. Depending on the experimental setup, the periodic drive in Eq. (11) comes either from periodic modulations of $\Omega_\rho(t)$, as in the experiments of Engels et al. [24], or from periodic modulations of the scattering length, as in the experiments of Pollack et al. [29], both approaches yielding, in fact, a periodic modulation of the effective nonlinear interaction. The stability analysis of Mathieu equations indicates that for small values of $b(k,\omega)$ the unstable solutions of Eq. (11) correspond to a series of lobes centered around $a(k,\omega) = n^2$, where n is a positive integer, the most unstable one corresponding to $a(k,\omega) = 1$.

Engels et al. [24] have observed in their experiments the density waves which correspond to $a(k,\omega) = 1$ and $a(k,\omega) = 4$. The former waves are referred to as Faraday waves because parametrically excited waves with a frequency equal to half that of the drive have been first observed by Michael Faraday in 1831 [31], while the later are referred to as resonant waves as the two frequencies (namely that of the drive and that of the excited wave) are equal [25]. A peculiarity of the experiments of Engels et al. [24] is that for Faraday waves ($a(k,\omega) = 1$) there is almost no bulk dynamics of the condensate, while for resonant waves ($a(k,\omega) = 4$)

there is only transverse bulk dynamics, therefore in most analytical descriptions of these waves the longitudinal dynamics of the condensate is usually neglected. As we will show in the next section, the recent experimental results of Pollack et al. [29] on the dynamics of a cigar-shaped ^{7}Li Bose-Einstein condensate subject to periodic modulation of the scattering length open investigations into a distinct regime where a quadrupole collective mode is excited along with a high-amplitude resonant density wave which gives the condensate a fragmented appearance. Finally, let us note that the fragmentation of a BEC has been proposed and studied as a mechanism suitable for future atom lasers (coherent matter wave emitters) [32–34].

3 Results and Discussion

We investigate the dynamics of cigar-shaped ^{7}Li Bose-Einstein condensates by solving numerically the GPE

$$i\hbar\psi_t = \left[-\frac{\hbar^2}{2m}\nabla^2 + g(t)\,|\psi|^2 + V\right]\psi \tag{12}$$

in an experimental setup identical with that described in Ref. [29] (see Refs. [8, 9] for a detailed description of the numerical algorithms and codes). We consider a ^{7}Li Bose-Einstein condensate of 3×10^5 atoms in the $|1, 1\rangle$ hyperfine state loaded in a magnetic trap with frequencies $\Omega_\rho = (2\pi)235\,\text{Hz}$ and $\Omega_z = (2\pi)4.85\,\text{Hz}$. Neglecting nonlinear effects one can easily show using Eqs. (5)–(8) that the frequency of the lowest-lying quadrupole mode is given by $\omega_Q = \sqrt{5/2}\omega_z$ (in the high-density limit) and $\omega_Q = 2\omega_z$ (in the non-interacting limit), while a more detailed analysis which includes also nonlinear effects (see Ref. [11] for a detailed discussion) shows that $\omega_Q \approxeq (2\pi)8.2\,\text{Hz}$. The dynamics is induced by the modulation of the scattering length which is of the form

$$a(t) = a_{\text{av}} + \delta a \sin(\Omega t). \tag{13}$$

We start our investigation by solving the GPE for a modulation with $a_{\text{av}} = 3a_0$, where a_0 is the Bohr radius, $\delta a = 3a_0$ and $\Omega = (2\pi)9\,\text{Hz}$. The drive is applied for 500 ms, after which time the condensate evolves freely in the magnetic trap. As Ω is very close to the frequency of the lowest-lying quadrupole mode and the amplitude of the drive is quite large, the quadrupole mode is excited relatively fast while the drive is still on. After the quadrupole mode sets in the numerical simulations show that the condensate develops a series of density maxima and minima in the longitudinal direction and eventually appears to fragment in smaller condensates. Given the strength of the drive and, in particular, the uncommon fragmented appearance, we looked at the time evolution of the longitudinal extent of the condensate to see how well can the GPE capture its dynamics.

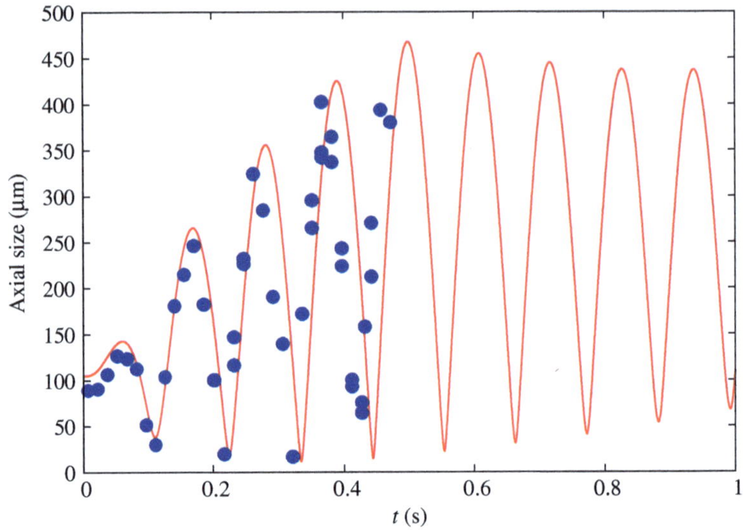

Fig. 1 The time evolution of the axial extent of the condensate for a modulation with $a_{av} = 3a_0$, $\delta a = 3a_0$ and $\Omega = (2\pi)9$ Hz. The full line corresponds to our Gross-Pitaevskii simulation while the *dots* represent the experimental data reported in Fig. 4 of Ref. [29]. The modulation is switched off after 500 ms and afterward the condensate evolves freely in the magnetic trap. Note that after switching off the drive the oscillations continue with very little damping. Also note that, while the drive is on, there is a clear signature of the "bouncing ball behavior" typical for strongly-driven systems with a $1/w_z^4(t)$ singular term, a behavior which gradually fades out after the drive is switched off

The dynamics of the longitudinal extent of the condensate shown in Fig. 1 is in remarkably good agreement with the experimental findings reported in Ref. [29]. The "bouncing ball behavior" visible in Fig. 1 is typical for strongly-driven collective modes due to the $1/w_z^4(t)$ singular term which appears in Eqs. (5)–(8). The very small discrepancy seen in Fig. 1 between the experimental data and the result of our numerical investigation can be traced back to the fact that each experimental point represents a distinct realization of the condensate and that the number of atoms fluctuates between values as low as 1.5×10^5 and values as high as 5×10^5 (Private communication from Randall Hulet), while in our numerical simulations we use the average value of 3×10^5 atoms.

Next, we plot in Figs. 2 and 3 the density and phase profile of the condensate before and after switching off the modulation of the scattering length. Given the energy transferred to the condensate by the drive while it was still on, the dynamics of the condensate suffers no major changes after switching off the drive, though there is an overall damping which is clearly visible in Fig. 1. The three density profiles shown in Fig. 2 suggest either a train of (bright or dark) solitons similar to those studied by Tsuzuki in Ref. [35] or a density wave [25]. The phase of the condensate depicted in Fig. 3 shows, however, only jumps of 2π and therefore indicates that the observed train of maxima and minima is not soliton-like.

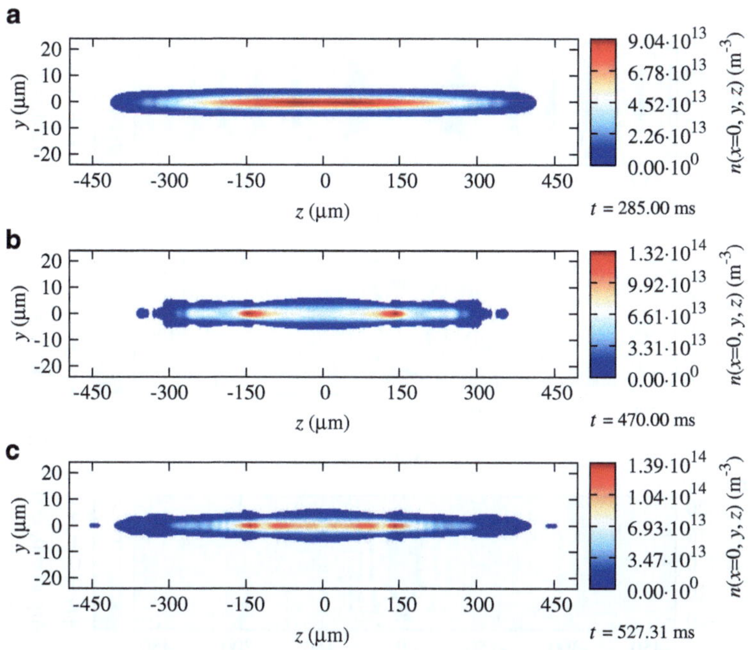

Fig. 2 The density profile of the condensate for a modulation with $a_{\mathrm{av}} = 3a_0$, $\delta a = 3a_0$ and $\Omega = (2\pi)9\,\mathrm{Hz}$. As before, the modulation is switched off after 500 ms and afterward the condensates evolves freely in the magnetic trap. The *upper panel* depicts the density profile at $t = 285\,\mathrm{ms}$, the *middle panel* depicts the profile at $t = 470\,\mathrm{ms}$, while the *lower panel* depicts the density profile just after switching off the drive at $t = 527.31$ ms. Note that the condensate reaches the fragmented appearance before the drive is switched off. Also note the imperfect periodicity of the four maxima visible in the lower panel which is due to the longitudinal inhomogeneity of the condensate

A close inspection at the radial mode of the condensate shows, as expected, that it has a frequency equal to $2\Omega_\rho$ while the drive is on and that the mode continues with negligible damping after the drive is switched off. Because the frequency of the longitudinal mode is some two orders of magnitude lower than that of the radial mode, the longitudinal extent for the condensate $w_z(t)$ and the nonlinear interaction $g(t)$ can be considered constant over a couple of periods of the radial mode and the above mechanism for generating density waves can be safely applied. In other words, for time scales of the order of a period of the radial mode the longitudinal density waves are described by

$$\dot{v}(t) = -u(t)\left(\frac{k^2}{2} + \frac{Ng^*}{\sqrt{2}\pi^{3/2}w_\rho^2(t)w_z^*}\right), \tag{14}$$

$$\dot{u}(t) = \frac{k^2}{2}v(t), \tag{15}$$

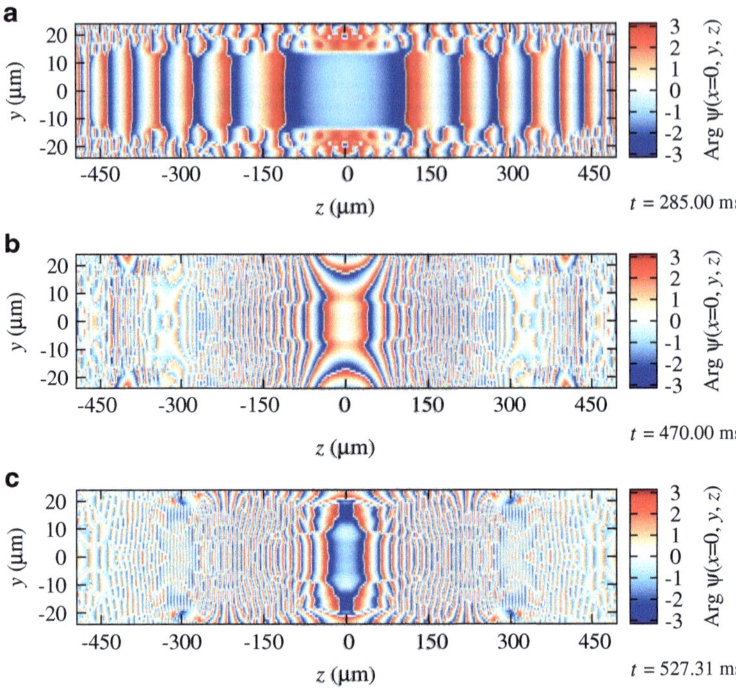

Fig. 3 The phase profile of the condensate for a modulation with $a_{\rm av} = 3a_0$, $\delta a = 3a_0$ and $\Omega = (2\pi)9\,$Hz. As before, the modulation is switched off after 500 ms and afterward the condensates evolves freely in the magnetic trap. The *upper panel* depicts the phase profile at $t = 285\,$ms, the *middle panel* depicts the profile at $t = 470\,$ms, while the *lower panel* depicts the phase profile just after switching off the drive at $t = 527.31\,$ms. Note that all phase jumps visible in the three panels are multiples of 2π, therefore the observed fragmentation is not soliton-like

where w_z^* and g^* are the average values of the longitudinal width and nonlinear interaction in a vicinity of t. Solving numerically the GPE we see that the $1/w_\rho^2(t)$ term exhibits modulations on a frequency equal to $2\Omega_\rho$ and thereby constitute the drive term in Eq. (11). We show in Fig. 4 that the density in the center of the trap shows large amplitude oscillations on a frequency equal to ω_Q and small amplitude oscillations on a frequency equal to $2\Omega_\rho$, therefore the induced density wave oscillates on the same frequency with that of the drive and represents a resonant wave [25].

The peculiarity of this wave is that, unlike that observed by Engels et al. [24], it emerges in a system where the longitudinal density is not constant but oscillates slowly in time (due to the excitation of the quadrupole mode), therefore the spatial period of the wave will not be constant in time and will, in fact, exhibit oscillations on a frequency equal to that of the longitudinal mode.

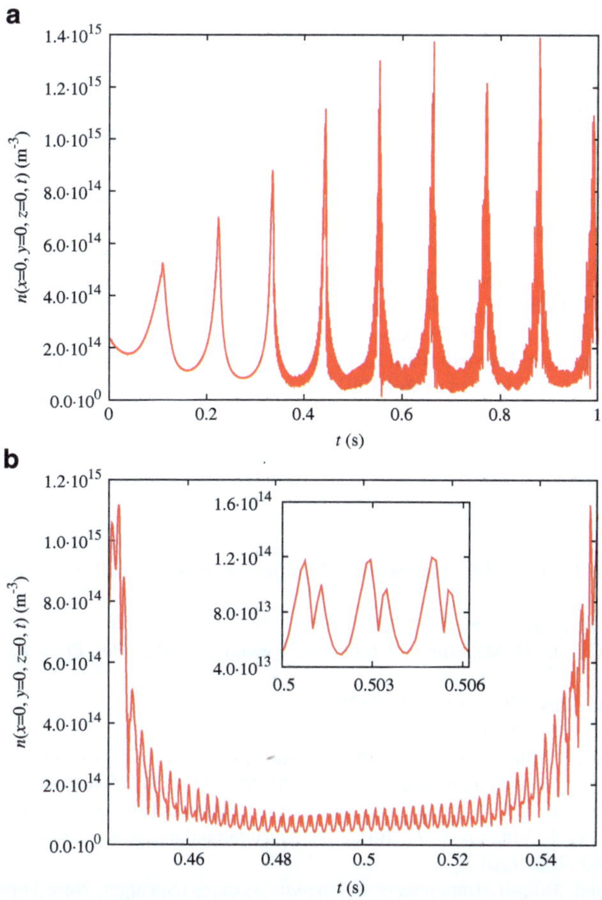

Fig. 4 Time evolution of the density of the condensate in the center of the trap. Note the large amplitude nonlinear oscillations on a frequency equal to ω_Q (*upper panel*) and small amplitude (weakly nonlinear) oscillations on a frequency equal to $2\Omega_\rho$ (*lower panel*). The *inset* in the *lower panel* shows the high frequency oscillations of the density and one easily sees that these are not simple sinusoidal oscillations as the contribution of at least the first harmonic is clearly visible

4 Conclusions

In this paper we have revisited recent experimental results on the dynamics of a cigar-shaped ^{7}Li Bose-Einstein condensate subject to periodic modulation of the scattering length and have shown by numerical means that for resonant drives of large amplitude the condensate reaches a fragmented state. We have shown that the fragmented state is a hybrid state of the condensate which consists of a quadrupole mode on which a longitudinal resonant density wave is grafted. The peculiarity of this hybrid state is that the resonant density wave is dynamically coupled to the quadrupole mode, the spatial period of the wave exhibiting oscillations on a frequency equal to that of the longitudinal mode. As related topics for future research we mention the emergence of density waves for modulations of the scattering length of small and moderate amplitudes and the dynamics of the condensate under the joint modulation of the scattering length and the magnetic trap.

Acknowledgements A. B. acknowledges financial support from the Ministry of Education, Science and Technological Development of the Republic of Serbia under projects No. ON171017 and NAI-DBEC, from DAAD – German Academic and Exchange Service under project NAI-DBEC, and from the European Commission under EU FP7 projects PRACE-2IP, PRACE-3IP, HP-SEE, and EGI-InSPIRE, while A. I. N. was supported by CNCS-UEFISCDI through PN-II-ID-PCE-2011-3-0972 and by ANCS through project PN 09370104/2012. The authors thank Randall Hulet, Vanderlei Bagnato, Paul Dyke and Dan Dries for insightful discussions concerning the experimental results reported in Ref. [29] and for providing the experimental data from Fig. 1.

References

1. D.S. Jin, J.R. Ensher, M.R. Matthews, C.E. Wieman, E.A. Cornell, Phys. Rev. Lett. **77**, 420 (1996)
2. S. Stringari, Phys. Rev. Lett. **77**, 2360 (1996)
3. V.M. Pérez-García, H. Michinel, J.I. Cirac, M. Lewenstein, P. Zoller, Phys. Rev. Lett. **77**, 5320 (1996)
4. S.K. Adhikari, Phys. Rev. E **62**, 2937 (2000)
5. S.K. Adhikari, Phys. Lett. A **265**, 91 (2000)
6. S.K. Adhikari, P. Muruganandam, J. Phys. B At. Mol. Opt. **35**, 2831 (2002)
7. S.K. Adhikari, P. Muruganandam, J. Phys. B: At. Mol. Opt. **36**, 2501 (2003)
8. P. Muruganandam, S.K. Adhikari, Comput. Phys. Commun. **180**, 1888 (2009)
9. D. Vudragović, I. Vidanović, A. Balaž, P. Muruganandam, S.K. Adhikari, Comput. Phys. Commun. **183**, 2021 (2012)
10. B.A. Malomed, *Soliton Management in Periodic Systems* (Springer, New York 2005)
11. I. Vidanović, A. Balaž, H. Al-Jibbouri, A. Pelster, Phys. Rev. A **84**, 013618 (2011)
12. J.J. García-Ripoll, V.M. Pérez-García, P. Torres, Phys. Rev. Lett. **83**, 1715 (1999)
13. C. Tozzo, M. Krämer, F. Dalfovo, Phys. Rev. A **72**, 023613 (2005)
14. I. Vidanović, H. Al-Jibbouri, A. Balaž, A. Pelster, Phys. Scr. **T149**, 014003 (2012)
15. K. Staliunas, S. Longhi, G.J. de Valcárcel, Phys. Rev. Lett. **89**, 210406 (2002)
16. K. Staliunas, S. Longhi, G.J. de Valcárcel, Phys. Rev. A **70**, 011601(R) (2004)
17. A.I. Nicolin, R. Carretero-González, P.G. Kevrekidis, Phys. Rev. A **76**, 063609 (2007)
18. P.G. Kevrekidis, D.J. Frantzeskakis, R. Carretero-González (eds.), *Emergent Nonlinear Phenomena in Bose-Einstein Condensates* (Springer, New York, 2008)
19. R. Carretero-González, D.J. Frantzeskakis, P.G. Kevrekidis, Nonlinearity **21**, R139 (2008)
20. L. Salasnich, A. Parola, L. Reatto, Phys. Rev. A **65**, 043614 (2002)
21. A. Muñoz Mateo, V. Delgado, Phys. Rev. A **77**, 013617 (2008)
22. A.I. Nicolin, Rom. Rep. Phys. **61**, 641 (2009)
23. A.I. Nicolin, M.C. Raportaru, Physica A **389**, 4663 (2010)
24. P. Engels, C. Atherton, M.A. Hoefer, Phys. Rev. Lett. **98**, 095301 (2007)
25. A.I. Nicolin, Phys. Rev. E. **84**, 056202 (2011)
26. A.I. Nicolin, Physica A **391**, 1062 (2012)
27. A. Balaž, A.I. Nicolin, Phys. Rev. A **85**, 023613 (2012)
28. K. Lakomy, R. Nath, L. Santos, Phys. Rev. A **86**, 023620 (2012)

29. S.E. Pollack, D. Dries, R.G. Hulet, K.M.F. Magalhães, E.A.L. Henn, E.R.F. Ramos, M.A. Caracanhas, V.S. Bagnato, Phys. Rev. A **81**, 053627 (2010)
30. C.J. Pethick, H. Smith, *Bose-Einstein Condensation in Dilute Gases* (Cambridge University Press, Cambridge, 2008)
31. M. Faraday, Philos. Trans. R. Soc. Lond. **121**, 299 (1831)
32. M.I. Rodas-Verde, H. Michinel, V.M. Perez-Garcia, Phys. Rev. Lett. **95**, 153903 (2005)
33. L.D. Carr, J. Brand, Phys. Rev. A **70**, 033607 (2004)
34. H. Michinel, A. Paredes, M.M. Valado, D. Feijoo, Phys. Rev. A **86**, 013620 (2012)
35. T. Tsuzuki, J. Low Temp. Phys. **4**, 441 (1971)

Contact Intensity and Extended Hydrodynamics in the BCS-BEC Crossover

Luca Salasnich

Abstract In the first part of this chapter we analyze the contact intensity C, which has been introduced by Tan (Ann Phys 323:2952 (2008)) and appears in several physical observables of the strongly correlated two-component Fermi gas. We calculate the contact C in the full BCS-BEC crossover for a uniform superfluid Fermi gas by using an efficient parametetrization of the ground-state energy. In the case of harmonic confinement, within the Thomas-Fermi approximation, we derive analytical formulas of C in the three relevant limits of the crossover. In the second part of this chapter we discuss the extended superfluid hydrodynamics we have recently proposed to describe static and dynamical collective properties of the Fermi gas in the BCS-BEC crossover. In particular we show the relation with the effective theory for the Goldstone field derived by Son and Wingate (Ann Phys 321, 197 (2006)) on the basis of conformal invariance. By using our equations of extended hydrodynamics we determine nonlinear sound waves, static response function and structure factor of a generic superfluid at zero temperature.

1 Contact Intensity

It has been proved by Tan [1] that the momentum distribution $\rho_\sigma(k)$ in an arbitrary system consisting of fermions in two spin states ($\sigma = \uparrow, \downarrow$) with a large scattering length has a tail that falls off as

$$\rho_\sigma(k) \sim \frac{C}{k^4} \tag{1}$$

L. Salasnich (✉)
Dipartimento di Fisica e Astronomia "Galileo Galilei" and CNISM, Università di Padova, Via Marzolo 8, 35122 Padova, Italy
e-mail: luca.salasnich@unipd.it

R. Carretero-González et al. (eds.), *Localized Excitations in Nonlinear Complex Systems*, Nonlinear Systems and Complexity 7, DOI 10.1007/978-3-319-02057-0_6,

for $k \to \infty$, where C is the so-called contact intensity [1]. Here large s-wave scattering length a means that $|a| \gg r_0$, where r_0 is the effective interaction radius. Under this condition Tan [1] has shown that C is related to the total energy E of the Fermi system by the rigorous expression

$$C = \frac{4\pi m a^2}{\hbar^2} \frac{dE}{da}, \tag{2}$$

where the derivative is taken under constant entropy and, in general, C depends on the number N of fermions, the scattering length a and the parameters of the trapping potential [2,3]. Remarkably, Eqs. (1) and (2) work also at finite temperature and in this case C will be a function of T [2, 4]. Tan has also derived, for finite scattering lengths, a generalized virial theorem and a generalized pressure relation where the contact C appears [3]. The contact intensity C appears also in other physical observables of the strongly correlated Fermi system. For instance, the radio-frequency spectroscopy shift is proportional to C [5–8], and the same happens to the photoassociation rate [9]. Very recently, it has been shown that the contact C gives the asymptotic tail behavior of the shear viscosity as a function of the frequency [10].

Using the methods of quantum field theory, Braaten and Platter have rederived [11, 12] the Tan's universal relations [1–3]. In addition, they have shown that the contact intensity can be written as

$$C = \int g^2 \langle \hat{\psi}_\uparrow^+(\mathbf{r}) \hat{\psi}_\downarrow^+(\mathbf{r}) \hat{\psi}_\downarrow(\mathbf{r}) \hat{\psi}_\uparrow(\mathbf{r}) \rangle \; d^3\mathbf{r},$$

where $\hat{\psi}_\sigma(\mathbf{r})$ is the fermionic field operator of spin σ and $g = 4\pi a/(1-(2ak_{cut}/\pi))$ is the bare coupling constant of the Fermi pseudo-potential interaction, with k_{cut} the ultraviolet wavenumber cutoff [11, 13]. Braaten and Platter have also shown that the number $\mathcal{N}_{pair}(\mathbf{r})$ of pairs of fermions with opposite spins in a small ball of volume $4\pi s^3/3$ centered at the point $\mathbf{r}$ scales as $s^4\mathcal{C}(\mathbf{r})/2$ for $s \to 0$, where $\mathcal{C}(\mathbf{r}) = g^2 \langle \hat{\psi}_\uparrow^+(\mathbf{r}) \hat{\psi}_\downarrow^+(\mathbf{r}) \hat{\psi}_\downarrow(\mathbf{r}) \hat{\psi}_\uparrow(\mathbf{r}) \rangle$ is the contact density [11, 13].

Explicit expressions of the universal quantity C have been derived by Tan [1] for a uniform superfluid Fermi gas at zero temperature only in three limits: the Bardeen-Cooper-Schrieffer (BCS) limit of weakly bound Cooper pairs, the unitarity limit of infinite scattering length, and the Bose-Einstein condensate (BEC) limit of weakly-interacting bosonic molecular pairs.

In this section we calculate the contact C as a function of the inverse scattering parameter $1/(k_F a)$ for a uniform superfluid Fermi gas in the full BCS-BEC crossover. To perform this calculation we use an efficient fitting formula of the ground-state energy [25, 26] and the Tan's equation (2). We find that the contact C has a maximum close to the unitarity limit of infinite scattering length. We also consider the interacting Fermi system under harmonic confinement. For this superfluid Fermi cloud we derive analytical formulas of the contact C in the three relevant limits of the crossover.

1.1 Uniform Superfluid Fermi Gas at Zero Temperature

In the case of a zero-temperature uniform two-component superfluid Fermi gas of total density $n = n_\uparrow + n_\downarrow$ ($n_\uparrow = n_\downarrow$), large scattering length a ($a \gg r_0$) in a volume V, the energy density can be written as

$$\frac{E}{V} = \frac{3}{5} n \epsilon_F f\left(\frac{1}{k_F a}\right), \tag{3}$$

where $f(y)$ is a universal function of inverse interaction parameter $y = 1/(k_F a)$, with $\epsilon_F = \hbar^2 k_F^2/(2m)$ the Fermi energy and $k_F = (3\pi^2 n)^{1/3}$ the Fermi wave number[4]. We observe that at finite temperature T the function $f(y)$ is substituted by a more general universal function $\Phi(y,t)$, where $t = T/\epsilon_F$ is the scaled temperature. $\Phi(y,t)$ has been studied with Monte Carlo methods by Bulgac, Drut, and Migierski [14, 15], but only in the unitarity limit ($y = 0$).

It is straightforward to derive from Eq. (2) the expression of the contact density

$$\frac{C}{V} = -\frac{6\pi}{5} k_F n \frac{df}{dy}. \tag{4}$$

The behavior of $f(y)$ is well known in three relevant limits:

$$f(y) = \begin{cases} 1 + \dfrac{10}{9\pi}\dfrac{1}{y} + O(1/y^2)\,, & y \ll -1 \\ \xi - \zeta y + O(y^2)\,, & -1 \ll y \ll 1 \\ \dfrac{5\mathcal{P}}{18\pi}\frac{1}{y} + O(1/y^{5/2})\,, & y \gg 1 \end{cases} \tag{5}$$

In fact, in the weakly attractive limit ($y \ll -1$) one expects a BCS Fermi gas of weakly bound Cooper pairs. As the superfluid gap correction is exponentially small, the function $f(y)$ follows the Fermi-gas expansion [16, 17]. In the so-called unitarity limit ($y = 0$) one expects that the energy per particle is proportional to that of a non-interacting Fermi gas with the universal constant ξ given by $\xi \simeq 0.42$ [18]. Note that more recent auxiliary-field Monte Carlo results [19] predict a smaller value for ξ, namely $\xi \simeq 0.38$, while the experiment performed at Ecole Normale Superieure [20] suggests $\xi \simeq 0.40$. The first correction to this behavior, shown in Eq. (5), has been estimated from Monte Carlo data with $\zeta \simeq 1$ [21]. In the weak BEC limit ($y \gg 1$) one expects a weakly repulsive Bose gas of dimers. Such Bose-condensed molecules of mass $m_M = 2m$ and density $n_M = n/2$ interact with a positive scattering length $a_M = \mathcal{P}a$ with $\mathcal{P} \simeq 0.6$, as predicted by Petrov et al. [22]. In this regime, after subtraction of the molecular binding energy, the function $f(y)$ follows the Bose-gas expansion [23]. It is easy to obtain the contact intensity C by using Eqs. (2) and (5) in the relevant limits of the crossover. One finds

$$\frac{C}{V} = \begin{cases} \frac{4}{3} k_F^3 n a^2 + O(a^3), & y \ll -1 \\ \frac{6\pi}{5} k_F n \zeta + O(1/a), & -1 \ll y \ll 1 \\ \frac{\mathcal{P}}{3} k_F^3 n a^2 + O(a^{7/2}), & y \gg 1 \end{cases} \tag{6}$$

in agreement with the previous determinations of Tan [1] and Werner, Tarruell and Castin [9]. Notice that in the BEC limit we have removed the binding energy of molecules. Moreover, very recently finite temperature corrections to Eq. (6) were given in Ref. [24].

1.2 Contact Intensity in the BCS-BEC Crossover

Now we want to calculate the behavior of C in the full BCS-BEC crossover. In 2005 we have proposed [25] the following analytical fitting formula

$$f(y) = \alpha_1 - \alpha_2 \arctan\left(\alpha_3\, y\, \frac{\beta_1 + |y|}{\beta_2 + |y|}\right), \tag{7}$$

interpolating the Monte Carlo energy per particle [18] and the limiting behaviors for large and small $|y|$. Equation (7) is very reliable [25,26] and it has been successfully used by various authors for studying grund-state and collective properties of this superfluid Fermi system [27–31]. The parameter α_1 is fixed by the value ξ of $f(y)$ at $y = 0$, the parameter α_2 is fixed by the value of $f(y)$ at $y = \infty$, and α_3 is fixed by the asymptotic $1/y$ coefficient of $\epsilon(y)$ at large $|y|$. The ratio β_1/β_2 is determined by the linear behavior ζ of $\epsilon(y)$ near $y = 0$. The value of β_1 is then determined by minimizing the mean square deviation from the Monte-Carlo data. Of course, we have considered two different set of parameters: one set in the BCS region ($y < 0$) and a separate set in the BEC region ($y > 0$) [25]. Table 1 of [25] reports the values of these parameters, with $\zeta = \zeta_- = 1$ in the BCS region but $\zeta = \zeta_+ = 1/3$ in the BEC region.

Here we use Eq. (7) to calculate the contact density given by Eq. (4), but contrary to [25], we choose $\zeta_+ = \zeta_- = \zeta = 1$ to ensure the continuity of $f'(y)$ at $y = 0$. Notice that the recent experimental results obtained at Ecole Normale Superieure [20] indeed suggest the continuity of $f'(y)$ at $y = 0$. In this way, in the BEC region $\beta_2 = 0.1517$ while β_1 is unchanged (see Table 1 of [25]). In the upper panel of Fig. 1 we plot the ground-state energy E while in the lower panel we plot the contact C, both as a function of the inverse interaction parameter $y = 1/(k_F a)$. For comparison, in addition to the data obtained with our method (solid lines), we insert also the results (dashed lines) one obtains using the Pade parametrization of $f(y)$ proposed by Kim and Zubarev [32]. The figure shows that solid and

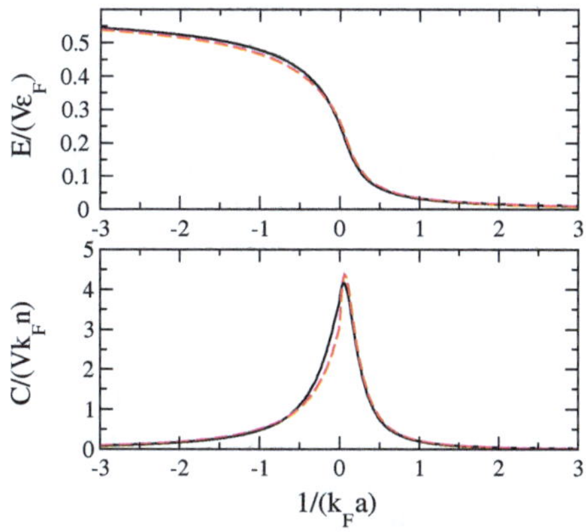

Fig. 1 *Upper panel*: Scaled ground-state energy $E/(V\epsilon_F)$ of the uniform Fermi gas as a function of the inverse interaction parameter $1/(k_F a)$ in the BCS-BEC crossover. *Lower panel*: Scaled contact intensity $C/(V k_F n)$ of the Fermi system as a function of the inverse interaction parameter $1/(k_F a)$. Here V and $n = N/V$ are the volume and the density of the Fermi gas, $k_F = (3\pi^2 n)^{1/3}$ is the Fermi wave number and $\epsilon_F = \hbar^2 k_F^2/(2m)$ is the Fermi energy. Two different parametrization of the universal function $f(y)$: *solid lines* are obtained with Eq. (7), while *dashed lines* are calculated using the Pade approximant of [32]

dashed lines are always close each other, apart for $-1 \ll y \leq 0$ where our fitting formula is smoother (and closer to the Monte Carlo data [18]). Moreover, the scaled contact $C/(V k_F n)$ as a function of $1/(k_F a)$ has its maximum near to the unitarity limit $1/(k_F a) = 0$: the position of the maximum is located at $1/(k_F a) \simeq 0.05$. Remarkably, the scaled contact has a behavior quite similar to the Landau's critical velocity v_c (at which there is the breaking of superfluid motion), calculated along the BCS-BEC crossover. In fact, also v_c goes to zero for $y \to \pm\infty$ and it has a peak at $y \simeq 0.08$ [33]. Clearly, the contact C exhibits a maximum close to unitarity because we have subtracted the molecular binding energy contribution, given by y^2 in the BEC side ($y > 0$). Including this energy term it is easy to show that C increases monotonically from the BCS side to the BEC side, according the definition in Eq. (2). Nevertheless, as previously stated, the radio-frequency spectroscopy shift is proportional to C, and its maximum around unitarity has been shown in Ref. [6], where the same artificial subtraction has been adopted.

1.3 Trapped Superfluid Fermi Gas

Let us now consider the superfluid Fermi gas under an external harmonic confinement

$$U(\mathbf{r}) = \frac{1}{2} m\omega^2 (x^2 + y^2 + z^2). \tag{8}$$

In the limit of a large number N of fermions we can use the local density (Thomas-Fermi) approximation [34–41] and the energy of the system can be written as

$$E = \int \left\{ \frac{3}{5} n(\mathbf{r}) \epsilon_F(\mathbf{r}) f\left(\frac{1}{k_F(\mathbf{r}) a} \right) + n(\mathbf{r}) U(\mathbf{r}) \right\} d^3\mathbf{r}, \tag{9}$$

where $\epsilon_F(\mathbf{r}) = \hbar^2 k_F(\mathbf{r})^2/(2m)$ is the local Fermi energy and $k_F = (3\pi^2 n(\mathbf{r}))^{1/3}$ is the local Fermi wave number. As in the uniform case, this expression is not very useful without the knowledge of the universal function $f(y)$.

The numerical calculation of the contact intensity C for a harmonically trapped Fermi superfluid in the full BCS-BEC crossover by using Eqs. (2) and (7) is very demanding. Consequently, we calculate C only in the three relevant limits of the BEC-BEC crossover. In these limits we obtain useful analytical expressions for the contact C.

BCS limit. In the BCS limit ($a \to 0^-$) from Eqs. (5) and (9) we find

$$\frac{dE}{da} = \frac{3}{5} \frac{10}{9\pi} \int \{ \epsilon_F(\mathbf{r}) n(\mathbf{r}) k_F(\mathbf{r}) \} \, d^3\mathbf{r}, \tag{10}$$

where $n(\mathbf{r})$ is the density profile of the ideal Fermi gas in the harmonic potential (8), given by Salasnich [42]

$$n(\mathbf{r}) = \frac{2\sqrt{2}}{3\pi^2 a_H^3} (3N)^{1/2} \left(1 - \frac{r^2}{r_F^2} \right)^{3/2},$$

where $r_F = \sqrt{2}(3N)^{1/6} a_H$ is the Fermi radius of the cloud, with $a_H = \sqrt{\hbar/(m\omega)}$ the characteristic harmonic length. Inserting this density profile into Eq. (10) and using Eq. (7) we get the contact intensity in the BCS limit:

$$C = \frac{4{,}096\sqrt{2}}{2{,}835\pi} \frac{1}{a_H} \left(\frac{a}{a_H} \right)^2 (3N)^{3/2}.$$

Unitarity limit. In the unitarity limit ($a \to \pm\infty$) from Eqs. (5) and (9) we have

$$\frac{dE}{da} = \frac{3}{5} \frac{\zeta}{a^2} \int \left\{ \epsilon_F(\mathbf{r}) n(\mathbf{r}) \frac{1}{k_F(\mathbf{r})} \right\} d^3\mathbf{r}, \tag{11}$$

where $n(\mathbf{r})$ is the density profile of the unitary Fermi gas in the potential (8), namely [34]

$$n(\mathbf{r}) = \frac{2\sqrt{2}}{3\pi^2 a_H^3 \xi^{3/4}} (3N)^{1/2} \left(1 - \frac{r^2}{r_F^2} \right)^{3/2},$$

where $r_F = \sqrt{2}\xi^{1/4}(3N)^{1/6}a_H$ is the Fermi radius of the unitary cloud. Inserting this density profile into Eq. (11) and using Eq. (7) we obtain the contact in the unitarity limit:

$$C = \frac{512\sqrt{2}}{525}\frac{\zeta}{\xi^{1/4}}\frac{1}{a_H}(3N)^{7/6}.$$

BEC limit. In the BEC limit ($a \to 0^+$) it is straightforward to calculate the contact C. In fact, the explicit formula of the ground-state energy of the dilute BEC is well known [43] and for a BEC of molecules it is given by

$$E = \frac{5}{7}\frac{\hbar\omega}{2}\left(\frac{15\mathcal{P}a}{a_H}\right)^{2/5}\left(\frac{N}{2}\right)^{7/5},$$

where $\mathcal{P} \simeq 0.6$ and $N/2$ is the number of molecules. Then from Eq. (2) the contact intensity reads

$$C = \frac{2\pi}{7}\frac{1}{a_H}\left(\frac{15\mathcal{P}}{2}\right)^{2/5}\left(\frac{aN}{a_H}\right)^{7/5}.$$

2 Extended Superfluid Hydrodynamics

In this section we discuss the extended Lagrangian density of superfluids we have proposed few years ago [34] and applied to study mainly the unitary Fermi gas [34–41]. The internal energy density of this Lagrangian contains a term proportional to the kinetic energy of a uniform non interacting gas of fermions, plus a gradient correction of the von-Weizsacker form $\lambda\hbar^2/(8m)(\nabla n/n)^2$ [44]. This approach has been adopted for studying the quantum hydrodynamics of electrons by March and Tosi [45] and Zaremba and Tso [46]. In the context of the BCS-BEC crossover, the gradient term is quite standard [25, 32, 47–53]. In particular we show the relation between our approach and the effective theory for the Goldstone field derived by Son and Wingate [54], and improved by Manes and Valle [55], on the basis of conformal invariance. Finally, by using our equations of extended superfluid hydrodynamics at zero temperature we calculate sound waves, static response function and structure factor of a generic superfluid.

The extended Lagrangian density of superfluids is given by

$$\mathcal{L} = -\hbar\,\dot{\theta}\,n - \frac{\hbar^2}{2m}(\nabla\theta)^2\,n - \mathcal{E}(n, \nabla n) - U(\mathbf{r})\,n, \tag{12}$$

where $n(\mathbf{r}, t)$ is the local density, $U(\mathbf{r})$ is the external potential acting on particles, and m the mass of superfluid particles. In the case of superfluid bosons $\theta(\mathbf{r}, t)$ is the phase of the condensate order parameter, while in the case of superfluid fermions $\theta(\mathbf{r}, t)$ is half of the phase of the condensate order parameter (of Cooper pairs). $\mathcal{E}(n, \nabla n)$ is the internal energy density of the system. Note that we are supposing that this equation of state $\mathcal{E}(n, \nabla n)$ can depend not only on the local density $n(\mathbf{r}, t)$ but also on its space derivatives. For this reason we call (12) the extended superfluid Lagrangian. We stress that in the context of the BCS-BEC crossover the extended internal energy density could be written as

$$\mathcal{E}(n, \nabla n) = \mathcal{E}_0(n) + \lambda \frac{\hbar^2}{8m} \frac{(\nabla n)^2}{n},$$

where

$$\mathcal{E}_0(n) = \frac{3}{5} n \epsilon_F f\left(\frac{1}{k_F a}\right) \tag{13}$$

is the energy density discussed in the previous section (see Eq. (3)) which depends on the universal function $f(y)$ of the BCS-BEC crossover, while the second term is the gradient correction of the von-Weizsacker form [35]. In the BCS-BEC crossover we expect that $1/6 \leq \lambda \leq 1/4$, where $\lambda = 1/6$ is the appropriate value in BCS regime of weakly-interacting superfluid fermions of mass m [45,46], while $\lambda = 1/4$ is the appropriate value in the deep BEC regime of weakly-interacting superfluid bosonic dimers of mass $2m$ [35].

By using the Lagrangian density (12) the Euler-Lagrange equation

$$\frac{\partial \mathcal{L}}{\partial \theta} - \frac{\partial}{\partial t}\frac{\partial \mathcal{L}}{\partial \dot{\theta}} - \nabla \cdot \frac{\partial \mathcal{L}}{\partial (\nabla \theta)} = 0$$

gives

$$\frac{\partial n}{\partial t} + \frac{\hbar}{m} \nabla \cdot (n\, \nabla \theta) = 0\,. \tag{14}$$

The Euler-Lagrange equation

$$\frac{\partial \mathcal{L}}{\partial n} - \frac{\partial}{\partial t}\frac{\partial \mathcal{L}}{\partial \dot{n}} - \nabla \cdot \frac{\partial \mathcal{L}}{\partial (\nabla n)} = 0$$

gives instead

$$\hbar\, \dot{\theta} + \frac{\hbar^2}{2m} (\nabla \theta)^2 + U(\mathbf{r}) + X(n, \nabla n) = 0, \tag{15}$$

where

$$X(n, \nabla n) = \frac{\partial \mathcal{E}}{\partial n} - \nabla \cdot \frac{\partial \mathcal{E}}{\partial (\nabla n)} \tag{16}$$

is the local chemical potential of the system (see also [54] and [55]).

The local field velocity $\mathbf{v}(\mathbf{r}, t)$ of the superfluid is related to the phase $\theta(\mathbf{r}, t)$ of the condensate by

$$\mathbf{v}(\mathbf{r}, t) = \frac{\hbar}{m} \nabla \theta(\mathbf{r}, t). \tag{17}$$

This definition ensures that the velocity is irrotational, i.e. $\nabla \wedge \mathbf{v} = \mathbf{0}$. By using the definition (17) in both Eqs. (14) and (15) and applying the gradient operator ∇ to Eq. (15) one finds the extended hydrodynamic equations of superfluids

$$\frac{\partial n}{\partial t} + \nabla \cdot (n\, \mathbf{v}) = 0\,. \tag{18}$$

$$m \frac{\partial \mathbf{v}}{\partial t} + \nabla \left[\frac{1}{2} m \mathbf{v}^2 + U(\mathbf{r}) + X(n, \nabla n) \right] = \mathbf{0}. \tag{19}$$

We stress that in the presence of an external confinement $U(\mathbf{r})$ the chemical potential μ of the system does not coincide with the local chemical potential $X(n, \nabla n)$. The chemical potential μ can be obtained from Eq. (15) setting $\theta(\mathbf{r}, t) = -\mu t / \hbar$ and $\mathbf{v}(\mathbf{r}, t) = \mathbf{0}$, such that

$$U(\mathbf{r}) + X(n_0, \nabla n_0) = \mu,$$

where $n_0(\mathbf{r})$ is the ground-state local density.

The Lagrangian density (12) depends on the dynamical variables $\theta(\mathbf{r}, t)$ and $n(\mathbf{r}, t)$. The conjugate momenta of these dynamical variables are then given by

$$\pi_\theta = \frac{\partial \mathcal{L}}{\partial \dot{\theta}} = -\hbar\, n, \tag{20}$$

$$\pi_n = \frac{\partial \mathcal{L}}{\partial \dot{n}} = 0, \tag{21}$$

and the corresponding Hamiltonian density reads

$$\mathcal{H} = \pi_\theta\, \dot{\theta} + \pi_n\, \dot{n} - \mathcal{L} = -\hbar\, n\, \dot{\theta} - \mathcal{L},$$

namely

$$\mathcal{H} = \frac{\hbar^2}{2m} (\nabla \theta)^2\, n + \mathcal{E}(n, \nabla n) + U(\mathbf{r})\, n,$$

which is the sum of the flow kinetic energy density $\hbar^2(\nabla\theta)^2 n/(2m) = (1/2)mv^2 n$, the internal energy density $\mathcal{E}(n,\nabla n)$, and the external energy density $U(\mathbf{r})n$.

2.1 Extended Hydrodynamics in Terms of Goldstone Field

Note that taking into account Eq. (15) one immediately finds

$$X(n,\nabla n)\, n = -\hbar\, \dot{\theta}\, n - \frac{\hbar^2}{2m}(\nabla\theta)^2\, n - U(\mathbf{r})\, n.$$

Consequently the Lagrangian density (12) can be rewritten as

$$\mathcal{L} = X(n,\nabla n)\, n - \mathcal{E}(n,\nabla n)\ .$$

Remarkably

$$P(n,\nabla n) = X(n,\nabla n)\, n - \mathcal{E}(n,\nabla n) \tag{22}$$

is the local pressure of the system as a function of the density and its spatial derivatives, which can be written as a function of the local chemical potential X and its spatial derivatives, namely

$$\mathcal{L} = P(X,\nabla X). \tag{23}$$

This result, based on a Legendre transformation, is clearly illustrated in the book of Popov [56] and used in the recent papers of Son and Wingate [54] and Manes and Valle [55]. Finally one can introduce the Goldstone field $\phi(\mathbf{r},t)$ as

$$\phi(\mathbf{r},t) = \theta(\mathbf{r},t) + \frac{\mu}{\hbar}t\ .$$

In this way, by using again Eq. (15), one can write

$$X = \mu - \hbar\dot{\phi} - \frac{\hbar^2}{2m}(\nabla\phi)^2 - U(\mathbf{r})\ ,$$

Thus, the Lagrangian density (23) actually depends only on the Goldstone field $\phi(\mathbf{r},t)$. This is exactly the main message of the paper of Son and Wingate [54], which however traces back to the older results of Popov [56].

2.2 Application: The Unitary Fermi Gas

Let us now suppose that the equation of state of the superfluid at zero temperature is that of the unitary Fermi gas, i.e.

$$\mathcal{E}(n, \nabla n) = \frac{3}{5}\frac{\hbar^2}{2m}(3\pi^2)^{2/3}\xi\, n^{5/3} + \lambda\frac{\hbar^2}{8m}\frac{(\nabla n)^2}{n} \,,$$

where $\xi \simeq 0.4$ and $\lambda \simeq 0.25$ [37, 41]. It follows from Eq. (16) that

$$X(n, \nabla n) = \frac{\hbar^2}{2m}(3\pi^2)^{2/3}\xi\, n^{2/3} - \lambda\frac{\hbar^2}{8m}\frac{(\nabla n)^2}{n^2}, \tag{24}$$

by taking into account that the surface terms give zero contribution. In addition we get from Eq. (22) that

$$P(n, \nabla n) = \frac{2}{5}\frac{\hbar^2}{2m}(3\pi^2)^{2/3}\xi\, n^{5/3} - \lambda\frac{\hbar^2}{4m}\frac{(\nabla n)^2}{n},$$

The Lagrangian density of Eq. (23) is then obtained by finding n and ∇n as functions of X and ∇X by inverting Eq. (24). This can be done in terms of a derivative expansion. One gets $n = (2m\xi)^{3/2}X^{3/2}/(3\pi^2\hbar^3)$ and

$$\mathcal{L} = \mathcal{L}_{LO} + \mathcal{L}_{NLO} \tag{25}$$

where

$$\mathcal{L}_{LO} = c_0\,\frac{m^{3/2}}{\hbar^3}\,X^{5/2},$$

with

$$c_0 = \frac{2^{5/2}}{15\pi^2\xi^{3/2}},$$

is the Lagrangian density at the leading order, and

$$\mathcal{L}_{NLO} = c_1\,\frac{m^{1/2}}{\hbar}\,\frac{(\nabla X)^2}{\sqrt{X}},$$

with

$$c_1 = -\lambda\frac{3\cdot 2^{1/2}}{8\pi^2\xi^{3/2}},$$

is the next-to-leading contribution to the Lagrangian density. The Lagrangian density (25) is the same of that derived by Son and Wingate [54] from general coordinate invariance and conformal invariance. Actually, at the lext-to-leading order Son and Wingate have found an additional term [54], which has been questioned by Manes and Valle [55] and is absent in our approach.

2.3 Nonlinear Sound Waves, Static Response Function and Structure Factor

In this subsection we consider the following zero-temperature equation of state of a generic superfluid

$$\mathcal{E}(n, \nabla n) = \mathcal{E}_0(n) + \lambda \frac{\hbar^2}{2m} \frac{(\nabla n)^2}{4n}.$$

Here the internal energy is the sum of two contributions: a generic internal energy $\mathcal{E}_0(n)$ which depends only of the local density $n(\mathbf{r}, t)$ (for instance that of Eq. (13)) plus the gradient correction of the von Weizsaćker type, where the coefficient λ can be a function of the interaction strength.

The equation of motion (19) becomes

$$m \frac{\partial \mathbf{v}}{\partial t} + \nabla \left[\frac{1}{2} m \mathbf{v}^2 + U(\mathbf{r}) + X_0(n) - \lambda \frac{\hbar^2}{2m} \frac{\nabla^2 \sqrt{n}}{\sqrt{n}} \right] = \mathbf{0},$$

with

$$X_0(n) = \frac{\partial \mathcal{E}_0}{\partial n}.$$

We are interested on the propagation of sound waves in superfluids. For simplicity we set

$$U(\mathbf{r}) = 0,$$

and consider a small perturbation $\tilde{n}(\mathbf{r}, t)$ around a uniform and constant configuration n_0, namely

$$n(\mathbf{r}, t) = n_0 + \tilde{n}(\mathbf{r}, t).$$

Neglecting quadratic terms in $\tilde{n}$ and v we derive the linearized equations of extended superfluid hydrodynamics

$$\frac{\partial \tilde{n}}{\partial t} + n_0 \nabla \cdot \mathbf{v} = 0, \tag{26}$$

$$n_0 \frac{\partial \mathbf{v}}{\partial t} + c_s^2 \nabla \tilde{n} - \lambda \frac{\hbar^2}{4m^2} \nabla \left(\nabla^2 \tilde{n} \right) = \mathbf{0}, \tag{27}$$

where c_s is the sound velocity, given by

$$c_s^2 = \frac{n_0}{m} \frac{\partial X_0(n_0)}{\partial n} = \frac{n_0}{m} \frac{\partial^2 \mathcal{E}_0(n_0)}{\partial n^2}.$$

Applying the operator $\frac{\partial}{\partial t}$ to Eq. (26) and the operator ∇ to Eq. (27) and subtracting the two resulting equations we get

$$\left(\frac{\partial^2}{\partial t^2} - c_s^2 \nabla^2 + \lambda \frac{\hbar^2}{4m^2} \nabla^4 \right) \tilde{n}(\mathbf{r}, t) = 0.$$

This is the wave equation of the small perturbation $\tilde{n}(\mathbf{r}, t)$ around the uniform density n_0. We stress that the effect of the gradient term in the equation of state is the presence of quartic spatial derivatives in this wave equation.

It is straightforward to show that the wave equation admits the real solution

$$\tilde{n}(\mathbf{r}, t) = A \, e^{i(\mathbf{q}\cdot\mathbf{r} - \omega_q t)} + A^* \, e^{-i(\mathbf{q}\cdot\mathbf{r} - \omega_q t)},$$

where the frequency ω_q and the wave vector $\mathbf{q}$ are related by the Bogoliubov-like dispersion formula

$$\hbar \omega_q = \sqrt{\frac{\hbar^2 q^2}{2m} \left(\lambda \frac{\hbar^2 q^2}{2m} + 2mc_s^2 \right)},$$

or equivalently

$$\omega_q = c_s q \sqrt{1 + \alpha q^2} = c_s q \left(1 + \frac{\alpha}{2} q^2 - \frac{\alpha^2}{8} q^4 + \ldots \right),$$

with $\alpha = \lambda \hbar^2 / (4m^2 c_s^2)$. Thus, the dispersion relation ω_q is linear in $q = |\mathbf{q}|$ only for small values of the wavenumber q and becomes quadratic for large values of q.

We observe that, for a generic many-body system, the dispersion relation can be written as [57]

$$\hbar \omega_q = \sqrt{\frac{m_1(q)}{m_{-1}(q)}}, \tag{28}$$

where $m_n(q)$ is the n moment of the dynamic structure function $S(q,\omega)$ of the many-body system under investigation, i.e.

$$m_n(q) = \int_0^\infty d\omega \; S(q,\omega) \; (\hbar\omega)^n .$$

Note that Eq. (28) is not exact and is valid under the approximation of a single-mode density excitation. It therefore only gives an upper bound for the dispersion relation. This is important for large q, for which quasiparticles other than density excitations may contribute. In our problem we have

$$m_1(q) = \frac{\hbar^2 q^2}{2m}$$

and

$$m_{-1}(q) = \frac{1}{\lambda \frac{\hbar^2 q^2}{2m} + 2mc_s^2} .$$

In general, the static response function $\chi(q)$ is defined as [57]

$$\chi(q) = -2\, m_{-1}(q),$$

and in our problem it reads

$$\chi(q) = -\frac{2}{\lambda \frac{\hbar^2 q^2}{2m} + 2mc_s^2},$$

or equivalently

$$\chi(q) = -\frac{1}{mc_s^2}\frac{1}{1+\alpha q^2} = -\frac{1}{mc_s^2}\left(1 - \alpha q^2 + \alpha^2 q^4 + \ldots\right),$$

where again $\alpha = \lambda\hbar^2/(4m^2c_s^2)$.

The static structure factor $S(q)$ is instead defined as [57]

$$S(q) = m_0(q) = \int_0^\infty d\omega \; S(q,\omega),$$

but it can be approximated by the expression

$$\tilde{S}(q) = \sqrt{m_1(q)\, m_{-1}(q)},$$

which gives an upper bound of $S(q)$, i.e. $\tilde{S}(q) \geq S(q)$ [57]. In our problem we immediately find

$$\tilde{S}(q) = \sqrt{\frac{\frac{\hbar^2 q^2}{2m}}{\lambda \frac{\hbar^2 q^2}{2m} + 2mc_s^2}},$$

or equivalently

$$\tilde{S}(q) = \frac{\hbar q}{2mc_s} \frac{1}{\sqrt{1+\alpha q^2}} = \frac{\hbar q}{2mc_s}\left(1 - \frac{1}{2}\alpha q^2 + \frac{3}{8}\alpha^2 q^4 + \ldots\right).$$

Our results clearly indicate that it should be possible to observe experimentally the effect of the dispersive von-Weizsacker-like gradient term from sound-wave measurements.

3 Conclusions

In the first part of this contribution we have calculated the contact C as a function of the inverse scattering parameter $1/(k_F a)$ for a uniform superfluid Fermi gas in the full BCS-BEC crossover at zero temperature. We have found that the contact C has a maximum close to the unitarity limit of infinite scattering length, in analogy with the behavior of the Landau's critical velocity v_c, at which there is the breaking of superfluid motion [33]. We have also considered the interacting Fermi system under harmonic confinement. In this case, we have derived analytical formulas of the contact intensity C in the three relevant limits of the crossover. Our results can be experimentally tested with ultracold atomic clouds by measuring one of the quantities which are directly related to the contact intensity C: the tail of the momentum distribution, the derivative of the total energy with respect to the scattering length, the radio-frequency spectroscopy shift, or the photoassociation rate. In the second part we have analyzed some properties of the extended superfluid hydrodynamics [34]; in particular, we have shown its strict relation with the low-energy effective field theory built on the Goldstone mode. Finally, by using the extended hydrodynamics we have calculated, for generic superfluid in the absence of external confinement, the nonlinear dispersion relation of sound waves, and, as a by-product, both static response function and structure factor.

References

1. S. Tan, Ann. Phys. **323**, 2952 (2008)
2. S. Tan, Ann. Phys. **323**, 2971 (2008)
3. S. Tan, Ann. Phys. **323**, 2987 (2008)
4. S. Zhang, A.J. Leggett, Phys. Rev. A **79**, 023601 (2009)
5. M. Punk, W. Zwerger, Phys. Rev. Lett. **99**, 170404 (2007)
6. G. Baym, C.J. Pethick, Z. Yu, M.W. Zwierlein, Phys. Rev. Lett. **99**, 190407 (2007)

7. P. Pieri, A. Perali, G.C. Strinati, Nat. Phys. **5**, 736 (2009)
8. J.P. Gaebler, J.T. Stewart, T.E. Drake, D.S. Jin, A. Perali, P. Pieri, G.C. Strinati, Nat. Phys. **6**, 569 (2010)
9. F. Werner, L. Tarruell, Y. Castin, Eur. Phys. J. B **68**, 401 (2009)
10. E. Braaten, in *The BCS-BEC Crossover and the Unitary Fermi Gas*, ed. by W. Zwerger (Springer, Berlin, 2012)
11. E. Braaten, L. Platter, Phys. Rev. Lett. **100**, 205301 (2008)
12. E. Braaten, D. Kang, L. Platter, Phys. Rev. A **78**, 053606 (2008)
13. E. Braaten, L. Platter, Laser Phys. **19**, 550 (2009)
14. A. Bulgac, J.E. Drut, P. Magierski, Phys. Rev. Lett. **96**, 090404 (2006)
15. A. Bulgac, J.E. Drut, P. Magierski, Phys. Rev. Lett. **99**, 120401 (2007)
16. K. Huang, C.N. Yang, Phys. Rev. **105**, 767 (1957)
17. T.D. Lee, C.N. Yang, Phys. Rev. **105**, 1119 (1957)
18. G.E. Astrakharchik, J. Boronat, J. Casulleras, S. Giorgini, Phys. Rev. Lett. **93**, 200404 (2004)
19. J. Carlson, S. Gandolfi, K.E. Schmidt, S. Zhang, Phys. Rev. A **84**, 061602(R) (2011)
20. N. Navon, S. Nascimbene, F. Chevy, C. Salomon, Science **328**, 729 (2010)
21. A. Bulgac, G.F. Bertsch, Phys. Rev. Lett. **94**, 070401 (2005)
22. D.S. Petrov, C. Salomon, G.V. Shlyapnikov, Phys. Rev. Lett. **93**, 090404 (2004)
23. T.D. Lee, K. Huang, C.N. Yang, Phys. Rev. **106**, 1135 (1957)
24. Z. Yu, G.M. Bruun, G. Baym, Phys. Rev. A **80**, 023615 (2009)
25. N. Manini, L. Salasnich, Phys. Rev. A **71**, 033625 (2005)
26. G. Diana, N. Manini, L. Salasnich, Phys. Rev. A **73**, 065601 (2006)
27. T.N. De Silva, E.J. Mueller, Phys. Rev. A **72**, 063614 (2005)
28. Yu. Zhou, G. Huang, Phys. Rev. A **75**, 023611 (2007)
29. T.K. Ghosh, Phys. Rev. A **76**, 033602 (2007)
30. S.K. Adhikari, Phys. Rev. A **77**, 045602 (2008)
31. S.K. Adhikari, Phys. Rev. A **79**, 023611 (2009)
32. Y.E. Kim, A.L. Zubarev, Phys. Rev. A **70**, 033612 (2004)
33. R. Combescot, M.Yu. Kagan, S. Stringari, Phys. Rev. A **74**, 042717 (2006)
34. L. Salasnich, F. Toigo, Phys. Rev. A **78**, 053626 (2008)
35. L. Salasnich, Laser Phys. **19**, 642 (2009)
36. F. Ancilotto, L. Salasnich, F. Toigo, Phys. Rev. A **79**, 033627 (2009)
37. S.K. Adhikari, L. Salasnich, New J. Phys. **11**, 023011 (2009)
38. L. Salasnich, F. Ancilotto, F. Toigo, Laser Phys. Lett. **7**, 78 (2010)
39. L. Salasnich, EPL **96**, 40007 (2011)
40. F. Ancilotto, L. Salasnich, F. Toigo, Phys. Rev. A **85**, 063612 (2012)
41. L. Salasnich, Few-Body Syst. doi:10.1007/s00601-012-0442-y
42. L. Salasnich, J. Math. Phys. **41**, 8016 (2000)
43. F. Dalfovo, S. Giorgini, L.P. Pitaevskii, S. Stringari, Rev. Mod. Phys. **71**, 463 (1999)
44. C.F. von Weizsäcker, Z. Phys. **96**, 431 (1935)
45. N.H. March, M.P. Tosi, Proc. R. Soc. A **330**, 373 (1972)
46. E. Zaremba, H.C. Tso, Phys. Rev. B **49**, 8147 (1994)
47. M.A. Escobedo, M. Mannarelli, C. Manuel, Phys. Rev. A **79**, 063623 (2009)
48. E. Lundh, A. Cetoli, Phys. Rev. A **80**, 023610 (2009)
49. G. Rupak, T. Schäfer, Nucl. Phys. A **816**, 52 (2009)
50. S.K. Adhikari, Laser Phys. Lett. **6**, 901 (2009)
51. W.Y. Zhang, L. Zhou, Y.L. Ma, EPL **88**, 40001 (2009)
52. A. Csordas, O. Almasy, P. Szepfalusy, Phys. Rev. A **82**, 063609 (2010)
53. S.N. Klimin, J. Tempere, J.P.A. Devreese, J. Low Temp. Phys. **165**, 261 (2011)
54. D.T. Son, M. Wingate, Ann. Phys. **321**, 197 (2006)
55. J.L. Manes, M.A. Valle, Ann. Phys. **324**, 1136 (2009)
56. V.N. Popov, *Functional Integrals in Quantum Field Theory and Statistical Physics* (Reidel, Dordrecht, 1983)
57. F. Dalfovo, A. Lastri, L. Pricaupenko, S. Stringari, J. Treiner, Phys. Rev. B **52**, 1193 (1995)

N-Soliton Interactions for the Manakov System: Effects of External Potentials

V.S. Gerdjikov and M.D. Todorov

Abstract We analyze the dynamical behavior of the N-soliton train in adiabatic approximation of the Manakov system (MS) perturbed by three types of external potentials: periodic, quadratic and quartic ones. We show that the dynamics of the N-soliton train is modeled by a perturbed complex Toda chain for certain choices of the train parameters and for small magnitudes of the intensities of the potentials. Possible applications of these results for Bose-Einstein condensates are discussed.

1 Introduction

The Gross-Pitaevski (GP) equation and its multicomponent generalizations are important tools for analyzing and studying the dynamics of the Bose-Einstein condensates (BEC) , see the monographs [25, 29, 42] and the numerous references therein among which we mention [4–6, 15, 21, 22, 26, 31, 36, 38, 41]. In the 3-dimensional case these equations can be analyzed solely by numerical methods. If we assume that BEC is quasi-one-dimensional then the GP equations mentioned above reduce to to the nonlinear Schrödinger (NLS) equation

$$iu_t + \frac{1}{2}u_{xx} + |u|^2 u(x,t) = 0, \tag{1}$$

V.S. Gerdjikov (✉)
Institute for Nuclear Research and Nuclear Energy, Bulgarian Academy of Sciences,
72 Tzarigradsko chaussee, Blvd., 1784 Sofia, Bulgaria
e-mail: gerjikov@inrne.bas.bg

M.D. Todorov
Department of Applied Mathematics and Computer Science, Technical University of Sofia,
8 Kliment Ohridski, Blvd., 1000 Sofia, Bulgaria
e-mail: mtod@tu-sofia.bg

R. Carretero-González et al. (eds.), *Localized Excitations in Nonlinear Complex Systems*, Nonlinear Systems and Complexity 7, DOI 10.1007/978-3-319-02057-0_7,

and its vector generalizations (VNLS)

$$i\boldsymbol{u}_t + \frac{1}{2}\boldsymbol{u}_{xx} + (\boldsymbol{u}^\dagger, \boldsymbol{u})\boldsymbol{u}(x,t) = 0. \tag{2}$$

perturbed with external potential.

Until now two types of integrable VNLS equations are known. The first and oldest one is known as the Manakov model (MM) [37] – two-component VNLS which is easily generalized also to any number of components. Most of our results below concern this model.

The second VNLS was discovered by Kulish and Sklyanin [34]. Its importance for the spin-1 and spin-2 BEC was discovered in [26,27,40,48,49].

The Lax representation, N-soliton solutions and the fundamental properties of the MM have been well known for long time now [37], see also [1, 19] and the numerous references therein.

The Kulish-Sklyanin model [34], though less popular have also been thoroughly investigated. Its Hamiltonian properties, N-soliton solutions and their interactions have also been derived [9,23,32,33].

The analytical approach to the N-soliton interactions was proposed by Zakharov and Shabat [39,51] for the scalar NLS. They treated the case of the exact N-soliton solution where all solitons had different velocities. They calculated the asymptotics of the N-soliton solution for $t \to \pm\infty$ and proved that both asymptotics are sums of N 1-soliton solutions with the same sets of amplitudes and velocities. The effects of the interaction were shifts in the relative center of masses and phases of the solitons.

Similar results for the N-soliton interactions were obtained also for the Kulish-Sklyanin model [9,23,30,32,33]. Again we proved that if all solitons have different velocities their interaction is purely elastic and results in shifting of the relative center of masses and phases. The same approach, however, is not applicable to the MM, because the asymptotics of the soliton solution for $t \to \pm\infty$ do not commute.

In the present paper we in fact prove the hypothesis in [8] and extend the results in [10, 16–18, 20, 35, 43, 50] concerning the model of soliton interactions of VNLS (2) in adiabatic approximation.

Next, we consider the perturbed VNLS

$$i\boldsymbol{u}_t + \frac{1}{2}\boldsymbol{u}_{xx} + (\boldsymbol{u}^\dagger, \boldsymbol{u})\boldsymbol{u}(x,t) = iR[\boldsymbol{u}]. \tag{3}$$

with the special choice of the perturbation

$$iR[\boldsymbol{u}] = V(x)\boldsymbol{u}(x,t), \tag{4}$$

due to the external potential $V(x)$ and analyze the effects of external potentials on the N-soliton interactions in the same adiabatic approximation.

The corresponding vector N-soliton train is a solution of the VNLS determined by the initial condition:

$$u(x,t=0) = \sum_{k=1}^{N} u_k(x,t=0) \tag{5}$$

with

$$\begin{aligned} u_k(x,t) &= \frac{2\nu_k e^{i\phi_k}}{\cosh(z_k)} n_k, \\ z_k &= 2\nu_k(x-\xi_k(t)), \qquad \xi_k(t) = 2\mu_k t + \xi_{k,0}, \\ \phi_k &= \frac{\mu_k}{\nu_k} z_k + \delta_k(t), \qquad \delta_k(t) = 2(\mu_k^2 + \nu_k^2)t + \delta_{k,0}. \end{aligned} \tag{6}$$

where the constant polarization vector n_k is normalized by

$$(n_k^\dagger, n_k) = 1, \qquad \sum_{s=1}^{n} \arg n_{k;s} = 0.$$

More precisely we derive a generalized version of the Complex Toda Chain (CTC) (see Eqs. (20) and (21) below) as a model describing the behavior of the N-soliton trains of the vector NLS (2). This generalized CTC includes dependence on the polarization vectors n_k. It allows us to analyze how the changes of the polarization vectors influence the soliton interactions.

In Sect. 2 we outline how the variational approach developed in [2] can be used to derive the perturbed CTC (PCTC) model [10, 18]. In Sect. 3 we briefly treat the N-soliton interactions of the MM without external potential. Obviously in order to determine the N-soliton train for the MM, along with the usual sets of solitons parameters ν_k, μ_k, ξ_k and δ_k we need also the set of polarization vectors n_s. In [10] we proposed that a condition on n_s should be imposed in order the adiabatic approximation to be valid for this soliton train. Here we put this condition to a numeric test and find that it could be substantially relaxed. More precisely we analyze the effects of the polarization vectors n_s on the soliton interaction. In Sect. 4 we derive the effects of the external potentials on the soliton interactions. This is a perturbed form of the CTCT (PCTC) for generic potentials of the form (40). Section 5 is dedicated to the comparison between the numeric solutions of the perturbed VNLS (3) with the predictions of the PCTC model. We plot trajectories of the center of masses for each of the solitons from the numeric solution of the perturbed VNLS and compare them with the (numeric solutions) for $\xi_k(t)$ of the PCTC. Thus we are able to analyze the effects of the external potentials on the soliton interactions. We treat separately the periodic, the quadratic and the quartic potentials, and for a number of cases demonstrate excellent matches between the PCTC and the relevant numeric solutions of the perturbed MM. More precisely we solve the vector Schrödinger equation numerically by using a Crank-Nicolson-like conservative difference scheme with inner iterations in complex arithmetic and plot the maxima of $(u^\dagger, u)$ as the trajectories of the solitons [7, 45–47]. The solutions of the relevant PCTC have been obtained using Maple. Finally, Sects. 6 and 7 contain discussion and conclusions.

2 Preliminaries

Here we briefly remind the derivation of the CTC as a model describing the N-soliton interactions of more general multicomponent NLS (MNLS) systems using the variational approach [2].

We apply it to the vector NLS equation (2) where $\boldsymbol{u}$ is an n-component vector-valued function and the scalar product $\langle \boldsymbol{u}, \boldsymbol{w}\rangle \equiv \sum_{s=1}^{n} \boldsymbol{u}_s^* \boldsymbol{w}_s$. For $n = 2$ Eqs. (2) and (3) go into the Manakov model (MM) and the perturbed MM respectively.

2.1 Derivation of the CTC as a Model for the Soliton Interaction of VNLS Systems

The vector NLS equation (2) allows Hamiltonian formulations with the Poisson brackets

$$\{\boldsymbol{u}_j(x,t), \boldsymbol{u}_k^*(y,t)\} = \delta_{jk}\delta(x-y) \tag{7}$$

and the Hamiltonian

$$H = \int_{-\infty}^{\infty} dx \left[\frac{1}{2}\langle \boldsymbol{u}_x, \boldsymbol{u}_x\rangle - \frac{1}{2}\langle \boldsymbol{u}^{\dagger}, \boldsymbol{u}\rangle^2 \right]. \tag{8}$$

It also admits Lagrangian formulation. Indeed, choose as the Lagrangian

$$\mathcal{L} = \int_{-\infty}^{\infty} dt \, \frac{i}{2} [\langle \boldsymbol{u}, \boldsymbol{u}_t\rangle - \langle \boldsymbol{u}_t, \boldsymbol{u}\rangle] - H. \tag{9}$$

Then the Lagrange equations of motion

$$\frac{d}{dt}\frac{\delta \mathcal{L}}{\delta \boldsymbol{u}_t^{\dagger}} - \frac{\delta \mathcal{L}}{\delta \boldsymbol{u}^{\dagger}} = 0, \tag{10}$$

will coincide with (2).

It is well known that the vector NLS equation is solvable by the inverse scattering method applied to a generalized Zakharov–Shabat system, see [1, 37]. This allows one to derive its N-soliton solutions.

In what follows we will analyze the large time behavior of the N-soliton train determined as the solution of the vector NLS by the initial condition (5) and (6). We remind that the polarization vectors $\boldsymbol{n}_k$ are normalized by $\langle \boldsymbol{n}_k, \boldsymbol{n}_k\rangle \equiv \sum_{s=1}^{n} \kappa_k^{(s)2} = 1$, and the phases ϕ_k are fixed uniquely by requesting that $\sum_{s=1}^{n} \gamma_k^{(s)} = 0$.

Note that the N-soliton train can not be considered as an N-soliton solution evaluated, say for $t = 0$. The N-soliton train is a generic solution of the vector NLS whose spectra contain, besides N pairs of eigenvalues $\tilde{\lambda}_k^{\pm}$, also a small but nontrivial contribution from the continuous spectrum of L.

Using the variational approach of [2] we insert the anzatz (5) into the Lagrangian and integrate over x. Skipping the details we obtain:

$$\mathcal{L} = \sum_{k=1}^{N} \mathcal{L}_k + \sum_{k=1}^{N} \sum_{n=k\pm 1} \tilde{\mathcal{L}}_{kn}, \tag{11}$$

$$\mathcal{L}_k = 4\nu_k \left(\frac{i}{2} \left(\langle \boldsymbol{n}_k, \boldsymbol{n}_{k,t} \rangle - \langle \boldsymbol{n}_{k,t}, \boldsymbol{n}_k \rangle \right) + 2\mu_k \frac{d\xi_k}{dt} - \frac{d\delta_k}{dt} - 2\mu_k^2 + \frac{2\nu_k^2}{3} \right), \tag{12}$$

$$\mathcal{L}_{kn} = 16\nu_0^3 e^{-\Delta_{kn}} (R_{kn} + R_{kn}^*) + \mathcal{O}(\epsilon^{3/2}), \tag{13}$$

$$R_{kn} = e^{i(\tilde{\delta}_n - \tilde{\delta}_k)} \langle \boldsymbol{n}_k, \boldsymbol{n}_n \rangle, \quad \tilde{\delta}_k = \delta_k - 2\mu_0 \xi_k, \quad \Delta_{kn} = 2s_{kn}\nu_0(\xi_k - \xi_n), \tag{14}$$

where $s_{k,k+1} = 1$ and $s_{k,k-1} = -1$.

All the integrals for obtaining $\mathcal{L}_k$ can be explicitly calculated, see [18]. As a result, we get terms of three different orders of magnitude: (i) terms of order $\Delta_{kn}^2 \exp(-\Delta_{kn})$; (ii) terms of order $\Delta_{kn} \exp(-\Delta_{kn})$ and (iii) terms of order $\exp(-\Delta_{kn})$. More detailed considerations show that the terms of types (i) and (ii) are multiplied by factors that are of the order of $\exp(-\Delta_{kn})$ due to the evolution equations for the soliton parameters. Finally, neglecting all the terms of order $\epsilon^{3/2}$ and higher, we arrive at the following set of dynamical equations for the soliton parameters:

$$\frac{d\xi_k}{dt} = 2\mu_k, \tag{15}$$

$$\frac{d\delta_k}{dt} = 2\mu_k^2 + 2\nu_k^2, \tag{16}$$

$$\frac{d\nu_k}{dt} = 8\nu_0^3 \sum_n e^{-\Delta_{kn}} i (R_{kn} - R_{kn}^*), \tag{17}$$

$$\frac{d\mu_k}{dt} = -8\nu_0^3 \sum_n e^{-\Delta_{kn}} (R_{kn} + R_{kn}^*), \tag{18}$$

$$\frac{d\boldsymbol{n}_k}{dt} = 4\nu_0^2 i \sum_{n=k\pm 1} e^{-\Delta_{kn}} \left[e^{i(\tilde{\delta}_n - \tilde{\delta}_k)} \boldsymbol{n}_n - R_{kn} \boldsymbol{n}_k + e^{i(\tilde{\delta}_n - \tilde{\delta}_k)} \boldsymbol{n}_n + R_{kn}^* \boldsymbol{n}_k \right] + C_k \boldsymbol{n}_k. \tag{19}$$

where the constants C_k are fixed up by the constraint $\langle \boldsymbol{n}_k, \boldsymbol{n}_k \rangle = 1$ for all t.

Next, we estimate the order of magnitude of the different terms in $\mathcal{L}$ and drop out those of them that are of order $\epsilon^{3/2}$ and higher.

Thus, from Eqs. (17) and (18) we get:

$$\frac{d(\mu_k + i\nu_k)}{dt} = 4\nu_0 \left[\langle \boldsymbol{n}_k, \boldsymbol{n}_{k-1} \rangle e^{q_k - q_{k-1}} - \langle \boldsymbol{n}_{k+1}, \boldsymbol{n}_k \rangle e^{q_{k+1} - q_k} \right], \tag{20}$$

where

$$\begin{aligned} q_k &= -2\nu_0 \xi_k + k \ln 4\nu_0^2 - i(\delta_k + \delta_0 + k\pi - 2\mu_0 \xi_k), \\ \nu_0 &= \frac{1}{N} \sum_{s=1}^{N} \nu_s, \qquad \mu_0 = \frac{1}{N} \sum_{s=1}^{N} \mu_s, \qquad \delta_0 = \frac{1}{N} \sum_{s=1}^{N} \delta_s. \end{aligned} \tag{21}$$

Besides, from (15) and (16) there follows (see [13]):

$$\frac{dq_k}{dt} = -4\nu_0(\mu_k + i\nu_k). \tag{22}$$

and

$$\frac{d^2 q_k}{dt^2} = 16\nu_0^2 \left[\langle \boldsymbol{n}_{k+1}, \boldsymbol{n}_k \rangle e^{q_{k+1} - q_k} - \langle \boldsymbol{n}_k, \boldsymbol{n}_{k-1} \rangle e^{q_k - q_{k-1}} \right], \tag{23}$$

which proves the statement in [8]. Equation (23), combined with the system of equations for the polarization vectors (19) provides the proper generalization of the CTC model for the MNLS.

The equations for the polarization vectors are nonlinear. So the whole system of equations for q_k and $\boldsymbol{n}_k$ seems to be nonintegrable. However, all terms in the right hand sides of the evolution equations for $\boldsymbol{n}_k$ are of the order of ϵ. This allows us to neglect in the leading approximation the evolution of $\boldsymbol{n}_k$ and to approximate them with their initial values. As a result we obtain that the N-soliton interactions for the vector NLS in the adiabatic approximation are modeled by the CTC , see Sect. 3.

It is easy to see, that if all $\langle \boldsymbol{n}_{k+1}, \boldsymbol{n}_k \rangle \neq 0$ then formally we can define

$$\tilde{q}_k = q_k + \sum_{s=1}^{k-1} \ln m_{0s}, \qquad m_{0s} = \langle \boldsymbol{n}_{k+1}, \boldsymbol{n}_k \rangle, \tag{24}$$

and rewrite the CTC (23) in its standard form:

$$\frac{d^2 \tilde{q}_k}{dt^2} = 16\nu_0^2 \left[e^{\tilde{q}_{k+1} - \tilde{q}_k} - e^{\tilde{q}_k - \tilde{q}_{k-1}} \right]. \tag{25}$$

Obviously, $\mathrm{Re}(\tilde{q}_{k+1} - \tilde{q}_k) = 2\nu_0(\xi_{k+1} - \xi_k) + \ln |m_{0k}|$, so the effect of $\ln |m_{0k}|$ can be viewed as changing the distance between the vector solitons.

Note also that the CTC models the soliton interactions for the VNLS with *any number of components*. The effect of the polarization vectors on the interaction comes into CTC only through the scalar products $m_{0s} = \langle \boldsymbol{n}_{k+1}, \boldsymbol{n}_k \rangle$. It is well known, that a gauge transformation $\boldsymbol{u} \to g_0 \boldsymbol{u}$ with any constant unitary g_0 leaves the VNLS, Eq. (2) invariant. Such transformation will change all polarization vectors simultaneously $\boldsymbol{n}_k \to g_0 \boldsymbol{n}_k$ but preserves their scalar products, and so will not influence the soliton interaction. Obviously, our CTC model is invariant under such transformations.

Our numerical checks for the validity of CTC and the perturbed CTC are done only for the Manakov model. Due to the above arguments our choice of the initial values of $\boldsymbol{n}_{k0}$ can be changed into $g_0 \boldsymbol{n}_{k0}$ with no effect on the interaction. That is why we specify only the scalar products m_{k0} for our runs, which we have chosen to be real.

3 Effects of the Polarization Vectors on the Soliton Interaction

In what follows we shall compare the CTC and the PCTC models with the numeric solutions of the corresponding (perturbed) MM. In doing this, to have better base for comparison we will keep fixed some of the initial parameters of the soliton trains:

$$\nu_k = \frac{1}{2}, \qquad \mu_k = 0, \qquad \delta_k = k\pi. \tag{26}$$

The other parameters may vary from run to run; their particular values will be specified in the captions of the figures. To avoid any confusion we mark the CTC and the PCTC solutions by solid lines, and the numeric solutions of the MM and the perturbed MM by dashed lines. Also, we plot the centers of solitons and track their trajectories.

Remark 1. The CTC and its perturbed versions PCTC are derived in the framework of the adiabatic approximation. Therefore they are expected to be adequate only up to times of the order of $1/\epsilon$. If the distance between neighboring solitons is 8, then $\epsilon \simeq 0.01$, so one might expect that the CTCT and the PCTC would be valid up to $t \simeq 100$. Rather surprisingly, see most of the figures below, we find that the models work well until $t \simeq 1{,}000$ or even longer. We also assume that $\xi_k < \xi_{k+1}$.

Remark 2. Note that all polarization vectors $\boldsymbol{n}_s$ are normalized by the condition $\langle \boldsymbol{n}_s{}^\dagger, \boldsymbol{n}_s \rangle = 1$. That means that all $|m_{0s}| = |\langle \boldsymbol{n}_{s+1}{}^\dagger, \boldsymbol{n}_s \rangle| \le 1$. The special case $|m_{0s}| = 1$ is obtained when all polarization vectors are equal: $\boldsymbol{n}_1 = \boldsymbol{n}_2 = \cdots = \boldsymbol{n}_N = \boldsymbol{n}_0$. Then by a simple gauge transformation we can always consider $\boldsymbol{n}_0 = (1, 0, \ldots, 0)^T$. Thus we conclude that the vector NLS with this special choice for all $\boldsymbol{n}_s$ becomes simply the scalar NLS, which has been already analyzed in details in [3, 11–14, 28].

To make it simple we assume that all $\boldsymbol{n}_s$ have only real components and that all m_{0s} are equal (say to 0.7), see Fig. 1. The left panel shows the soliton train of scalar

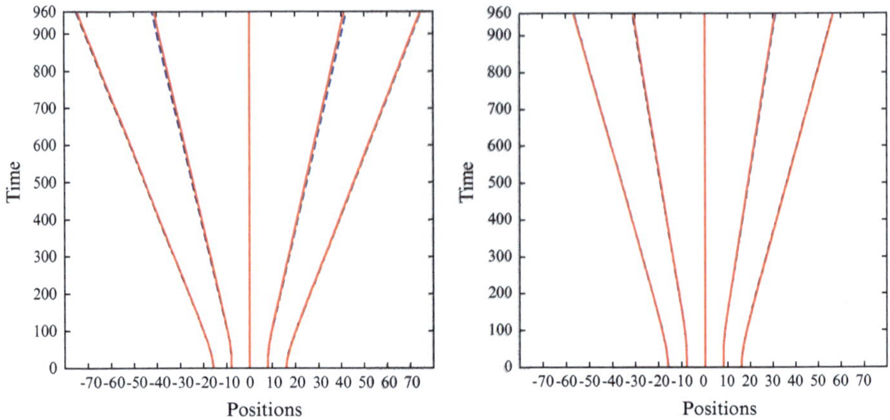

Fig. 1 The initial soliton parameters as like in (26) with $\xi_k = 8(k-3)$ – numerics of MM vs. CTC. *Left panel*: scalar 5-soliton train; *Right panel*: vector 5-soliton train $m_{0s} = 0.707$

solitons; the right panel obviously shows that the repulsion of the vector solitons is smaller. In this special case both the scalar and the vector soliton trains are in the same asymptotically free regime.

It was already noted above, that $|m_{0s}|$ effectively change the distance between the solitons. Since $|m_{0s}| < 1$, its effect will be as if the distance between the $(s+1)$-st and s-th solitons has increased by $-\ln|m_{0s}|$, which of course, makes the interaction between them weaker. Thus we may use $|m_{0s}|$ as 'switch' that turns on and off the soliton interactions. However, if we choose all $|m_{0s}| = 0$, i.e., all neighboring solitons have mutually orthogonal polarizations, then the CTC simplifies to a set of uncoupled linear equations:

$$\frac{\partial^2 q_k^{(0)}}{\partial t^2} = 0. \tag{27}$$

As a consequence we conclude, that the vector solitons generically interact weaker than the scalar solitons. This is illustrated by Fig. 1. The left panel shows 5-soliton train of the scalar NLS, while the right panel shows vector 5-soliton train with $|m_{0s}| = 0.707$. It is obvious that the spread of the scalar train at $t = 960$ is from -74.4 to 74.4; the spread of the vector train at $t = 960$ is from -56.6 to 56.6 which is roughly 75 % of the one for the scalar case.

However, if we choose m_{0s} to be different we may expect to encounter additional problems. Two or more of the solitons may come very close to each other which might violate the adiabatic condition, see Fig. 2. Therefore the first obvious move would be to assume that $|m_{0s}| - |m_{0s-1}| \simeq \sqrt{\epsilon}$ [10]. Contrary to our expectation we see, that even for very small values of $|m_{02}|$ the CTC gives rather adequate description of the soliton interactions, see Fig. 2. Only for $m_{02} = 0$ we see substantial discrepancies between the numerics and the CTC, see Fig. 3.

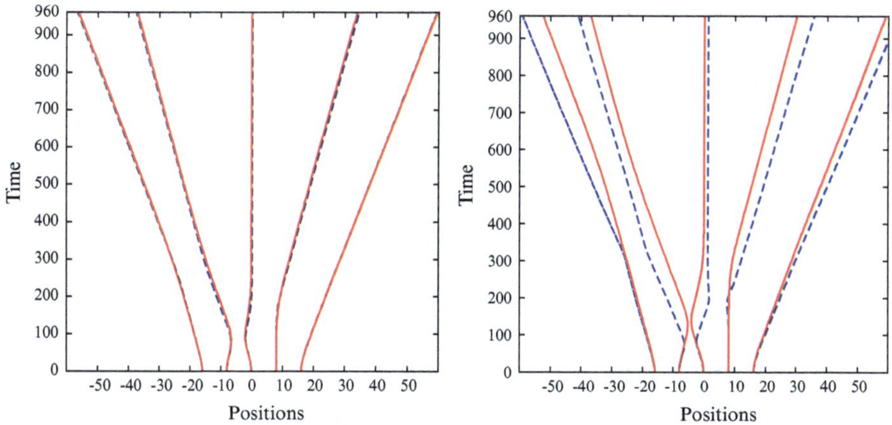

Fig. 2 Numerics of MM vs. CTC. *Left panel*: vector 5-soliton train with $m_{01} = m_{03} = m_{04} = 0.8$ and $m_{02} = 0.031$; *Right panel*: vector 5-soliton train with $m_{01} = m_{03} = m_{04} = 0.8$ and $m_{02} = 0.001$

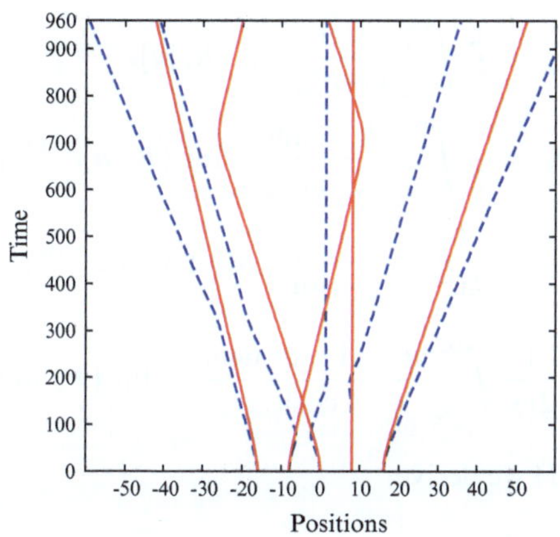

Fig. 3 Numerics of MM vs. CTC. Vector 5-soliton train with $m_{01} = m_{03} = m_{04} = 0.8$ and $m_{02} = 0.0$

4 Derivation of the PCTC as a Model for the Soliton Interactions

We will consider specific choices for the perturbations of the type $iR[u] \equiv V(x)u$. Such perturbed VNLS (or, the perturbed Manakov model) also allows Hamiltonian formulation with:

$$H_{\text{pert}} = H - \int_{-\infty}^{\infty} V(x)\langle \boldsymbol{u}^{\dagger}, \boldsymbol{u}\rangle \tag{28}$$

which means that we have to replace $\mathcal{L}_k$ in Eq. (12) by

$$\mathcal{L}_{k,\mathrm{pert}} = \mathcal{L}_k - 2\nu_k \int_{-\infty}^{\infty} \frac{dx\ V(x)}{\cosh^2(z_k)}. \tag{29}$$

while $\mathcal{L}_{kn}$ remains unchanged. Thus, in the adiabatic approximation the dynamics of the soliton parameters can be modeled by the following PCTC system:

$$\frac{d\lambda_k}{dt} = -4\nu_0 \left(e^{Q_{k+1}-Q_k} - e^{Q_k - Q_{k-1}} \right) + M_k + i N_k, \tag{30}$$

$$\frac{d\xi_k}{dt} = 2\mu_k + \Xi_k, \qquad \frac{d\delta_k}{dt} = 2(\mu_k^2 + \nu_k^2) + X_k, \tag{31}$$

where $\lambda_k = \mu_k + i\nu_k$ and $X_k = 2\mu_k \Xi_k + D_k$. To distinguish the different kinds of perturbations in this section we use the notation Q_k instead of q_k for the unperturbed situation. The right hand sides of Eqs. (30)–(31) are determined by $R_k[\boldsymbol{u}] \equiv iV(x)\boldsymbol{u}_k$ through:

$$N_k = \frac{1}{2} \int_{-\infty}^{\infty} \frac{dz_k}{\cosh z_k} \operatorname{Re} \left(R_k[\boldsymbol{u}] e^{-i\phi_k} \right), \tag{32}$$

$$M_k = \frac{1}{2} \int_{-\infty}^{\infty} \frac{dz_k\ \sinh z_k}{\cosh^2 z_k} \operatorname{Im} \left(R_k[\boldsymbol{u}] e^{-i\phi_k} \right), \tag{33}$$

$$\Xi_k = \frac{1}{4\nu_k^2} \int_{-\infty}^{\infty} \frac{dz_k\ z_k}{\cosh z_k} \operatorname{Re} \left(R_k[\boldsymbol{u}] e^{-i\phi_k} \right), \tag{34}$$

$$D_k = \frac{1}{2\nu_k} \int_{-\infty}^{\infty} \frac{dz_k\ (1 - z_k \tanh z_k)}{\cosh z_k} \operatorname{Im} \left(R_k[\boldsymbol{u}] e^{-i\phi_k} \right). \tag{35}$$

Using (30) and (31) we derive:

$$\begin{gathered} \frac{dQ_k}{dt} = -4\nu_0\lambda_k + \frac{2k}{\nu_0}\mathcal{N}_0 + 2i\xi_k \left(\mathcal{M}_0 + i\mathcal{N}_0 \right) + i\left(2\lambda_0 \Xi_k - X_k - \mathcal{X}_0 \right), \\ \mathcal{N}_0 = \frac{1}{N}\sum_{j=1}^{N} N_j, \qquad \mathcal{M}_0 = \frac{1}{N}\sum_{j=1}^{N} M_j, \qquad \mathcal{X}_0 = \frac{1}{N}\sum_{j=1}^{N} X_j. \end{gathered} \tag{36}$$

In deriving Eq. (36) we have kept terms of the order $\Delta\nu_k \simeq \mathcal{O}(\sqrt{\epsilon_0})$ and neglected terms of the order higher than $\mathcal{O}(\epsilon_0)$. These perturbations result in that ν_0 and μ_0 may become time-dependent. Indeed, from (30) we get:

$$\frac{d\mu_0}{dt} = \mathcal{M}_0, \qquad \frac{d\nu_0}{dt} = \mathcal{N}_0. \tag{37}$$

The small parameter ϵ_0 can be related to the initial distance $r_0 = |\xi_2 - \xi_1|_{t=0}$ between the two solitons. Assuming $\nu_{1,2} \simeq \nu_0$ we find:

$$\epsilon_0 = \int_{-\infty}^{\infty} dx \left| u_1^{1s}(x,0) u_2^{1s}(x,0) \right| \simeq 8\nu_0 r_0 e^{-2\nu_0 r_0}. \tag{38}$$

In particular, Eq. (38) means that $\epsilon_0 \simeq 0.01$ for $r_0 \simeq 8$ and $\nu_0 = 1/2$.

We assume that the initial configurations of solitons is ordered in such a way that $\xi_{k+1} - \xi_k \simeq r_0$. One can check [11, 13] that $N_k \simeq M_k \simeq \exp(-2\nu_0|k-p|r_0)$. Therefore the interaction terms between the k-th and $(k \pm 1)$-st solitons will be of the order of $\exp(-2\nu_0 r_0)$; the interaction terms between k-th and $(k \pm 2)$-nd soliton will be of the order of $\exp(-4\nu_0 r_0) \ll \exp(-2\nu_0 r_0)$ and must be neglected.

The terms $\Xi_k^{(0)}$, $X_k^{(0)}$ are of the order of $r_0^a \exp(-2\nu_0 r_0)$, where $a = 0$ or 1. However, they can be neglected as compared to $\tilde{\mu}_k$ and $\tilde{\nu}_k$, where

$$\tilde{\mu}_k = \mu_k - \mu_0 \simeq \sqrt{\epsilon_0}, \qquad \tilde{\nu}_k = \nu_k - \nu_0 \simeq \sqrt{\epsilon_0}. \tag{39}$$

It remains to insert into the above formulae the special choice of the perturbation (4), where the potential $V(x)$ is taken in the form:

$$V(x) = \sum_{k=0}^{4} V_k x^k + A\cos(\Omega x + \Omega_0). \tag{40}$$

Thus, in order to derive the relevant PCTC model we need to evaluate several integrals of the type shown in the Appendix. In order to simplify the expressions we also put $\nu_k = \nu_0 = 1/2$ and thus obtain:

$$\begin{aligned}
N_k[u] &- 0, \qquad \Xi_k[u] = 0, \\
M_k[u] &= -\frac{1}{4\nu_k}\left(V_1 + 2V_2\xi_k + 3V_3\xi_k^2 + 4V_4\xi_k^3\right) - \frac{\pi^2}{32\nu_k^2}(V_3 + 3V_4\xi_k) \\
&\quad + \frac{\pi A\Omega^2}{8\nu_k \sinh Z_k}\sin(\Omega\xi_k + \Omega_0), \\
D_k[u] &= -\frac{1}{2}V(\xi_k) + \frac{\pi^2}{96\nu_k^2}\left(V_2 + 3V_3\xi_k + 6V_4\xi_k^2\right) + \frac{7\pi^4}{169\cdot 16\nu_k^4}V_4 \\
&\quad - \frac{\pi^2 A\Omega^2}{16\nu_k^2}\frac{\cosh Z_k}{\sinh^2 Z_k}\cos(\Omega\xi_k + \Omega_0),
\end{aligned} \tag{41}$$

where $Z_k = \Omega\pi/(4\nu_k)$.

The corrections to $N_k, \ldots,$ coming from the terms linear in u depend only on the parameters of the k-th soliton; i.e., they are 'local' in k.

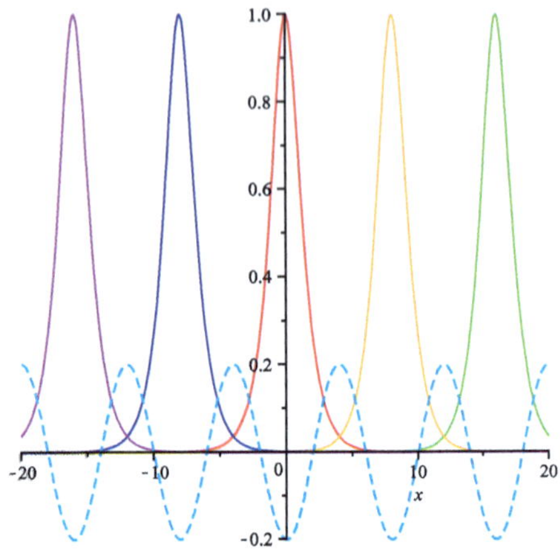

Fig. 4 Five-soliton train with periodic potential, $r_0 = 8$, $\Omega = \pi/4$ with intensity $A = -0.02$

5 Comparison of the PCTC with the Perturbed Manakov Model

5.1 The Periodic Potentials

In this Subsection we consider periodic potentials of the form:

$$V(x) = A\cos(\Omega x), \tag{42}$$

and impose special initial conditions for the vector soliton train, such that the centers of the solitons are located at the minima of the potential, i.e. $r_0 = 2\pi/\Omega$, see Fig. 4.

Each soliton of the train experiences confining force of the periodic potential and repulsive force of neighboring solitons. Therefore, one does not expect the equilibrium positions of solitons to coincide with the minima; instead they perform small amplitude oscillations around these minima, provided that the strength of the potential is big enough to keep the solitons confined. If the potential is weak then it is unable to confine solitons, and the repulsive forces between neighboring solitons (at phase difference π) induce unbounded expansion of the train.

Such behavior is observed on the left panel of Fig. 5 where the intensity $A = -0.0001$ is not strong enough to confine the solitons. On the right panel of the same Fig. 5 the value of $A = -0.0005$ exceeds the first critical value and the three central solitons are confined, while the two lateral solitons 'run away'.

On Fig. 6 the values of A ($A = -0.0025$ for the left panel and $A = -0.0075$ for the right panel) are greater than the second critical value and as a result all five solitons form a stable bound state.

Thus we have demonstrated that the periodic potential plays stabilizing role for the vector soliton trains. Obviously similar results hold also for trains with arbitrary number of solitons.

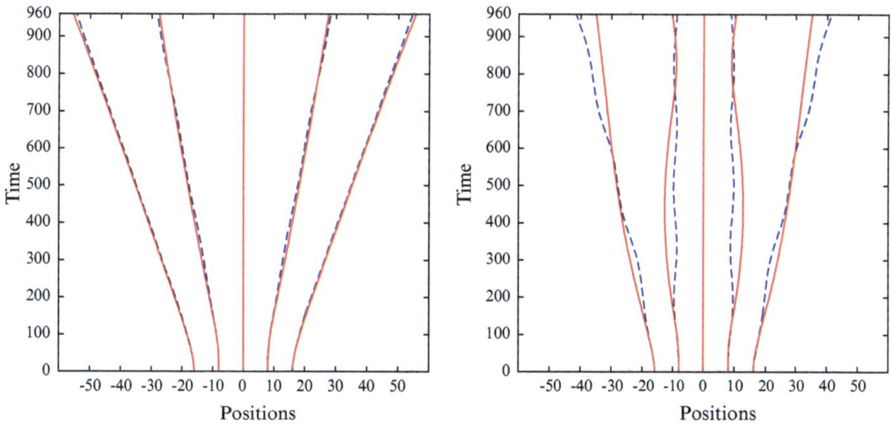

Fig. 5 Periodic potential on 5-soliton trains with $r_0 = 8$, $\Omega = \pi/4$ with subcritical intensity $A = -0.0001$ (*left panel*) and $A = -0.0005$ (*right panel*)

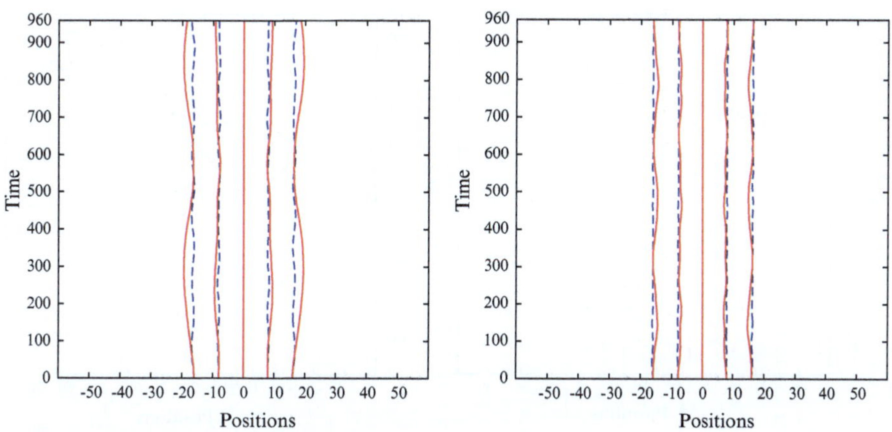

Fig. 6 Periodic potential on 5-soliton trains with $r_0 = 8$, $\Omega = \pi/4$ with supercritical intensity $A = -0.0025$ (*left panel*) and $A = -0.0075$ (*right panel*)

5.2 The Quadratic Potentials

Next we consider quadratic potentials of the form:

$$V(x) = a(x - w)^2. \tag{43}$$

Its effect on the vector N-soliton train with parameters (26) is to balance the repulsive interaction between the solitons, so that they remain bounded by the potential, as illustrated in the right panels of the Figs. 7 and 8. The quadratic

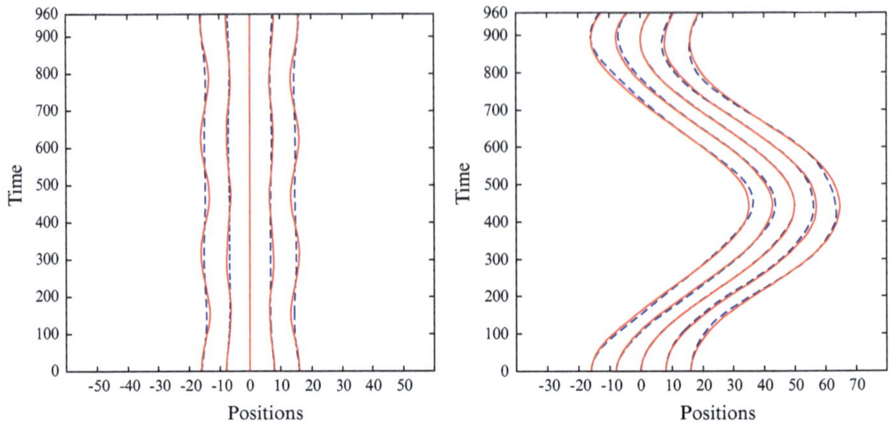

Fig. 7 Five-soliton train in a quadratic potential (43) with $m_{0s} = 0.309$, $a = 0.000025$ and $w = 0$ (*left panel*) and $w = 25$ (*right panel*)

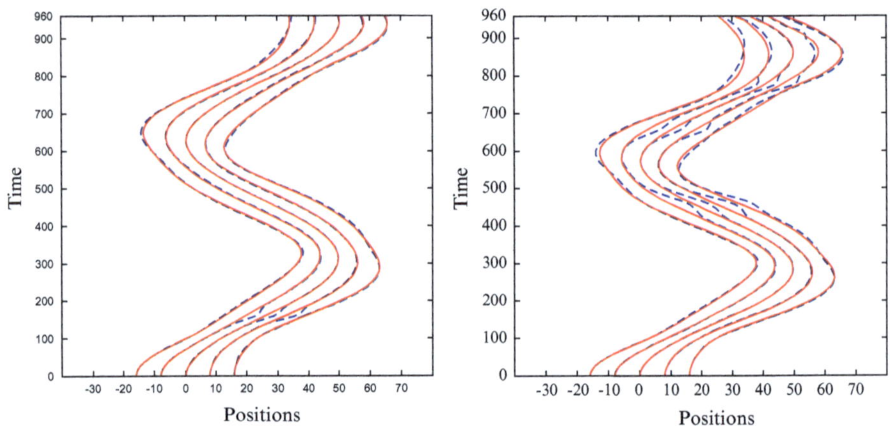

Fig. 8 Five-soliton train in a quadratic potential (43) with $m_{0s} = 0.309$, $a = 0.00005$ and $w = 25$ (*left panel*) and $a = 0.00006$, $w = 25$ (*right panel*)

potentials are supposed to be weak, i.e., we choose a so that

$$a\xi_N^2(0) \leq \nu_0, \qquad a\xi_1^2(0) \leq \nu_0. \tag{44}$$

It is reasonable to assume also that $a \simeq \mathcal{O}(\epsilon_0/N)$; this ensures the possibility to have the N-soliton train 'inside' the potential. It also means that both the exponential terms and the correction terms M_k are of the same order of magnitude. From Eqs. (41) there follows that $d\nu_0/dt = 0$ and:

$$\frac{d\mu_0}{dt} = -a\xi_0 - aw, \qquad \frac{d\xi_0}{dt} = 2\mu_0, \tag{45}$$

where μ_0 is the average velocity and $\xi_0 = \frac{1}{N}\sum_{j=1}^{N} \xi_j$, is the center of mass of the N-soliton train . The natural initial conditions are

$$\mu_0(0) = 0, \qquad \xi_0(0) = \xi_{00}. \tag{46}$$

The system of equations (45) for $a > 0$ has a simple solution

$$\begin{aligned} \mu_0(t) &= -\sqrt{\frac{a}{2}}(w + \xi_{00}) \sin(\sqrt{2a}t), \\ \xi_0(t) &= (w + \xi_{00}) \cos(\sqrt{2a}t) - w, \end{aligned} \tag{47}$$

ξ_{00} is the initial position of the center of mass of the soliton train with vanishing average initial velocity. Therefore the overall effect of such quadratic potential will be to induce a slow periodic motion of the train as a whole.

Figures 7 and 8 show good agreement between the PCTC model and the numerical solution of the perturbed VNLS equation (2). They also show two types of effects of the quadratic potential on the motion of the N-soliton train: (i) the train performs contracting and expanding oscillations if its center of mass coincides with the minimum of the potential, (ii) the train oscillates around the minimum of the potential as a whole if its center of mass is shifted. In the last case contracting and expanding motions of the soliton train are superimposed to the center of mass dynamics. As one can see on the figures the period of this motion matches very well the one predicted by formula (47). Indeed, from Eq. (47) it follows that the period of the center of mass motion is $T = 2\pi/\sqrt{2a}$. For the parameters on Fig. 7 we have $T \simeq 888.6$ and $T \simeq 628.3$ – on Fig. 8 in good agreement with the numerical simulations.

The direct simulations of the VNLS equation (2) show that stronger parabolic trap may cause merging of individual solitons at times of contraction, and restoring of the original configuration when the train is expanded. This behavior reminds the phenomenon of "missing solitons" observed in the experiment [44]. However, this situation is beyond the validity of the PCTC approach.

5.3 *The Quartic Potentials*

The periodic and harmonic potentials considered in the previous two subsections are among the simplest ones that are important for BEC. One can consider, however, more complicated potentials which could lead to richer dynamics of the solitons. The idea for the quartic potentials

$$W(x) = a(x^2 - w_0^2)(x^2 - w_1^2) \tag{48}$$

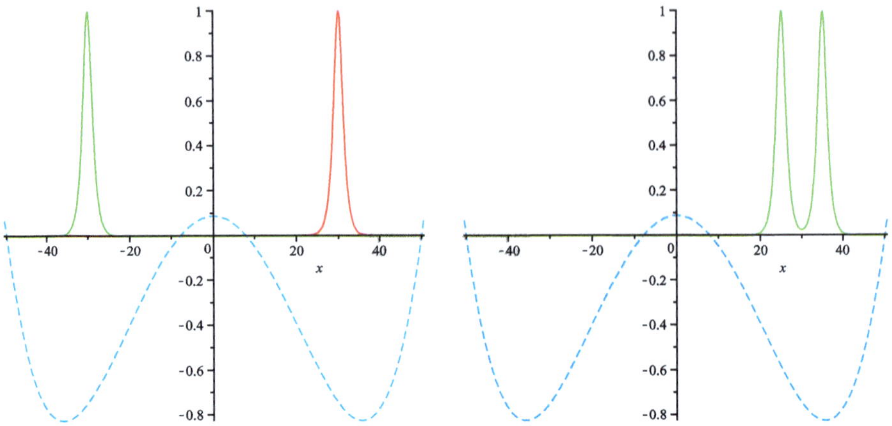

Fig. 9 Two solitons in a quartic potentials $W(x)$ with $a = 1/1\,600{,}000$, $w_0 = 8$, $w_1 = 50$, with soliton centers located at -30.0 and 30.0 (*left panel*) and at 25.0 and 35.0 (*right panel*)

with two minima at $x^{\pm}_{\rm min} = \pm\sqrt{(w_0^2 + w_1^2)/2}$ was suggested to us by R. Goodman's paper [24]. Such anharmonic potential is much stronger than the harmonic one. To be able to treat it as perturbation we must impose conditions similar to (44):

$$a\xi_N^2(0) \leq \nu_0, \qquad a\xi_1^2(0) \leq \nu_0. \tag{49}$$

Unlike the two-well potentials considered in [24] the quartic potentials (48) grow quickly with x and confine the soliton trains around the two wells. Now we can consider two types of soliton trains: one, that is initially located inside the well around $x^{-}_{\rm min}$ and another one located around $x^{+}_{\rm min}$. One may consider also two-soliton trains, one at each of the minima. In all considerations below we choose a to be large enough so that the conditions (49) to be satisfied and fix up $w_0 = 8$ and $w_1 = 50$. The latter ensures enough space to insert up to 2-soliton trains in each of the wells (see the simplest possible initial configurations on Fig. 9).

In the next two figures we compare the predictions of the PCTC with the numerical solutions of the perturbed MM for 2-soliton trains located at one of the wells of the potential. Figure 10 shows that one of the solitons may periodically 'jump over' the barrier between the two wells, while the second one slightly oscillates inside the same well and does not leave it.

The left panel of Fig. 11 shows a 2-soliton train oscillating inside the right well. Due to the fact that here we have chosen a to be large enough we have very good match between the PCTC and the numerics of PMM. Of course, this match is much worse when we choose the solitons to have orthogonal polarizations. As we mentioned above in this case the PCTC requires that each soliton is driven by the external potential only. So we have the solid lines that intersect each other. The direct numerical simulation of MM, however, shows that the trajectories of the solitons do

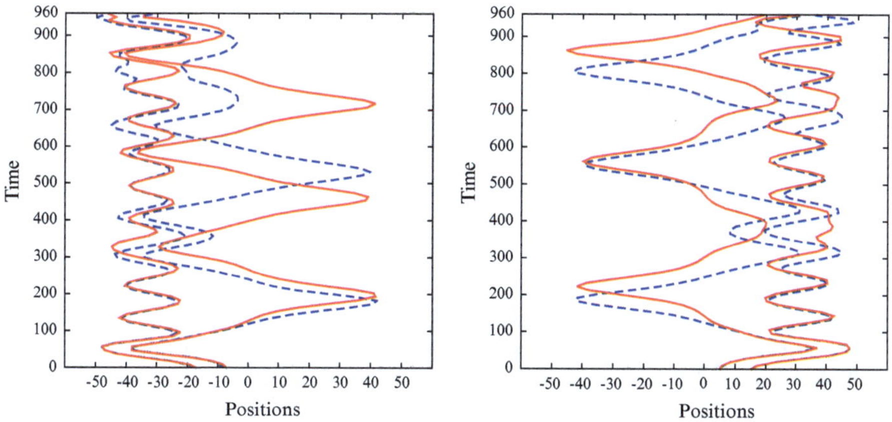

Fig. 10 Two-soliton train in quartic potential (48) with $a = 1/1\,600{,}000$, $w_0 = 8$, $w_1 = 50$, $m_{0s} = 0.951$, $\xi_k = -17.0, -7.0$ (*left panel*) and $\xi_k = 5.0, 15.0$ (*right panel*)

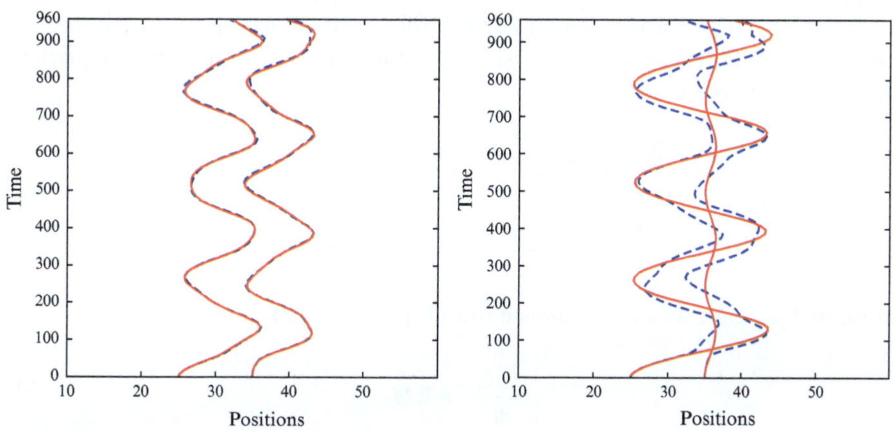

Fig. 11 Two-soliton train in quartic potential (48) with $a = 1/16{,}000{,}000$, $w_0 = 8$, $w_1 = 50$, $m_{0s} = 0.951$, $\xi_k = 25.0, 35.0$ (*left panel*) and $m_{0s} = 0.0$, $\xi_k = 25.0, 35.0$ (*right panel*)

not intersect. Even in this case we see qualitative match between the two models except at regions around the intersections.

The last Fig. 12, illustrates the fact, that the PCTC can be used only if the quartic potential can be viewed as a perturbation. If the potential is strong and the threshold condition (49) is violated the numeric simulation of the PMM is very close to chaos.

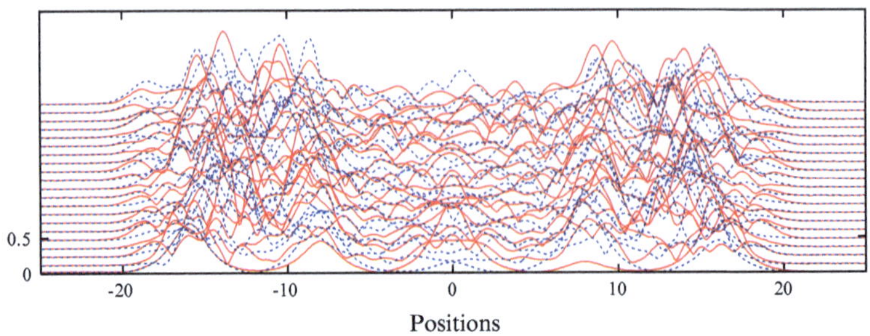

Fig. 12 Five-soliton trains with stronger quartic potential are quickly destroyed and turned into chaos; the initial train configuration is plotted at the lower-most stage. The final time is 960 (upper-most stage), $m_{0s} = 0.309$, $w_0 = 8$, $w_1 = 64$, $a = 1/10{,}000$, $\xi_k = 0.0; \pm 16.0; \pm 8.0$

6 Discussion

Here we would also outline possible further investigations. An important question concerns the vector solitons with nearly orthogonal polarizations. In the limit of $m_{0s} \to 0$ the PCTC system for N solitons splits into N disjoint sets of equations for the parameters of each soliton:

$$\begin{aligned} \frac{\partial \mu_k}{\partial t} &= M_k, & \frac{\partial \nu_k}{\partial t} &= 0, \\ \frac{\partial \xi_k}{\partial t} &= 2\mu_k, & \frac{\partial \delta_k}{\partial t} &= 4\nu_0 \nu_k + D_k, \end{aligned} \tag{50}$$

for $k = 1, \dots, N$. As a consequence from Eq. (50) we get

$$\frac{\partial^2 \xi_k}{\partial t^2} = 2M_k. \tag{51}$$

Thus the external potential onto soliton train with pair-wise orthogonal polarizations drives each of the solitons separately. Note that M_k depends only on ξ_k and ν_k. But, since ν_k are time independent (see Eq. (50)) we can replace them by their initial values. In this way Eq. (51) reduces a second order nonlinear equation for ξ_k, which obviously admits $d\xi_k/dt$ as an integration multiplier and can be integrated. More details about the solutions of these equations for various special choices of $V(x)$ will be published elsewhere. Here we just note that:

- The trajectory of each soliton is fixed up by its initial condition ξ_{k0};
- The different solitons do not interact between themselves and their trajectories can intersect, see for example the solid lines on the right panel of Fig. 11.

However, the numerical simulation of the vector NLS for such train *does not* show intersection of soliton trajectories – the dashed lines on the right panel of Fig. 11.

Notice, that even in this case the PCTC and the numerical solution for the VNLS practically coincide except in small regions around the 'intersection points'.

Of course we have investigated only a few possible types of soliton trains of the perturbed MM. Full examination requires a lot more studies which can not be squeezed into one or few papers.

Similar results for the second type VNLS responsible for the spin-1 and spin-2 BEC are much harder to obtain. The difficulty is that their one-soliton solutions are more complicated and therefore the evaluation of the relevant integrals providing $\mathcal{L}_{kn}$ in Eq. (11) has not been done yet.

7 Conclusions

We showed that the PCTC describes adequately the behavior of the vector N-soliton trains in several types of small intensity external potentials: periodic, harmonic and anharmonic ones. Obviously with the growth of the intensity of the potentials they can not be viewed any more as perturbations. Then the adiabatic conditions are violated and as a result the initial N-soliton train turns into a chaotic mode, see Fig. 12.

In studying the periodic potentials we used a special but important configuration of a 5-soliton trains when the center of each soliton initially is located at a minimum of the potential. One can study also, say 3- or 4-soliton trains obtaining similar results.

It is also natural to expect that the quartic anharmonic potential will influence substantially the solitons even when their intensity a is of the order of 10^{-6} or less.

Acknowledgements This paper is dedicated to the late Professor Evgeni Doktorov with whom this topic was started. One of us (V.S.G.) is grateful also to Professor Jesus Cuevas Maraver and to the Organizing Committee of LENCOS-2 for the chance to participate in the conference. We are obliged to Professor Anca Visinescu and Dr. Assen Kyuldjiev for useful discussions and help, Finally, we thank the anonymous referees for careful reading of the manuscript.

This work is supported in part by the National Science Foundation of the Bulgarian Ministry of Youth, Education and Science under grant DDVU02/71.

Appendix

Several Important Integrals

Here we list several important integrals, needed in calculating the additional terms in the PCTC.

Table 1 Values of the integrals

p	0	1	2	3	4	5	6
$\mathcal{J}_p$	1	0	$\frac{\pi^2}{12}$	0	$\frac{7\pi^4}{240}$	0	$\frac{31\pi^6}{1,344}$
$\mathcal{K}_p$	0	$\frac{1}{2}$	0	$\frac{\pi^2}{8}$	0	$\frac{7\pi^4}{96}$	0
$\mathcal{L}_p$	$\frac{1}{2}$	0	$-\frac{\pi^2}{24}$	0	$\frac{-7\pi^4}{160}$	0	$-\frac{155\pi^6}{2,688}$

$$\begin{aligned} \mathcal{J}_2(a) &= \int_{-\infty}^{\infty} \frac{dz\, e^{iaz}}{2\cosh^2 z} = \frac{\pi a}{2\sinh(\pi a/2)}, \\ \mathcal{K}_2(a) &= \int_{-\infty}^{\infty} \frac{dz\, \sinh z\, e^{iaz}}{2\cosh^2 z} = \frac{\pi a}{2\sinh(\pi a/2)}, \\ &\int_{-\infty}^{\infty} \frac{dz z^s e^{iaz}}{2\cosh^2 z} = \lim_{a\to 0} \frac{1}{i^s} \frac{\partial^s \mathcal{J}_2(a)}{\partial a^s} \end{aligned} \tag{52}$$

and as consequence, putting

$$\begin{aligned} \mathcal{J}_p &= \int_{-\infty}^{\infty} \frac{dz z^p}{2\cosh^2 z}, \quad \mathcal{K}_p = \int_{-\infty}^{\infty} \frac{dz z^p \sinh z}{2\cosh^2 z}, \\ \mathcal{L}_p &= \int_{-\infty}^{\infty} \frac{dz z^p (1 - z\tanh z)}{2\cosh^2 z}, \end{aligned} \tag{53}$$

we get the results in Table 1.

We need them to calculate the effects of the external potentials.

References

1. M.J. Ablowitz, B. Prinari, A.D. Trubatch, *Discrete and Continuous Nonlinear Schrödinger Systems*. London Mathematical Society Lecture Notes Series, vol. 302 (Cambridge University Press, Cambridge/New York, 2004)
2. D. Anderson, M. Lisak, Nonlinear asymmetric self-phase modulation and self-steepening of pulses in long optical waveguides. Phys. Rev. A **27**, 1393–1398 (1983); D. Anderson, M. Lisak, T. Reichel, Approximate analytical approaches to nonlinear pulse propagation in optical fibers: a comparison. Phys. Rev. A **38**, 1618–1620 (1988)
3. J.M. Arnold, Complex Toda lattice and its application to the theory of interacting optical solitons. J. Opt. Soc. Am. A **15**, 1450–1458 (1998); 979–986 (1999); J.M. Arnold, Stability of solitary wave trains in Hamiltonian wave systems. Phys. Rev. E **60**, 979–986 (1999)
4. V.A. Atanasov, V.S. Gerdjikov, G.G. Grahovski, N.A. Kostov, Fordy-Kulish models and spinor Bose-Einstein condensates. J. Nonlinear Math. Phys. **15**(3), 29-1-0298 (2008). ArXiv:0802.4405 [nlin.SI]

5. S.M. Baker, J.N. Elgin, J. Gibbons, Polarization dynamics of solitons in birefringent fibers. Phys. Rev. E **62**, 4325–4332 (1999)
6. Th. Busch, J.R. Anglin, Dark-bright solitons in inhomogeneous Bose-Einstein condensates. Phys. Rev. Lett. **87**, 010401 (2001)
7. C.I. Christov, S. Dost, G.A. Maugin, Inelasticity of soliton collisions in systems of coupled NLS equations. Phys. Scr. **50**, 449–454 (1994)
8. V.S. Gerdjikov, N-soliton interactions, the Complex Toda Chain and stability of NLS soliton trains, in *Proceedings of the International Symposium on Electromagnetic Theory*, Aristotle University of Thessaloniki, vol. 1, ed. by E. Kriezis (1998), pp. 307–309
9. V.S. Gerdjikov, On soliton interactions of vector nonlinear Schrödinger equations, in *AMiTaNS'11*, Albena, vol. 1404, ed. by M.D. Todorov, C.I. Christov (AIP CP, Melville, 2011), pp. 57–67
10. V.S. Gerdjikov, Modeling soliton interactions of the perturbed vector nonlinear Schrödinger equation. Bulg. J. Phys. **38**, 274–283 (2011)
11. V.S. Gerdjikov, I.M. Uzunov, Adiabatic and non-adiabatic soliton interactions in nonlinear optics. Physica D **152–153**, 355–362 (2001)
12. V.S. Gerdjikov, D.J. Kaup, I.M. Uzunov, E.G. Evstatiev, Asymptotic behavior of N-Soliton trains of the nonlinear Schrödinger equation. Phys. Rev. Lett. **77**, 3943–3946 (1996)
13. V.S. Gerdjikov, I.M. Uzunov, E.G. Evstatiev, G.L. Diankov, Nonlinear Schrödinger equation and N–soliton interactions: generalized Karpman–Soloviev approach and the Complex Toda Chain. Phys. Rev. E **55**(5), 6039–6060 (1997)
14. V.S. Gerdjikov, E.G. Evstatiev, D.J. Kaup, G.L. Diankov, I.M. Uzunov, Stability and quasi-equidistant propagation of NLS soliton trains. Phys. Lett. A **241**, 323–328 (1998)
15. V.S. Gerdjikov, B.B. Baizakov, M. Salerno, Modelling adiabatic N-soliton interactions and perturbations. Theor. Math. Phys. **144**(2), 1138–1146 (2005)
16. V.S. Gerdjikov, B.B. Baizakov, M. Salerno, Modelling adiabatic N-soliton interactions and perturbations. Teor. i Matem. Fizika **144**(2), 302–312 (2005) (In Russian); English translation: Theor. Math. Phys. **144**(2), 1138–1146 (2005)
17. V.S. Gerdjikov, B.B. Baizakov, M. Salerno, N.A. Kostov, Adiabatic N-soliton interactions of Bose-Einstein condensates in external potentials. Phys. Rev. E. **73**, 046606 (2006)
18. V.S. Gerdjikov, E.V. Doktorov, N.P. Matsuka, N-soliton train and generalized Complex Toda Chain for Manakov system. Theor. Math. Phys. **151**(3), 762–773 (2007)
19. V.S. Gerdjikov, G. Vilasi, A.B. Yanovski, *Integrable Hamiltonian Hierarchies. Spectral and Geometric Methods*. Lecture Notes in Physics, vol. 748 (Springer, Berlin/Heidelberg/New York, 2008). ISBN: 978-3-540-77054-1
20. V.S. Gerdjikov, N.A. Kostov, E.V. Doktorov, N.P. Matsuka, Generalized Perturbed Complex Toda Chain for Manakov System and exact solutions of the Bose-Einstein mixtures. Math. Comput. Simul. **80**, 112–119 (2009). doi:10.1016/j.matcom.2009.06.013. Special Issue of the Journal: Nonlinear Waves: Computation and Theory
21. V.S. Gerdjikov, N.A. Kostov, T.I. Valchev, Solutions of multi-component NLS models and spinor Bose-Einstein condensates. Physica D **238**, 1306–1310 (2009). ArXiv:0802.4398 [nlin.SI]
22. V.S. Gerdjikov, N.A. Kostov, T.I. Valchev, Bose-Einstein condensates with $F = 1$ and $F = 2$. Reductions and soliton interactions of multi-component NLS models, in *International Conference on Ultrafast and Nonlinear Optics 2009, Proceedings of SPIE*, Burgas, vol. 7501, 7501W, ed. by S.M. Saltiel, A.A. Dreischuh, I.P. Christov (2009). arXiv: 1001.0168 [nlin.SI]
23. V.S. Gerdjikov, N.A. Kostov, T.I. Valchev, Bose-Einstein condensates with $F = 1$ and $F = 2$. Reductions and soliton interactions of multi-component NLS models, in *International Conference on Ultrafast and Nonlinear Optics 2009, Proceedings of SPIE*, Burgas, vol. 7501, 7501W, ed. by S.M. Saltiel, A.A. Dreischuh, I.P. Christov (2009). arXiv: 1001.0168 [nlin.SI]
24. R. Goodman, Hamiltonian Hopf bifurcations and dynamics of NLS/GP standing-wave modes J. Phys. A Math. Theor. **44**, 425101 (28pp) (2011). doi:10.1088/1751-8113/44/42/425101
25. A. Griffin, T. Nikuni, E. Zaremba, *Bose-Condensed Gases at Finite Temperatures* (Cambridge University Press, Cambridge, 2009)

26. T.-L. Ho, Spinor Bose condensates in optical traps. Phys. Rev. Lett. **81**, 742 (1998)
27. J. Ieda, T. Miyakawa, M. Wadati, Exact analysis of soliton dynamics in spinor Bose-Einstein condensates. Phys. Rev. Lett. **93**, 194102 (2004)
28. V.I. Karpman, V.V. Solov'ev, A perturbational approach to the two-solition systems. Physica D **3**, 487–502 (1981)
29. P.G. Kevrekidis, D.J. Frantzeskakis, R. Carretero-Gonzalez (eds.), *Emergent Nonlinear Phenomena in Bose-Einstein Condensates: Theory and Experiment*, vol. 45 (Springer, Berlin/New York, 2008)
30. N.A. Kostov, V.S. Gerdjikov, On the soliton interactions in NLS equation with external potentials, in *14-th International School on Quantum Electronics: Laser Physics and Applications. Proceedings of SPIE*, Sunny Beach, Bulgaria, vol. 6604, 66041S, ed. by P.A. Atanasov, T.N. Dreischuh, S.V. Gateva, L.M. Kovachev (2007). doi: 10.1117/12.730712
31. N.A. Kostov, V.Z. Enol'skii, V.S. Gerdjikov, V.V. Konotop, M. Salerno, On two-component Bose-Einstein condensates in periodic potential. Phys. Rev. E. **70**, 056617 (2004)
32. N.A. Kostov, V.A. Atanasov, V.S. Gerdjikov, G.G. Grahovski, On the soliton solutions of the spinor Bose-Einstein condensate, in *14-th International School on Quantum Electronics: Laser Physics and Applications, Proceedings of SPIE*, Sunny Beach, Bulgaria, vol. 6604, 66041T, ed. by P.A. Atanasov, T.N. Dreischuh, S.V. Gateva, L.M. Kovachev (2007). doi: 10.1117/12.730712
33. N.A. Kostov, V.S. Gerdjikov, T.I. Valchev, Exact solutions for equations of Bose-Fermi mixtures in one-dimensional optical lattice, in *SIGMA 3* paper 071, 14p (2007). nlin.SI/0703057. http://www.emis.de/journals/SIGMA/
34. P.P. Kulish, E.K. Sklyanin, $O(N)$-invariant nonlinear Schrodinger equation – a new completely integrable system. Phys. Lett. **84A**, 349–352 (1981)
35. T.I. Lakoba, D.J. Kaup, Perturbation theory for the Manakov soliton and its applications to pulse propagation in randomly birefringent fibers. Phys. Rev. E **56**, 6147–6165 (1997)
36. B.A. Malomed, Variational methods in nonlinear fiber optics and related fields. Prog. Opt. **43**, 71–193 (2002)
37. S.V. Manakov, On the theory of two-dimensional stationary self-focusing of electromagnetic waves. Zh. Eksp. Teor. Fiz. **65**, 1392 (1973) (in Russian). (English translation) Sov. Phys. JETP **38**, 248 (1974)
38. M. Modugno, F. Dalfovo, C. Fort, P. Maddaloni, F. Minardi, Dynamics of two colliding Bose-Einstein condensates in an elongated magnetostatic trap. Phys. Rev. A **62**, 063607 (2000)
39. S.P. Novikov, S.V. Manakov, L.P. Pitaevski, V.E. Zakharov, *Theory of Solitons, the Inverse Scattering Method* (Consultant Bureau, New York, 1984)
40. T. Ohmi, K. Machida, Bose-Einstein condensation with internal degrees of freedom in alkali atom gases. J. Phys. Soc. Jpn. **67**, 1822 (1998)
41. V.M. Perez-Garcia, H. Michinel, J.I. Cirac, M. Lewenstein, P. Zoller, Dynamics of Bose-Einstein condensates: variational solutions of the Gross-Pitaevskii equations. Phys. Rev. A **56**, 1424–1432 (1997)
42. L. Pitaevskii, S. Stringari, *Bose-Einstein Condensation* (Oxford University Press, Oxford, 2003)
43. V.S. Shchesnovich, E.V. Doktorov, Perturbation theory for solitons of the Manakov system. Phys. Rev. E **55**, 7626 (1997)
44. K.E. Strecker, G.B. Partridge, A.G. Truscott, R.G. Hulet, Formation and propagation of matter-wave soliton trains. Nature **417**, 150–153 (2002); K.E. Strecker et al., Tunable interactions in ultracold Bose gases. Adv. Space Res. **35**, 78–81 (2005)
45. M.D. Todorov, C.I. Christov, Conservative numerical scheme in complex arithmetic for coupled nonlinear Schrodinger equations. Discret. Contin. Dyn. Syst.- Suppl. 2007, 982–992 (2007)
46. M.D. Todorov, C.I. Christov, Impact of the large cross-modulation parameter on the collision dynamics of quasi-particles governed by vector NLSE. Math. Comput. Simul. **80**, 46–55 (2008)
47. M.D. Todorov, C.I. Christov, Collision dynamics of elliptically polarized solitons in coupled nonlinear Schrödinger equations. Math. Comput. Simul. **82**, 1221–1232 (2012)

48. M. Uchiyama, J. Ieda, M. Wadati, Multicomponent bright solitons in $F = 2$ spinor Bose-Einstein cndensates. J. Phys. Soc. Jpn. **76**(7), 74005 (2007)
49. M. Ueda, M. Koashi, Theory of spin-2 Bose-Einstein condensates: spin correlations, magnetic response, and excitation spectra. Phys. Rev. A **65**, 063602 (2002)
50. J. Yang, Suppression of Manakov soliton interference in optical fibers. Phys. Rev. E **65**, 036606 (2002)
51. V.E. Zakharov, A.B. Shabat, Exact theory of two-dimensional self-focusing and one-dimensional self-modulation of waves in nonlinear media. Sov. Phys.-JETP **34**, 62–69 (1972) (English translation)

Wave Modes Trapped in Rotating Nonlinear Potentials

Yongyao Li, Wei Pang, and Boris A. Malomed

Abstract We study modes trapped in a rotating ring with the local strength of the nonlinearity modulated as cos (2θ), where θ is the azimuthal angle. This modulation pattern may be of three different types: self-focusing (SF), self-defocusing (SDF), and alternating SF-SDF. The model, based on the nonlinear Schrödinger (NLS) equation with periodic boundary conditions, applies to the light propagation in a twisted pipe waveguide, and to a Bose-Einstein condensate (BEC) loaded into a toroidal trap, under the action of the rotating nonlinear *pseudopotential* induced by means of the Feshbach resonance in an inhomogeneous external field. This is the difference from the recently considered similar setting with the rotating linear potential. In the SF, SDF, and alternating regimes, four, three, and five different types of stable trapped modes are identified, respectively: even, odd, second-harmonic (2H), symmetry-breaking, and 2H-breaking ones. The shapes and stability of these modes, together with transitions between them, are investigated in the first rotational Brillouin zone. Ground-state modes are identified in each regime. Boundaries between symmetric and asymmetric modes are also found in an analytical form, by means of a two-mode approximation.

Y. Li
Faculty of Engineering, Department of Physical Electronics, School of Electrical Engineering, Tel Aviv University, Tel Aviv 69978, Israel

Department of Applied Physics, South China Agricultural University, Guangzhou 510642, China
e-mail: yongyaoli@gmail.com

W. Pang
Department of Experiment Teaching, Guangdong University of Technology, Guangzhou 510006, China
e-mail: pangwei@gdut.edu.cn

B.A. Malomed (✉)
Faculty of Engineering, Department of Physical Electronics, School of Electrical Engineering, Tel Aviv University, Tel Aviv 69978, Israel
e-mail: malomed@post.tau.ac.il

R. Carretero-González et al. (eds.), *Localized Excitations in Nonlinear Complex Systems*, Nonlinear Systems and Complexity 7, DOI 10.1007/978-3-319-02057-0_8,

1 Introduction

Optical and matter waves exhibit a plenty of dynamical scenarios under the action of effective nonlinear potentials (which may sometimes be combined with usual linear potentials) [1]. The dynamics of such systems is governed by the nonlinear Schrödinger equation (NLSE) in optical media, or Gross-Pitaevskii equation (GPE) in the context of Bose-Einstein condensates (BECs). In either case, the nonlinear *pseudopotential* [2] may be induced by a regular [3, 4] or singular [5] spatial modulation of the local nonlinearity. These systems have been studied theoretically in a variety of one- [3,5] and two-dimensional (1D and 2D) [4] settings, and recently reviewed in Ref. [1]. To the same general class belong models which predict that, in any dimension D, stable fundamental and vortex solitons can be supported by a purely self-defocusing (SDF) nonlinearity growing towards the periphery ($r \to \infty$) at any rate faster than r^D [6].

In optics, such nonlinear potentials may be designed using the mismatch between the nonlinearity of the host material and solid [7] or liquid [8] stuff filling voids of photonic-crystal-fiber waveguides. Another possibility to create the effective nonlinear potential in optics is offered by inhomogeneous distributions of dopants which induce the resonantly enhanced nonlinearity [9]. In particular, it is possible to use the Rhodamine B dopant added to the SU-8 polymer (a commonly used transparent negative photoresist) [10], or Pr^{3+} ions doping the Y_2SiO_5 host medium [11, 12].

In BEC, the pseudopotential can be created with the help of the Feshbach resonance controlled by nonuniform magnetic [13] or optical [14] fields. In particular, the necessary pattern of the spatial modulation of the scattering length, which determines the local strength of the cubic nonlinearity in the respective GPE, can be induced by appropriately designed magnetic lattices [15].

Another well-known tool for the creation of various dynamical states is provided by rotating potentials, which may trap optical and matter waves. Effects of the rotation have drawn a great deal of attention in the studies of BEC. A well-known result is that rotational stirring of the condensate with repulsive interactions leads to the formation of vortex lattices, see review [16]. Under special conditions (the compensation of the trapping by the centrifugal force), giant vortices can be produced too, which were studied in detail experimentally [17] and theoretically [18]. In a binary immiscible BEC, vortex streets were predicted to form, instead of the vortex lattices [19]. On the other hand, it was predicted that the rotation of self-attractive condensates gives rise to several species of localized modes, such as solitary vortices, mixed-vorticity states ("crescents"), and quasi-solitons [20].

It is also natural to consider the dynamics of matter-wave modes trapped in rotating lattices, which can be created in BEC by broad laser beams transmitted through a revolving mask [21]. Quantum states and vortex lattices have been studied in this setting [22], as well as the depinning of trapped and solitary vortices when the rotation rate exceeds a critical value [23]. The nucleation of vortices in the rotating lattice was demonstrated experimentally [24].

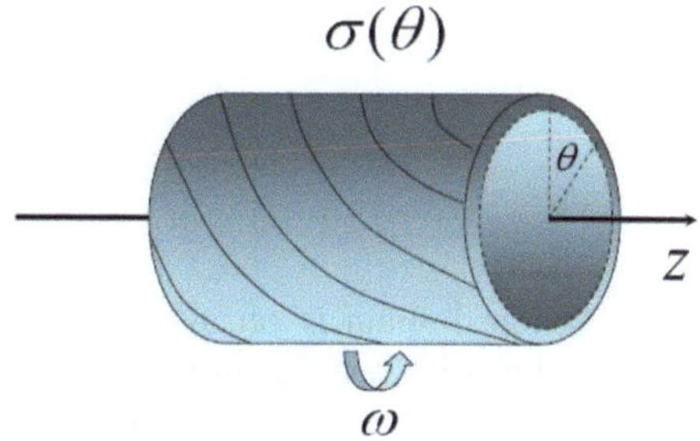

Fig. 1 The pipe nonlinear waveguide with the intrinsic nonlinear potential, twisted at rate ω

In optics, a setting similar to the rotating lattice can be realized in twisted photonic-crystal fibers [25, 26]. In plain optical fibers, the twist affects the polarization dynamics [27] and couples it to the transmission of temporal solitons [28]. In helical photonic-crystal fibers, modified Bragg reflection and enhancement of the mode conversion and transport have been studied [29], and the transformation of the linear moment of photons into the orbital angular momentum has been demonstrated recently [26].

The simplest version of the rotating lattice is represented by the revolving quasi-1D double-well potential (DWP) . It gives rise to azimuthal Bloch bands [30], and allows one to support solitons and solitary vortices even in the case of the SDF nonlinearity [31]. The generation of a vortex lattice in the rotating DWP was studied too [32]. Further, it is well known that the interplay of the DWP with the SF or SDF nonlinearity provides for the simplest setting for the study of the spontaneous symmetry breaking of even and odd states, respectively, in one dimension [33]. In this connection, a natural problem, which was recently considered in Ref. [34], is a modification of the symmetry-breaking scenarios in a rotating ring carrying the DWP potential, along with the nonlinearity.

While the dynamics of nonlinear waves trapped in rotating linear potentials has been studied in detail, previous works did not tackle modes pulled by rotating *nonlinear* (pseudo-) potentials. This setting, which may be implemented in optics and BEC alike, is the subject of the present work. To analyze the basic features of the respective phenomenology, we here concentrate on the 1D nonlinear potential on a rotating ring, as shown in Fig. 1. In optics, this system is realized as a pipe (hollow) waveguide with an azimuthal modulation of the local nonlinearity, $\sigma(\theta)$ (θ is the angular coordinate), *twisted* with pitch $2\pi/\omega$, where ω plays the role of the effective rotation speed. In BEC, a similar setting corresponds to a toroidal trap, which is available in the experiment [35], combined with the rotating nonlinear potential, that can be superimposed onto the trap [36]. This combination realizes a *rotating ring* [37] carrying the nonlinear potential.

It is necessary to define the sign of the nonlinear potential. In this work, we consider three distinct cases, namely, the self-focusing (SF), i.e., attractive, SDF (repulsive), and SF-SDF (alternating) nonlinearities, all subject to the spatial harmonic modulation, which is defined as follows:

$$\sigma_{\mathrm{SF}}(\theta) = -\sin^2\theta$$
$$\sigma_{\mathrm{SDF}}(\theta) = \cos^2\theta \tag{1}$$
$$\sigma_{\mathrm{SF-SDF}}(\theta) = \cos(2\theta)$$

The solution domain is set as $-\pi \leq \theta \leq +\pi$. Note that, in all the cases the local nonlinearity coefficient (1) has its maxima at points $\theta = 0$ and $\theta = \pm\pi$, and minima at $\theta = \pm\pi/2$. We aim to find basic types of modes trapped in rotating nonlinearity profiles (1), and establish their basic properties, such as symmetry/asymmetry, stability, and the identification of the respective ground states, varying two control parameters, viz., the rotation speed, ω, and the total power (norm), P, which is defined below in Eq. (4).

The paper is structured as follows. In Sect. 2, we formulate the system and methods of the stability analysis of the modes. In Sect. 3, we present numerical results for the basic modes and their stability in each type of the nonlinear potential (1). In Sect. 4, we present analytical results, obtained by means of a two-mode approximation, which explain a boundary between the symmetric and asymmetric modes in each case. The paper is concluded by Sect. 5.

2 The Model

The dynamics of the optical wave (or the BEC wave function) in the rotating ring is governed by the normalized one-dimensional NLSE (GPE), subject to the periodic boundary conditions [1]:

$$i\frac{\partial}{\partial z}\psi = \left[-\frac{1}{2}\frac{\partial^2}{\partial\theta^2} + \sigma(\theta - \omega z)|\psi|^2\right]\psi, \tag{2}$$

$\psi(\theta) \equiv \psi(\theta + 2\pi)$, where z is the propagation distance in the case of the optical waveguide, and the radius of the ring is scaled to be 1. It is more convenient to rewrite Eq. (2) in the rotating reference frame, with $\theta' \equiv \theta - \omega z$:

$$i\frac{\partial}{\partial z}\psi = \left[-\frac{1}{2}\frac{\partial^2}{\partial\theta'^2} + i\omega\frac{\partial}{\partial\theta'} + \sigma(\theta')|\psi|^2\right]\psi, \tag{3}$$

while the solution domain is defined as above, i.e., $-\pi \leq \theta' \leq +\pi$. For the BEC trapped in the rotating potential, the respective GPE differs by replacing z with time t. Equation (3) conserves the total power (norm) of the field and its Hamiltonian (energy),

$$P = \int_{-\pi}^{+\pi} \left|\psi(\theta')\right|^2 d\theta', \tag{4}$$

$$H = \frac{1}{2}\int_{-\pi}^{+\pi}\left[\left|\frac{\partial\psi}{\partial\theta'}\right|^2 + i\omega\left(\psi^*\frac{\partial\psi}{\partial\theta'} - \psi\frac{\partial\psi^*}{\partial\theta'}\right) + \sigma(\theta')|\psi|^4\right]d\theta', \quad (5)$$

with the asterisk standing for the complex conjugate. Stationary modes with real propagation constant $-\mu$ (in terms of BEC, μ is the chemical potential) are sought for as

$$\psi(\theta', z) = \exp(-i\mu z)\,\phi(\theta'), \quad (6)$$

with complex function $\phi(\theta')$ obeying equation

$$\mu\phi = \left[-\frac{1}{2}\frac{d^2}{d\theta'^2} + i\omega\frac{d}{d\theta'} + \sigma(\theta')|\phi|^2\right]\phi. \quad (7)$$

The periodic boundary conditions, $\sigma(\theta'+2\pi) = \sigma(\theta')$ and $\psi(\theta'+2\pi) = \psi(\theta')$, make Eq. (3) invariant with respect to the *boost transformation*, which allows one to change the rotation speed from ω to $\omega - N$ with arbitrary integer N:

$$\psi(\theta', z; \omega - N) = \psi(\theta', z; \omega)\exp\left[-iN\theta' + i\left(\frac{1}{2}N^2 - N\omega\right)z\right], \quad (8)$$

hence the speed may be restricted to interval $0 \leq \omega < 1$. Furthermore, Eq. (3) admits an additional invariance, relating solutions with opposite signs of the speed: $\psi(\theta', z; \omega) = \psi^*(\theta', -z; -\omega)$. If combined with boost $\omega \to \omega + 1$, the latter transformation demonstrates that the solutions with

$$\omega = 1/2 \pm \delta, \quad (9)$$

where $\delta < 1/2$, are tantamount to each other. Thus, the rotation speed may be eventually restricted to the fundamental interval,

$$0 \leq \omega \leq 1/2, \quad (10)$$

which plays the role of the first *rotational Brillouin zone*, cf. Ref. [30].

Comparing the scaled equation (3) with the GPE written in physical units for the condensate of ^{7}Li, one can estimate that $\omega = 1/2$ corresponds to the ring of radius $\sim$25 μm rotating at frequency $\sim$1 Hz. On the other hand, in terms of the NLSE for the optical propagation in the twisted pipe waveguide of radius $\sim$10 μm, $\omega = 0.5$ corresponds to the twist pitch $\sim$10 cm, which is easy to implement in the experiment, although it may not be easy to fabricate the waveguide long enough to accommodate many such rotations.

The dynamical stability of the stationary solutions has been investigated via numerical computation of eigenvalues for small perturbations, and verified by direct simulations of the perturbed evolution. The perturbed solutions are introduced as usual,

$$\psi = e^{-i\mu z}[\phi(\theta') + \varepsilon u(\theta')e^{i\lambda z} + \varepsilon v^*(\theta')e^{-i\lambda^* z}], \tag{11}$$

where ε is an infinitesimal amplitude of the disturbance, $u(\theta')$ and $v(\theta')$ are the corresponding eigenmodes, and λ the eigenfrequency. The substitution of ansatz (11) into Eq. (3) and linearization leads to the linear eigenvalue problem,

$$\begin{pmatrix} \mu - \hat{h} - i\omega\frac{\partial}{\partial\theta'} & -\sigma\phi^2 \\ \sigma(\phi^*)^2 & -\mu + \hat{h} - i\omega\frac{\partial}{\partial\theta'} \end{pmatrix} \begin{pmatrix} u \\ v \end{pmatrix} = \lambda \begin{pmatrix} u \\ v \end{pmatrix}, \tag{12}$$

where $\hat{h} = -(1/2)\partial^2/\partial(\theta')^2 + 2\sigma|\phi|^2$ is the single-particle Hamiltonian. The underlying solution ϕ is stable if Eq. (12) generates solely real eigenvalues. Lastly, stationary equation (7) was solved using numerical code "PCSOM" borrowed from Ref. [38].

3 Numerical Results

3.1 The Classification of Trapped Modes

For all types of the nonlinear potentials defined in Eq. (1), the numerical solution of stationary equation (7) with different inputs (initial guesses) makes it possible to identify five basic species of stationary states, which are listed in Table 1.

Following the symmetry of the nonlinearity-modulation patterns in Eq. (1), the modes are identified as symmetric (alias even) and antisymmetric (alias odd), with respect to the central point, $\theta' = 0$. Another basic mode is one which breaks the symmetry of the even solution (the fourth entry in Table 1). The mode of the "second-harmonic" (2H) type, which is also even, refers to the dominant term in its Fourier decomposition, and the "2H-breaking" type originates as a result of the spontaneous symmetry breaking of the 2H mode. Note that all the inputs displayed in the table are real functions, while numerically found solutions of Eq. (7) with $\omega \neq 0$ are complex ones. Naturally, real parts of the solutions generated by the real inputs indicated in the table have the same parity (even/odd) as the inputs, while the imaginary parts feature the opposite parity (odd/even) (the input of the 2H-breaking type does not feature a certain parity). It is shown below too that maxima and minima of the local power (density) of the even mode defined in the table coincide with maxima and minima of the local nonlinearity, $\sigma(\theta')$, while for the odd and 2H modes the relation is the opposite, with the peak powers sitting in potential wells (see Eq. (1)), hence the odd and 2H modes tend to have lower values of the energy, and may play the role of the ground state, as confirmed below.

It is relevant to compare this set of possible modes with its counterpart found in the similar model with the linear rotating potential [34]. In that case, an additional species is possible, viz., an antisymmetry-breaking one.

Table 1 Different species of stationary modes, labeled by input waveforms which generate them as solutions of Eq. (7)

Inputs	Types of modes
$\cos\theta'$	Symmetric (even)
$\sin\theta'$	Antisymmetric (odd)
$\sin^2\theta'$	Second-harmonic (2H)
$b+\cos\theta'$, $0<b\leq 1$	Symmetry-breaking
$b+\sin\theta'$, $0<b\leq 1$	2H-breaking

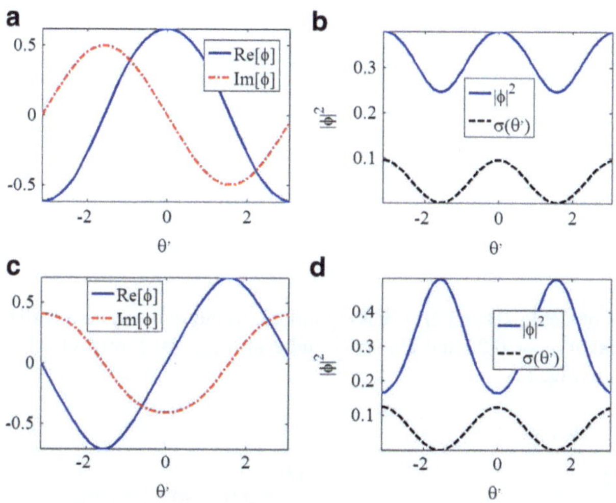

Fig. 2 Examples of stable even and odd modes, found in the system with self-focusing nonlinear potential (1a), at rotation speed $\omega = 0.25$, with total power $P_{\text{even}} = P_{\text{odd}} = 2$. Panels (**a**), (**c**) display, severally, real and imaginary parts of the even and odd modes, while (**b**), (**d**) show their local-power (density) profiles. The *dashed curves* in (**b**) and (**d**), and in similar panels displayed below, depict the corresponding nonlinearity-modulation profile, $\sigma(\theta')$; in the present case, it is $\cos^2\theta'$

3.2 The Self-Focusing Nonlinearity

The SF nonlinear potential, defined as per Eq. (1a), gives rise to four types of dynamically stable trapped modes, viz., even (symmetric), odd (antisymmetric), 2H and 2H-breaking ones (i.e., only the symmetry-breaking species is missing in this case). Typical examples of these stable modes are displayed in Figs. 2 and 3. In addition to the above-mentioned fact that the maxima and minima of the local power coincide with those of the nonlinear potential for the even mode, and, on the contrary, coincide with minima and maxima of the potential for the odd and 2H modes, suggesting that either of the latter modes may be a ground state, the figures demonstrate that the 2H-breaking mode has one maximum and one minimum of the local power, both sitting in nonlinear-potential wells (minima).

Results of the numerical analysis for the stability of the modes in the model with the SF nonlinear potential are summarized in Fig. 4, in the form of diagrams

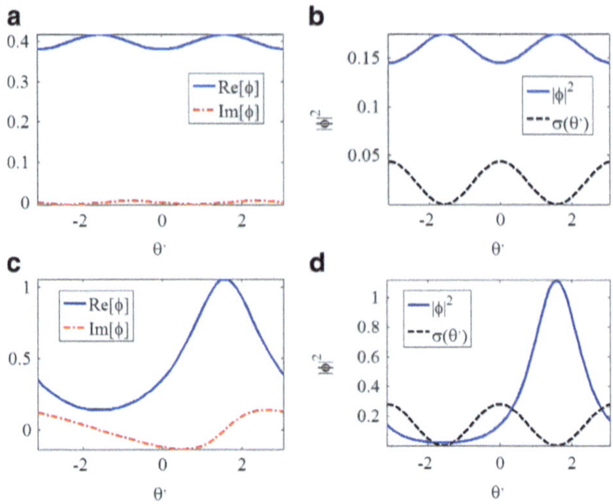

Fig. 3 Examples of stable 2H and 2H-breaking modes, found in the system with the self-focusing nonlinear potential at $\omega = 0.25$ and $P_{2H} = 1$ and $P_{2H-\text{break.}} = 2$, respectively. Panels have the same meaning as in Fig. 2

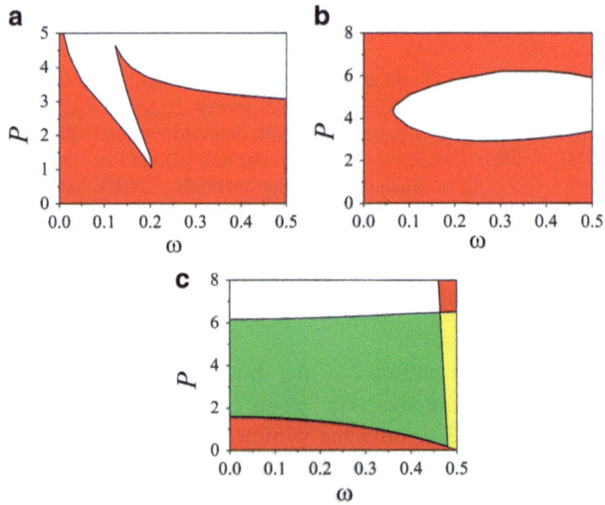

Fig. 4 (**a**) and (**b**): Stability diagrams for the even and odd (antisymmetric) modes, respectively, obtained in the system with of the self-focusing nonlinearity, in the plane of the rotation speed (ω) and total power (P). (**c**) The stability diagram for the set of the second-harmonic (2H) and 2H-breaking modes. In panels (**a**)–(**c**) the even, odd, and 2H modes are stable in the *red areas*, and unstable in the blank ones, respectively. In panel (**c**), the 2H-breaking mode is stable in the *green* (*middle*) *area*, and the bistability, i.e., coexistence of 2H and 2H-breaking stable modes, occurs in the *yellow* (*right edge*) *region*. In the blank area of panel (**c**), no 2H-breaking mode, stable or unstable one, is found

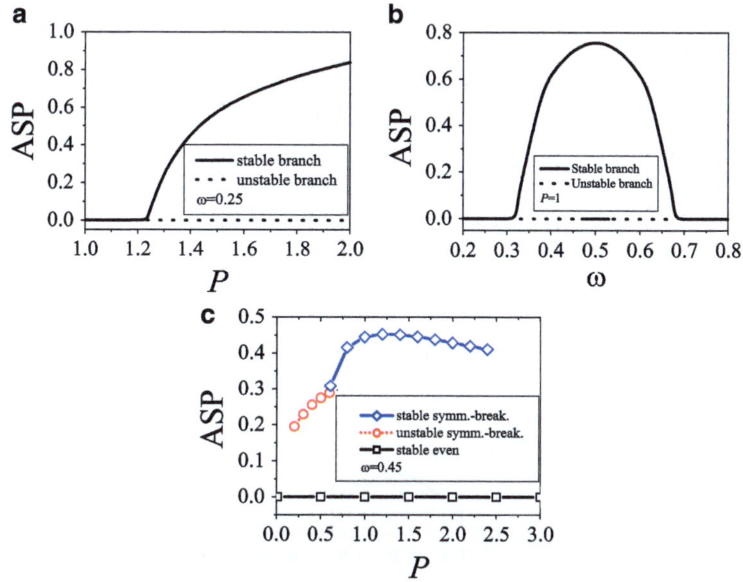

Fig. 5 (**a**) The ASP of the 2H and 2H-breaking modes, defined as per Eq. (13), as a function of the total power, P, with the rotation speed fixed at $\omega = 0.25$, in the system with the self-focusing nonlinearity (**b**). The same as a function of ω, at a fixed total power, $P = 1$. The plot includes regions symmetric with respect to $\omega = 0.5$, to stress the respective symmetry of modes in the present system. (**c**) The ASP of even and symmetry-breaking mode as a function of P for $\omega = 0.45$, in the case of the self-defocusing nonlinearity

drawn in the plane of (P, ω). They show that the stability regions of the even, odd, 2H and 2H-breaking modes strongly overlap between themselves. In particular, the asymmetry parameter of the 2H-breaking modes, which is defined as

$$\mathrm{ASP} \equiv \left| \int_0^{\pi} |\phi|^2 d\theta' - \int_{-\pi}^{0} |\phi|^2 d\theta' \right| / P, \tag{13}$$

is displayed in Fig. 5a, b. The plots demonstrate that the transition between 2H and 2H-breaking modes is of the *supercritical* type [39].

The multistability, which is obvious in Fig. 4, makes it necessary to compare energies of the coexisting dynamically stable modes, defined by Eq. (5), in order to identify the ground state that realizes the energy minimum. First, in Fig. 6a we show the results along horizontal cuts of all the three panels of Fig. 4, made at a constant value of the total power, $P = 0.5$, with the rotation speed varying in the interval of $0 \leq \omega \leq 0.4$. The *tristability* of the even, odd and 2H modes takes place along this segment. Further, Fig. 6b displays the results along vertical cuts of panel (c) of Fig. 4 made at $\omega = 0.5$, while the power is varying as $0.2 \leq P \leq 1.6$ (it is seen in panels (b) and (c) of Fig. 4 that the 2H-breaking mode also coexists with the stable odd one, but the energy of the odd mode is definitely larger, therefore it

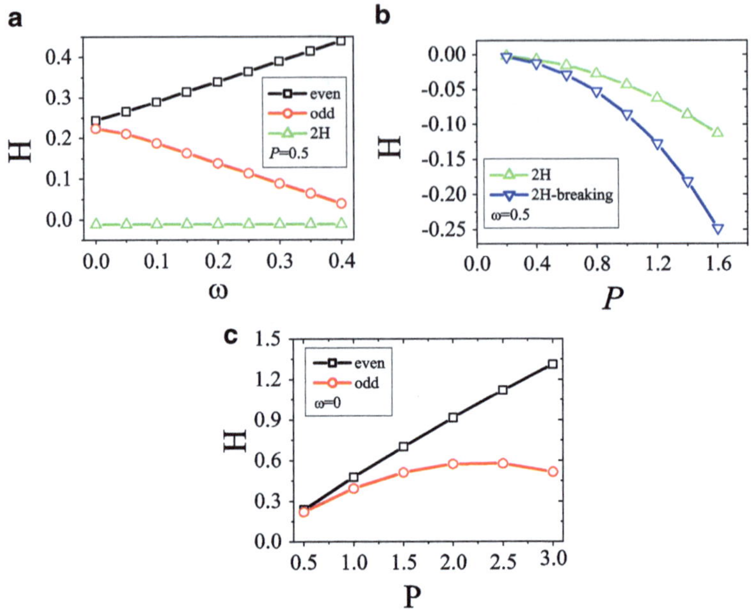

Fig. 6 (**a**) Energies of the even, odd, and second-harmonic modes in the system with the self-focusing nonlinearity, computed, as per Eq. (5), for $P = 0.5$ and $0 \leq \omega \leq 0.4$. (**b**) Energies of the 2H-breaking and second-harmonic modes, computed for $\omega = 0.5$ and $0.2 \leq P \leq 1.6$. (**c**) Energies of the even and odd modes along the vertical segment, with $\omega = 0$ and $0.5 \leq P \leq 3$

is not displayed in Fig. 6b). In panel (b) of Fig. 6, the two branches merge into one, with $H \to 0$, at $P \to 0$, as the *nonlinear* potential vanishes in this limit (this is another characteristic difference from the similar model [34] with the rotating linear potential).

From Fig. 6a, we conclude that $H_{2H} < H_{odd} < H_{even}$, while Fig. 6b shows that $H_{2H-break} < H_{2H}$. Calculations of the energy, performed along other horizontal and vertical cuts, demonstrate that the following relation between the energies of the different modes, suggested by these inequalities, is always correct:

$$H_{2H-break} < H_{2H} < H_{odd} < H_{even}. \tag{14}$$

Thus, the 2H-breaking mode, if it exists (recall that, in the plane of (ω, P) shown in Fig. 4c, it exists above the curve separating the bottom (red) and middle (green) areas), represents the ground state in the system with the self-focusing nonlinearity. If the latter mode does not exist, then the 2H state plays the same role.

In Fig. 6a we can see that, in the limit of $\omega = 0$, the energies of the even and odd modes are very close, which seems to be in contradiction with the fact that these two modes are essentially different, having the opposite parities. The reason is that the total power that we chose here ($P = 0.5$) is not large enough to show the energy difference between the two modes. To clarify the point, in Fig. 6c we display the energy curves for these two modes selected along the vertical segment with $\omega = 0$ and $0.5 \leq P \leq 3$. It shows that the energy difference indeed increases with the growth of P.

Lastly, it is relevant to compare these conclusions with those produced by the analysis of the system combining the rotating linear DWP potential and spatially uniform SF nonlinearity [34]. In that case, the 2H-breaking mode may also play the role of the ground state, but the one competing with it in this capacity is not the 2H state with the unbroken symmetry, but rather what is defined above as the symmetry-breaking mode, see Table 1.

3.3 The Self-Defocusing Nonlinearity

In the case of the SDF nonlinear potential, represented by Eq. (1b), the numerical solution of Eq. (7) reveals stable modes of three types, viz., even, symmetry-breaking, and 2H (recall that symmetry-breaking modes were not found in the system with the SF nonlinearity), while odd and 2H-breaking states do not exist in this case. Because the profiles of the even and 2H modes are quite similar to their counterparts presented above in Fig. 2, we here display, in Fig. 7, only a typical stable symmetry-breaking mode. The shape of this mode seems symmetric, centered at $\theta' = 0$; however, it is classified as an asymmetric mode, as the power profile of a true symmetric state would be double-humped, cf. Figs. 2b, d and 3b, while the present one has a single maximum, similar to the intensity distribution in the 2H-breaking state in Fig. 3d. Accordingly, the ASP (effective asymmetry measure) for the symmetry-breaking mode is introduced as follows, instead of the above definition (13):

$$(\mathrm{ASP})_{\mathrm{symm-break}} \equiv \left| \left[\int_{-\pi/2}^{+\pi/2} - \left(\int_{-\pi}^{-\pi/2} + \int_{+\pi/2}^{+\pi} \right) \right] |\phi|^2 d\theta' \right| / P, \qquad (15)$$

to stress the lack of the asymmetry between the central and peripheral parts of the mode.

The stability and energy diagrams for the even, symmetry-breaking, and 2H modes are displayed in Fig. 8 (the 2H mode exists and is stable in the entire plane of (P, ω), therefore it is not specially marked in panel Fig. 8a). In particular, it is observed that the symmetry-breaking mode exists near the right edge of the rotational Brillouin zone, i.e., it does not exist in the stationary system (with $\omega = 0$). The ASP of the even and symmetry-breaking modes, defined per Eq. (15), is displayed in Fig. 5c as a function of the total power. The absence of the linkage between the branches representing these two modes implies that they are not related by any bifurcation.

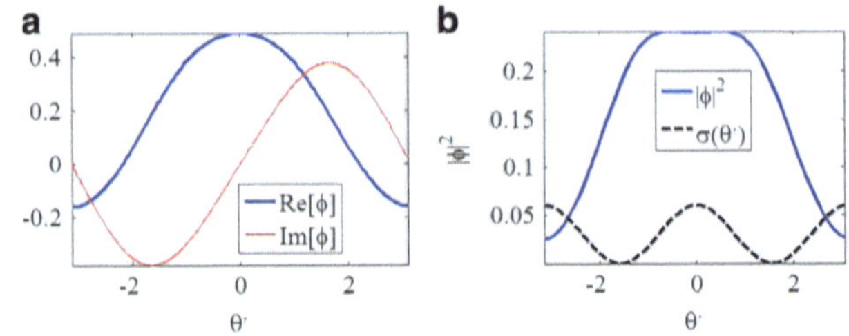

Fig. 7 A stable symmetry-breaking mode, found in the system with the self-defocusing nonlinearity, at $(\omega, P) = (0.45, 1)$: (**a**) its real and imaginary parts; (**b**) the power profile

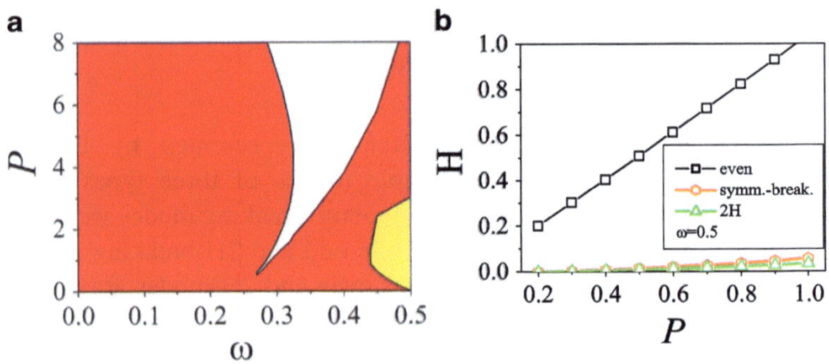

Fig. 8 (**a**) The stability diagram for the even and symmetry-breaking modes in the system with the self-defocusing nonlinearity. The *red* (largest) and *yellow* (smallest, near the right edge) areas designate, severally, the stability region of the even mode, and the region of the coexistence (bistability) of the even and symmetry-breaking modes. In the blank area, no stable modes of these types are found (in fact, an unstable even mode exists in that area). The second-harmonic mode exists and is stable in the entire plane. (**b**) Energies of the even, symmetry-breaking, and second-harmonic modes along the vertical cut of panel (**a**) at $\omega = 0.5$ and $0.2 \leq P \leq 1$

Figure 8b displays the comparison of energies of these three kinds of the modes (even, symmetry-breaking, and 2H ones) along the vertical cut made at $\omega = 0.5$, with the power varying in interval $0.2 \leq P \leq 1$. It demonstrates that the curve for the 2H mode goes close to but slightly lower than its counterpart for the symmetry-breaking mode. The analysis of more general data, produced by the numerical calculations for the system with the SDF nonlinear potential, demonstrates that energies of all the three dynamically stable modes existing in this case are ordered as follows:

$$H_{\mathrm{2H}} < H_{\mathrm{symm-break}} < H_{\mathrm{even}}, \tag{16}$$

cf. Eq. (14). Thus, the 2H mode plays the role of the ground state in the case of the SDF nonlinearity (recall this mode exists at all values of ω and P). A similar conclusion was obtained for the system combining the rotating linear DWP potential and spatially uniform SDF nonlinearity [34].

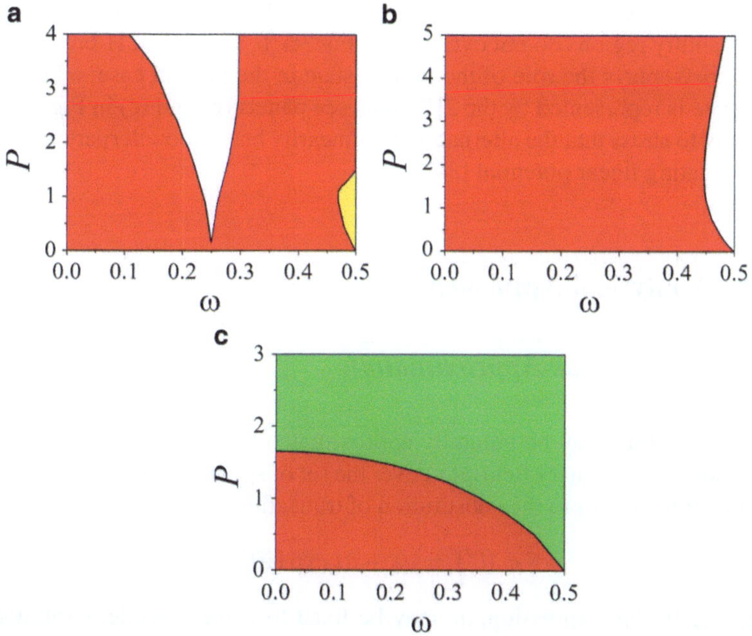

Fig. 9 Stability diagrams for the system with the alternating self-focusing – self-defocusing nonlinearity. (**a**) Even and symmetry-breaking modes; (**b**) the odd mode; (**c**) the 2H and 2H-breaking modes. In (**a**), the even mode is stable in the *red (bottom) area*; in the small *yellow region* it coexists with the symmetry-breaking one, and no stable modes of these types are found in the *blank area*. In (**b**) the odd mode is stable in the *red area*, and unstable in the *blank area*. In (**c**) the 2H mode is stable in the *red area*, while the 2H-breaking mode is stable in the *green region*. A clear border between the two latter areas suggest that the transition between the 2H and 2H-breaking modes is supercritical

3.4 The Alternating Self-Focusing: Self-Defocusing Nonlinear Potential

In the case of the alternating SF-SDF nonlinearity, defined as per Eq. (1), the numerical solutions reveal the existence of all the five types of stable trapped modes indicated in Table 1. Profiles of these modes are quite similar to those of their counterparts displayed above in Figs. 2, 3, and 7, therefore we do not show them again here. The respective stability diagrams in the (P, ω) plane are presented in Fig. 9. In particular, the absence of a bistability area in panel (c) of Fig. 9 suggests that the transition between the 2H and 2H-breaking modes is supercritical, like in the system with the SF nonlinearity, cf. Fig. 4c.

The comparison of the energies of the five types of the modes which may be stable in the case of the sign-alternating nonlinearity demonstrates the following ordering, cf. Eqs. (14) and (16):

$$H_{\text{2H-break}} < H_{\text{2H}} < H_{\text{symm.-break}},\ H_{\text{odd}} < H_{\text{even}}. \tag{17}$$

The energies of the symmetry-breaking and odd modes are not compared in Eq. (17), as their stability regions do not overlap, see Fig. 9a, b. Thus, the 2H-breaking mode, when it exists, plays the role of the ground state in the present case; otherwise, the ground state is represented by the 2H mode, see panels (b) and (c) in Fig. 9. Finally, it is worthy to stress that the alternating nonlinearity has no counterpart in the model with the rotating linear potential [34].

4 The Analytical Approach

4.1 The Two-Mode Approximation

The present setting may be naturally approximated by a finite-mode truncation of the expansion of stationary field $\phi(x)$ over the set of spatial harmonics. The simplest approximation reduces to the substitution of truncation

$$\phi\left(\theta'\right) = a_0 + a_1 \exp\left(i\theta'\right) \tag{18}$$

into Eq. (7). In this expression, a_0 may be fixed to be real, while amplitude a_1 is allowed to be complex. This approach is consistent if, in the linear approximation, each term in combination (18) is an exact solution of Eq. (7) for a common value of μ, hence the zeroth-order approximation exists, and the analysis of weakly nonlinear states can be developed around it. It is easy to see that such a case corresponds, in the zeroth approximation, to $\omega = 1/2$ (which is exactly the right edge of zone (10)) and $\mu = 0$ [34]. Then, weakly nonlinear modes can be constructed in an approximate analytical form, assuming that μ, a_0^2, $|a_1|^2$, and

$$\delta \equiv 1/2 - \omega \tag{19}$$

are all small quantities. To this end, ansatz (18) and a particular expression (1) for $\sigma\left(\theta'\right)$ are substituted into Eq. (7), and equations for amplitudes a_0 and a_1 are derived as balance conditions for the zeroth and first harmonics.

Ansatz (18) corresponds to the following approximation for the total power (4),

$$P = \int_{-\pi}^{+\pi} |\phi|^2 d\theta' = 2\pi(a_0^2 + |a_1|^2), \tag{20}$$

which will be used below too.

4.2 The Self-Focusing Nonlinearity

In the case of the SF nonlinear potential (1a), the two-mode approximation (18) leads to the following equations:

$$\mu = -\frac{1}{2}a_0^2 - |a_1|^2 + \frac{1}{4}a_1^2 \tag{21}$$

$$\mu - \delta = -a_0^2 - \frac{1}{2}|a_1|^2 + \frac{1}{4}a_0^2\frac{a_1^*}{a_1}. \tag{22}$$

As shown above by the numerical analysis, the SF nonlinearity gives rise, *inter alia*, to the 2H-breaking mode, which is generated from input $b + \sin\theta'$ in Table 1. To capture the part of the solution corresponding to $\sin\theta'$ in the input, we set

$$a_1 = ic, \tag{23}$$

Then, Eq. (22) yields

$$a_0^2 = (4/21)\,(5\delta - 3\mu) \tag{24}$$

$$c^2 = -\,(4/21)\,(2\delta + 3\mu). \tag{25}$$

Substituting solutions (24) and (25) into expression (20), we obtain a relation between the propagation constant and total power for the 2H-breaking mode,

$$\mu = \frac{\delta}{2} - \frac{7}{16\pi}P. \tag{26}$$

It demonstrates that this mode, as predicted by the analytical approximation, satisfied the Vakhitov-Kolokolov (VK) criterion, $d\mu/dP < 0$, which is a necessary stability condition for patterns supported by the SF nonlinearity [40].

Furthermore, Eq. (24) predicts the emergence of the 2H-breaking modes at $a_0^2 = 0$, i.e.,

$$\mu = 5\delta/3. \tag{27}$$

Since the modes with rotation speeds related by Eq. (9) are mutually tantamount, Eqs. (27) and (26) predict the coexistence of the odd and 2H-breaking modes at

$$P \geq (P_{\min}) = (8\pi/3)\,|\delta|. \tag{28}$$

This analytical result is compared with the corresponding numerical findings in Fig. 10a, which shows a reasonably good agreement.

4.3 *The Self-Defocusing Nonlinearity*

The substitution of the same ansatz (18) into Eq. (7) in the case of the SDF nonlinear potential , taken as per Eq. (1b), yields the following algebraic equations, instead of Eq. (21) and (22) derived above for the SF nonlinearity:

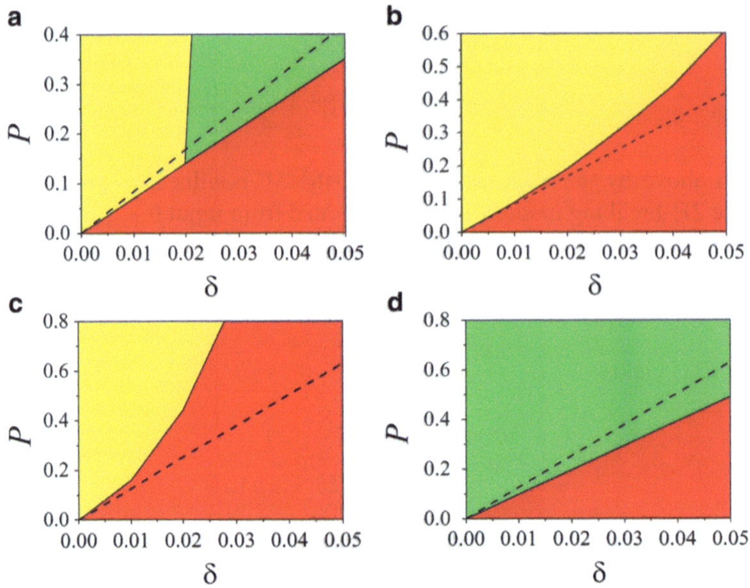

Fig. 10 The comparison between the theoretically predicted (*dashed lines*) and numerically found boundaries of the existence of the stable 2H-breaking (**a**, **d**) and symmetry-breaking (**b**, **c**) modes in the plane of (δ, P), in the interval of $0 \leq |\delta| \equiv |1/2 - \omega| \leq 0.05$. In the *red areas*, only even or 2H modes are produced by the numerical solution, while in the *yellow* (bistability) *regions* they coexist with the stable symmetry- or 2H-breaking modes, respectively. Panels (**a**) and (**b**) pertain, severally, to the SF and SDF nonlinear potentials , while (**c**) and (**d**) correspond to the alternating SF-SDF potential. These four panels are, actually, zoomed versions of the right bottom corners of Figs. 4b, 8a, and 9a, b, respectively, with analytically predicted lines (28) or (40) added to each panel

$$\mu = \frac{1}{2}a_0^2 + |a_1|^2 + \frac{1}{4}a_1^2 \tag{29}$$

$$\mu - \delta = a_0^2 + \frac{1}{2}|a_1|^2 + \frac{1}{4}a_0^2\frac{a_1^*}{a_1}. \tag{30}$$

The above numerical results for the SDF nonlinearity demonstrate the existence of the symmetry-breaking mode in this case, which is generated by input $b + \cos\theta'$ in Table 1. To capture this mode by means of ansatz (18), it is natural to set

$$a_1 = c \equiv \text{real}, \tag{31}$$

on the contrary to Eq. (23), where a_1 was imaginary. In this case, the algebraic system (30) yields

$$a_0^2 = (4/21)(3\mu - 5\delta) \tag{32}$$

$$c^2 = (4/21)(3\mu + 2\delta), \tag{33}$$

cf. Eq. (25). According to Eq. (20), the relations between the propagation constant and total power takes the following form for solution (33):

$$\mu = \frac{\delta}{2} + \frac{7}{16\pi}P, \tag{34}$$

cf. Eq. (26). This equation demonstrates that the asymmetric mode satisfies the *anti-VK* criterion, $d\mu/dP > 0$, which, as argued in Ref. [41], may play the role of a necessary stability condition for modes supported by the SDF nonlinearity.

The analytical approximation predicts the existence boundary for the symmetry-breaking states, $a_0^2 = 0$, in the same form (27) as it was obtained above for the SF nonlinearity. Consequently, the existence region for these states is predicted in the form coinciding with that given by Eq. (28). This result is compared with its numerical counterpart in Fig. 10b.

4.4 The Alternating Self-Focusing: Self-Defocusing Nonlinearity

In the case of the alternating SF-SDF nonlinear potential (1c), the substitution of ansatz (18) into Eq. (7) yields the algebraic equations in the form which is somewhat simpler than Eqs. (22) and (30) derived above for the "pure" SF and SDF nonlinearities:

$$\mu = \frac{1}{2}a_1^2 \tag{35}$$

$$\mu - \delta = \frac{1}{2}a_0^2\frac{a_1^*}{a_1}. \tag{36}$$

First, if we choose real a_1, as in Eq. (31), which refers to the symmetry-breaking mode, Eq. (36) yields

$$a_0^2 = 2(\mu - \delta) \tag{37}$$

$$c^2 = 2\mu. \tag{38}$$

The corresponding relation between μ and P is

$$\mu = \frac{\delta}{2} + \frac{P}{8\pi}, \tag{39}$$

which satisfies the above-mentioned anti-VK criterion, cf. Eq. (34), i.e., it is possible to assume that the stability of the symmetry-breaking mode is supported by the SDF part of the alternating nonlinear potential . Further, Eq. (38) predicts the emergence of the symmetry-breaking mode ($a_0^2 = 0$) at $\mu = \delta$. In terms of the total power related to μ by Eq. (20), this implies that this mode exists at

$$P \geq P_{\min} = 4\pi|\delta|, \tag{40}$$

cf. Eq. (28). The comparison of this analytical result with its numerical counterpart in shown in Fig. 10c.

The numerical results presented in the previous section demonstrate that the alternating SF-SDF nonlinear potential support the stable 2H-breaking mode too. To describe it in the framework of the two-mode approximation, we now assume a_1 to be imaginary, as in Eq. (23). In this case, Eq. (38) yields

$$a_0^2 = 2(\delta - \mu) \tag{41}$$

$$c^2 = -2\mu, \tag{42}$$

cf. Eq. (36), the respective relation between μ and P being

$$\mu = \frac{\delta}{2} - \frac{P}{8\pi}, \tag{43}$$

cf. Eq. (39). The latter relation satisfies the VK criterion, which implies that the stability of the 2H-breaking mode is supported by the SF part of the alternating nonlinear potential. The emergence of the 2H-breaking mode corresponds to $a_0^2 = 0$, i.e., again $\mu = \delta$, as in the case of the symmetry-breaking mode, under the same alternating nonlinear potential . Finally, this means that the existence of the stable 2H-breaking mode is predicted in the same region (40) as for its symmetry-breaking counterpart. The latter prediction is compared to the numerical findings in Fig. 10d.

5 Conclusions

This work aimed to study the existence and stability of modes trapped in the rotating nonlinear-lattice potentials of the SF, SDF (self-focusing and defocusing) and alternating SF-SDF types. The consideration was carried out for the first Brillouin zone in the rotating reference frame, i.e., for $0 \leq \omega \leq 1/2$, where ω is the rotation speed. The stability analysis was performed through the computation of eigenvalues for small perturbations, and verified by direct simulations. The model can be realized in the spatial domain, in terms of a twisted-pipe optical waveguide, with the built-in azimuthal modulation of the local Kerr coefficient, or, in the temporal domain, as the Gross-Pitaevskii equation for BEC loaded into a toroidal trap, under the action of a rotating optical or magnetic structure which affects the local value of the scattering length.

In the SF system, four types of different modes have been identified: even, odd, which are dominated by combinations of the fundamental and zeroth angular harmonics, and, in addition, the 2H (second-harmonic) and 2H-breaking states.

On the other hand, the SDF nonlinear potential supports three species of the trapped states: 2H, even, and symmetry-breaking ones, the latter existing only at the rotation speed close to the right edge of the Brillouin zone, $\omega = 1/2$, and in a limited interval of values of the total power, P. The alternating SF-SDF nonlinear potential supports all the five species of the trapped modes. Transitions between the 2H and 2H-breaking modes are of the supercritical types (subcritical transitions are possible in the system which combines the rotating linear DWP potential and spatially uniform SF nonlinearity). The energy comparison reveals that, in the SF and SF-SDF systems alike, the 2H-breaking mode, if it exists, represents the ground state; otherwise, this role is played by the 2H mode. The ground state of the SDF system is always represented by the 2H solution.

This work may be naturally developed in other directions. On the one hand, it is relevant to consider rotating nonlinear lattices with smaller azimuthal periods, $2\pi/n$, for integer $n > 1$, unlike the case of $n = 1$ investigated here. On the other hand, it may be interesting to consider a two-dimensional version of the present one-dimensional model (with an entire rotating plane, rather than the thin ring, which will include effects of the Coriliolis' force).

Acknowledgements We appreciate a valuable discussion with Dimitri J. Frantzeskakis, and help in the use of numerical methods provided by Nir Dror and Shenhe Fu. This work was supported by Chinese agencies NKBRSF (grant No. G2010CB923204) and CNNSF (grant No. 11104083,11204089,11205063), by the German-Israel Foundation through grant No. I-1024-2.7/2009, and by the Tel Aviv University in the framework of the "matching" scheme.

References

1. Y.V. Kartashov, B.A. Malomed, L. Torner, Rev. Mod. Phys. **83**, 247 (2011)
2. W.A. Harrison, *Pseudopotentials in the Theory of Metals* (Benjamin, New York, 1966)
3. F.K. Abdullaev, A. Gammal, L. Tomio, J. Phys. B **37**, 635 (2004); H. Sakaguchi, B.A. Malomed, Phys. Rev. E **72**, 046610 (2005); F.K. Abdullaev, J. Garnier, Phys. Rev. A **72**, 061605(R) (2005); G. Theocharis, P. Schmelcher, P.G. Kevrekidis, D.J. Frantzeskakis, ibid. **72**, 033614 (2005); D.L. Machacek, E.A. Foreman, Q.E. Hoq, P.G. Kevrekidis, A. Saxena, D.J. Frantzeskakis, A.R. Bishop, Phys. Rev. E **74**, 036602 (2006); P. Niarchou, G. Theocharis, P.G. Kevrekidis, P. Schmelcher, D.J. Frantzeskakis, Phys. Rev. A **76**, 023615 (2007); Z. Rapti, P.G. Kevrekidis, V.V. Konotop, C.K.R.T. Jones, J. Phys. A **40**, 14151 (2007); F. Abdullaev, A. Abdumalikov, R. Galimzyanov, Phys. Lett. A **367**, 149 (2007); F.K. Abdullaev, A. Gammal, M. Salerno, L. Tomio, Phys. Rev. A **77**, 023615 (2008); L.C. Qian, M.L. Wall, S. Zhang, Z. Zhou, H. Pu, ibid. **77**, 013611 (2008); A.S. Rodrigues, P.G. Kevrekidis, M.A. Porter, D.J. Frantzeskakis, P. Schmelcher, A.R. Bishop, ibid. **78**, 013611 (2008); F.K. Abdullaev, Y.V. Bludov, S.V. Dmitriev, P.G. Kevrekidis, V.V. Konotop, Phys. Rev. E **77**, 016604 (2008); Y.V. Kartashov, B.A. Malomed, V.A. Vysloukh, L. Torner, Opt. Lett. **34**, 3625 (2009); F.K. Abdullaev, R.M. Galimzyanov, M. Brtka, L. Tomio, Phys. Rev. E **79**, 056220 (2009); N. Dror, B.A. Malomed, Phys. Rev. A **83**, 033828 (2011); X.-F. Zhou, S.-L. Zhang, Z.-W. Zhou, B.A. Malomed, H. Pu, ibid. A **85**, 023603 (2012)
4. H. Sakaguchi, B.A. Malomed, Phys. Rev. E **73**, 026601 (2006); G. Fibich, Y. Sivan, M.I. Weinstein, Physica D **217**, 31 (2006); Y. Sivan, G. Fibich, M.I. Weinstein, Phys. Rev. Lett. **97**, 193902 (2006); Y.V. Kartashov, B.A. Malomed, V.A. Vysloukh, L. Torner, Opt. Lett. **34**,

770 (2009); Phys. Rev. A **80**, 053816 (2009); O.V. Borovkova, B.A. Malomed, Y.V. Kartashov, EPL **92**, 64001 (2010); N.V. Hung, P. Ziń, M. Trippenbach, B.A. Malomed, Phys. Rev. E **82**, 046602 (2010); T. Mayteevarunyoo, B.A. Malomed, A. Reoksabutr, J. Mod. Opt. **58**, 1977 (2011); H. Sakaguchi, B.A. Malomed, Opt. Lett. **37**, 1035 (2012)

5. T. Mayteevarunyoo, B.A. Malomed, G. Dong, Phys. Rev. A **78**, 053601 (2008); O.V. Borovkova, V.E. Lobanov, B.A. Malomed, ibid. A **85**, 023845 (2012); A. Acus, B.A. Malomed, Y. Shnir, Physica D **241**, 987 (2012)
6. O.V. Borovkova, Y.V. Kartashov, B.A. Malomed, L. Torner, Opt. Lett. **36**, 3088 (2011); O.V. Borovkova, Y.V. Kartashov, L. Torner, B.A. Malomed, Phys. Rev. E **84**, 035602 (R) (2011); W.-P. Zhong, M. Belić, G. Assanto, B.A. Malomed, T. Huang, Phys. Rev. A **84**, 043801 (2011); Y.V. Kartashov, V.A. Vysloukh, L. Torner, B.A. Malomed, Opt. Lett. **36**, 4587 (2011); V.E. Lobanov, O.V. Borovkova, Y.V. Kartashov, B.A. Malomed, L. Torner, ibid. **37**, 1799 (2012)
7. F. Luan, A.K. George, T.D. Hedley, G.J. Pearce, D.M. Bird, J.C. Knight, P.St.J. Russell, Opt. Lett. **29**, 2369 (2004); G. Bouwmans, L. Bigot, Y. Quiquempois, F. Lopez, L. Provino, M. Douay, Opt. Exp. **13**, 8452 (2005); A. Fuerbach, P. Steinvurzel, J.A. Bolger, A. Nulsen, B.J. Eggleton, Opt. Lett. **30**, 830 (2005)
8. T.T. Larsen, A. Bjarklev, D.S. Hermann, J. Broeng, Opt. Exp. **11**, 2589 (2003); F. Du, Y.Q. Lu, S.T. Wu, Appl. Phys. Lett. **85**, 2181 (2004); M.W. Haakestad, T.T. Alkeskjold, M.D. Nielsen, L. Scolari, J. Riishede, H.E. Engan, A. Bjarklev, IEEE Photon. Tech. Lett. **17**, 819 (2005); C.R. Rosberg, F.H. Bennet, D.N. Neshev, P.D. Rasmussen, O. Bang, W. Królikowski, A. Bjarklev, Y.S. Kivshar, Opt. Exp. **15**, 12145 (2007)
9. J. Hukriede, D. Runde, D. Kip, J. Phys. D **36**, R1 (2003)
10. J. Li, B. Liang, Y. Liu, P. Zhang, J. Zhou, S.O. Klimonsky, A.S. Slesarev, Y.D. Tretyakov, L.O. Faolain, T.F. Krauss, Adv. Mater. **22**, 1 (2010); M. Feng, Y. Liu, Y. Li, X. Xie, J. Zhou, Opt. Exp. **19**, 7222 (2011); Y. Li, B.A. Malomed, J. Wu, W. Pang, S. Wang, J. Zhou, Phys. Rev. A **84**, 043839 (2011); B. Liang, Y. Liu, J. Li, L. Song, Y. Li, J. Zhou, K.S. Wong, J. Micromech. Microeng. **22**, 035013 (2012)
11. Y. Li, B.A. Malomed, M. Feng, J. Zhou, Phys. Rev. A **82**, 063813 (2010); W. Pang, J. Wu, Z. Yuan, Y. Liu, G. Chen, J. Phys. Soc. Jpn. **80**, 113401 (2011); J. Wu, M. Feng, W. Pang, S. Fu, Y. Li, J. Nonlinear Opt. Phys. **20**, 193 (2011); Y. Li, W. Pang, S. Fu, B.A. Malomed, Phys. Rev. A **85**, 053821 (2012)
12. M. Fleischhauer, A. Imamoğlu, J.P. Marangos, Rev. Mod. Phys. **77**, 633 (2005); H. Schmidt, A. Imamoğlu, Opt. Lett. **21**, 1936 (1996); H. Wang, D. Goorskey, M. Xiao, Phys. Rev. Lett. **87**, 073601 (2001)
13. S. Inouye, M.R. Andrews, J. Stenger, H.J. Miesner, D.M. Stamper-Kurn, W. Ketterle, Nature (London) **392**, 151 (1998); Ph. Courteille, R.S. Freeland, D.J. Heinzen, F.A. van Abeelen, B.J. Verhaar, Phys. Rev. Lett. **81**, 69 (1998); J.L. Roberts, N.R. Claussen, J.P. Burke, C.H. Greene, E.A. Cornell, C.E. Wieman, ibid. **81**, 5109 (1998)
14. P.O. Fedichev, Y. Kagan, G.V. Shlyapnikov, J.T.M. Walraven, Phys. Rev. Lett. **77**, 2913 (1996); M. Theis, G. Thalhammer, K. Winkler, M. Hellwig, G. Ruff, R. Grimm, J.H. Denschlag, ibid. **93**, 123001 (2004)
15. S. Ghanbari, T.D. Kieu, A. Sidorov, P. Hannaford, J. Phys. B **39**, 847 (2006); A. Abdelrahman, P. Hannaford, K. Alameh, Opt. Express **17**, 24358 (2009)
16. A.E. Fetter, Rev. Mod. Phys. **81**, 647 (2009)
17. P. Engels, I. Coddington, P.C. Haljan, V. Schweikhard, E.A. Cornell, Phys. Rev. Lett. **90**, 170405 (2003)
18. A.L. Fetter, Phys. Rev. A **64**, 063608 (2001); E. Lundh, ibid. **65**, 043604 (2002); K. Kasamatsu, M. Tsubota, M. Ueda, ibid. **66**, 053606 (2002); G.M. Kavoulakis, G. Baym, New J. Phys. **5**, 51 (2003); A. Aftalion, I. Danaila, Phys. Rev. A **69**, 033608 (2004); U.R. Fischer, G. Baym, Phys. Rev. Lett. **90**, 140402 (2003); A.D. Jackson, G.M. Kavoulakis, Phys. Rev. A **70**, 023601 (2004); T.K. Ghosh, ibid. **69**, 043606 (2004)
19. K. Kasamatsu, M. Tsubota, Phys. Rev. A **76**, 023606 (2009)

20. N.K. Wilkin, J.M.F. Gunn, R.A. Smith, Phys. Rev. Lett. **80**, 2265 (1998); B. Mottelson, ibid. **83**, 2695 (1999); C.J. Pethick, L.P. Pitaevskii, Phys. Rev. A **62**, 033609 (2000); E. Lundh, A. Collin, K.-A. Suominen, Phys. Rev. Lett. **92**, 070401 (2004); G.M. Kavoulakis, A.D. Jackson, G. Baym, Phys. Rev. A **70**, 043603 (2004); A. Collin, ibid. **73**, 013611 (2006); A. Collin, E. Lundh, K.-A. Suominen, Phys. Rev. A **71**, 023613 (2005); S. Bargi, G.M. Kavoulakis, S.M. Reimann, ibid. **73**, 033613 (2006); H. Sakaguchi, B.A. Malomed, ibid. **78**, 063606 (2008)
21. V. Schweikhard, I. Coddington, P. Engels, S. Tung, E.A. Cornell, Phys. Rev. Lett. **93**, 210403 (2004); S. Tung, V. Schweikhard, E.A. Cornell, ibid. **97**, 240402 (2006)
22. J.W. Reijnders, R.A. Duine, Phys. Rev. Lett. **93**, 060401 (2004); Phys. Rev. A **71**, 063607 (2005); H. Pu, L.O. Baksmaty, S. Yi, N.P. Bigelow, Phys. Rev. Lett. **94**, 190401 (2005); H. Pu, L.O. Baksmaty, S. Yi, N.P. Bigelow, ibid. **94**, 190401 (2005); R. Bhat, L.D. Carr, M.J. Holland, ibid. **96**, 060405 (2006); K. Kasamatsu, M. Tsubota, ibid. **97**, 240404 (2006); M.P. Mink, C. Morais Smith, R.A. Duine, Phys. Rev. A **79**, 013605 (2009)
23. H. Sakaguchi, B.A. Malomed, Phys. Rev. A **75**, 013609 (2007); **79**, 043606 (2009)
24. R.A. Williams, S. Al-Assam, C.J. Foot, Phys. Rev. Lett. **104**, 050404 (2010)
25. G. Kakarantzas, A. Ortigosa-Blanch, T.A. Birks, P.St.J. Russell, L. Farr, F. Couny, B.J. Mangan, Opt. Lett. **28**, 158 (2003)
26. G.K.L. Wong, M.S. Kang, H.W. Lee, F. Biancalana, C. Conti, T. Weiss, P.St.J. Russell, Science **337**, 446 (2012)
27. R.Y. Chiao, Y.S. Wu, Phys. Rev. Lett. **57**, 933 (1986); A. Tomita, R.Y. Chiao, ibid. **57**, 937 (1986); V.I. Kopp, A.Z. Genack, Opt. Lett. **28**, 1876 (2003); M. Ornigotti, G. Della Valle, D. Gatti, S. Longhi, Phys. Rev. A **76**, 023833 (2007)
28. B.A. Malomed, Phys. Rev. A **43**, 410 (1991)
29. S. Jia, J.W. Fleischer, Phys. Rev. A **79**, 041804(R) (2009)
30. H. Saito, M. Ueda, Phys. Rev. Lett. **93**, 220402 (2004); S. Schwartz, M. Cozzini, C. Menotti, I. Carusotto, P. Bouyer, S. Stringari, New J. Phys. **8**, 162 (2006)
31. Y.V. Kartashov, B.A. Malomed, L. Torner, Phys. Rev. A **75**, 061602(R) (2007)
32. L. Wen, H. Xiong, B. Wu, Phys. Rev. A **82**, 053627 (2010)
33. G.J. Milburn, J. Corney, E.M. Wright, D.F. Walls, Phys. Rev. A **55**, 4318 (1997); I. Zapata, F. Sols, A.J. Leggett, ibid. **57**, R28 (1998); E.A. Ostrovskaya, Y.S. Kivshar, M. Lisak, B. Hall, F. Cattani, D. Anderson, ibid. **61**, 031601(2000); R. D'Agosta, B.A. Malomed, C. Presilla, Phys. Lett. A **275**, 424 (2000); R.K. Jackson, M.I. Weinstein, J. Stat. Phys. **116**, 881 (2004); K.W. Mahmud, H. Perry, W.P. Reinhardt, Phys. Rev. A. **71**, 023615 (2005); H.T. Ng, P.T. Leung, ibid. **71**, 013601 (2005); T. Schumm, S. Hofferberth, L.M. Andersson, S. Wildermuth, S. Groth, I. Bar-Joseph, J. Schmiedmayer, P. Krüger, Nat. Phys. **1**, 57 (2005); M. Albiez, R. Gati, J. Fölling, S. Hunsmann, M. Cristiani, M.K. Oberthaler, Phys. Rev. Lett. **95**, 010402 (2005)
34. Y. Li, W. Pang, B.A. Malomed, Phys. Rev. A **86**, 023832 (2012)
35. S. Gupta, K.W. Murch, K.L. Moore, T.P. Purdy, D.M. Stamper-Kurn, Phys. Rev. Lett. **95**, 143201 (2005); A.S. Arnold, C.S. Garvie, E. Riis, Phys. Rev. A **73**, 041606(R) (2006); I. Lesanovsky, W. von Klitzing, ibid. **99**, 083001 (2007); C. Ryu, M.F. Andersen, P. Cladé, V. Natarajan, K. Helmerson, W.D. Phillips, Phys. Rev. Lett. **99**, 260401 (2007); A. Ramanathan, K.C. Wright, S.R. Muniz, M. Zelan, W.T. Hill III, C.J. Lobb, K. Helmerson, W.D. Phillips, G.K. Campbell, ibid. **106**, 130401 (2011)
36. L. Salasnich, A. Parola, L. Reatto, Phys. Rev. A **74**, 031603(R) (2006); R. Kanamoto, H. Saito, M. Ueda, ibid. **73**, 033611 (2006); M. Modugno, C. Tozzo, F. Dalfovo, ibid. **74**, 061601 (R) (2006); A.V. Carpentier, H. Michinel, EPL **78**, 10002 (2007); P. Mason, N.G. Berloff, Phys. Rev. A **79**, 043620 (2009); J. Brand, T.J. Haigh, U. Zülicke, ibid. **80**, 011602 (R) (2009); P. Capuzzi, D.M. Jezek, J. Phys. B: At. Mol. Opt. Phys. **42**, 145301 (2009); A. Aftalion, P. Maso, ibid. **81**, 023607 (2010); J. Smyrnakis, M. Magiropoulos, G.M. Kavoulakis, A.D. Jackson, ibid. **81**, 063601 (2010); S. Zöllner, G.M. Bruun, C.J. Pethick, S.M. Reimann, Phys. Rev. Lett. **107**, 035301 (2011); Z.-W. Zhou, S.-L. Zhang, X.-F. Zhou, G.-C. Guo, X. Zhou, H. Pu, Phys. Rev. A **83**, 043626 (2011); M. Abad, M. Guilleumas,

R. Mayol, M. Pi, D.M. Jezek, ibid. **84**, 035601 (2011); X. Zhou, S. Zhang, Z. Zhou, B.A. Malomed, H. Pu, ibid. **85**, 023603 (2012); S.K. Adhikari, ibid. **85**, 053631 (2012)
37. S. Schwartz, M. Cozzini, C. Menotti, I. Carusotto, P. Bouyer, S. Stringari, New J. Phys. **8**, 162 (2006); J. Smyrnakis, S. Bargi, G.M. Kavoulakis, M. Magiropoulos, K. Kärkkäinen, S.M. Reimann, Phys. Rev. Lett. **103**, 100404 (2009)
38. J. Yang, T.I. Lakoba, Stud. Appl. Math. **118**, 153 (2007); **120**, 265 (2008)
39. G. Iooss, D.D. Joseph, *Elementary Stability and Bifurcation Theory* (Springer, New York, 1980)
40. M. Vakhitov, A. Kolokolov, Izvestiya VUZov Radiofizika **16**, 1020 (1973) (in Russian; English translation: Radiophys. Quantum. Electron. **16**, 783 (1973)); L. Bergé, Phys. Rep. **303**, 259 (1998); E.A. Kuznetsov, F. Dias, ibid. **507**, 43 (2011)
41. H. Sakaguchi, B.A. Malomed, Phys. Rev. A **81**, 013624 (2010)

Light Propagation in Two Dimensional Plasmonic Arrays

Danhua Wang and Alejandro B. Aceves

Abstract We present results on the dynamics of light beams propagating in two dimensional dielectric/metallic waveguides. We show how different configurations provide a rich dynamics of localization, solitary wave formation and instabilities.

1 Introduction

Light propagation in one dimensional coupled nonlinear waveguide arrays is a mature area of research with numerous experimental and theoretical discoveries on localization, discrete soliton formation and modulational instability dynamics to name some [11]. The pioneering experiment demonstrating light localization [5] triggered efforts by many groups which lead to a large body of work in a variety of arrays in different optical media (fibers, liquid crystals, etc.). This, together with the parallel theoretical studies on the discrete nonlinear Schrödinger equation (DNLSE) and the long-wavelength integrable NLSE approximations, has advanced our understanding of the dynamics and possible applications in this area. Amongst the advances that followed from one dimensional arrays in large measure due to the emergence of photonics crystals was that of two dimensional arrays, for which spatio-temporal localization and optical bullet formation first proposed in [2] and later experimentally demonstrated in [14]. For a long time, the models studied were uniform and only recently interesting scenarios that depart from uniformity in the array have been explored. Two examples are: disordered arrays where Anderson localization has been demonstrated [18] and binary arrays, where either the array consists of dielectric waveguides of different size and/or unequal spacing

D. Wang · A.B. Aceves (✉)
Department of Mathematics, Southern Methodist University, Dallas, TX 75275, USA
e-mail: danhuaw@mail.smu.edu; aaceves@mail.smu.edu

R. Carretero-González et al. (eds.), *Localized Excitations in Nonlinear Complex Systems*, Nonlinear Systems and Complexity 7, DOI 10.1007/978-3-319-02057-0_9,

[15, 19] or plasmonic arrays of alternating metallic/dielectric waveguides [3]. In both cases, where the coupled mode theory applies, the resulting system of equations break into two pairs which together with their long wavelength continuum approximations depart from NLSE turning into models such as the coupled mode equations (with gap soliton solutions) or the Dirac equations [1]. Distinct features that result from these models include the presence of gaps and singular (diabolical) points in the dispersion relation [1, 16]. These models also predict the existence (and co-existence) of bright and dark [10] solitary waves. By now the type of one dimensional binary arrays continues to be extensively studied [4, 6–9, 12, 13, 17].

In this work we report results on two dimensional binary (metallic/dielectric) nonlinear waveguides (see Fig. 1), which can be seen as a two dimensional tension of the work in [3]. We discuss dispersion properties both in the discrete and in the continuum approximation, modulational instability dynamics. Results obtained also include the existence and stability of solitary-wave solutions.

2 Theoretical Model and Dispersion Properties: Case 1

We consider two dimensional structures where we alternate dielectric and metallic waveguides and extend the work in [3] to a geometry. Specifically we have studied the configurations shown in Fig. 1. For simplicity we briefly present the derivation of the coupled mode equations for one dimensional plasmonic waveguides. As with similar systems, we model light propagation in this array according to the coupled mode theory in which the complex electric field envelope propagation in the N-th waveguide $E_N(z)$ is governed by the system of equations

$$iE_{N\,z} + \beta_N\, E_N + C_{N-1}\, E_{N-1} + C_{N+1}\, E_{N+1} + \gamma_N\, |E_N|^2\, E_N = 0\,, \tag{1}$$

where β_N is the propagation constant of each individual waveguide and C_{N-1}, C_{N+1} are the coupling coefficients with the $N-1$ and $N+1$ waveguide respectively. These coefficients are proportional to the overlap of the N-th guiding mode with its neighbors so the coupling strength depends on their separation. The non-uniformity in the arrays accounts for the explicit N-dependence. In these particular arrays of alternating metallic/dielectric waveguides, this non-uniformity emerges from the fact that while the fundamental guiding mode in the dielectric is even, it is odd for the metallic waveguide. As a consequence, the pairs C_{N-1}, C_{N+1} are positive, for N even and negative for N odd. Effectively, Eq. (1) can be re-written as a system for $E_{2N} = A_n, E_{2N-1} = B_n$ [3]. A this time, by applying the same principles, we present the governing equations for a *two* dimensional array of alternating dielectric/metallic waveguides

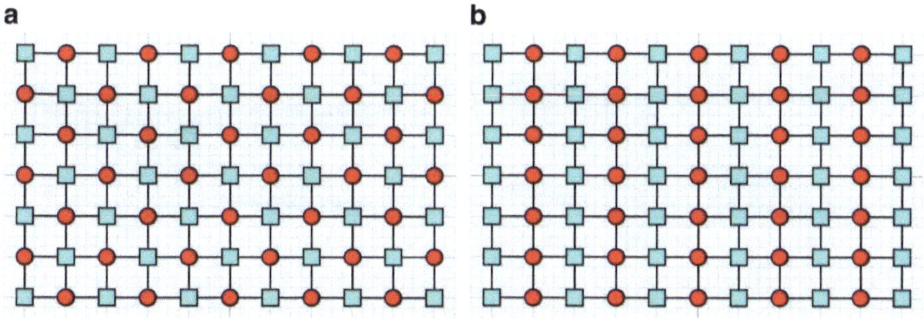

Fig. 1 (**a**) The metal and dielectric are alternating in both horizontal and vertical direction. (**b**) The metal and dielectric are alternating only in one direction, either in horizontal or in vertical direction

$$\begin{aligned} iA_{n,m\,z} + \frac{\Delta\beta}{2} A_{n,m} + C_{n,m} B_{n,m} + C_{n+1,m} B_{n+1,m} + C_{n,m+1} B_{n,m+1} \\ + C_{n-1,m} B_{n-1,m} + \gamma_1 |A_{n,m}|^2 A_{n,m} = 0, \\ iB_{n,m\,z} - \frac{\Delta\beta}{2} B_{n,m} + C_{n,m} A_{n,m} + C_{n+1,m} A_{n+1,m} + C_{n,m-1} A_{n,m-1} \\ + C_{n-1,m} A_{n-1,m} + \gamma_2 |B_{n,m}|^2 B_{n,m} = 0. \end{aligned} \tag{2}$$

Here we first consider arrays designed in such a way that the coupling between successive waveguides switches periodically from C to $-C\ (1+\varepsilon)$ reflecting the fact that the metal and dielectric guides are alternating in both the horizontal and vertical directions as shown in Fig. 1a. Without loss of generality, we take $\Delta\beta = 0$ of (2) everywhere in this paper.

We start by studying the dispersion properties in the linear regime. For this, we assume plane-wave solutions, $(A_{n,m},\ B_{n,m}) \propto e^{i(n\,k_x + m\,k_y + k_z\,z)}$, whose linear dispersion relation is

$$k_z^2 = \left(\frac{\Delta\beta}{2}\right)^2 + 4 - 2\cos(2k_x) - 2\cos k_y - 2\cos(k_x + k_y) + 2\cos(k_x - k_y)\,. \tag{3}$$

As we will see in the longwave analysis below, this dispersion relation has a gap near $(k_x, k_y) = (0, 0)$ as shown in Figs. 2a and 3a.

The longwave limit $(k_x, k_y \approx 0)$ can be better analysed if we use the continuum approximation given by the expansions

$$A_{n\pm1,m}(z) = u(x,y,z) \pm u_x(x,y,z) + \frac{1}{2}u_{xx}(x,y,z) + \cdots,$$

$$A_{n,m\pm1}(z) = u(x,y,z) \pm u_y(x,y,z) + \frac{1}{2}u_{yy}(x,y,z) + \cdots,$$

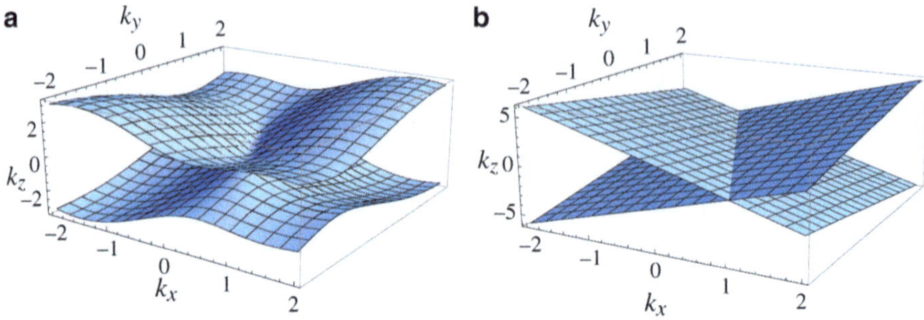

Fig. 2 (**a**) Linear dispersion relation (3). (**b**) Linear dispersion relation (5)

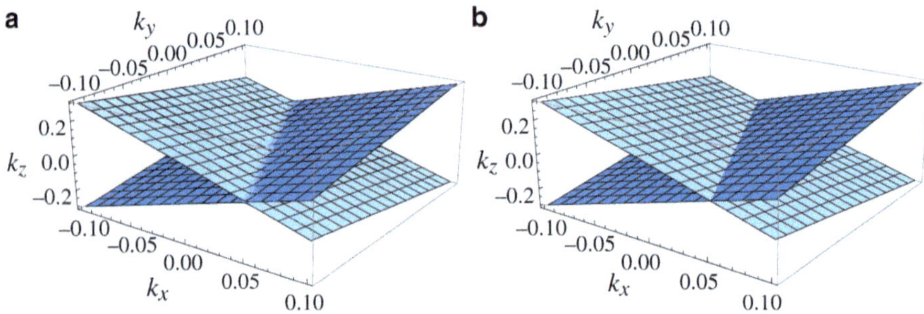

Fig. 3 Zoom in Fig. 2 at $(k_x, k_y) = (0, 0)$. (**a**) Linear dispersion relation (3). (**b**) Linear dispersion relation (5)

$$B_{n\pm1,m}(z) = w(x, y, z) \pm w_x(x, y, z) + \frac{1}{2}w_{xx}(x, y, z) + \cdots,$$

$$B_{n,m\pm1}(z) = w(x, y, z) \pm w_y(x, y, z) + \frac{1}{2}w_{yy}(x, y, z) + \cdots,$$

obtaining (as a first-order approximation)

$$\begin{aligned} i\,u_z + \frac{\Delta\beta}{2}u + 2w_x + w_y + \varepsilon\, w + \gamma_1|u|^2\, u = 0\,, \\ i\,w_z - \frac{\Delta\beta}{2}w - 2u_x - u_y + \varepsilon\, u + \gamma_2|w|^2\, w = 0\,, \end{aligned} \tag{4}$$

where $C_{n,m} = -1 + \varepsilon$. Observe that the continuum limit preserves (modulus a trivial scaling) as it should the 90° rotational invariance of the array. In the linear case if we set $\Delta\beta = 0$ and $\varepsilon = 0$ the dispersion relation of Eqs. (4) reads

$$k_z = \pm|2k_x + k_y| \tag{5}$$

which has a degenerate (diabolic) line at $2k_x + k_y = 0$ as shown in Figs. 2b and 3b. In Fig. 3 we highlight that as expected, in the longwave limit ($k_x, k_y \approx 0$) both the discrete model and its continuous approximation have the same singular-like linear dispersion relation.

To see how this dispersion relation is regularized by the addition of nonlinearity in Eqs. (4) (i.e., $\gamma_1, \gamma_2 \neq 0$) we insert the plane wave solution form leading to the system

$$-k_z A + iB(2k_x + k_y) + \gamma_1 A^3 = 0\,, \tag{6}$$

$$-k_z B - iA(2k_x + k_y) + \gamma_2 B^3 = 0\,. \tag{7}$$

We rewrite Eqs. (6) and (7) as a system

$$\begin{pmatrix} -k_z + \gamma_1 A^2 & i(2k_x + k_y) \\ -i(2k_x + k_y) & -k_z + \gamma_2 B^2 \end{pmatrix} \begin{pmatrix} A \\ B \end{pmatrix} = \mathbf{0}\,.$$

for which the determinant of the coefficient matrix is zero, if,

$$(k_z - \gamma_1 A^2)(k_z - \gamma_2 B^2) - (2k_x + k_y)^2 = 0\,. \tag{8}$$

The main task here is we want to find the explicit form of k_z depends on k_x and k_y from Eq. (8), i.e., $k_z = f(k_x, k_y)$. Let $B = \alpha A$, where α is a constant. Then Eq. (8) becomes

$$k_z^2 - (\gamma_1 + \gamma_2\alpha^2)A^2 k_z + \gamma_1\gamma_2\alpha^2 A^4 - (2k_x + k_y)^2 = 0\,. \tag{9}$$

From Eq. (6), we have

$$-\,k_z + \gamma_1 A^2 + \alpha i(2k_x + k_y) = 0\,. \tag{10}$$

In Eqs. (9) and (10), by eliminating α, we obtain

$$k_z^2 - \left(\gamma_1 - \gamma_2\frac{(-k_z + \gamma_1 A^2)^2}{(2k_x + k_y)^2}\right)A^2 k_z - \gamma_1\gamma_2\frac{(-k_z + \gamma_1 A^2)^2}{(2k_x + k_y)^2}A^4 - (2k_x + k_y)^2 = 0\,.$$

Finally, if we define $K = 2k_x + k_y$, the above equation reads

$$k_z^2 - \left(\gamma_1 - \gamma_2\frac{(-k_z + \gamma_1 A^2)^2}{K^2}\right)A^2 k_z - \gamma_1\gamma_2\frac{(-k_z + \gamma_1 A^2)^2}{K^2}A^4 - K^2 = 0\,. \tag{11}$$

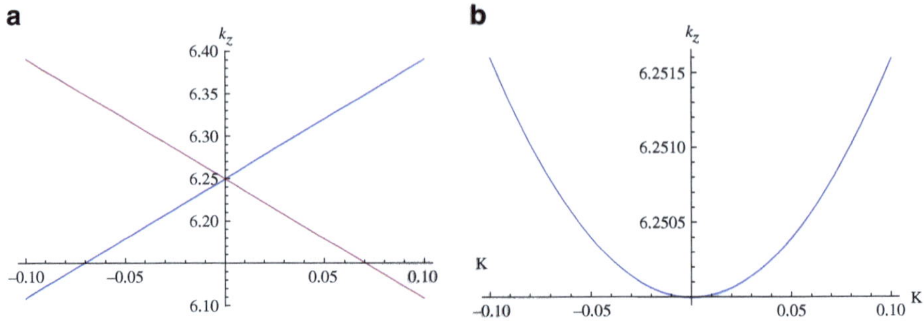

Fig. 4 $A = 2.5$, $\gamma_1 = 1$, $\gamma_2 = -0.5$. (**a**) Nonlinear dispersion relation $k_z \sim \gamma_1 A^2 \pm \sqrt{-\frac{\gamma_1}{\gamma_2}} K$ if $\gamma_1\gamma_2 < 0$. (**b**) Nonlinear dispersion relation $k_z \sim \gamma_1 A^2 + \frac{1}{\gamma_1 A^2} K^2$ if $\gamma_1\gamma_2 > 0$

So we have found $k_z = f(K)$ here. We are interested in the behavior of k_z for K approaching 0 by use of asymptotic perturbation methods. We find that $k_z \sim \gamma_1 A^2 \pm \sqrt{-\frac{\gamma_1}{\gamma_2}} K$ if $\gamma_1\gamma_2 < 0$ and $k_z \sim \gamma_1 A^2 + \frac{1}{\gamma_1 A^2} K^2$ if $\gamma_1\gamma_2 > 0$ (see Fig. 4). Interestingly, while the case $\gamma_1\gamma_2 < 0$ is regular for all amplitudes A, it is not the case for $\gamma_1\gamma_2 > 0$. Here we need analyse separately the expansion near $A = 0$. If $A = 0$, from Eq. (11), we know $k_z^2 = K^2$, i.e., $k_z = \pm K$. So we can assume $k_z(A) \sim \pm K + O(A^p)$ without any loss of generality, where p is to be determined by inserting this approximation into (11). We find that $p = 2$ and $k_z(A) \sim \pm K + \frac{\gamma_1-\gamma_2}{2} A^2$ for $\gamma_1 \neq \gamma_2$. If $\gamma_1 = \gamma_2$, $p = 4$ and $k_z(A) \sim \pm K \pm \frac{3\gamma_1^2}{2K} A^4$. The complete picture of the nonlinear dispersion relation for $\gamma_1\gamma_2 > 0$ for all amplitudes would require matching the small A expansion with the finite A formula.

3 Theoretical Model and Dispersion Properties: Case 2

In this section, we study an alternative array geometry as shown in Fig. 1b where the metal and dielectric waveguides are alternating only in one direction, which we take it to be Case 2. Here we consider the horizontal direction. Then the governing equations read as the following two sets of coupled equations with constant coefficients $C_{n,m}$:

$$\begin{aligned} iA_{n,m\,z} + \frac{\Delta\beta}{2} A_{n,m} + C_{n,m}\, B_{n,m} + C_{n,m+1}\, B_{n,m+1} + C_{n-1,m}\, A_{n-1,m} \\ + C_{n+1,m}\, A_{n+1,m} + \gamma_1 \left|A_{n,m}\right|^2 A_{n,m} = 0\,, \\ iB_{n,m\,z} - \frac{\Delta\beta}{2} B_{n,m} + C_{n,m}\, A_{n,m} + C_{n,m-1}\, A_{n,m-1} + C_{n-1,m}\, B_{n-1,m} \\ + C_{n+1,m}\, B_{n+1,m} + \gamma_2 \left|B_{n,m}\right|^2 B_{n,m} = 0\,. \end{aligned} \tag{12}$$

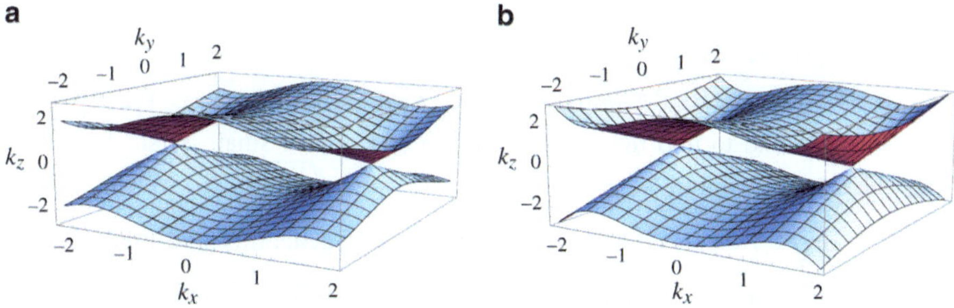

Fig. 5 (**a**) Linear dispersion relation (13). (**b**) Linear dispersion relation (15)

In this case, the signs of coefficients in each row are positive and negative alternately. As before, in the n-th row, we usually set $C_{n,m} = -1$ without loss of generality, then $C_{n,m\pm 1} = 1$. In the vertical direction, we have to set the same signs of coefficients in each column and the signs of two adjacent columns are opposite since one column is for $A_{n,m}$ and the other is for $B_{n,m}$. Here we assume the sign of the coefficients of columns for $A_{n,m}$ are negative, i.e., $C_{n\pm 1,m} = -1$ in the first equation of (12) and $C_{n\pm 1,m} = 1$ in the second equation of (12). In this case, the linear dispersion relation of Eqs. (12) whose plot is shown in Fig. 5a is

$$k_z^2 = \left(\frac{\Delta\beta}{2}\right)^2 + 4 - 2\cos k_x\, \Delta\beta + 2\cos(2k_x) - 2\cos k_y\,. \tag{13}$$

Here, the longwave approximation leads to the continuous system of Eqs. (12) is

$$\begin{aligned} i\, u_z + \frac{\Delta\beta}{2} u + w_y - 2u - \delta u_{xx} + \varepsilon\,(u + w) + \gamma_1 |u|^2\, u = 0\,, \\ i\, w_z - \frac{\Delta\beta}{2} w - u_y + 2w + \delta w_{xx} + \varepsilon\,(u + w) + \gamma_2 |w|^2\, w = 0\,, \end{aligned} \tag{14}$$

where $C_{n,m} = -1 + \varepsilon$. Here in the vertical (i.e., x-) direction we consider the first two terms of the Taylor expansion. We introduce a parameter δ to allow for different scalings in each direction. If we set $\delta = 0$ we effectively return to the one dimensional models (see equation in [3]) but with a different phase mismatch.

For the case where $\epsilon = 0$ and $\delta = 1$, the linear dispersion relation (see Fig. 5b) reads

$$k_z = \pm\sqrt{(2 - k_x^2)^2 + k_y^2} \tag{15}$$

Now we want to find the nonlinear dispersion relation of Eqs. (14) (i.e., $\gamma_1, \gamma_2 \neq 0$), then we have

$$-k_z A + i k_y B - 2A + k_x^2 A + \gamma_1 A^3 = 0\,, \tag{16}$$

$$-k_z B - i k_y A + 2B - k_x^2 B + \gamma_2 B^3 = 0\,. \tag{17}$$

The same technique as in Sect. 2 applies; namely we study the behavior of k_z for k_y approaching 0. We obtain $k_z \sim \gamma_1 A^2 - 2 \pm \sqrt{-\frac{\gamma_1 A^2 - 4}{\gamma_2 A^2}}\, k_y$ if $\frac{\gamma_1 A^2 - 4}{A^2 \gamma_2} < 0$, otherwise $k_z \sim \gamma_1 A^2 - 2 + \frac{1}{\gamma_1 A^2 - 4} k_y^2$ (similar to Fig. 4). As in the previous case, For $\frac{\gamma_1 A^2 - 4}{A^2 \gamma_2} < 0$, we need to find to find the right expansion of k_z for A small. First, if $A = 0$, after calculation, we find, $k_z^2 = 4 + k_y^2$, i.e., $k_z = \pm\sqrt{4 + k_y^2}$. So we can assume $k_z(A) \sim \pm\sqrt{4 + k_y^2} + O(A^p)$ without any loss of generality and find p. After substitution, we get $p = 2$ and $k_z(A) \sim \pm\sqrt{4 + k_y^2} + \frac{\gamma_1 + \gamma_2}{2} A^2$ for $\gamma_1 + \gamma_2 \neq 0$. If $\gamma_1 + \gamma_2 = 0$, we obtain $p = 4$, then $k_z(A) \sim \pm\sqrt{4 + k_y^2} + \frac{3\gamma_1^2}{4} A^4$. In all instances the role of nonlinearity is to remove the singular behavior and opened a gap of the linear dispersion relation.

4 Modulation Instability

Modulational instabilities of continuous wave (CW) solutions are of interest as they can develop into solitary wave formation. As with other nonlinear discrete systems, one gains much insight by studying the more manageable continuum limit, understanding that comparisons are only meaningful of instabilities occurring at long wavelengths.

Plane wave solutions of the continuum PDE approximations take the form

$$\begin{pmatrix} u(x,y,z) \\ w(x,y,z) \end{pmatrix} = \begin{pmatrix} A e^{i\gamma_1 A^2 z} \\ B e^{i\gamma_2 B^2 z} \end{pmatrix}.$$

for both configurations. Looking first at the case $\epsilon = 0$ we linearize about these CW solutions,

$$\begin{pmatrix} u(x,y,z) \\ w(x,y,z) \end{pmatrix} = \begin{pmatrix} (A + \delta_1(x,y,z)) e^{i\gamma_1 A^2 z} \\ (B + \delta_2(x,y,z)) e^{i\gamma_2 B^2 z} \end{pmatrix}, \tag{18}$$

obtaining for Case 1 (i.e., Eq. (4))

$$\begin{aligned} i\,\delta_{1z} + 2\delta_{2x} + \delta_{2y} + \gamma_1 A^2(\delta_1 + \delta_1^*) &= 0\,, \\ i\,\delta_{2z} - 2\delta_{1x} - \delta_{1y} + \gamma_2 B^2(\delta_2 + \delta_2^*) &= 0\,, \end{aligned}$$

whose solutions take the general form

$$\begin{pmatrix} \delta_1(x,y,z) \\ \delta_2(x,y,z) \end{pmatrix} = \begin{pmatrix} \alpha_1 \\ \beta_1 \end{pmatrix} \mathrm{e}^{i(x\rho_x + y\rho_y - z\rho_z)} + \begin{pmatrix} \alpha_2 \\ \beta_2 \end{pmatrix} \mathrm{e}^{-i(x\rho_x + y\rho_y - z\rho_z)} . \tag{19}$$

leading to the system

$$\begin{pmatrix} \rho_z + \gamma_1 A^2 & \gamma_1 A^2 & i(2\rho_x + \rho_y) & 0 \\ \gamma_1 A^2 & -\rho_z + \gamma_1 A^2 & 0 & -i(2\rho_x + \rho_y) \\ -i(2\rho_x + \rho_y) & 0 & \rho_z + \gamma_2 B^2 & \gamma_2 B^2 \\ 0 & i(2\rho_x + \rho_y) & \gamma_2 B^2 & -\rho_z + \gamma_2 B^2 \end{pmatrix} \begin{pmatrix} \alpha_1 \\ \alpha_2 \\ \beta_1 \\ \beta_2 \end{pmatrix} = \mathbf{0} . \tag{20}$$

Non-trivial solutions require the determinant of the coefficient matrix in (20) to be zero, i.e.,

$$\rho_z^4 + 2(2\rho_x + \rho_y)^2(A^2\gamma_1 + B^2\gamma_2)\rho_z - (2\rho_x + \rho_y)^4 = 0 . \tag{21}$$

Modulations instabilities (MI) arise when ρ_z is complex, which is the case in the small wave-number regime as shown in Fig. 6a. This is verified by an asymptotic analysis done of (21) which proves that the root $\rho = 0$ of multiplicity 4 at $2\rho_x + \rho_y = 0$ bifurcates into two complex and two real roots

$$\begin{aligned} &\text{I. } C = A^2\gamma_1 + B^2\gamma_2 > 0 : \rho_1 = \frac{\mu^2}{2} C \,,\ \rho_2 = 2(\mu^2 C)^{\frac{1}{3}} , \\ &\qquad \rho_3 = 2(\mu^2 C)^{\frac{1}{3}} e^{i\pi/3} \,,\ \rho_4 = 2(\mu^2 C)^{\frac{1}{3}} e^{5i\pi/3} . \\ &\text{II. } C = A^2\gamma_1 + B^2\gamma_2 < 0 : \rho_1 = \frac{\mu^2}{2} C \,,\ \rho_2 = 2(\mu^2 |C|)^{\frac{1}{3}} , \\ &\qquad \rho_3 = 2(\mu^2 |C|)^{\frac{1}{3}} e^{2i\pi/3} \,,\ \rho_4 = 2(\mu^2 |C|)^{\frac{1}{3}} e^{4i\pi/3} . \end{aligned} \tag{22}$$

where $\mu = 2\rho_x + \rho_y$.

While we will not be presenting a more detailed parameter space analysis of MI regimes, we are certain that instabilities are generic for wide ranges of wavenumbers. The numerical simulations in the section that follows are a good representation that shade some light into the dynamics that develops. If instead we consider possible modulation instabilities of (4) with $\varepsilon \neq 0$ for Case 1 where the solution can be written as

$$\begin{pmatrix} u(x,y,z) \\ w(x,y,z) \end{pmatrix} = \begin{pmatrix} A\mathrm{e}^{i\gamma_1 Bz} \\ C\mathrm{e}^{i\gamma_2 Dz} \end{pmatrix} ,$$

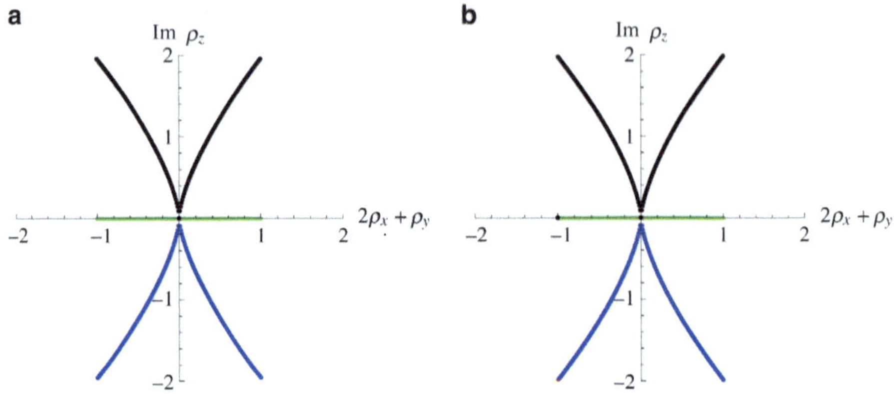

Fig. 6 Imaginary part of roots ρ_z. (**a**) (21). (**b**) (24)

with A, B, C, D satisfying $AC(\gamma_1 A^2 - \gamma_2 C^2) - \varepsilon(A^2 - C^2) = 0$ and $\gamma_1 B - \gamma_2 D = 0$.

Linearizing

$$\begin{pmatrix} u(x,y,z) \\ w(x,y,z) \end{pmatrix} = \begin{pmatrix} (A + \delta_1(x,y,z))e^{i\gamma_1 Bz} \\ (C + \delta_2(x,y,z))e^{i\gamma_2 Dz} \end{pmatrix}, \tag{23}$$

leads to the characteristic polynomial

$$\rho_z^4 + 2(2\rho_x + \rho_y)^2(A^2\gamma_1 + C^2\gamma_2 + 2\varepsilon)\rho_z - (2\rho_x + \rho_y)^4 = 0\,. \tag{24}$$

Observe that if $(2\rho_x + \rho_y) \approx 0$ an asymptotic balance of this equation gives two roots $\rho_z \approx \pm i(2\rho_x + \rho_y)$ indicating instabilities arise in the long wavelength $(\rho_x, \rho_y \approx 0)$ regime as shown in Fig. 6b.

For Case 2 with $\varepsilon = 0$, the solution of can be written as

$$\begin{pmatrix} u(x,y,z) \\ w(x,y,z) \end{pmatrix} = \begin{pmatrix} Ae^{i(\gamma_1 A^2 - 2)z} \\ Be^{i(\gamma_2 B^2 + 2)z} \end{pmatrix}.$$

for which the linearized equations are

$$\begin{aligned} i\,\delta_{1z} + \delta_{2y} - 2\delta_1 - \delta_{1xx} + (\gamma_1 A^2 - 2)(\delta_1 + \delta_1^*) &= 0\,, \\ i\,\delta_{2z} - \delta_{1y} + 2\delta_2 + \delta_{2xx} + (\gamma_2 B^2 + 2)(\delta_2 + \delta_2^*) &= 0\,. \end{aligned} \tag{25}$$

Use of the solution form (19) in (25) results in the system

$$\begin{pmatrix} \rho_z + \rho_{xx}^2 + \gamma_1 A^2 - 4 & \gamma_1 A^2 - 2 & i\rho_y & 0 \\ \gamma_1 A^2 - 2 & -\rho_z + \rho_{xx}^2 + \gamma_1 A^2 - 4 & 0 & -i\rho_y \\ -i\rho_y & 0 & \rho_z - \rho_{xx}^2 + \gamma_2 B^2 + 4 & \gamma_2 B^2 + 2 \\ 0 & i\rho_y & \gamma_2 B^2 + 2 & -\rho_z - \rho_{xx}^2 + \gamma_2 B^2 + 4 \end{pmatrix} \begin{pmatrix} \alpha_1 \\ \alpha_2 \\ \beta_1 \\ \beta_2 \end{pmatrix} = \mathbf{0}. \qquad (26)$$

For the determinant of the coefficient matrix in (26) to be zero, ρ_z satisfies the quartic polynomial

$$\rho_z^4 - 2[12 - 8\rho_x^2 + \rho_x^4 + A^2\gamma_1(-2+\rho_x^2) - B^2\gamma_2(-2+\rho_x^2) + \rho_y^2]\rho_z^2$$
$$-[-12 + 8\rho_x^2 - \rho_x^4 + 2B^2\gamma_2(-2+\rho_x^2) - \rho_y^2][12 - 8\rho_x^2$$
$$+\rho_x^4 + 2A^2\gamma_1(-2+\rho_x^2) + \rho_y^2] = 0.$$

Finally Case 2 with $\varepsilon \neq 0$ has as its characteristic equation,

$$\rho_z^4 - 2[12 - 8\rho_x^2 + \rho_x^4 + A^2\gamma_1(-2+\rho_x^2) - C^2\gamma_2(-2+\rho_x^2) + \rho_y^2]\rho_z^2$$
$$-[-12 - 4\varepsilon + 2(4+\varepsilon)\rho_x^2 - \rho_x^4 + 2C^2\gamma_2(-2+\rho_x^2) - \rho_y^2][12 - 4\varepsilon$$
$$+2(-4+\varepsilon)\rho_x^2 + \rho_x^4 + 2A^2\gamma_1(-2+\rho_x^2) + \rho_y^2] = 0.$$

Again, long wavelength instabilities are generic. To determine if this is true for the discrete systems, we performed numerical studies discussed in the next section.

4.1 Numerical Simulations

The theoretical results presented in the previous sections consider an infinite domain. Numerical studies must include boundaries and as such, it is of interest to consider realistic conditions that better reflect experimental situations. Here, we present numerical simulations for initial conditions representing amplitude values where plane waves are modulationally unstable, if one considers an infinite medium. Examples are shown for both Eqs. (2) and (12) again for $\Delta\beta = \epsilon = 0$. The simulations we highlight describe the fields $A_{n,m}(z)$ and $B_{n,m}(z)$ in the domain $(n,m,z) \in [1,N] \times [1,M] \times [0,\infty)$ for finite values of N and M and we use periodic boundary conditions, i.e. $A_{1,m}(z) = A_{N+1,m}(z)$, $B_{1,m}(z) = B_{N+1,m}(z)$,

$A_{n,1}(z) = A_{n,M+1}(z)$ and $B_{n,1}(z) = B_{n,M+1}(z)$. The initial conditions we consider are of two types:

(i) $A_{n,m}(0) = A$ and $B_{n,m}(0) = B$ for $n = 1, 2, \cdots, N-1, N, m = 1, 2, \cdots, M-1, M$.
(ii) $A_{n,m}(0) = A$ and $B_{n,m}(0) = B$ for $n = 4, 5, \cdots, N-4, N-3$, $m = 4, 5, \cdots, M-4, M-3$,
and $A_{n,m}(0) = B_{n,m}(0) = 0$ for $n = 1, 2, 3, N-2, N-1, N$, $m = 1, 2, 3, M-2, M-1, M$,

where A and B are constants. The difference between case (i) and (ii) is that for the second case we consider the edge of the array to have zero values for the field. All together, while these numerical simulations approximate the evolution of Eqs. (2) and (12) for CW initial conditions, it is clear that boundary conditions play a role, which at this time we cannot explain, in triggering instabilities different than those analysed in the previous section. Having said this, we think it is of interest to showcase these more realistic simulations where we chose a suitable number N spatial grid points for n and M spatial grid points for m, thus $n = 1, 2, \cdots, N-1, N$ and $m = 1, 2, \cdots, M-1, M$. Letting Δz be the propagation step size and $z_k = k\Delta z$, with $K = Z/\Delta z$ we highlight the solutions at $Z = 2, 20$, where we used $\Delta z = 0.002$. We take as initial condition $A^k_{n,m} = A^{(\mathrm{num})}(n, m, z_k)$ and $B^k_{n,m} = B^{(\mathrm{num})}(n, m, z_k)$ for $n = 1, 2, \cdots, N-1, N, m = 1, 2, \cdots, M-1, M$, and $k = 0, 1, \cdots, K-1, K$. The superscript "(num)" denotes the numerical solution. Since $A_{n,m}(z)$ and $B_{n,m}(z)$ are complex functions, we only present their magnitude. Briefly on the numerical approach, we split the fields into real and imaginary parts $A_{n,m}(z) = \mathrm{Re}(A_{n,m}(z)) + i\ \mathrm{Im}(A_{n,m}(z))$ and $B_{n,m}(z) = \mathrm{Re}(B_{n,m}(z)) + i\ \mathrm{Im}(B_{n,m}(z))$, obtaining for each index pair (n, m) a system of four ODEs. Results of the numerical simulations for these two types of initial conditions, for Eqs. (2) are shown in Figs. 7 and 8. Similarly, the results for initial conditions Type (i) for Eqs. (12) are shown in Fig. 9.

In all instances we can clearly see that initially longwavelength instabilities develop ($z = 2$) but for longer propagation distances ($z = 20$) and apparently in only one direction, short wavelength instabilities not captured by the continuum approximation appear. We also notice that for the long wavelength modulations, coherent structures seem to emerge (in Figs. 7 and 9 at $z = 2$, the localized mode in n is seen by 'gluing' both ends of the grid in that direction). That solitary wave solutions exist is demonstrated in the next section. Whether instabilities lead to their formation as suggested in Case 1, will not be presented in this work.

5 Solitary Wave Solutions

The simulations shown in the previous section reveal in some instances modulations along one direction suggesting they may lead to discrete one dimensional solitary waves. It may even be the case that such waves can be well described by the

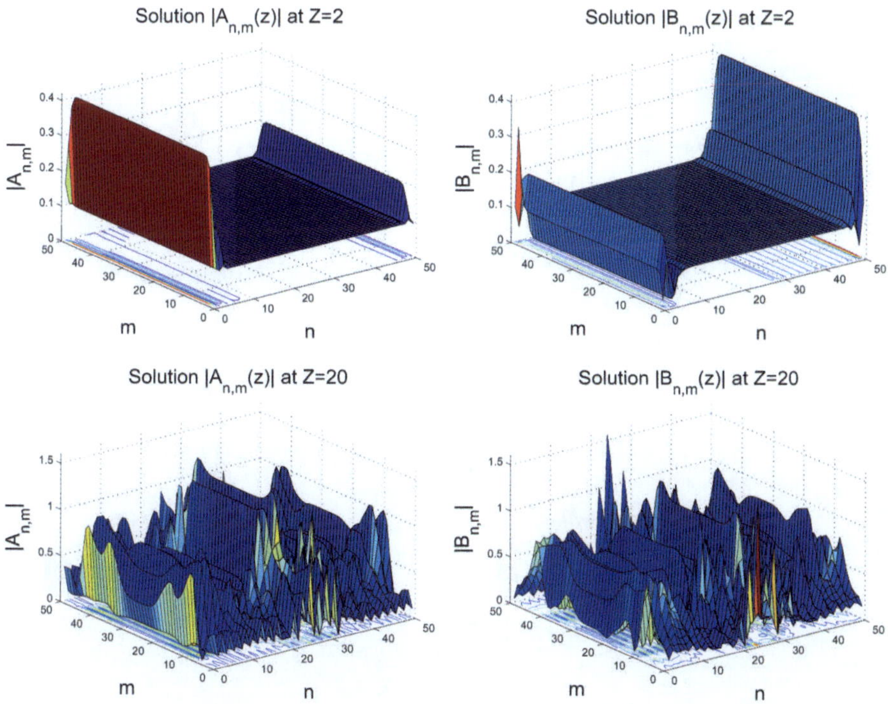

Fig. 7 $A = B = 0.1, N = M = 50$. The numerical solution of (2) with the initial conditions Type (i)

continuum approximation where their existence as we briefly discuss now is easy to proof.

Following the approach in Ref. [3], which only applies for Case 1, we seek solutions of the system (2) in the form

$$u(x, y, z) = \tfrac{1}{2}[K_1\, g_1(\xi) + i\, K_2\, g_2(\xi)]\, \exp(i\, \psi\, \cos\, Q)\,, \tag{27}$$

$$w(x, y, z) = \tfrac{1}{2i}[K_1\, g_1(\xi) - i\, K_2\, g_2(\xi)]\, \exp(i\, \psi\, \cos\, Q)\,, \tag{28}$$

$$\xi = \tfrac{\eta(x,y)+v\, z}{\sqrt{1-v^2}}\,, \qquad \psi = \tfrac{v\, \eta(x,y)+z}{\sqrt{1-v^2}}\,,$$

$$K_1 = (\tfrac{1+v}{1-v})^{\frac{1}{4}}\,, \qquad K_2 = (\tfrac{1-v}{1+v})^{\frac{1}{4}}\,,$$

where

$$\eta(x, y) = \begin{cases} \frac{1}{2}\lambda x + (1-\lambda) y, & x \neq 0,\ y \neq 0; \\ y, & x = 0,\ y \neq 0; \\ \frac{1}{2} x, & x \neq 0,\ y = 0; \end{cases}$$

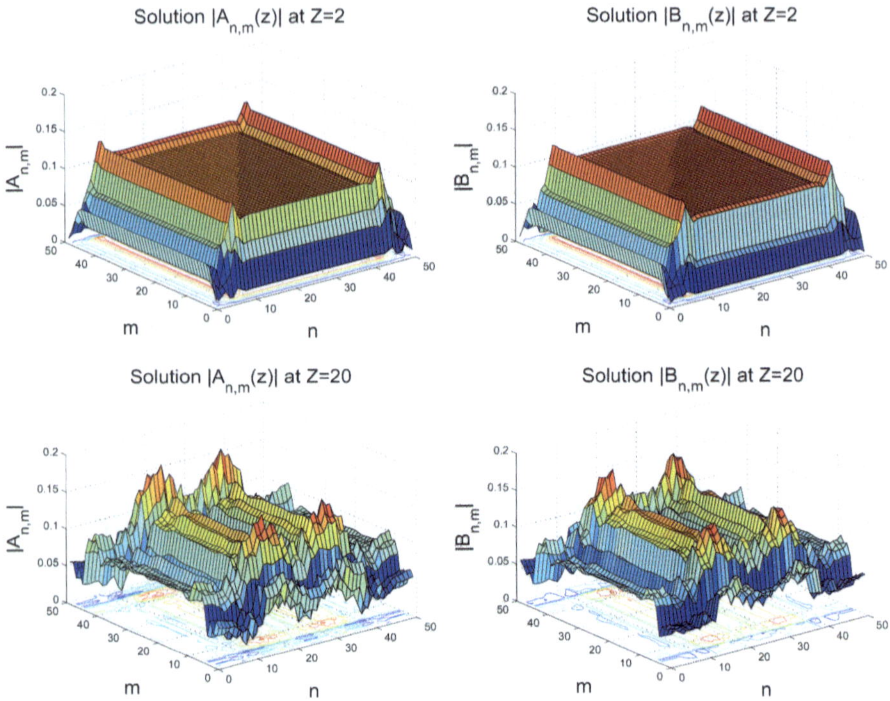

Fig. 8 $A = B = 0.1, N = M = 50$. The numerical solution of (2) with the initial conditions Type (ii)

with $g_{1,2}$ two arbitrary complex functions, $-1 \leq v \leq 1$, $0 \leq Q \leq \pi$ and $\lambda \neq 0, 1$. Observe that the invariance $\frac{1}{2}x \leftrightarrow y$ preserves the 90° rotational invariance of the array. For the simplicity, we set $\Delta\beta = 0$, substitute of (27) and (28) into Eqs. (4) gives ($s = \gamma_1 + \gamma_2$, $d = \gamma_1 - \gamma_2$)

$$
\begin{aligned}
&-\dot{g}_1 + i g_1 \cos Q + i\varepsilon g_2 + \tfrac{s}{8i}(K_1^4|g_1|^2 g_1 + 2|g_2|^2 g_1 - g_2^2 g_1^*) \\
&\quad -\tfrac{d}{8}(-K_2^2|g_2|^2 g_2 - 2K_1^2|g_1|^2 g_2 + K_1^2 g_1^2 g_2^*) = 0\,, \qquad (29)\\
&\dot{g}_2 + i g_2 \cos Q + i\varepsilon g_1 + \tfrac{s}{8i}(K_2^4|g_2|^2 g_2 + 2|g_1|^2 g_2 - g_1^2 g_2^*) \\
&\quad -\tfrac{d}{8}(K_1^2|g_1|^2 g_1 + 2K_2^2|g_2|^2 g_1 - K_2^2 g_2^2 g_1^*) = 0\,. \qquad (30)
\end{aligned}
$$

These equations imply the invariant $P = |g_1|^2 - |g_2|^2$. In the case $P = 0$, we have $|g_1|^2 = |g_2|^2$ and $g_{1,2}(\xi) = f(\xi)\exp[i\theta_{1,2}(\xi)]$. Therefore, by $\mu = f^2$ and $\nu = \theta_1 - \theta_2$, we obtain

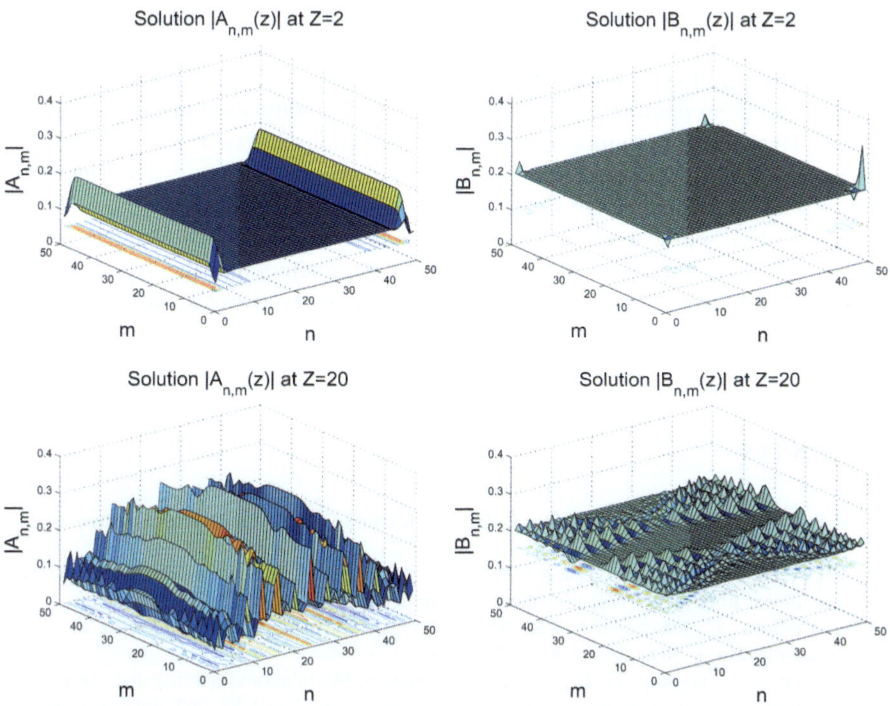

Fig. 9 $A = 0.1, B = 0.2, N = M = 50$. The numerical solution of (12) with the initial conditions Type (i)

$$\dot{\mu} = -\frac{\partial H}{\partial \nu}, \qquad \dot{\nu} = \frac{\partial H}{\partial \mu},$$

$$H = 2\mu(\varepsilon \cos\nu + \cos Q) - \frac{s}{8}\mu^2\left(\frac{K_1^4}{2} + \frac{K_2^4}{2} + 2 - \cos(2\nu)\right) - \frac{d}{4}\mu^2(K_1^2 + K_2^2)\sin\nu\,. \tag{31}$$

Equations (31) represent a one-dimensional integrable Hamiltonian system from which solitary wave solutions (including the case $\epsilon = 0$) are obtained [3]. In this case, given that the spatial variable ξ is a linear combination of the triad (x, y, z) these solutions are effectively one dimensional.

6 Conclusion

We have analyzed in some detail two possible models describing light propagation in two dimensional plasmonic arrays. These models could also describe different physical settings besides plasmonic arrays; examples include Bragg or photonic crystal waveguides. A proper description results in coupled mode systems for which as can be expected results in nonlinear dispersion relations having gaps and/or singular diabolic like features. We discuss continuous wave solutions in both the discrete and continuum approximation models with modulation instabilities being generic. Here we find a unique feature of coexisting long and short wave instabilities which may lead to the development of one dimensional solitary waves that were obtained from a reduced ODE Hamiltonian formulation.

Future work will extend on these results with a more detailed characterization of dynamical regimes over a broad range of parameters and the search of two dimensional discrete solitary waves and one should include losses which have been ignored in the present work

References

1. M.J. Ablowitz, S.D. Nixon, Y. Zhu, Conical diffraction in honeycomb lattices. Phys. Rev. A **79**, 053830 (2009)
2. A.B. Aceves, G.G. Luther, C. De Angelis, A.M. Rubenchik, S.K. Turitsyn, Energy localization in nonlinear fiber arrays: collapse-effect compressor. Phy. Rev. Lett **75**, 73 (1995)
3. M. Conforti, C. De Angelis, T.R. Akylas, Energy localization and trans port in binary waveguide arrays. Phys. Rev. A **83**, 043822 (2011)
4. M. Conforti, C. De Angelis, T.R. Akylas, A.B. Aceves, Modulational stability and gap solitons of gapless systems: continuous versus discrete limits. Phys. Rev. A **85**, 063836(1–4) (2012)
5. H.S. Eisenberg, Y. Silberberg, R. Morandotti, A.R. Boyd, J.S. Aitchison, Discrete spatial optical solitons in waveguide arrays. Phys. Rev. Lett. **81**, 3383 (1998)
6. N.K. Efremidis, P. Zhang, Z. Chen, D.N. Christodoulides, C.E. Rüter, D. Kip, Wave propagation in waveguide arrays with alternating positive and negative couplings. Phys. Rev. A **81**, 053817 (2010)
7. M. Guasoni, A. Locatelli, C. De Angelis, Peculiar properties of photonic crystal binary waveguide array. J. Opt. Soc. Am. B **25**, 1515 (2008)
8. M. Guasoni, M. Conforti, C. De Angelis, Light propagation in nonuniform plasmonic subwavelength waveguide arrays. Opt. Commun. **283**, 1161 (2010)
9. K. Hizanidis, Y. Kominis, N.K. Efremidis, Interlaced linear-nonlinear optical waveguide arrays. Opt. Express **22**, 18296 (2008)
10. A. Kanshu, Ch. E. Ruter, D. Kip, J. Cuevas, P. Kevrekidis, Dark lattice solitons in one-dimensional waveguide arrays with defocusing saturable nonlinearity and alternating couplings. Eur. Phys. J. D **66**, 182 (2012)
11. F. Lederer, G.I. Stegeman, D.N. Christodoulides, G. Assanto, M. Segev, Y. Silberberg, Discrete solitons in optics. Phys. Rep. **463**, 1 (2008)
12. S. Longhi, Multiband diffraction and refraction control in binary arrays of periodically curved waveguides. Opt. Lett. **31**, 1857–1859 (2006)
13. S. Longhi, Photonic analog of Zitterbewegung in binary waveguides arrays. Phys. Rev. B **81**, 075102 (2010)
14. S. Minardi, F. Eilenberger, Y.V. Kartashov, A. Szameit, U. Ropke, J. Kobelke, K. Schuster, H. Bartelt, S. Nolte, L. Torner, F. Lederer, A. Tunnermann, T. Pertsch, Three-dimensional light bullets in arrays of waveguides. Phys. Rev. Lett. **105**, 263901 (2010)
15. R. Morandotti, D. Mandelik, Y. Silberberg, J.S. Aitchison, M. Sorel, D.N. Christodoulides, A.A. Sukhorukov, Y.S. Kivshar, Observation of discrete gap solitons in binary waveguide arrays. Opt. Lett. **29**, 2890–2892 (2004)

16. S.H. Nam, A.J. Taylor, A. Efimov, Diabolical point and conical-like diffraction in periodic plasmonic nanostructures. Opt. Express **18**, 10120 (2010)
17. S.H. Nam, E. Ulin-Avila, G. Bartal, X. Zhang, Deep subwavelength surface modes in metal-dielectric metamaterials. Opt. Lett. **35**, 1847 (2010)
18. Th. Pertsch, U. Peschel, J. Kobelke, K. Schuster, H. Bartelt, S. Nolte, A. Tünnermann, F. Lederer, Nonlinearity and disorder in fiber arrays. Phys. Rev. Lett. **93**, 053901 (2004)
19. A.A. Sukhorukov, Y.S. Kivshar, Discrete gap solitons in modulated waveguide arrays. Opt. Lett. **27**, 2112–2114 (2002)

Part III
Nonlinear Dynamical Lattices

Localized Excitations and Anisotropic Directional Ordering in a Two-Dimensional Morse Lattice Model of Cuprate Layers

Manuel G. Velarde, Werner Ebeling, and Alexander P. Chetverikov

Abstract We study here localized excitations in a Morse lattice model of two-dimensional CuO_2 layers (cuprates in short). The Cu-atoms are positioned motionless in a square lattice and the oxygen atoms are able to oscillate around originally equilibrium positions on another superposed square lattice. After studying regular oxygen lattices we investigate lattices with bonds which are weakly distorted. We estimate the density of compressions (strain density) in dependence of the misfit of the Cu–O-bonds. We show that with increasing misfit the nonlinear oscillations of the O-atoms yield stripes hence patches exhibiting anisotropic directional ordering with an overall yet transient tessellated structure.

1 Introduction

The model system which we study here is a square lattice with lattice constant σ (equilibrium distance between lattice units), formed by oxygen atoms, which is embedded or superposed into another square lattice, consisting of copper atoms.

M.G. Velarde (✉)
Instituto Pluridisciplinar, Universidad Complutense, Paseo Juan XXIII, 1, Madrid-28040, Spain

Fundación Universidad Alfonso X El Sabio, Villanueva de la Cañada-28691, Madrid, Spain
e-mail: mgvelarde@uax.es,mgvelarde@pluri.ucm.es

W. Ebeling
Instituto Pluridisciplinar, Universidad Complutense, Paseo Juan XXIII, 1, Madrid-28040, Spain

Institut für Physik, Humboldt-Universität Berlin, Newtonstrasse 15, Berlin-12489, Germany
e-mail: ebeling@physik.hu-berlin.de

A.P. Chetverikov
Instituto Pluridisciplinar, Universidad Complutense, Paseo Juan XXIII, 1, Madrid-28040, Spain

Department of Physics, Saratov State University, Astrakhanskaya 83, Saratov-410012, Russia
e-mail: chetverikovap@info.sgu.ru

R. Carretero-González et al. (eds.), *Localized Excitations in Nonlinear Complex Systems*, Nonlinear Systems and Complexity 7, DOI 10.1007/978-3-319-02057-0_10,

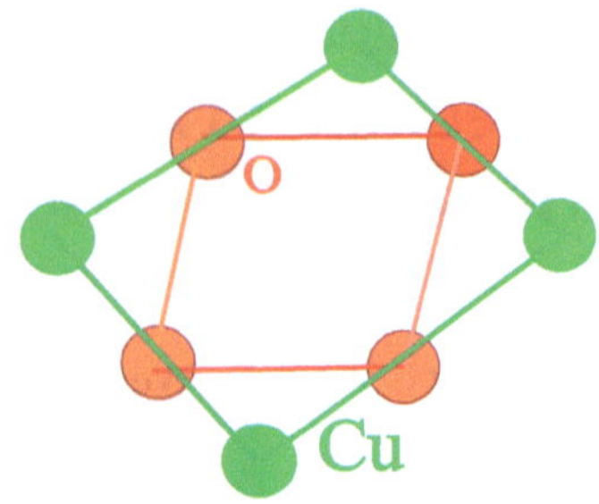

Fig. 1 Model CuO_2 cuprate layer. Two superposed square lattices, one with oxygens (*red color dots*) and another with coppers (*green color dots*)

We assume that the interaction potential parameters are such that the distances between minima of Cu and of O – atoms are related like the geometrical distances as $1/\sqrt{2}$ [1, 2]. This case is denoted as "regular"; otherwise, if the distances deviate from such value we speak about "irregular" cases. The case of a regular lattice exhibiting breathers was already considered by Russell and colleagues [3–5]. For other works on breathers for two-dimensional lattices see Refs. [6–11]. Here we shall explore stripe structures and soliton-like localized excitations rather than breathers. In the case of heated lattices, irregular configurations may appear as time proceeds, so the term regular and irregular refers only to the initial *equilibrium* situation. Note that we give only the initial square lattice configuration and then we allow the O-units to freely move, but the Cu-units are held fixed in the time scale considered here which agrees with available experimental observations [12, 13].

Figure 1 provides a schematic representation of our model two-lattice problem. The motivation of our study comes from arguments proposed by Müller [12, 13], Busmann-Holder [14–16], Newns and Tsuei [17, 18] and Alexandrov [19] and the need of interpreting experimentally observed facts. Long ago Müller [12] when discussing significant features of high-Tc superconductors (HTSC) underlined the possible role of anharmonic phonons. He proposed a model of anharmonic vibrations of apical oxygens interacting with carriers in CuO_2 planes. Recent topographic and other studies give us a lot of detailed information about the microscopic charge distributions [20–26]. Typically "stripe" patterns are observed whose structure depends on charge doping. Accordingly, we assume that below the HTSC regime exists a regime of stripes (electronic glass). Here we show that the long range uniform order may be broken leading to a kind of dynamic tessellation patterns.

In particular we consider here a model for multistable lattices first introduced by Bussmann-Holder [14–16] and similar to the fluctuating bonds model proposed by Newns and Tsuei [17, 18]. In this, our first approach to the above posed experimental evidence, no excess added electrons are considered here as our aim is not the study of the superconducting regime itself, but merely to analyze the role and influence of lattice nonlinearities leading to localized excitations and anisotropic directional ordering along certain axes of the cuprate lattice layer.

We start with the case of nonlinear oscillations of cuprate planar lattices with regular structure (see also [3–5]). Then we include qualitatively the expected effect

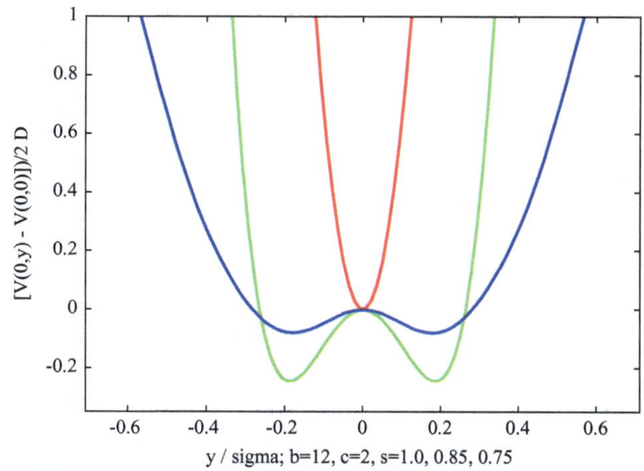

Fig. 2 Possibilities (minima) offered to an oxygen atom by a quartic potential (1). The single minimum curve corresponds to no doping. As charge doping enters we have two symmetric stable equilibria. The abcissa axis is orthogonal to the *Cu–Cu* line with units are proportional to the *O–O* distance divided by σ

of some weak doping by assuming that the bond length *Cu–O* is increased due to added, excess free charges. Following Newns and Tsuei [17, 18] the charges lead to instability of the originally equilibrium regular O-positions between two copper atoms which are replaced by two new, equally available, stable positions. This is locally accounted by a fourth-order nonlinearity in the energy,

$$E_O \simeq E_0^{'} - g_2 y^2 + g_4 y^4, \tag{1}$$

where y is a coordinate perpendicular to the direct connection between two copper atoms and $g_i (i = 2, 4)$ are parameters of the potential. Then depending on the sign of g_2 the double-well quartic potential could accommodate bistable oxygen configurations (Fig. 2). However we will not use (1) but rather a Morse potential locally exhibiting the same feature. The Morse interatomic potential is the first quantum based one. It does not differ significantly from the (12,6) Lennard-Jones potential of much current use [2, 27, 28]. It differs from the Toda potential [29] only in the attractive component which is unphysical in the Toda case. The Toda dynamic problem is exactly solvable in the one-dimensional (1d) case, and it is known that differences in the actual form of its solutions relative to the numerical solutions of the Morse dynamic problem are within 10 % deviation [30, 31]. Accordingly, we shall use the explicitly known analytical solutions of the Toda lattice to guide us in the computer simulations with the Morse potential.

Let us emphasize that the energy landscape (1) proposed by Newns and Tsuei [17, 18] comes from an ab initio quantum mechanical calculation for small lattice displacements. Yet in order to allow free oxygen motions, as stated above, we need an interaction potential which, for our molecular dynamics computational algorithm, has free asymptotics at large inter-unit distances. This is possible with the Morse or the Lennard-Jones potential. However, we prefer the former since it corresponds better to quantum-mechanical repulsion at short inter-unit distances, which is the region of significance in our approach here. The quantum mechanical calculations made by Newns, Tsuei and collaborators [18] use "frozen" configurations to obtain (1). Locally, our Morse potential is chosen to fit their data thus ensuring the correct polynomial expansion around the potential minimum while being more physically acceptable at quite short (and also long) inter unit distances.

For simplicity we have placed all *Cu*-atoms and *O*-atoms superposed into a single and same plane (Fig. 1). This is not completely realistic, so our *O*- and *Cu*-atoms are a kind of suitable projections of the real atoms. The players in our nonlinear dynamics are the *O*-atoms which are oscillating atoms interacting with six neighbors (in the initial configuration). Four of the neighbors are also *O*-atoms and the two other are *Cu*-atoms which as earlier noted will be motionless. The *Cu*-sub-lattice gives the frame for our game, the *O*-atoms are quite free and are looking for the "best place" according to the Morse interactions which locally correspond also to the double-well quartic potential (1).

2 Model Hamiltonian and Visualization of Local and Collective Atom (Lattice Units) Motions

We write the lattice Hamiltonian as

$$H_a = \sum_n \frac{m}{2} v_n^2 + \frac{1}{2} \sum_{n,k'} V_{Cu-O}^M \left(|r_n - r_{k'}|\right) + \frac{1}{2} \sum_{n,j} V_{O-O}^M \left(|r_n - r_j|\right), \quad (2)$$

where the subscripts locate atoms sitting near to lattice sites and for the summations over n and j run from 1 to N_O while the summations over k' from 1 to N_{Cu} with subscripts O and Cu referring to oxygen and copper, respectively (N denotes number of units). All masses are taken equal: $m_n = m$ for all n. The Morse potential we use is

$$V^M(r) = D\left[\exp(-2b(r-\sigma)) - 2\exp(-b(r-\sigma))\right], \quad (3)$$

where D, b, σ denote the corresponding potential-well depth, lattice stiffnesses and initial inter unit equilibrium distances (D_{O-O}, b_{O-O}, σ_{O-O} for oxygen-oxygen interactions, and D_{O-Cu}, b_{O-Cu}, σ_{O-Cu} for oxygen-copper interactions). The free space variable r will be subsequently considered in the two dimensional (x, y) plane with complex coordinates, and hence r would be replaced by $Z = x + iy$.

Putting the *Cu*-atoms and *O*-atoms together and giving full freedom to the *O*-atoms, as time proceeds we are bound to obtain a complex multi-stable potential landscape. In order to observe the rather complex dynamics in such "regular" or

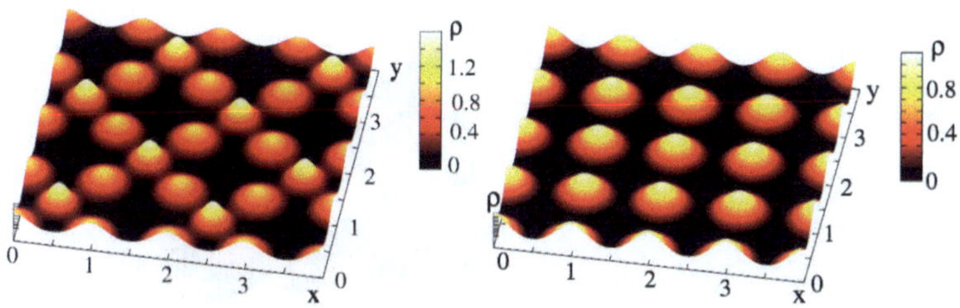

Fig. 3 *Left panel*: Visualization of copper atoms (narrower peaks) and oxygen atoms (broader peaks) (compound of both terms in Eq. (4)) by considering that atoms may be described as Gaussian densities of their core electrons in a regular lattice with bond lengths in the $\sqrt{2}$ ratio as required by the regular geometry at zero T ($\eta = 1/\sqrt{2}$; $\lambda_O = 0.25$, $\lambda_{Cu} = 0.177$, $T = 0.0002 D_{O-O}$). *Right panel* shows only the oxygen units (first term in Eq. (4))

"irregular" lattices we need an appropriate method of visualization. We model the lattice units or atoms as points on a plane which are surrounded by little spheres formed by the atomic electrons. We will assume that these atomic electrons may be represented by a Gaussian distribution centered on each lattice site. Accordingly, in Fig. 3 the left panel indicates how to experimentally observe atom motions and eventually moving local excitations by means of, e.g., Scanning Tunneling Microscopy (STM) or a more appropriately faster scanning methodology [32]. When two atoms approach each other like in a strong enough compression the superposition of their corresponding electron Gaussian densities yield, e.g., an enhanced single maximum out of the composition of the original two separated maxima. For a thorough discussion of this methodology and earlier results in lattice systems the reader is referred to [33–38]. The compound electron density is then:

$$\rho(Z,t) = \sum\nolimits_{n;|Z-Z_n(t)|<1.5} \exp\left[-\frac{|Z-Z_n(t)|^2}{2\lambda_O^2}\right],$$
$$\sum\nolimits_{k',|Z-Z_{k'}|<1.5} \exp\left[-\frac{|Z-Z_{k'}|^2}{2\lambda_{Cu}^2}\right]. \tag{4}$$

Here the Z_n are the complex oxygen coordinates, rescaled using as unit σ_{O-O}. We will use different λ-parameters for oxygen and for copper, (λ_O and λ_{Cu}, respectively) taking into account their different atom size. Note in Fig. 3 that copper-peaks are narrower than oxygen peaks. Needless to say when only oxygens are displayed there is no contribution from the second term in the r.h.s. of (4).

We denote by σ_{O-O} and σ_{O-Cu} the actual distances between two oxygen units and between an oxygen unit and a copper unit in the regular geometrical configuration. Then we define $\eta \equiv \sigma_{O-Cu}/\sigma_{O-O}$. Note that σ_{Cu-Cu} is of no relevance as it remains with a fixed constant and the same value at all times. Besides σ_{O-O} is the reference value and, consequently, only σ_{O-Cu} relative to σ_{O-O} is allowed to change.

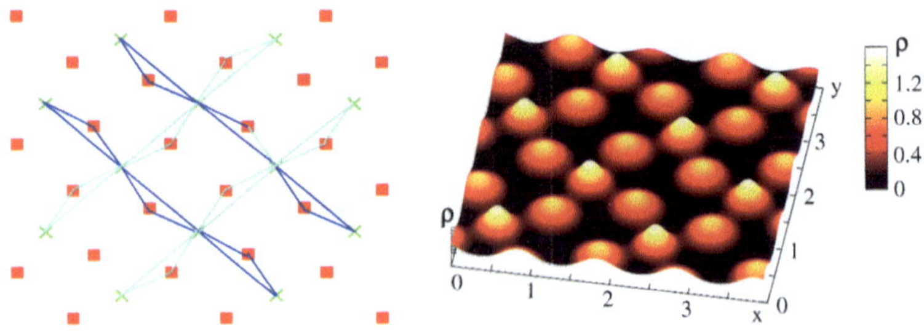

Fig. 4 Two-lattice model cuprate layer. *Left panel*: CuO_2 lattice illustrating one among the two possible oxygen configurations for η slightly different from $1/\sqrt{2}$ (see also the single minimum in Fig. 2). *Right panel*: Snapshot of an equivalent visualization of *left panel* for $\eta = 0.74$ in a rather cold lattice ($T = 2.10^{-4} D_{O-O}$). Enhanced (narrower) peaks correspond to copper units. Parameter values: $(b\sigma)_{O-O} = (b\sigma)_{O-Cu} = 10$, $D_{O-Cu}/D_{O-O} = 1.5$, $\lambda_O = 0.2$, $\lambda_{Cu} = \lambda_O \eta$

From the Hamiltonian (2) with (3) we can get the corresponding Newton's equations of motions. However, as we want to play with temperature we introduce heating by introducing in the dynamic system an appropriate thermal bath. Hence the evolution of the oxygen atoms is governed by the Langevin equations:

$$d^2 Z_n/dt^2 = \sum_k F_{nk}^{O-O}\left(|Z_{nk}|\right) z_{nk} + \sum_{k'} F_{nk'}^{O-Cu}\left(|Z_{nk'}|\right) z_{nk'} \\ - \gamma dZ_n/dt + \sqrt{2D_v}\left(\xi_{nx} + i\xi_{ny}\right), \tag{5}$$

where now, for better clarity, n and k denote lattice positions of the oxygens and k' those of coppers; $Z_{nk} = Z_n - Z_k$ and $z_{nk} = Z_{nk}/|Z_{nk}|$. γ, D_v and ξ_s account for friction, noise strength and noise fluctuating variables. Temperature is thus defined using Einstein's relation $T = D_v/\gamma$ (in computations we use $\gamma = 0.2$ or 0.02).

The forces are

$$F_{nk}^{O-O}\left(|Z_{nk}|\right) = -dV_{O-O}^{M}(r)/\,dr|_{r=|Z_{nk}|}, \tag{6}$$

and

$$F_{nk'}^{O-Cu} = -dV_{O-Cu}^{M}(r)/\,dr|_{r=|Z_{nk'}|}. \tag{7}$$

The quantity V^M accounts for both terms in (2). All computer simulations start with a lattice equilibrium distribution. Then after integrating Eq. (5) we wait for a *steady state* corresponding to the a priori fixed temperature.

Figure 4 shows the Cu-atoms as enhanced peaks in a landscape of O-densities (right panel). With these assumptions we study the oxygen dynamics seeing the fixed copper positions only in enlarged zoomed pictures. For simplicity and illustration of results we set $\lambda_{Cu} \equiv \lambda_O \eta$ with parameter λ to be allowed to vary at will. Note that we use σ_{O-O} and $2D_{O-O}$ as distance and energy units, respectively.

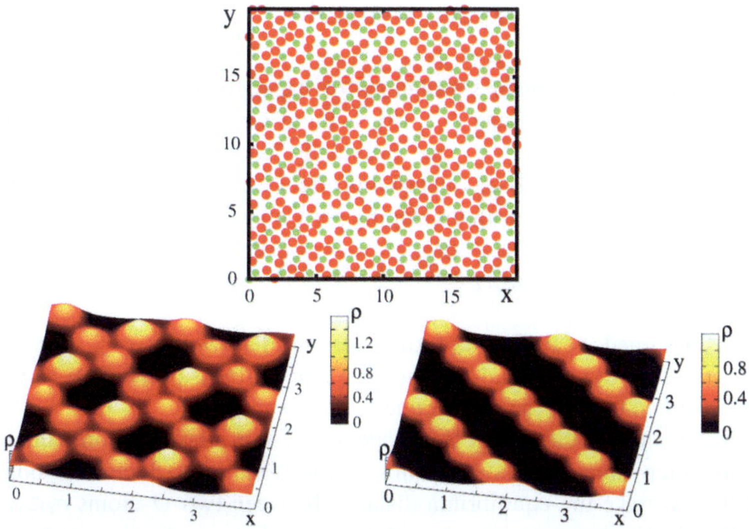

Fig. 5 Snapshots of a cuprate layer slightly heated. *Upper panel*: Snapshot of the cuprate layer with irregular bonds. Most of the oxygen units (*red dots*) are off their original equilibrium centers of the connecting lines between the copper units (see Fig. 1). *Left panel below*: Snapshot of an enlarged view of a portion of the *upper panel*. Such a portion shows one among many other patches where we see in the patch rather long range anisotropic liquid-crystalline directional ordering as further illustrated in the *lower right panel*, where, extracted from the *lower left panel*, only the oxygen atoms are displayed. Parameter values and temperature as in Fig. 4, save $\eta = 0.95$

Besides $(b\sigma)_{O-O} = (b\sigma)_{O-Cu} = 5$ or 10 in our computations. In order to represent the bonds between *Cu–O* and their evolution we play with the free lambda-parameter looking for values which allow an easy representation of bond structure.

Note that as oxygen motions get stronger and stronger the apparent symmetry shown in the Fig. 4 will be altered with the possibility of configurations with many more oxygen units aligned together albeit always within patches, as displayed in Fig. 5.

Two cases are of interest:

(i) **Case 1**: In the absence of added, excess free electrons (no doping), the O-atoms are sitting in equilibrium on a square lattice where they are located half-way between the Cu-atoms. When heating the system the O-atoms perform nonlinear oscillations. In an ideal regular lattice $\eta = 1/\sqrt{2}$; the effective binding radii are in $\sqrt{2}$ ratio as indicated in the Introduction. According to Pauling [1, 2] the actual ratio between lattice radii is around $0.96/1.40 \simeq 0.69$ which is near to $1/\sqrt{2} = 0.707$.

(ii) **Case 2**: At stronger but still undercritical doping we have according to the fluctuating bond effect [17, 18] a small increase of the bond length $\eta(n) > 1/\sqrt{2}$ where the parameter n denotes doping percentage due to the added, excess free electrons. This is expected to lead to the formation of "stripes" which we are going to study in the following section.

3 Simulation of the Oxygen Dynamics

We start with some computer simulations of the regular or nearly regular case. According to Newns and Tsuei [17, 18], a finite amount of doping changes the effective binding length $Cu - O$ and eventually leads to bistable configurations. For a qualitative (mean field) account of this effect we assume that the length ratio, η, increases nearly linearily with the doping percentage, n, and hence

$$\eta(n) = 1/\sqrt{2} + \alpha n + \ldots,$$

where n is assumed weak. Let us further assume that $(b\sigma)_{O-O} = 5$ or 10 is the stiffness of the $O - O$ potential and $(b\sigma)_{O-Cu}$ the stiffness of the $Cu - O$ potential which is assumed to have the same value as the former.

We investigate now the energy landscape of a system admitting some small misfit of the parameter values and show that this may lead indeed to fluctuating bonds. We remind that we take the equilibrium distance between two O-atoms σ_{O-O} as the unit of length. Furthermore in our model the equilibrium distance of $Cu - O$ is $\sqrt{2}\sigma_{O-O}/2$ and the $Cu - Cu$ equilibrium distance is $\sigma_{O-O}\sqrt{2}$.

Let us analyze the consequences of small changes of the bond length. We expect that an increased bond length $Cu - O$ will push the oxygens off the initial axes and will create a positive or negative elongation (Fig. 4, left panel). Recall that we consider one O-atom initially in a "cage" surrounded by four O-atoms and 2 Cu-atoms. The potential in the center of the cage may be represented by 6 contributions from the 6 neighbors and shows bistability indeed for appropriate parameter values.

The "fluctuating bond" condition is that the extremum connected with the regular position is higher than the global minimum what means in effect that the regular position is unstable. The condition that the side minima are deeper than the central extremum may be fulfilled for a large set of parameter values. The search for an appropriate set is quite complicated. Various cases are illustrated in Figs. 5 and 6.

4 Discussion and Conclusions

We have studied here the localized excitations in a Morse model of regular and irregular CuO_2 (cuprate) lattices under different conditions. Irregular lattices are obtained by introducing some small misfit of bond lengths. In the framework of the Newns-Tsuei fluctuating bond model [17, 18] this may be interpreted as due to electron doping in the cuprate layer i.e. due to adding excess free electrons, but a discussion of this latter point is beyond the scope of our interest here.

Consequently, the anharmonic oxygen motions, originally suggested by Müller [12] appear as a possible explanation for the observed stripe structures in cuprates [20–26]. We assumed here that such changes due to doping have the physical meaning that the ratio of the length of the $Cu - O$ bond to the $O - O$ bond may vary.

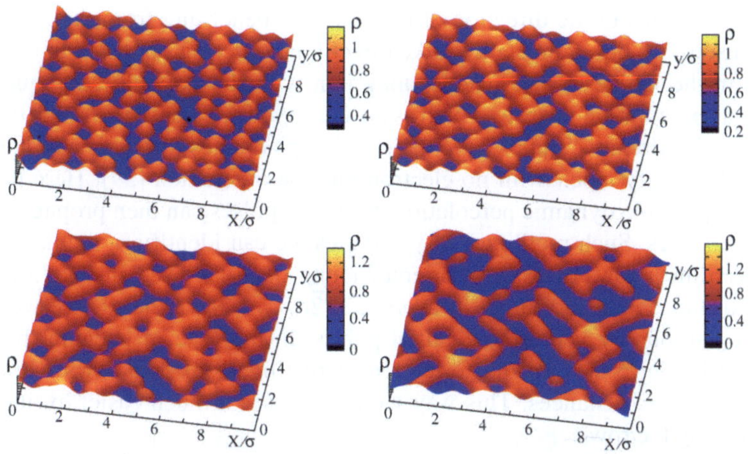

Fig. 6 Snapshots of a cuprate layer at a higher temperature than in Fig. 5. *Upper left panel*: $\eta = 0.74$; *upper right panel*: $\eta = 0.80$; *lower left panel*: $\eta = 0.85$; and *lower right panel*: $\eta = 0.95$. Note the difference between the latter case with the case of Fig. 5 for $\eta = 0.95$. For purpose of easier visualization only the oxygen densities are provided as also done in the *right panel* of Figs. 3 and 5. Note that the long range anisotropic directional order is eventually weakened as temperature grows. Parameter values: $(b\sigma)_{O-O} = (b\sigma)_{O-Cu} = 5$, $D_{O-Cu}/D_{O-O} = 1$, $\lambda_O = 0.35$, $T = 0.2D_{O-O}$ and $\gamma = 0.02$

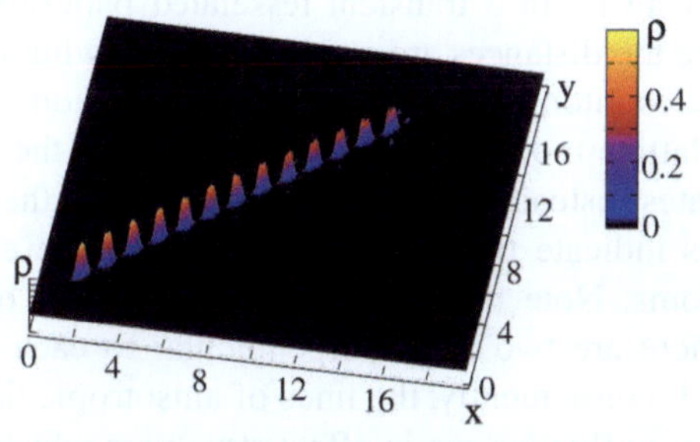

Fig. 7 Cumulative picture of a sequence of snapshots (all plotted in the same picture) to illustrate the trajectory of a soliton-like localized compression density in a lattice of oxygen atoms in a regular cuprate lattice launched by a strong pulse (momentum given) to one O-atom in the direction of the already ordered O-line. Parameter values: $\eta = 1/\sqrt{2}$, $\lambda_O = 0.25$, and all other parameters have the same values as in Fig. 6

In order to check this idea we played with the bond length of the original system where this ratio is approximately given by $1/\sqrt{2}$. The result is that the changes of the bond length are correlated with the nonlinear oxygen oscillations in the lattice and the multistability of configurations.

What can we say about solitonic excitations? From our present knowledge we can say that we can successfully launch solitons as shown in Fig. 7 by, e.g., giving momentum (giving a sudden kick) to one oxygen atom in a row of rectilinearly arranged O-atoms in the direction of the row thus benefiting from the rather long range anisotropic liquid crystalline-like directional ordering observed in one many patches in the lattice. Those patches may join with the possible consequence of creating a "percolation" transport path as the density of "patches" with such ordering is increased. Already "static" percolation via "plaquettes" of ordering has been suggested recently albeit with no electron-phonon interaction [39]. Here however we offer a possible dynamic percolation. Solitonic pulses can then propagate along the oxygen lines. Such rectilinear oxygen lines we can identify in regular lattices. In the rectilinear rows of oxygen, oriented along $Cu - Cu$ axes, in Figs. 3 or 4 the oxygen distances are too large, in average $\sqrt{2}\sigma_{O-O}$. Nevertheless we were able to excite solitons by very strong initial pulses. In the case of irregular bonds as illustrated in Fig. 5 we see the formation of rectilinear oxygen rows with rather small interatomic distances. This way we can foresee a much easier excitation of solitons along the rows.

Our basic parameter is the ratio of the radii $\eta = \sigma_{O-Cu}/\sigma_{O-O}$. For a regular lattice the ideal ratio is $1/\sqrt{2}$. For higher values of this ratio (irregular lattices) we observe bistability and the potential reminds then the quartic type potential along the axis perpendicular to the $Cu - Cu$-bond calculated by Newns and Tsuei [17, 18].

The Cu-lattice gives the static frame for our computer runs and the O-atoms are quite free and are looking for the "best place" according to the Morse interactions (3) leading locally to the minima of the quartic potential (1). Note that at equilibrium, in regular geometry each O-unit has four other O-atoms as neighbors and two other Cu-units a bit further away. In a transient tesselated pattern some O-units form rectilinear rows where the distances are smaller thus allowing lattice compressions eventually leading to soliton-like moving local excitations (Fig. 7). A special property of the two-lattice model is multistability, that is the existence of a high number of ground states instead of a single ground state in the regular lattice. Our computer simulations indicate the basic role played by the cristallographic axes formed by the O-atoms. Note that we see always patches of excited pieces of the O-axes. Since there are two axes, perpendicular to each other, the direction of the oscillations and, consequently, the lines of anisotropic liquid crystalline-like ordering, may change, so that we see in effect structures which are like tessellated patterns with patches of directional ordering. The $Cu - O$-bonds, where presumably most of the doping electrons tend to accumulate, are located right and left from the $O - O - O, \ldots$ axes. So we may imagine the existence of some "streets of local excitations" right and left besides the ordered O-axes along which added, excess electrons may travel bound to the local soliton-like excitations. Then eventually percolation paths could be established if an external electric field is also added.

In summary we may conclude that in our rather simple model we observe quite complex motions of the O-atoms oscillating in the field of the other O-atoms and in the field of the Cu-atoms. The structure mimics the effect of finite (but small) doping multistable potential landscape which is responsable for the existence of multi stable positions of the O-atoms. The field of parameters and significant parameter values is so big that we could explore so far only a small subset of all possibilities of nonlinear excitations in particular those leading to stripe-like structures. We have seen here several types of such structures for different choices of the misfit parameter $\eta(n)$. Since the hopping transport is in zeroth approximation from copper to copper position, it remains to clarify how possible superconducting-like electronic paths are related to the above mentioned "streets of local excitations" found in our computer simulations.

Acknowledgements The authors acknowledge fruitful discussions with A. A. Alexandrov (deceased), L. Bascones, E. Brändas, L. Brizhik, A. Bussmann-Holder, L. Cruzeiro, Ch. Eilbeck, W. A. Goddard III, V. Lakhno, S. Larsson, G. Röpke, F. M. Russell, J. Tahir-Kheli and E.G. Wilson. Part of the research here reported was done when MGV was a Del Amo Visiting Scholar with the University of California at Irvine. This research was supported by the Spanish *Ministerio de Economía y Competitividad*, under Grant MAT2011-026221. A.P.C. acknowledges support from the Ministry of Education and Science of the Russian Federation under Grant 14.B37.21.0751 (2009–2013).

References

1. L. Pauling, *General Chemistry* (Dover reprint, New York, 1988)
2. T. Kihara, S. Ichimaru, *Intermolecular Forces* (Wiley, New York, 1978)
3. F.M. Russell, D.R. Collins, Phys. Lett. A **216**, 197 (1996)
4. J.L. Marín, J.C. Eilbeck, F.M. Russell, Phys. Lett. A **248**, 225 (1998)
5. J.L. Marín, F.M. Russell, J.C. Eilbeck, Phys. Lett. A **281**, 21 (2001)
6. S. Flach, K. Kladko, C.R. Willis, Phys. Rev. E **50**, 2293 (1994)
7. P.G. Kevrekidis, K.O. Rasmussen, A.R. Bishop, Phys. Rev. E **61**, 2006 (2000)
8. X. Yi, J.A.D. Wattis, H. Susanto, L.J. Cummings, J. Phys. A: Math. Theor. **42**, 355207 (2009)
9. I.A. Butt, J.A.D. Wattis, J. Phys. A: Math. Gen. **39**, 4955 (2006)
10. I.A. Butt, J.A.D. Wattis, J. Phys. A: Math. Theor. **40**, 1239 (2007)
11. Y. Yamayose, Y. Kinoshita, Y. Doi, A. Nakatani, T. Kitamura, EPL **80**, 40008 (2007)
12. K.A. Müller, Z. Phys. B: Condens. Matt. **80**, 193 (1990)
13. J. Bednorz, K.A. Müller (eds.), *Superconductivity* (Springer, Berlin, 1990)
14. A. Bussmann-Holder, A. Simon, H. Büttner, Phys. Rev. B **39**, 207 (1989)
15. A. Bussmann-Holder, A.R. Bishop, Phys. Rev. B **44**, 2853 (1991)
16. A. Bussmann-Holder, A.R. Bishop, Ferroelectrics **128**, 99 (1992)
17. D.M. Newns, C.C. Tsuei, Nat. Phys. **3**, 184 (2007)
18. R.A. Nistor, G.J. Martyna, D.M. Newns, C.C. Tsuei, M.H. Müser, Phys. Rev. B **83**, 144503 (2011)
19. A.S. Alexandrov, Phys. Scripta **83**, 038301 (2011)
20. J.M. Tranquada, B.J. Sternlieb, J.D. Axe, Y. Nakamura, S. Uchida, Nature **375**, 561 (1995)
21. V.J. Emery, S.A. Kivelson, J.M. Tranquada, PNAS **96**, 8814 (1999)
22. D. Reznik, L. Pintschovius, M. Ito, S. Likubo, M. Sato, H. Goka, M. Fujita, K. Yamada, G.D. Gu, J.M. Tranquada, Nature **440**, 1170 (2006)
23. Y. Kohsaka, C. Taylor, K. Fujita, A. Schmidt, C. Lupien, T. Hanaguri, M. Azuma, M. Takano, H. Eisaki, H. Takagi, S. Uchida, J.C. Davis, Science **315**, 1380 (2007)
24. Y. Kohsaka, C. Taylor, P. Wahl, A. Schmidt, J. Lee, K. Fujita, J.W. Alldredge, K. McElroy, J. Lee, H. Eisaki, S. Uchida, D.-H. Lee, J.C. Davis, Nature **454**, 1072 (2008)
25. K.A. Moler, Nature **468**, 643 (2010)
26. J. Zaanen, Nature **440**, 1118 (2006)
27. P.M. Morse, Phys. Rev. **34**, 57 (1929)
28. M. Rieth, *Nano-Engineering in Science and Technology* (World Scientific, Singapore, 2003)

29. M. Toda, *Theory of Nonlinear Lattices*, 2nd edn. (Springer, Berlin, 1988)
30. J. Dancz, S.A. Rice, J. Chem. Phys. **67**, 1418 (1977)
31. T.J. Rolfe, S.A. Rice, J. Dancz, J. Chem. Phys. **70**, 26 (1979)
32. A.H. Zewail, *4D Electron Microscopy: Imaging in Space and Time* (Imperial College Press, London, 2010)
33. A.P. Chetverikov, W. Ebeling, M.G. Velarde, Eur. Phys. J. B **70**, 217 (2009)
34. A.P. Chetverikov, W. Ebeling, M.G. Velarde, Physica D **240**, 1954 (2011)
35. A.P. Chetverikov, W. Ebeling, M.G. Velarde, Wave Motion **48**, 753 (2011)
36. A.P. Chetverikov, W. Ebeling, M.G. Velarde, Eur. Phys. J. B **80**, 137 (2011)
37. M.G. Velarde, W. Ebeling, A.P. Chetverikov, C. R. Mecanique **340**, 910 (2012)
38. A.P. Chetverikov, W. Ebeling, M.G. Velarde, Eur. Phys. J. B **85**, 291 (2012)
39. J. Tahir-Kheli, W.A. Goddard III, J. Phys. Chem. Lett. **2**, 2326 (2011)

Variational Approximation to Electron Trapping by Soliton-Like Localized Excitations in One-Dimensional Anharmonic Lattices

Luis A. Cisneros-Ake, Antonmaria A. Minzoni, and Manuel G. Velarde

Abstract Electron trapping by soliton-like (traveling) localized excitations in one-dimensional anharmonic lattices is discussed with particular emphasis on the case of an initially completely delocalized electron.

1 Introduction

Let us consider a one-dimensional (1d) lattice with units interacting with anharmonic interactions. Then for appropriate anharmonic interactions it can be shown that the lattice can exhibit traveling solitary waves or periodic nonlinear waves, with soliton features as defined by Zabusky and Kruskal [1]. One particular such lattice is the Toda lattice where interactions are of exponential form for the repulsive component [2]. Its soliton solutions are known exactly as the system is integrable.

For our purpose here we shall consider Morse interactions [3] which are not significantly different from Toda's in their repulsive component while offering a physically justified attractive component. The latter is physically meaningless in the Toda case. Moreover, though the Morse lattice is non-integrable its computer

L.A. Cisneros-Ake
Department of Mathematics, ESFM, Instituto Politécnico Nacional, Unidad Profesional Adolfo López Mateos, Edificio 9, México DF-07738, México
e-mail: cisneros@esfm.ipn.mx

A.A. Minzoni
FENOMEC, Department of Mathematics and Mechanics, IIMAS, Universidad Nacional Autónoma de México, Apdo. 20-726, México, DF-01000, México
e-mail: tim@mym.iimas.unam.mx

M.G. Velarde (✉)
Instituto Pluridisciplinar, Universidad Complutense, Paseo Juan XXIII, 1, Madrid-28040, Spain

Fundación Universidad Alfonso X El Sabio, Villanueva de la Cañada-28691, Madrid, Spain
e-mail: mgvelarde@uax.es; mgvelarde@pluri.ucm.es

R. Carretero-González et al. (eds.), *Localized Excitations in Nonlinear Complex Systems*, Nonlinear Systems and Complexity 7, DOI 10.1007/978-3-319-02057-0_11,

solutions differ little from the solutions of the Toda lattice [4–6] and hence we shall benefit from such significantly useful approximation. Little has been established about the quantization of the Toda or other anharmonic lattices and, moreover, for the purpose of our present communication it is not needed here.

The next item to be introduced in our mathematical model is an electron. As a quantum mechanical object, its (space and time) evolution obeys the corresponding (linear) Schrödinger equation. We shall consider the electron evolution on a lattice using the tight binding approximation (TBA) . If we disregard the lattice dynamics and hence consider a free evolving electron on the discrete lattice space, if we place it at a given site with maximum probability density, then as time proceeds the electron probability density spreads "uniformly" all over the lattice sites though the probability remains normalized to unity. Such delocalization of the electron can be considered as a form of "dust", tiny spots on the lattice sites.

The electron-phonon interaction or, in more general terms the electron-lattice dynamics interaction, is nonlinear and eventually is an exponential nonlinearity [7,8]. Having in mind electron (charge) transfer processes in biomolecules, Davydov [9] was able, using suitable approximations, to reduce the electron-lattice phonon dynamics to soliton-bearing evolution equations. Then he was able to show that the real electric carrier was the compound he called electrosoliton . Unfortunately there is numerical evidence that such electrosolitons do not survive beyond 10 K. This temperature is far from physiological temperatures ca. 300 K [10–13].

Recently, the electron-lattice dynamics was considered for anharmonic (Toda, Morse, etc.) lattices [14–16]. Then the mixed quantum-classical (electron-lattice) evolution problem possesses two distinct nonlinearities. On the one hand, there is the electron-lattice interaction and, on the other hand, we have the lattice intrinsic anharmonicity . The latter already offers a soliton carrier and hence the compound electrosoliton (Davydov concept generalized to anharmonic lattices) together with the lattice soliton defines a dynamic (space and time evolving) bound state of an electron to a lattice soliton which is a new electron (charge) carrier that has been denoted solectron [17]. Thus the latter concept (and quasiparticle) appears as a natural generalization of the original (Landau-Pekar) polaron (for harmonic lattices) and the (Davydov) electrosoliton (also for harmonic lattices) to the case of initially anharmonic lattices, with, let us emphasize it, the genuinely new ingredient of offering a possible charge carrier before injecting, e.g., the electron. Davydov and Zolotaryuk and colleagues also elaborated some results about the same concept including supersonic motions [9, 18–20].

One interesting feature and the motivation for this communication is the following. It has been shown that placing an electron, using TBA, at any lattice site, and a soliton at the same or at another lattice site, as time proceeds the electron is eventually trapped and transported (electron surfing) by the lattice soliton. Moreover, if as noted above the electron is completely delocalized in the lattice as probability density "dust" spots, following Schrödinger equation alone, then as soon as the electron-lattice soliton is switched-on the latter is able to gather the electron probability density thus "reconstruding" the electron (or better said, its maximum probability density) around the soliton itself and trapping it with subsequent transport, as the soliton is a (generally) supersonic, always moving localized excitation. This phenomenon has been denoted as the "vacuum cleaner effect" [14, 15]. The above phenomena have been shown to occur for temperatures from zero K up to the physiological range (ca. 300 K; for materials other than biomolecules the actual temperature range may be different) [21].

In the present communication we take up the problem from a variational perspective starting with the Lagrangian embracing coupled together the lattice classical dynamics and the quantum (discrete) Schrödinger electron dynamics in the tight binding approximation.

2 Formulation of the Problem and Numerical Results

Let us consider a 1d lattice of N units all with equal mass, M, and interacting with a Morse potential [3] (like equal massless springs operating beyond Hooke's linear elasticity), to which an excess electron is added as a Gaussian probability density centered at site n. Having in mind the TBA the function $c_n(t)$ (or better said $|c_n|^2$) gives the probability of finding the electron at site n, while the function $\xi_n(t)$ gives the distortion of the lattice caused by the polarization of lattice atom-units. The model assumes a hopping probability from site n to site $n+1$ of the form $e^{\alpha(\xi_{n+1}-\xi_n)}$ [7,8]. The coupling parameter α is determined by a quantum mechanical calculation and can in principle be modified by the further electron doping of the material. In this case we will have α to be a given function of site the n, that is $\alpha = \alpha(n)$, which describes the doping. Clearly, such n-dependent coupling between the lattice and the electron changes the probability of hopping. In its absence this probability does not depend on the lattice deformation and the deformation just acts as a trapping potential. The model Lagrangian is given in the form:

$$\begin{aligned} L = \textstyle\sum_{n=-\infty}^{\infty} \frac{1}{2}\frac{p}{V_0}\dot{q}_n^2 + \frac{i}{2}\left(\dot{c}_n c_n^* - \dot{c}_n^* c_n\right) \\ +pe^{\alpha(q_{n-1}-q_n)}\left(c_{n-1}c_n^* + c_{n-1}^* c_n\right) - \frac{p}{2V_0}\left(1-e^{q_{n-1}-q_n}\right)^2, \end{aligned} \tag{1}$$

where we set

$$\begin{aligned} \xi_n = \frac{1}{\beta}q_n, \quad \tau = \omega_M t = t\sqrt{\frac{2D\beta^2}{M}}, \\ \tilde{\alpha} = \frac{\alpha}{\beta}, \quad \tilde{V}_0 = \frac{V_0}{2D}. \end{aligned}$$

The quantities β and D parameterize the Morse lattice. V_0 and α account for electron-lattice coupling strength and their tilde has been removed from (1) in the new variables for practical purposes. According to the physical values [9–11,22,23] the non dimensional parameter $p = 2D\tilde{V}_0/(\hbar\omega_M) = V_0/(\hbar\omega_M) \approx 10$ is the ratio between the electronic and mechanical time scales.

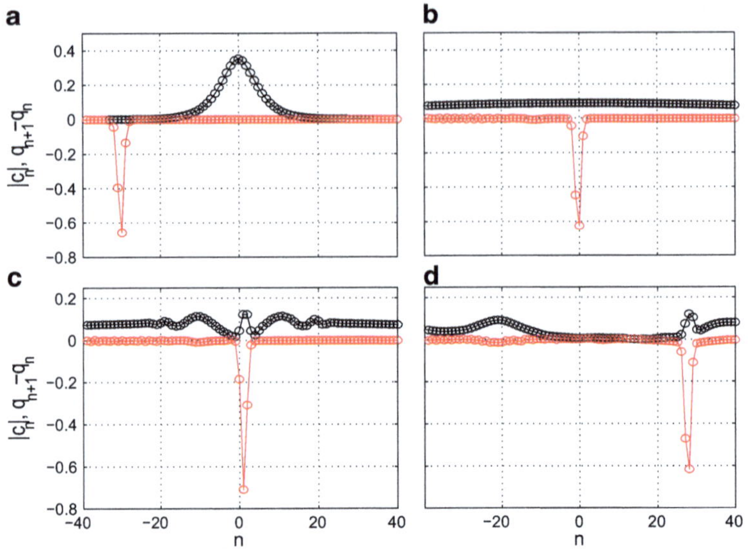

Fig. 1 Numerical evolution of system (2) and (3) for $w = 4$, $v = 0$, $\phi = 0$, $\xi_0 = 0$ and $k = 1.35$, $A = 0.29$, $\xi_1 = -30$, $\xi = 1.5$. Parameter values: $V_0 = 0.5$, $p = 10$ $\alpha = 0$ for $t \le 20$ and $\alpha = 1.75$ for $t > 20$ in (**a**) $t = 0$, (**b**) $t = 20$, (**c**) $t = 21$ and (**d**) $t = 40$ (*red lines*: soliton; *black lines*: electron)

We thus obtain from (1), and its associated Euler-Lagrange equations , the corresponding non dimensional equations of motion in the form:

$$\frac{dc_n}{d\tau} = ip\left(e^{\alpha(q_{n-1}-q_n)}c_{n-1} + e^{\alpha(q_n-q_{n+1})}c_{n+1}\right), \tag{2}$$

$$\begin{aligned} \frac{d^2q_n}{d\tau^2} &= (1-e^{q_n-q_{n+1}})\,e^{q_n-q_{n+1}} - (1-e^{q_{n-1}-q_n})\,e^{q_{n-1}-q_n} \\ &\quad +\alpha V_0\Big[\left(c_{n+1}c_n^* + c_{n+1}^*c_n\right)e^{\alpha(q_n-q_{n+1})} \\ &\quad -\left(c_{n-1}c_n^* + c_{n-1}^*c_n\right)e^{\alpha(q_{n-1}-q_n)}\Big]. \end{aligned} \tag{3}$$

In Refs. [14, 15] the authors considered as initial condition both a localized and a delocalized electron separated from a (supersonic) compression wave (a localized soliton-like excitation) which was allowed to travel along the lattice. The computer experiment starts with α set to zero, hence no interaction between electron and lattice dynamics. By the time the compression wave reaches the center of the lattice the original electron Gaussian distribution has decreased to 25 % of the original amplitude before α is switched to a nonzero value. Very rapidly a coherent compound localized structure is formed from the small amplitude delocalized electron wave function, which is swept away by the compression wave (Fig. 1). This is the above mentioned "vacuum cleaner effect". In Fig. 1 we reproduce this process which was obtained integrating numerically Eqs. (2) and (3) with initial conditions:

$$c_n(0) = A\text{sech}\left(\frac{n-\xi}{w}\right)\exp\left[i\phi + iv(n-\xi_0)\right], \tag{4}$$

$$q_n(0) = \ln\frac{1+e^{2k(n-1-\xi_1)}}{1+e^{2k(n-\xi_1)}}, \tag{5}$$

thus confirming the results of [14, 15]. Note the solectron detaching from the background in Fig. 1c.

It was shown in [16] that the equations

$$i\frac{dc_n}{d\tau} = \mu\left(q_{n+1}-q_n\right)c_n - p\left(c_{n+1}+c_{n-1}\right), \tag{6}$$

$$K\frac{d^2q_n}{d\tau^2} = \left(q_{n-1}-2q_n+q_{n+1}\right)\left(1-\beta\left(q_{n+1}-q_{n-1}\right)\right) + \mu\left(|c_n|^2 - |c_{n-1}|^2\right), \tag{7}$$

which are the Davydov analog to (2) and (3) with cubic anharmonicity in the lattice [9], have supersonic (generalized) polaron solutions similar in shape to the ones of (2) and (3) shown in Fig. 1. A similar "vacuum cleaner effect" was observed integrating (6) and (7) using same initial condition as (4) and a similar condition as (5) in the form:

$$q_n(0) = \frac{B}{2}\left[1-\tanh\left(\frac{n-\frac{1}{2}-\xi}{w}\right)\right].$$

Typical results found for a wide range of parameter values and initial conditions (4) and (5) are displayed in Figs. 2–5. We first consider the possibility of vacuum cleaning for the (linear) Davydov model ($\beta = 0$). In Fig. 2 we observe the formation of a right-to-left moving polaron due to the large Peierls-Nabarro (PN) potential (a signature of the lattice discreteness) which stops the left-to-right going waves. The PN potential is obtained at second order approximation in the Poisson summation formula. It is proportional to $\sinh^{-1}(\pi^2/k)$ for Morse interactions and $\sinh^{-1}(\pi^2/w)$ for cubic anharmonicities, according to the wave profile initial conditions (i.c.). PN terms raise due to the discreteness and non-integrability of a problem. In Fig. 3 we show the evolution of the initial conditions and we see no polaron formation but rather a bouncing back of the electron wave function caused by the peaks in the potential. In Fig. 4, very weak vacuum cleaning occurs as compared to the TBA. In Fig. 5 we see new phenomena where the interaction produces a left-to-right traveling compression and a right-to-left moving localized structure. This is due to the fact that for $p = 1$ there is a stronger PN potential which prevents the compression wave trapping process. In the supersonic case solectron formation is found.

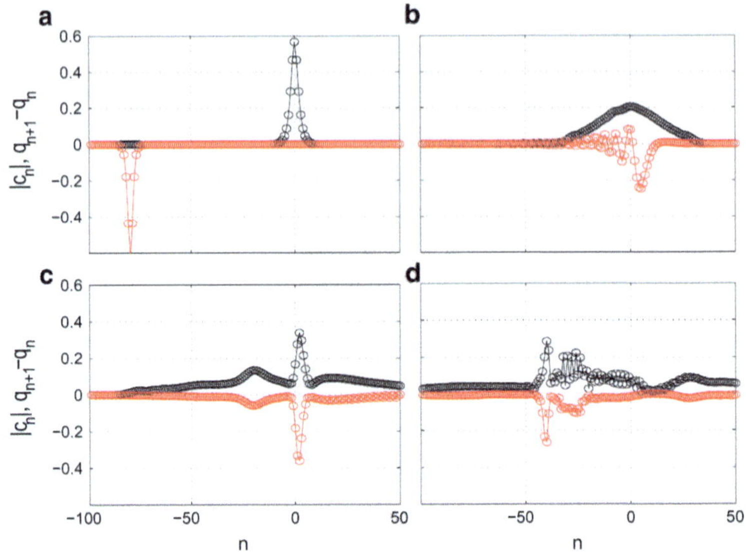

Fig. 2 Numerical evolution of system (6) and (7) for the harmonic (linear) lattice $\beta = 0$ and $p = 1$. The coupling parameter μ is switched on at $t = 14$ from 0 to 1.75. Wave parameter values: $B = 2$, $w = 1.56$, $v = 0$ and $\dot{\xi} = 1$. (**a**) $t = 0$, (**b**) $t = 15$, (**c**) $t = 40$ and (**d**) $t = 70$ (*red lines*: soliton; *black lines*: electron)

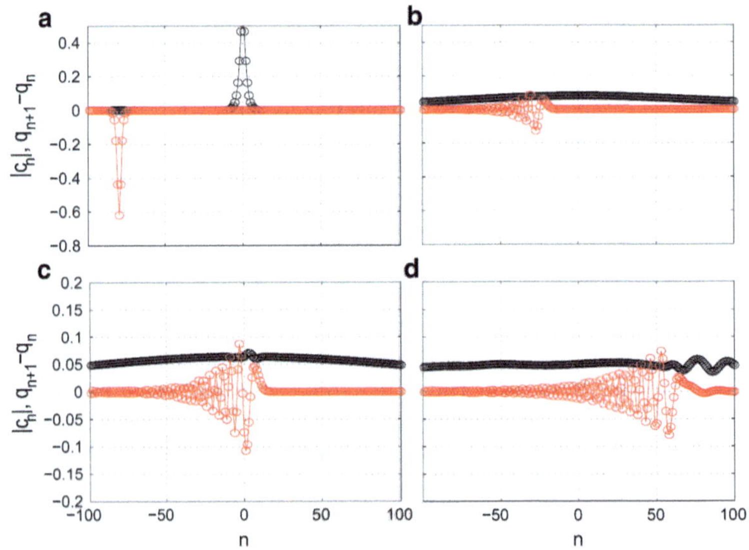

Fig. 3 Numerical evolution of system (6) and (7) for the harmonic (linear) lattice $\beta = 0$ and $p = 10$. The coupling parameter μ is switched on at $t = 14$ from 0 to 1. Wave parameter values as in Fig. 2: (**a**) $t = 0$, (**b**) $t = 15$, (**c**) $t = 25$ and (**d**) $t = 35$ (*red lines*: soliton; *black lines*: electron)

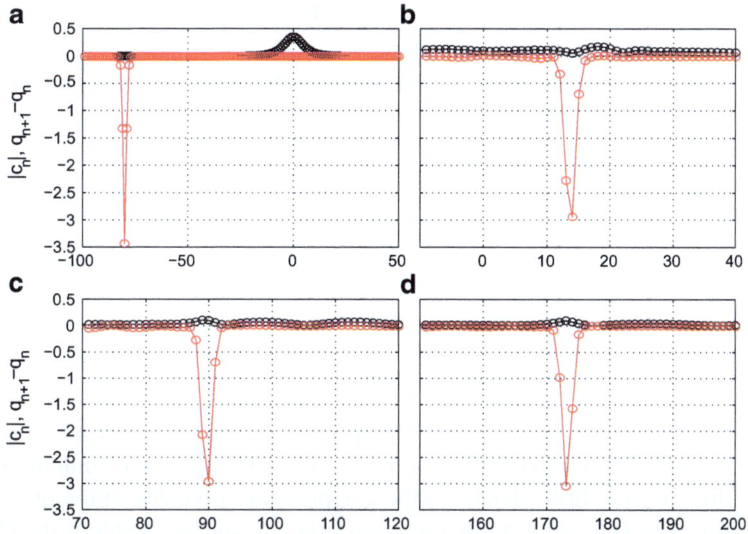

Fig. 4 Numerical evolution of system (6) and (7) for the anharmonic lattice $\beta = 0.5$ with $p = 10$. The coupling parameter μ is switched on at $t = 10$ from 0 to 1.75. Wave parameter values: $B = 6.407$, $w = 0.8385$, $\eta = 4$, $v = 0$ and $\dot{\xi} = 8.48$. **(a)** $t = 0$, **(b)** $t = 11$, **(c)** $t = 20$ and **(d)** $t = 30$ (*red lines*: soliton; *black lines*: electron)

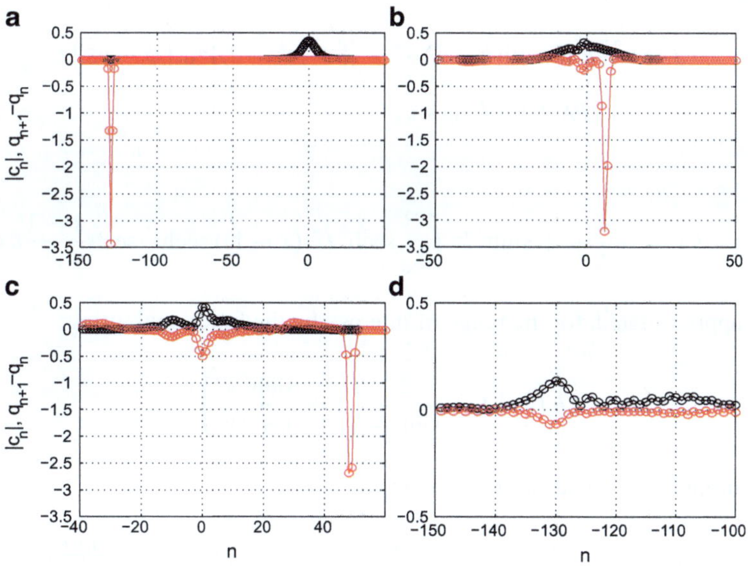

Fig. 5 Numerical evolution of system (6) and (7) for the anharmonic lattice $\beta = 0.5$ with $p = 1$. The coupling parameter μ is switched on at $t = 15$ from 0 to 1.75. Wave parameter values as in Fig. 4: **(a)** $t = 0$, **(b)** $t = 16$, **(c)** $t = 21$ and **(d)** $t = 100$ (*red lines*: soliton; *black lines*: electron)

3 Traveling Polaron/Solectron Solutions

Static polaron solutions were found numerically in [23]. To find traveling polarons/solectrons we proceed as in [16] averaging the Lagrangian in non dimensional variables (1) for the special case of $\alpha = 1$ using the trial functions:

$$c_n = A\text{sech}\left(\frac{n-\xi}{w}\right)\exp\left[i\phi + iv\left(n-\xi\right)\right], \tag{8}$$

$$q_n = \ln\frac{1+e^{2k(n-1-\xi)}}{1+e^{2k(n-\xi)}}, \tag{9}$$

$$q_n - q_{n+1} = \ln\left[1+\sinh^2 k\,\text{sech}^2\left(k\left(n-\xi\right)\right)\right]. \tag{10}$$

Here we take, unlike in [24], the kink/topological soliton/shock wave in $q_n(t)$ to satisfy the amplitude width relation for the Toda shock since in the numerics (unlike in [25]) there is very little change in the compression wave due to the presence of the electron. The velocity $\dot{\xi}$ of the localized structure and the width of the electron wave function will be determined as a function of the amplitude of the electron wave function (which is, in turn, given by the initial normalization to probability unity) and the compression width.

We now substitute (8)–(10) in the Lagrangian (1) and to leading order in the Poisson sum we obtain:

$$\begin{aligned} L = {} & \tfrac{2P}{V}k\left(k\coth k - 1\right)\dot{\xi}^2 - \tfrac{2P}{3V}\tfrac{\sinh^4 k}{k} - 2A^2 w\left(\dot{\phi} - v\dot{\xi}\right) \\ & + 2PA^2 h\left(w,k\right)\cos v, \end{aligned} \tag{11}$$

where

$$h\left(w,k\right) = \frac{2}{\sinh\frac{1}{w}} + \sinh^2 k\int_{-\infty}^{\infty}\text{sech}^2\left(k\left(x-1\right)\right)\text{sech}\frac{x}{w}\text{sech}\frac{x-1}{w}dx,$$

is well approximated, for the range of interest, by the functional fitting:

$$h\left(w,k\right) = \frac{2}{\sinh\frac{1}{w}}\left(1+\frac{\sinh^2 k}{1+kw}\right). \tag{12}$$

The modulation equations take the form:

$$\begin{aligned} \delta k : {} & \tfrac{2P}{V}\left(2k\coth k - \tfrac{k^2}{\sinh^2 k} - 1\right)\dot{\xi}^2 - \tfrac{2P}{3V}\tfrac{\sinh^3 k}{k}\left(4\cosh k - \tfrac{\sinh k}{k}\right) \\ & + 2PA^2 h_k\cos v = 0, \end{aligned} \tag{13}$$

which determines the velocity of the compression wave $\dot{\xi}$ as a function of the shock width and the amplitude of the electron wave function.

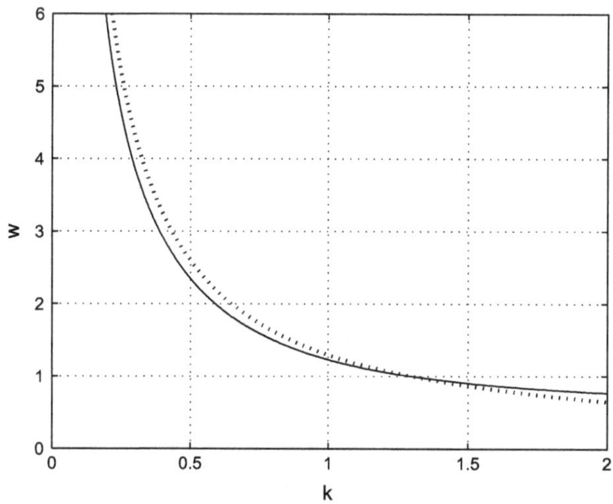

Fig. 6 Dispersion relation $w = w(k)$ obtained from equation (16). *Dotted line*: $w = 1.3/k$

To determine the width of the trapped electron wave function we take the equations:

$$\delta w: \; -2A^2\left(\dot{\phi} - v\dot{\xi}\right) + 2PA^2 h_w \cos v = 0, \tag{14}$$

$$\delta A: \; -4Aw\left(\dot{\phi} - v\dot{\xi}\right) + 4PAh\cos v = 0, \tag{15}$$

and obtain,

$$h_w = \frac{h}{w}, \tag{16}$$

an implicit equation which gives w as a function of k as shown in Fig. 6. It is shown in Fig. 6 that $w = 1.3/k$ is also a good approximation to the solution.

The variation with respect to v gives:

$$\dot{\xi} = \frac{P}{w} h \sin v. \tag{17}$$

We take $\frac{\pi}{2} \leq v < \pi$ which gives $\dot{\xi} > 0$. Solving for v as a function of $\dot{\xi}$ in (17) and using this solution into (13) we obtain a quadratic dependence of $\dot{\xi}$ as a function of k in the form:

$$\begin{aligned} &\tfrac{\sinh^3 k}{3k}\left(4\cosh k - \tfrac{\sinh k}{k}\right) - \left(2k\coth k - \tfrac{k^2}{\sinh^2 k} - 1\right)\dot{\xi}^2 \\ &+VA^2 h_k \sqrt{1 - \left(\tfrac{w\dot{\xi}}{Ph}\right)^2} = 0. \end{aligned} \tag{18}$$

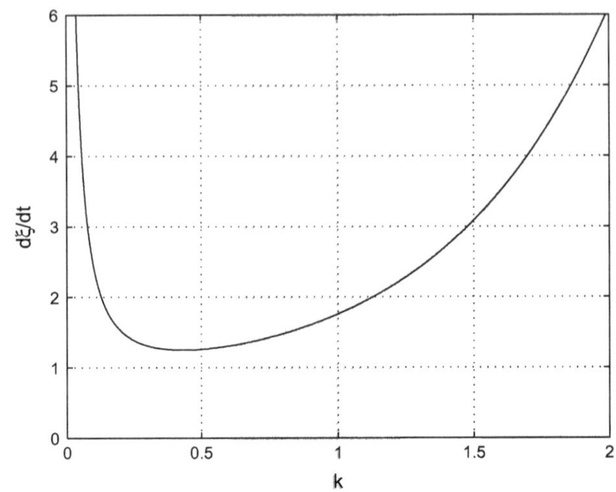

Fig. 7 Zero level curve for velocity $\dot{\xi}$ obtained from (18) for $A = 0.25$ and $V_0 = 0.5$ with $p = 10$.

Notice that for each given pair, A^2 and k, there is only one branch of solutions which represents a polaron, as illustrated in Fig. 7. Unlike the anharmonic one there are no multiple solutions and no sonic solution. This is due to the very small feedback of the electron to the distortion since $h\,(w, k)$ becomes very small as k is large compared to the $\sinh^3 k$ term.

4 Conclusions

When an electron is free to evolve along a lattice, obeying Schrödinger equation, after sometime it becomes completely deslocalized with its probability density uniformly distributed as "dust", tiny spots all over the lattice sites. Needless to say their integration gives probability one. If, however, the electron is coupled to lattice excitations in the form of solitons then we have shown that the latter are able to gather together all probability density spots thus reconstructing to a significantly large extent a localized electron which is carried with the soliton. This is the dynamic process we have denoted as "vacuum cleaner effect". The solution shown here illustrates the trapping of the electron as the consistent ground state for the potential. This in turn shows the supersonic compression wave giving the coherent structure.

Acknowledgements The authors acknowledge fruitful discussions with A.A. Alexandrov (deceased), L. Brizhik, A.P. Chetverikov, L. Cruzeiro, W. Ebeling, and D. Hennig. L.A.C.-A. gratefully acknowledges support from COFAA-IPN and project IPN-CGPI-20120651. M.G.V.

acknowledges support from the Del Amo Foundation for a visit to the University of California at Irvine. This research has been supported in part by the Spanish *Ministerio de Economía y Competitividad* under Grant MAT2011-026221.

References

1. N.J. Zabusky, M.D. Kruskal, Phys. Rev. Lett. **15**, 57 (1965)
2. M. Toda, *Theory of Nonlinear Lattices*, 2nd. edn. (Springer, Berlin, 1988)
3. P.M. Morse, Phys. Rev. **34**, 57 (1929)
4. J. Dancz, S.A. Rice, J. Chem. Phys. **67**, 1418 (1977)
5. T.J. Rolfe, S.A. Rice, J. Dancz, J. Chem. Phys. **70**, 26 (1979)
6. A.P. Chetverikov, W. Ebeling, M.G. Velarde, Int. J. Bifurcation Chaos **16**, 1613 (2006)
7. J.C. Slater, *Quantum Theory of Molecules and Solids*, vol. 4 (McGraw-Hill, New York, 1974)
8. A.A. Voityuk, J. Jortner, M. Bixon, N. Rösch, Chem. Phys. Lett. **324**, 430 (2000)
9. A.S. Davydov, *Solitons in Molecular Systems*, 2nd. edn. (Reidel, Dordrecht, 1991). (And references therein)
10. A.C. Scott, Phys. Rep. **217**, 1 (1992)
11. P.L. Christiansen, A.C. Scott (Eds.), *Davydov's Soliton Revisited. Self-Trapping of Vibrational Energy in Protein* (Plenum, New York, 1983)
12. P.S. Lomdahl, W.C. Kerr, Phys. Rev. Lett. **55**, 1235 (1985)
13. L. Cruzeiro-Hansson, S. Takeno, Phys. Rev. E **56**, 894 (1997). (And references therein)
14. D. Hennig, A.P. Chetverikov, M.G. Velarde, W. Ebeling, Phys. Rev. E **76**, 046602 (2007)
15. M.G. Velarde, W. Ebeling, A.P. Chetverikov, D. Hennig, Int. J. Bifurcation Chaos **18**, 521 (2008)
16. L.A. Cisneros-Ake, A.A. Minzoni, Phys. Rev. E **85**, 021925 (2012)
17. M.G. Velarde, J. Comput. Appl. Math. **233**, 1432 (2010)
18. A.V. Zolotaryuk, K.H.S. Spatschek, A.V. Savin, Europhys. Lett. **31**, 531 (1995)
19. Yu.B. Gaididei, P.L. Christiansen, S.F. Mingaleev, Physica Scripta **51**, 289 (1995)
20. A.V. Zolotaryuk, K.H.S. Spatschek, A.V. Savin, Phys. Rev. B **54**, 266 (1996)
21. M.G. Velarde, W. Ebeling, A.P. Chetverikov, Int. J. Bifurcation Chaos **18**, 3815 (2008)
22. E.M. Conwell, Proc. Natl. Acad. Sci. U.S.A. **102**, 8795 (2005)
23. J.-H. Park, H.-Y. Choi, E.M. Conwell, J. Phys. Chem. B **108**, 19483 (2004)
24. L.A. Cisneros, A.A. Minzoni, Stud. Appl. Math. **120**, 333 (2008)
25. A.B. Aceves, L.A. Cisneros-Ake, A.A. Minzoni, Discret. Contin. Dyn. Syst. – Ser. S. **4**, 975 (2011)

Stability of Mode-Locked Kinks in the AC Driven and Damped Sine-Gordon Lattice

Yaroslav Zolotaryuk

Abstract Kink dynamics in the underdamped and strongly discrete sine-Gordon lattice that is driven by the oscillating force is studied. The investigation is focused mostly on the properties of the mode-locked states in the *overband* case, when the driving frequency lies above the linear band. With the help of Floquet theory it is demonstrated that the destabilizing of the mode-locked state happens either through the Hopf bifurcation or through the tangential bifurcation. It is also observed that in the overband case the standing mode-locked kink state maintains its stability for the bias amplitudes that are by the order of magnitude larger than the amplitudes in the low-frequency case.

Main Abbreviations

- SG: Sine-Gordon
- FK: Frenkel-Kontorova
- JJ: Josephson junctions
- DSG: Discrete sine-Gordon
- PN: Peierls-Nabarro
- JJA: Josephson junction arrays

1 Introduction

The discrete sine-Gordon (DSG) equation, also known as the Frenkel-Kontorova (FK) model, is ubiquitous in condensed matter physics [1, 2]. It has a wide range of applications in the dislocation theory [3], weak superconductivity [4, 5] and magnetism [6]. Among the intensively discussed problems for the DSG dynamics

Y. Zolotaryuk (✉)
Bogolyubov Institute for Theoretical Physics, National Academy of Sciences of Ukraine, vul. Metrologichna 14-B 03680 Kiev, Ukraine
e-mail: yzolo@bitp.kiev.ua

R. Carretero-González et al. (eds.), *Localized Excitations in Nonlinear Complex Systems*, Nonlinear Systems and Complexity 7, DOI 10.1007/978-3-319-02057-0_12,

the problem of the topological soliton (fluxon) response to the ac (time-periodic) bias, remains to be important. This interest is caused in particular by the number of technological applications based on the Josephson junction arrays (JJAs), which are successively modeled by the DSG equation. Properties of the small ac-biased Josephson junctions have been extensively studied both experimentally (starting from the pioneering papers of Shapiro [7]) and theoretically (with the focus on the phase-locking [8] and chaotic regimes [9, 10]). In particular, the rf-biased Josephson junctions have been used as a voltage standard [10, 11].

It is well-known [12] that contrary to the continuous sine-Gordon (SG) equation the DSG equation is non-integrable, and, moreover, it does not possess moving kink solutions. The ac-driven DSG lattice has two independent sources of non-integrability: the external drive (bias) and the discreteness. Interplay of these two sources has led to a number of interesting effects: mode-locking to the frequency of the external drive [13] and kink mobility [14, 15] (including its experimental detection in periodically modulated Josephson junctions [16]), various regimes of the dynamical chaos [13, 18], biharmonically driven discrete kink ratchet [17, 18] to name a few. However, these studies have been performed mostly in the adiabatic, subband (the driving frequency lies in the gap of the linear spectrum) or resonant (the driving frequency lies in the linear band) cases. The high-frequency limit when the driving frequency exceeds the linear wave frequency by several orders of magnitude has been studied in Refs. [19, 20]. In these papers the inversion of the ground state that is based on the Kapitza pendulum effect has been reported. The intermediate *overband* case when the bias frequency exceeds the linear wave frequency, but remains approximately of the same order of magnitude, requires a special attention. This frequency range is the natural bridge between the cases, studied in the papers, mentioned above. The dynamics of topological solitary waves (kinks), especially their linear stability in the intermediately high-frequency regime is the main aim of this paper.

The paper is organized as follows. The model, the equations of motion and the usage of the Floquet method for the linear stability studies are described in the next section. In the Sect. 3 we present the main properties of the standing kinks in the driven DSG lattice. The discussion and the main conclusions are given in the last section.

2 The Model and Equations of Motion

2.1 The DSG Equation

The periodically driven and damped discrete sine-Gordon (DSG) equation is introduced in a dimensionless form as follows:

$$\ddot{\phi}_n - \kappa \Delta \phi_n + \sin \phi_n + \alpha \dot{\phi}_n + A \cos(\omega t) = 0,\ n = \overline{1, N}\,. \tag{1}$$

Here $\Delta\phi_n \equiv \phi_{n+1} - 2\phi_n + \phi_{n-1}$ is the discrete Laplacian and the dot represents the time differentiation. The physical meaning of the field variable ϕ_n depends on the underlying physical system. In the dislocation theory it stands for the particle displacement from its equilibrium position. In the JJ theory [21] ϕ_n corresponds to the phase difference of the wave functions at the nth junction.[1]

Only the periodic boundary conditions

$$\phi_{n+N}(t) = \phi_n(t) + 2\pi Q,\ \dot{\phi}_{n+N}(t) = \dot{\phi}_n(t), \tag{2}$$

which correspond to the circular JJAs, are to be considered. The topological charge Q is an integer constant that stands for the net number of kinks trapped in the lattice. Further on only we will study only the case of one kink ($Q = 1$). The experiments with annular JJAs have been performed for typical lengths $N \sim 8 - 30$ (see Refs. [4,5,22]). In the following, we consider the case of an array (lattice) with $N = 30$.

The dispersion law for the linear excitations (phonons) reads

$$\omega_L(q) = \sqrt{1 + 4\kappa\ \sin^2\frac{q}{2}}\ . \tag{3}$$

Due to finiteness of the array, the wavenumber $q \in [0, 2\pi)$ attains only the discrete set of values $q_m = 2\pi m/N, m = \pm 1, \ldots, \pm N$.

The regular kinks, mode-locked to the frequency of the external bias correspond to the limit cycles of Eq. (1). On these orbits, the average kink velocity is expressed as $\langle v\rangle = k\omega/(2\pi l)$, where the winding numbers k and l are integer. Thus, the kink travels k sites during the time $lT = 2\pi l/\omega$, so that, except for a shift in space, its profile is completely reproduced after this time interval (in the pendulum analogy, this orbit corresponds to k full rotations of the pendulum during l periods of the external drive).

2.2 Linear Stability and the Floquet Theory

In this article our main focus will be on the kinks locked to the external drive. In order to understand better their properties, we will focus on their linear stability. The fluxon periodic orbit is computed by finding zeroes of the map

$$\hat{\mathcal{I}}_{kl}(T)\mathbf{X} = \mathbf{X}, \tag{4}$$

---

[1] The coupling constant $\kappa = \sqrt{\Phi_0/(2\pi I_c L)}$ measures the discreteness of the array, where Φ_0 is the magnetic flux quantum, L is inductance of an elementary cell, and I_c is the critical current of an individual junction. The dimensionless dissipation parameter is then $\alpha = \Phi_0/(2\pi I_c R)$, where R is the resistance of an individual junction, and the time is normalized to the inverse Josephson plasma frequency $1/\omega_0 = \sqrt{C\Phi_0/(2\pi I_c)}$ with C being the junction capacitance.

where the vector $\mathbf{X}$ consists of the dynamical variables $\{\phi_n, \dot{\phi}_n\}_{n=1}^N$. The operator $\hat{\mathcal{I}}_{kl}$ stands for the integration of the equations of motion (1) during the time lT and afterwards the shift of the final solution by k sites forward if $k < 0$ or backward if $k > 0$. The case $k = 0$ corresponds to the fluxon pinned to a lattice site.

A fixed point of the map (4) is a mode-locked solution $\{\phi_n^{(0)}(t), \dot{\phi}_n^{(0)}(t)\}_{n=1}^N$ which reproduces itself after the time lT with the space shift by k lattice sites backward or forward. Next, we substitute the expansion

$$\phi_n(t) = \phi_n^{(0)}(t) + \varepsilon_n(t) , \tag{5}$$

into Eq. (1). For the case of *standing* kink ($k = 0$) after keeping only the linear terms, we obtain the following set of linear ODEs with periodic coefficients:

$$\ddot{\varepsilon}_n = -\alpha\dot{\varepsilon}_n + \kappa\Delta\varepsilon_n - \cos[\phi_n^{(0)}(t)]\varepsilon_n , \quad n = 1, 2, \ldots, N. \tag{6}$$

The map

$$\begin{bmatrix} \boldsymbol{\varepsilon}(lT) \\ \dot{\boldsymbol{\varepsilon}}(lT) \end{bmatrix} = \hat{M}(T) \begin{bmatrix} \boldsymbol{\varepsilon}(0) \\ \dot{\boldsymbol{\varepsilon}}(0) \end{bmatrix} \tag{7}$$

is constructed from the solutions of the system (6). It relates the small perturbations $\boldsymbol{\varepsilon}(t) = \{\varepsilon_n(t)\}_{n=1}^N$ at the time moments $t = 0$ and $t = lT$. The $2N \times 2N$ Floquet (monodromy) matrix $\hat{M}$ contains all the necessary information about the linear stability of the system. If this matrix has at least one eigenvalue with $|\Lambda_n| > 1$ ($n = 1, 2, \ldots, 2N$), then the system is unstable. If for all eigenvalues $|\Lambda_n| \leq 1$, the system is stable. It is well-known [23] that these eigenvalues come in quadruples, so that if Λ_n is an eigenvalue, then Λ_n^*, R/Λ_n and R/Λ_n^* (here $R = e^{-l\alpha\pi/\omega}$, see, for example, Refs. [24, 25]) are also eigenvalues. Thus, the Floquet multipliers lie either on the circle of the radius R (will be referred to as a *R-circle* throughout the paper) or may depart from it after collisions. The notable difference of the ac-driven case from the dc-driven (autonomous) case is the absence of the degeneration with respect to time shifts, which manifests itself in the absence of the eigenvalue $\Lambda = 1$ [26].

Collision of the Floquet eigenvalues on the real axis signals the tangential (saddle-node) bifurcation if it happens at $\arg\Lambda = 0$ or the period-doubling bifurcation if $\arg\Lambda = \pi$. Eigenvalue collision away from the real axis means that the Hopf bifurcation is taking place.

3 Kinks in the High-Frequency Driven DSG Equation

In this paper, we plan to compute the mode-locked limit cycle that corresponds to the standing kink and to path-follow it while a control parameter is changed until the cycle becomes unstable or completely disappears. By monitoring the Floquet eigenvalues Λ_n one can obtain the information about the underlying bifurcations and, consequently, about the unlocking process.

3.1 The Numerical Scheme

The scheme of the numerical studies can be described in the following way. As a starting iteration in the anti-continuum limit ($\kappa \equiv 0$) we consider the kink state that can be described by the following coding sequence

$$\{\underbrace{00\cdots 00}_{n=1,\ldots,N_0},\ \underbrace{2\pi\cdots 2\pi}_{n=N_0+1,\ldots,N}\}\,. \tag{8}$$

This means that we start with the kink which is centered between the N_0th and N_0+1th sites. In should be noted that due to the translational invariance the position of the kink center (defined by N_0) does not influence on the kink properties. The above choice of the coding sequence is defined by the well-known (see [2, 12]) fact that such configuration, sometimes referred as a bond-centered kink, is stable for the DSG equation, in contrary to the site-centered kinks which are unstable. Then the initially stable (in the $\kappa = 0$ limit) mode-locked state is continued numerically. For the numerical computation of the fixed point of Eq. (4) we use the Newton-Raphson iterative method. With this method it is possible to compute numerically the respective mode-locked limit cycle for the given period lT with a desired computer precision. For details one might consult Ref. [28], Chap. 6.1. The advantage of this approach is that not only attractors, but also repellers, can be computed. Also, wrong conclusions which can be made due to sensitivity to initial conditions can be avoided. Once the fixed point of (4) is computed, $\phi_n^{(0)}$ is plugged into Eqs. (6) and the Floquet matrix is computed and diagonalized with the standard numerical methods.

3.2 The Existence Diagram

It is well known that discreteness causes kink pinning to the lattice [12]. The kink in a lattice can be approximately described as a particle that moves in the spatially periodic potential (the PN potential) with the period that coincides with the lattice spacing. If the bias amplitude is weak enough, the kink will oscillate around the minima of the PN potential with the bias period. In other words, the kink oscillations are locked to the external drive. If the amplitude (or, alternatively, the coupling) is increased till some critical value, the external drive becomes strong enough to unlock the kink and to destroy the mode-locked state. Intuitively, it is not hard to understand that on the parameter plane (κ, A) one can draw a curve

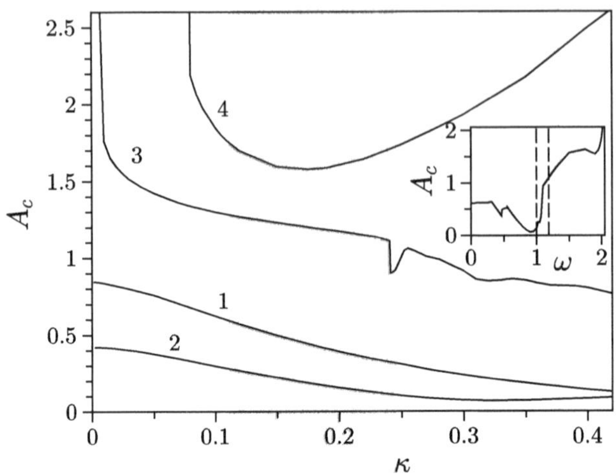

Fig. 1 Existence diagram of standing mode-locked kinks for $\alpha = 0.1$, $\omega = 0.25$ (curve *1*), $\omega = 0.7$ (curve *2*), $\omega = 1.3$ (curve *3*) and $\omega = 2$ (curve *4*). The *inset* shows the $A_c(\omega)$ at $\kappa = 0.1$, the *dashed vertical bars* mark the edges of the linear band, $\omega_L(0)$ and $\omega_L(\pi)$

$A_c = A_c(\kappa)$ that marks the loss of stability of a stable mode-locked standing kink. Suppose κ is fixed and the mode-locked state is continued while the amplitude A is increased until this state loses stability at $A = A_c$. At $A > A_c$, depending on the type of the destabilizing bifurcation and on the system parameters, different dynamical regimes can take place such as diffusively moving kinks, mode-locked moving kinks, quasiperiodic kinks, another standing mode-locked kink states with the different shape or even the chaotic dynamics of the whole array (when individual kinks cannot be identified). The latter case corresponds to the *non-existing* area and will not be discussed in this paper. The nature of the kink dynamics in the non-pinned area also strongly depends on the driving frequency ω.

The existence diagrams on the (κ, A) plane for the different values of ω are shown in Fig. 1. The critical dependence $A_c(\kappa)$ is defined by the first bifurcation that makes the mode-locked standing kink unstable. In the case of *subband* $(0 < \omega < \omega_L(q))$ frequencies the dependence $A_c(\kappa)$ is almost monotonic (see the curves 1–2 in Fig. 1) with the two well-defined limiting cases. In the limit $\kappa \to \infty$, the effects of discreteness disappear, thus A_c should decrease. On the other hand, the decrease of κ means that the PN barrier becomes stronger and thus a larger amplitude is necessary to overcome it. As a result, A_c increases when $\kappa \to 0$. The exit from the pinning area (below the curve $A_c(\kappa)$) can lead to different scenarios depending on the direction of the exit. The issue of discrete kink unlocking (depinning) in the DSG lattice driven by the subband frequencies has been studied in Refs. [13,18]. In Fig. 2a the kink dynamics just above the A_c value at $\kappa = 0.1$ and $\omega = 1.3 > \omega_L(q)$ is shown. For these parameters the first destabilizing bifurcation takes place at $A_c \simeq 1.301145$, while the driving amplitude in Fig. 2a corresponds to the slightly larger value $A = 1.3012$. It can be clearly observed that initially the

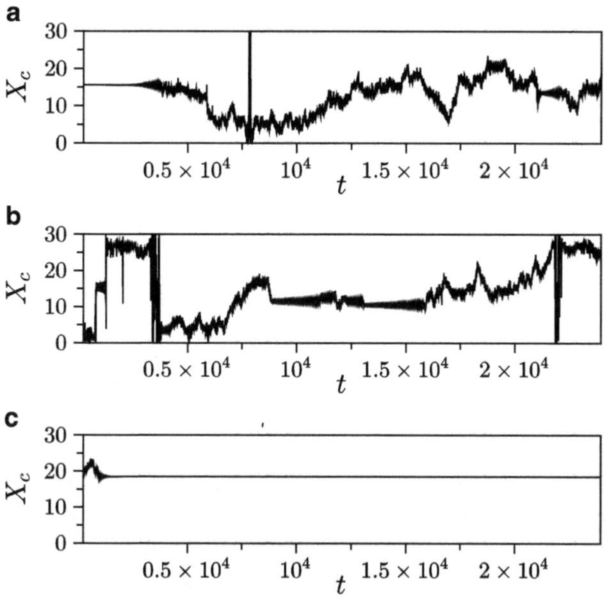

Fig. 2 Time evolution of the kink center of mass $X_c = (4\pi)^{-1} \sum_{n=1}^{N} n(\phi_{n+1} - \phi_{n-1})$ at $\alpha = 0.1$, $\omega = 1.3$, $\kappa = 0.1$, and $A = 1.3012$ (**a**), $A = 1.3$ (**b**) and $A = 1.295$ (**c**)

kink stays pinned, but at $t \sim 5{,}000$ it unlocks and begins to move chaotically. At this point we should remark that at $A < A_c$ the standing mode-locked kink is not the unique solution. Typically, the non-uniqueness is manifested by the existence of hysteresis loops in the neighborhood of A_c (for the subband case see Ref. [13]) . This situation is demonstrated in Fig. 2b, c. After crossing the critical value A_c the driving amplitude is decreased back, and the chaotic moving solution is followed [2] to the values $A < A_c$ (Fig. 2b) until it falls back to the mode-locked state as shown in Fig. 2c. The hysteresis loop appears to be quite narrow, constituting less then 1 % of A_c. At $\omega = 0.35$ a different scenario has been observed, when the stability loss at A_c leads to the complete kink destruction and chaotic dynamics of the whole lattice. If this chaotic solution is followed while A is decreased, it does not return to the mode-locked kink state. Instead, at some $A_* < A_c$ the system falls back to the complex mode-locked state that includes the kink and one or several breathers, placed at the different lattice sites.

The dependence on the driving frequency $A_c = A_c(\omega)$ is non-monotonic (see the inset in Fig. 1). The main resonance with the linear band can be clearly identified with the minimum at $\omega \simeq \omega_L(0) \equiv 1$. In this frequency range a linear wave around

[2]It should be noted that the persistence of the diffusive kink solution at $A < A_c$ has been checked for the times $t \sim 10^6$ which are much larger then in Fig. 2b. For the sake of clearness of the figure these data were not plotted.

the kink can be excited at a rather small driving amplitude. Two significantly shallow minima at $\omega \simeq 0.5$ and $\omega \simeq 2$ appear due to the subharmonic half-frequency resonance and the resonance with the double frequency of the linear spectrum, respectively.

The case of *overband* frequencies (lying above the linear band, $\omega > \omega_L(q)$) is not well studied yet. Let us now focus on the curves 3 and 4 in the Fig. 1. It should be noted that they are not monotonous, and, in addition they show sharp growth at $\kappa \to 0$. In order to understand this behavior we study the Floquet spectrum and the nature of the destabilizing bifurcations.

3.3 Floquet Spectrum and the Destabilizing Bifurcations

In the subband (low-frequency) case the stability loss leads to the unlocking of the standing kink. Typically it takes place through the tangential bifurcation [13] which leads to the disappearance of the mode-locked state. After this bifurcation the kink starts to move in a chaotic way. More precisely, its regime belongs to the type-I intermittency. If the coupling is weak enough, the instability may lead to the destruction of the kink state and to chaotic motion of the whole lattice. In the neighborhood of the main resonance $\omega \simeq 1$ the first destabilizing bifurcation is also tangential and it takes place at rather small values of the amplitude. It is caused by the resonance with the linear band and transforms the spatially monotonic standing kink into the standing kink with the oscillating background. Further increase of A leads to the second destabilizing bifurcation after which the kink undergoes either depinning transition or destruction.

The case with $\omega = 1.3$ that corresponds to the curve 3 in the Fig. 1, can be considered either as overband or as a resonant depending on the value of κ, which defines the upper edge $\omega_L(\pi) = \sqrt{1 + 4\kappa}$ of the linear spectrum (3). Evolution of the Floquet spectrum for this case for the values of $\kappa = 0.1$ and $\kappa = 0.25$ is shown in Fig. 3. In the first case ($\kappa = 0.1$) the bias frequency lies above the linear spectrum. In the anticontinuum limit, all the eigenvalues sit in one point and with the growth of κ they separate, forming two distinct groups: the modes associated with the linear spectrum and the internal mode(s) [27]. The linear band extends with κ according to the dispersion law (3), while the localized eigenmode is distinctly detached from the linear band, as can be clearly seen in Fig. 3a, c.

The destabilizing bifurcation takes place at $A \simeq 1.31$ and it can be clearly seen from the Fig. 3a that this is a Hopf bifurcation in which the localized mode (for the respective eigenvector shape see the inset in the panel (a)) is involved. Stability loss in this case causes the kink to unlock and to start moving chaotically along the lattice.

A different bifurcation scenario is observed when $\kappa = 0.25$ (Fig. 3c, d). Now the resonance with the linear spectrum occurs and the destabilizing eigenvector is delocalized as shown in the Fig. 3g. Obviously it is associated with the linear spectrum. Indeed, one can clearly observe multiple collisions of the eigenvalues

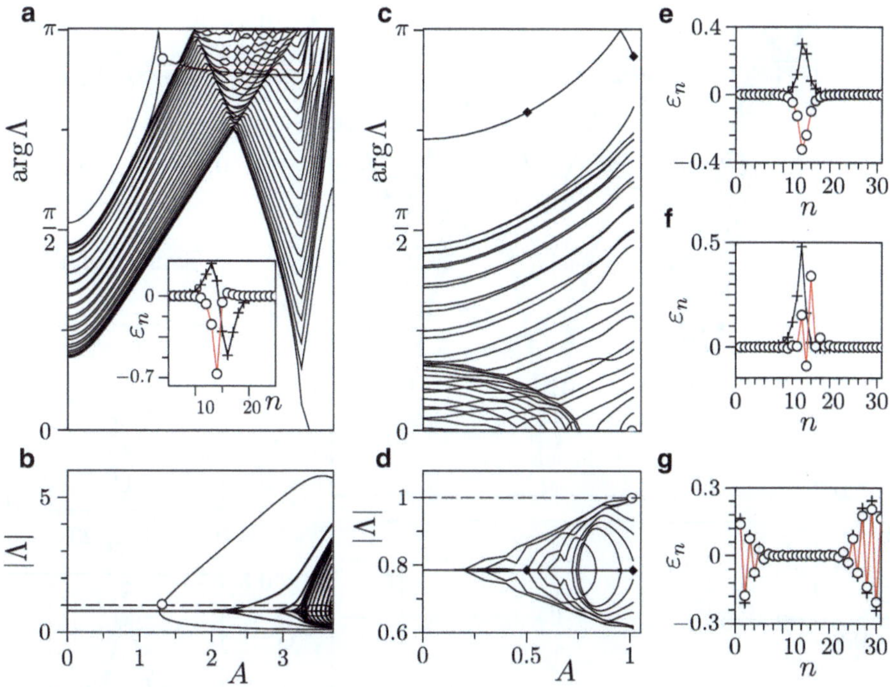

Fig. 3 Phases and moduli of the Floquet multipliers as a function of the bias amplitude for $\omega = 1.3$, $\kappa = 0.1$ (**a** and **b**) and $\kappa = 0.25$ (**c** and **d**). The inset in the panel (**a**) shows the unstable eigenvector at $A = 1.31$. The *dashed horizontal line* in the panels (**b**) and (**d**) corresponds to the radius of the unit circle. The unstable eigenvalues are marked by ∘. Panels (**e**) and (**f**) show the shape of the localized eigenvector at $A = 0.5$ and $A = 1.015$, respectively. The respective eigenvalues are marked by ♦. Panel (**g**) shows the unstable eigenvector at $A = 1.008$. In panels (**e**)–(**g**) + corresponds to $\mathrm{Re}\epsilon_n$ and ∘ corresponds to $\mathrm{Im}\epsilon_n$. The rest of the parameters is the same as in Fig. 1

from the linear band on the positive side of real axis. These collisions are represented as "bubbles" in the $|\Lambda_n(A)|$ dependences Fig. 3d. One of these "bubbles" expands beyond the value $|\Lambda_n| = 1$ at $A = 1.008$ and causes the instability of the mode-locked state. The further small increase of the driving amplitude (to the value $A = 1.095$) brings the kink to the regime of chaotic diffusion. The localized eigenvalue (Fig. 3e, f) stays distinctly detached from the linear band.

The resonance with the linear waves explains the small dip and some oscillations in the $A_c(\kappa)$ dependence (curve 3 in the Fig. 1). In the intervals $\kappa \in [0.241, 0.253]$ and $\kappa \in [0.33, 0.335]$ the first destabilizing bifurcation is the tangential (due to the resonance with the linear waves) and the Hopf bifurcation occurs later, while everywhere outside this interval it is the Hopf bifurcation which comes first. This tangential bifurcation occurs at slightly smaller values of A, therefore the respective interval on the κ axis is marked by the small dip. We remind that due to the finiteness of the lattice the linear spectrum is discrete and the external ac bias resonates with

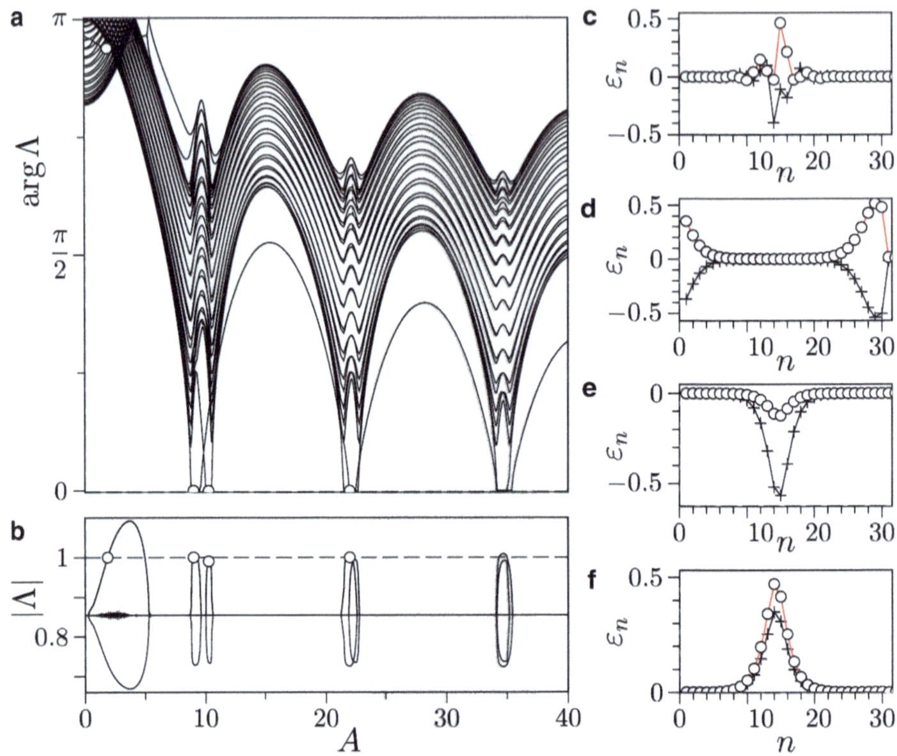

Fig. 4 Phases (**a**) and moduli (**b**) of the Floquet multipliers as a function of the bias amplitude for $\kappa = 0.1$, $\omega = 2$ and $\alpha = 0.1$. The *dashed horizontal line* in the panel (**b**) corresponds to the radius of the unit circle. The unstable eigenvectors ($\mathrm{Re}\varepsilon_n$ (+) and $\mathrm{Im}\varepsilon_n$ (∘)) are shown in the panels (**c**) ($A = 1.85$), (**d**) ($A = 9$), (**e**) ($A = 10.25$) and (**f**) ($A = 22$). The respective eigenvalues are marked by ∘ in the panel (**a**)

the particular cavity modes. That is why the tangential bifurcation takes place not for all $\kappa \geq (\omega^2 - 1)/4$, but in the certain intervals of κ. After the tangential bifurcation the kink shape changes as it becomes surrounded by the non-decaying phonon tail.

Now we increase the bias frequency up to the value $\omega = 2$. The behavior of the Floquet multipliers as the bias amplitude is increased is shown in Fig. 4. Similarly to the previous case of $\omega = 1.3$ the Hopf bifurcation, driven by the spatially localized perturbation (see Fig. 4c) makes the standing kink unstable. The instability leads to the appearance of a small localized mode on the top of the kink. The whole lattice is no longer in the mode-locked state but in the quasiperiodic, although the kink remains pinned. Further increase of A leads to the second Hopf bifurcation at $A \simeq 5.25$ that restores the mode-locked standing kink state. Inside this interval the dynamic is quasiperiodic and no dynamical chaos has been detected. It appears that this Hopf bifurcation can be controlled by changing the system parameters. In particular, this can be done by varying the coupling constant κ, the

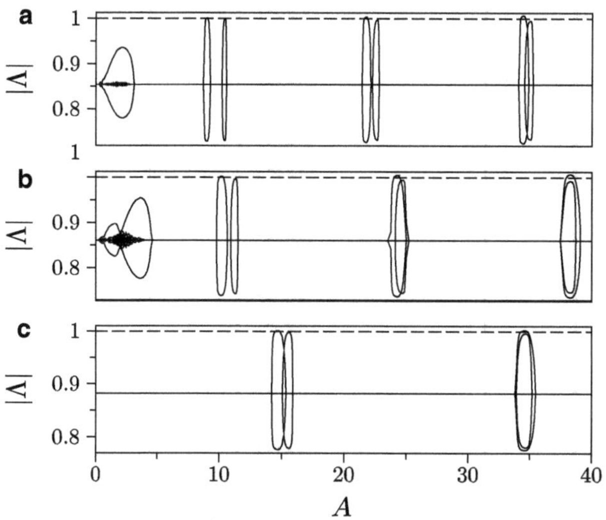

Fig. 5 Moduli of the Floquet multipliers as a function of the bias amplitude for $\kappa = 0.05$, $\omega = 2$ (**a**); $\kappa = 0.1$, $\omega = 2.1$ and $\kappa = 0.1$, $\omega = 2.5$ (**c**). The *dashed horizontal line* marks the radius of the unit circle

bias frequency ω or the damping constant α. After the Hopf bifurcation a number of tangential bifurcations take place at the much larger values of A. Some of them are destabilizing and some are not. Before discussing them we turn our attention on the control of the first (Hopf) bifurcation.In the Fig. 5 we show how the moduli of the Floquet eigenvalues evolve with the increase of A for the different parameter values. The first case, shown in the Fig. 5a corresponds to the same frequency $\omega = 2$, but the coupling is reduced to $\kappa = 0.05$. As a result, the bubble associated with the destabilizing Hopf bifurcation is significantly reduced, and, more importantly, the collision of the respective Floquet eigenvalues does not lead to the instability, because these eigenvalues remain inside the unit circle. We have monitored the intermediate cases $\kappa = 0.07, 0.08, 0.09$ and have observed the gradual increase of the bubble with the growth of κ within this interval. Now we can explain the sharp growth of the $A_c(\kappa)$ dependence when κ is decreased (see Fig. 1). It happens because the Hopf bifurcation does not longer lead to the kink instability, and the first destabilizing bifurcation is the tangential bifurcation. Similar control of the destabilizing bifurcation can be performed by increasing the damping parameter, which decreases R and removes the instability. Interestingly, the increase of ω also reduces the Hopf bifurcation, and can completely remove it. This is demonstrated in the Fig. 5b, c, where the $|\Lambda_n(A)|$ is plotted for $\omega = 2.1$ and $\omega = 2.5$. In the case $\omega = 2.1$ the “bubble” stays inside the unit circle, while in the case $\omega = 2.5$ it is almost unnoticeable.

Another interesting observation that comes from Figs. 3 to 5 is the possibility of the standing kink existence for the rather large bias values that exceed the values of

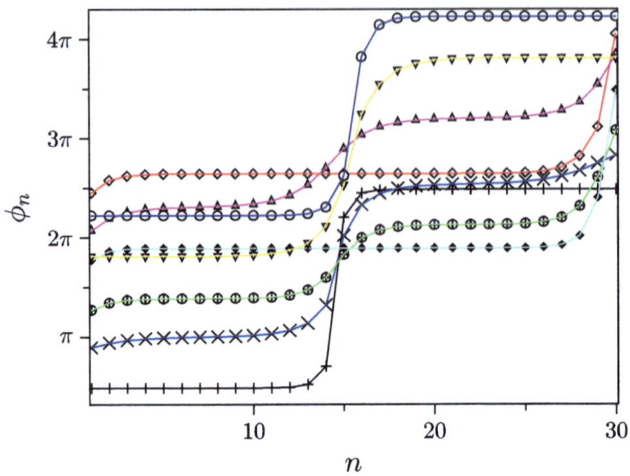

Fig. 6 Kink profiles for the case shown in Fig. 4 and for the different ac bias amplitudes: $A = 5$ (+), $A = 9\,(\times)$, $A = 10\,(\oplus)$, $A = 12$ (♦), $A = 20\,(\diamond)$, $A = 22\,(\Delta)$, $A = 23.5\,(\nabla)$ and $A = 28$ (∘). *Solid lines* are used as guides for an eye

A_c in the subband case by the order of magnitude. Indeed, if one forgets the first Hopf bifurcation (which can be controlled by the proper parameter choice anyway), the existence interval of the standing kink solution stretches along the A-axis interrupted only by narrow intervals where the Floquet multipliers exit the unit circle or approach it. These intervals are represented by the typical "bubbles" in the $|\Lambda_n(A)|$ dependencies in the Fig. 4b and in the Fig. 5. The amplitude of these bubbles can also be controlled by the increase of α so that the instability can be removed. These underlying bifurcations are the tangential ones and the respective unstable perturbation in the phase space is driven by the U-shaped eigenvector, as shown in Fig. 4d. It means that the instability develops in the following way: the core of the kink stays unchanged, while the tails tend to become deformed. The second bifurcation, on the contrary, causes the deformation in the kink core. In the Fig. 6 the change of the kink profiles with the growth of A is shown. At $A = 5$ the kink profile remains very much as for the usual strongly discrete kink. Further increase of A leads to the broadening of the kink core and to the small deformation away from the core. Note that these deformations in the tail are actually caused by the instabilities, described in the previous paragraph. More precisely, they are associated with the tangential bifurcation at $A \sim 9$ and the instability direction, illustrated in Fig. 4d, prescribes exactly the same: the tail deformation. As the amplitude increased further, the deformation in the tails grows and the core straightens up. As a result, the kink restores its strongly localized shape, but now it is centered between the sites $N - 1$ and N (shown by ♦ in Fig. 6). Also, the field variable has been increased by 2π. Further increase of the amplitude repeats the deformation scenario: in the neighborhood of the second set of tangential bifurcations (around $A \sim 22 - 23.5$) the tails deform and the core straightens up. Finally the kink re-emerges in as a strongly localized excitation with its center placed between the sites $n = 15$ and $n = 16$. Note that the initially at $A = 0$ the kink position was between the sites $n = 14$ and $n = 15$. The solution has been followed up to $A \simeq 50$ and such a transformation has been observed once again. Thus, the increase of the bias amplitude causes a sequence of the kink shape transformations that are driven by the tangential bifurcations and that shift the kink by $N/2$ sites along the lattice. Here we have reached the high-frequency limit, described in Ref. [19] where the fast-oscillating drive effectively makes (after the averaging over the fast variables) the sine-term in the DSG equation to look like $J_0\left(A/[\omega\sqrt{\omega^2+\alpha^2}]\right)\sin\phi_n$ in the limit $\omega \gg 1$, and where J_0 is the Bessel function. As a result, depending on the driving amplitude the stable ground may become unstable, and back again, perfectly explaining the intermediate structures which have flat plateau at $\phi_n \sim \pi$ ($A = 10$ ($\oplus$)) or at $\phi_n \sim 3\pi$ ($A = 23.5$ (∇)) in Fig. 6.

4 Discussion and Conclusions

To summarize, we have studied the kink stability in the ac-driven and damped sine-Gordon lattice. We have focused mainly on the overband regime where the bias frequency lies above the upper edge of the linear spectrum.

We have shown that the mode-locked high-frequency driven standing kink is much more stable with respect to the external ac drive comparing to the low-frequency driven kink. The critical bias amplitude at which the mode-locked state loses its stability may be several times larger in the high-frequency case comparing to the low-frequency case. For example, for the coupling $\kappa = 0.1$ we have $A_c \simeq 0.62$ for $\omega = 0.25$ and $A_c \simeq 1.85$ for $\omega = 2$. For $\kappa = 0.05$ the difference is even more drastic: $A_c \simeq 0.84$ ($\omega = 0.25$) against $A_c \simeq 8.89$ ($\omega = 2$). The instability of the mode-locked state is driven in the different frequency regime by the different bifurcations.

In the low-frequency case it is the tangential bifurcation, associated with the internal mode, while in the high-frequency case it is either Hopf bifurcation, associated again with the internal mode, or the tangential bifurcation, associated with the tail deformation. When the lattice is driven slowly, it is much easier to drive it away from the mode-locked state because different parts of the kink react in the different way to the perturbation: the kink may be depinned or even destroyed if the coupling is very small. Crossing the critical line $A_c(\kappa)$ in the low-frequency regime will almost probably lead either to the chaotic kink diffusion or to the complete kink destruction. The instability scenario in the high-frequency case is different. The Hopf bifurcation transforms the mode-locked periodic kink state into the quasiperiodic but still standing one with the small distortion in the core. Further increase of the driving amplitude may unlock the kink and it begins to travel chaotically along the lattice. If one takes even higher driving frequency and trace again the mode-locked states while the bias amplitude is increased, such an quasiperiodic state will not turn into the chaotic diffusive regime, but instead will be transformed back into the mode-locked state. Thus, the Hopf bifurcation can be controlled by the proper damping and/or frequency choice and the instability can be arrested.

Another interesting result is the kink structural deformation, driven by the tangential bifurcation at very large amplitudes ($A \gtrsim 10$). Here we reach the high-frequency limit, studied in Refs. [19, 20] where the ground state of the chain alternates between the $\phi = \pi$ and $\phi = 2\pi$. We would like to note that the DSG lattice in the above-mentioned papers has been studied at $\kappa \geq 4$ and $\omega \sim 10 - 100$ what can be considered as a high-frequency driven weakly discrete case, while in this article we deal with the strongly discrete lattices at $\kappa < 1$ and in the frequency range that exceeds the linear frequencies insignificantly: $\omega \sim 2$.

References

1. L.M. Floría, J. J. Mazo, Adv. Phys. **45**, 505 (1996)
2. O.M. Braun, Y.S. Kivshar, Phys. Rep. **306**, 2 (1998)
3. Ya. Frenkel, T. Kontorova, Phys. Z. Sowietunion **13**, 1 (1938)
4. S. Watanabe, H.S.J. van der Zant, S.H. Strogatz, T.P. Orlando, Physica D **97**, 429 (1996)
5. A.V. Ustinov, Physica D **123**, 315 (1998)
6. H.-J. Mikeska, M. Steiner, Adv. Phys. **40**, 191 (1991).
7. S. Shapiro, Phys. Rev. Lett. **11**, 80 (1963); S. Shapiro, A.R. Janus, S. Holly, Rev. Mod. Phys. **36**, 223 (1964)
8. R.L. Kautz, J. Appl. Phys. **52**, 3528 (1981)
9. R.L. Kautz, J. Appl. Phys. **52**, 6241 (1981); E. Ben-Jacob, Y. Braiman, R. Shainsky, Y. Imry, Appl. Phys. Lett. **38**, 822 (1981); R.L. Kautz, R. Monaco, J. Appl. Phys. **57**, 875 (1985)
10. R.L. Kautz, Rep. Prog. Phys. **59**, 935 (1996)
11. M.T. Levinsen, R.Y. Chiao, M.J. Feldman, B.A. Tucker, Appl. Phys. Lett. **31**, 776 (1977); R.L. Kautz, Appl. Phys. Lett. **36**, 386 (1980)
12. M. Peyrard, M.D. Kruskal, Physica D **14**, 88 (1984)
13. P.J. Martínez, F. Falo, J. Mazo, L.M. Floría, A. Sánchez, Phys. Rev. B **56**, 87 (1997)
14. L.L. Bonilla, B.A. Malomed, Phys. Rev. B **43**, 11539 (1991)
15. G. Filatrella, B.A. Malomed, J. Phys.: Condens. Matter **11**, 7103 (1999)
16. A.V. Ustinov, B.A. Malomed, Phys. Rev. B **64**, 020302(R) (2001)
17. Y. Zolotaryuk, M. Salerno, Phys. Rev. E **73**, 066621 (2006)
18. Y. Zolotaryuk, Phys. Rev. E **86**, 026604 (2012)
19. N. Grønbech-Jensen, Y.S. Kivshar, Phys. Lett. A **171** (1992) 338
20. Y.S. Kivshar, N. Grønbech-Jensen, M.R. Samuelsen, Phys. Rev. B **45** (1993) 7789
21. A. Barone, G. Paterno, *Physics and Applications of the Josephson Effect* (Wiley, New York, 1982)
22. P. Binder, D. Abraimov, A.V. Ustinov, S. Flach, Y. Zolotaryuk, Phys. Rev. Lett. **84**, 745 (2000)
23. V.I. Arnol'd, *Mathematical Methods of Classical Mechanics* (Springer, Berlin, 1989)
24. A.E. Miroshnichenko, S. Flach, M.V. Fistul, Y. Zolotaryuk, J.B. Page, Phys. Rev. E **64**, 066601 (2001)
25. J.L. Marín, F. Falo, P. Martínez, L.M. Floría, Phys. Rev. E **63**, 066603 (2001)
26. J.A. Sepulchre, R.S. MacKay, Nonlinearity **10**, 679 (1997)
27. C. Baesens, S. Kim, R. Mackay, Physica D Nonlinear Phenomena **113**, 242 (1998)
28. S. Flach, C.R. Willis, Phys. Rep. **295**, 182 (1998)

Spatio-Temporal Dynamics in a Ring of Coupled Pendula: Analogy with Bubbles

V.J. Sánchez-Morcillo, N. Jiménez, J. Chaline, A. Bouakaz, and S. Dos Santos

Abstract Many systems in nature, like drops, bubbles or some macromolecules present circular or spherical symmetry. Under the influence of some external force, such objects often develop surface patterns whose properties are greatly influenced by the underlying geometry. However, differently from the planar case, patterns in curved geometries have been much less explored. Despite the complexity of the particular physical problems, the basic dynamical features are often captured by simple models of coupled oscillators. Here we present a theoretical and experimental study of the spatial instabilities of circular ring of coupled pendula parametrically driven by a vertical harmonic force. Normal oscillation modes (breathing, dipole, quadrupole) and localized patterns of different types (breathers and kinks) are predicted and observed. The analogy between the considered discrete mechanical system and a gas bubble cavitating under the action of an acoustic field is established. On the basis of this analogy, the oscillation patterns and localized modes observed experimentally in acoustically driven bubbles are interpreted and discussed.

V.J. Sánchez-Morcillo (✉) · N. Jiménez
Instituto de Investigación para la Gestión, Integrada de las Zonas Costeras,
Universitat Politécnica de Valencia, c/ Paranimf no. 1, 46730 Grao de Gandia, Spain
e-mail: victorsm@upv.es

J. Chaline · A. Bouakaz
Unité Mixte de Recherche Inserm U930 "Imagerie et Cerveau, Université François-Rabelais
PRES Centre Val de Loire Université, 10 boulevard Tonnellé, 37032 Tours Cedex, France

S. Dos Santos
ENI Val de Loire, 3 Rue de la Chocolaterie BP 3410, F-41034 Blois cedex, France

Unité Mixte de Recherche Inserm U930 "Imagerie et Cerveau, Université François-Rabelais
PRES Centre Val de Loire Université, 10 boulevard Tonnellé, 37032 Tours Cedex, France

R. Carretero-González et al. (eds.), *Localized Excitations in Nonlinear Complex Systems*, Nonlinear Systems and Complexity 7, DOI 10.1007/978-3-319-02057-0_13,

1 Introduction

The spatio-temporal dynamics of a gas bubble in a fluid under the action of an external oscillating pressure field has been the subject of intense investigations during the last decades [1]. An acoustically forced bubble experiences different instabilities as the driving parameters (mainly the amplitude and the frequency of the external field) are changed. One of such instabilities leads to the formation of surface patterns, similar to the Faraday patterns observed in the free surface of vertically vibrated fluids, by means of parametric excitation of subharmonic modes [2,3]. Nonlinear patterns on the bubble surface have been however much less explored. Only recently, the existence of localized soliton-type surface waves has been theoretically considered [11]; their existence has been evidenced by different experimental observations. An interesting extension is the bubble with an artificial coating, so called ultrasound contrast agent (UCA), with current applications in medical imaging and different therapies. Besides free gas bubbles or UCAs, there exist many other systems, either natural or artificial, characterized by spherical (or more generally curved) geometries, with typical sizes ranging from micro to macro scale, which show evidences of complex spatio-temporal dynamics under external forcing [4]. Some problems of interest in this category are for example the vibrational modes of small liquid droplets or spherical molecules, or the buckling and rupture of UCAs.

We present here an experimental and numerical study of spatiotemporal dynamics in a parametrically driven system presenting a curved geometry. The system is a ring-shaped chain of coupled oscillators subjected to a vertical harmonic acceleration. The theoretical model is the well known Frenkel-Kontorova lattice (a discrete version of the sine-Gordon equation) with damping and parametric driving. A practical implementation of this chain (but in a straight configuration) was proposed in [5], where the existence of kinks was demonstrated. Later, the analysis was extended [6] to show the existence of breathers in this chain. Such a system, with the appropriate modifications, allows to investigate fundamental aspects not studied before, such as the effect of curvature in the existence and stability of the different localized solutions. Here we propose that a ring-type configuration of the linear 1D chain can be used as a toy model of acoustically driven UCA and related systems. We also investigate the effect of defects in the chain: nucleation of solitons, enhancement of stability. Thus, the aim of this work is twofold. On one hand, to study parametric instability leading to pattern formation and localized patterns (breathers and kinks) in a chain of coupled oscillators with circular symmetry (a ring) and, on the other hand, to show some formal analogies between this system and a spherical pulsating bubble. Our study could be useful to understand and predict localized solutions in bubbles, as certain localized modes observed experimentally [7], and to determine the optimal conditions for the excitation of particular patterns.

2 The Ring of Coupled Pendula

2.1 The Model

A driven-damped chain of coupled oscillators, despite its simplicity, is known to present a very rich dynamics, and is often used as a toy model to describe the behavior of other physical systems or processes, such as ladder networks of discrete Josephson junctions, charge-density wave-conductors, crystal dislocation in metals, DNA dynamics, and many others, see e.g. [8]. Most of these systems are well described by a straight chain. When disposed in the form of a circular ring, connecting first and last masses as we propose in this work, the chain forms the equivalent of a 2D bubble, and for small radius and/or large amplitudes the effect of curvature becomes non negligible. However, in the theoretical description below we will ignore for simplicity the curvature effects.

Let's consider a monoatomic pendulum lattice, formed by equal masses linearly coupled to their nearest neighbors. The whole chain is subjected to vertical oscillations (parametric driving) of amplitude A and angular frequency ω. The motion on the $n-th$ pendulum is described by its angle θ with respect to the vertical, which obeys the equation

$$
\begin{aligned}
& m l_n^2 \ddot{\theta}_n + \gamma l_n \dot{\theta}_n + m l_n \left[g + 4A\omega^2 \cos(2\omega t) \right] \sin(\theta_n) \\
&= k_2 \left(\theta_{n+1} - 2\theta_n + \theta_{n-1} \right) + k_4 \left[(\theta_{n+1} - \theta_n)^3 - (\theta_n - \theta_{n-1})^3 \right],
\end{aligned}
\tag{1}
$$

where l_n is the length of the $n-th$ pendulum, m its mass, k_2 and k_4 are the nearest coupling strengths of the harmonic and quartic interactions respectively, γ is the friction coefficient and g the gravity constant acceleration. We note (see below) that the anharmonic coupling is included since it is related to the existence of upper cutoff breathers which are actually observed in the experiments. In this section, a homogeneous chain with identical elements is assumed, with $l_n = L$, but the effect of inhomogeneities will be considered later. After normalization, the homogeneous chain can be written in the form of a driven-damped Frenkel-Kontorova (FK) equation:

$$
\begin{aligned}
& \ddot{\theta}_n + \beta \dot{\theta}_n + \left[\omega_0^2 + \eta \cos(2\omega t) \right] \sin(\theta_n) \\
&= c_2 \left(\theta_{n+1} - 2\theta_n + \theta_{n-1} \right) + c_4 \left[(\theta_{n+1} - \theta_n)^3 - (\theta_n - \theta_{n-1})^3 \right]
\end{aligned}
\tag{2}
$$

where $\eta = 4\omega^2 A/L$, $\omega_0 = \sqrt{g/L}$ is the natural frequency of the uncoupled system, $\beta = \gamma/ml$ and c_2, c_4 are the normalized damping and coupling terms, respectively. As stated, Eq. (2) does not take into account the curvature effect which are necessarily present in a ring geometry.

The type of localized solutions studied here (breathers and kinks) appear as a slow spatial modulation of the cutoff modes. In this case it is possible to derive equations describing the evolution of the envelopes. Following [5, 6] we consider solutions in the form

$$\theta_n(t) = \epsilon(\pm 1)^n \psi_\pm(\tau, x_n) \exp(-i\omega t) + c.c. \tag{3}$$

where ϵ is a smallness parameter and the positive (negative) sign describe modes existing close to the lower (upper) band edges, also called phase-matched and phase-mismatched modes respectively. Assuming the following scalings $\beta, \eta, (\omega^2 - \omega_\pm^2) \sim O(\epsilon^2)$, $\tau = \epsilon^2 t$, $x_n = \epsilon n a$, it follows that the amplitude of the envelope obeys the parametrically driven-damped nonlinear Schroedinger equation,

$$2i\omega_\pm \frac{\partial \psi_\pm}{\partial t} = c_2^2 \frac{\partial^2 \psi_\pm}{\partial x^2} + i\omega_\pm \beta \psi_\pm + \frac{\eta}{2}\psi_\pm^* + \chi_\pm \psi_\pm \left|\psi_\pm\right|^2 . \tag{4}$$

where $\omega_\pm = \{\omega_0, \sqrt{\omega_0^2 + 4c_2^2}\}$ and $\chi_\pm = \{\omega_0^2/2, \omega_0^2/2 - 96c_4^2\}$. Equation (4) has been derived for a variety of system in different disciplines (e.g. Faraday waves in vibrated fluid or granular layers, or pulses in optical fibers), differing only in the coefficients. Interestingly, it has been also derived in [11] to describe the azimuthal envelope of equatorial surface mode in bubbles. Thus, the ring of pendula is suitable for studying these solution which are characteristic of real bubbles. This and other analogies will be discussed also in Sect. 5.

2.2 Experimental Setup

The setup follows basically that proposed in [5], except for the circular geometry. The chain is formed by N spheres of mass m, suspended on a ring-shaped support of radius R by strings forming a "V" with vertical length L, Fig. 1. The coupling is obtained by overlapping the strings and fixing the crossing point with a knot at a distance d from the support. The whole system lies on a flat surface attached to a loudspeaker which provides the parametric driving. The images are taken with a camera located 1 m above the ring. For quantitative measurements of the amplitudes, optical correction was applied.

The linear (phonon) dispersion relation of the chain is bounded: it presents a lower cutoff frequency $\omega_l = \sqrt{g/L}$ and an upper cutoff frequency $\omega_u = \sqrt{g/(L-d)}$, which corresponds to the natural frequency of pendula oscillating either from the rigid support or from the knot, respectively. The dispersion relation can be obtained in the form

$$\left(\frac{\omega_n}{\omega_0}\right)^2 = \frac{1}{1-\eta}\left[1 - \eta \cos^2\left(\frac{\pi}{2}\frac{n}{N}\right)\right], \tag{5}$$

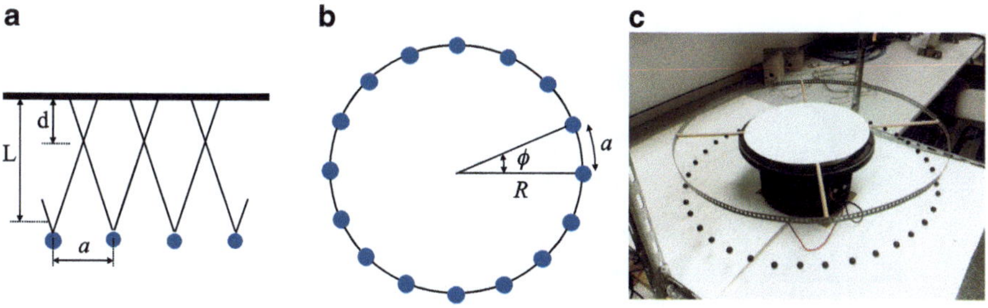

Fig. 1 Experimental setup: (**a**) side view (**b**) top view (**c**) photograph of the setup

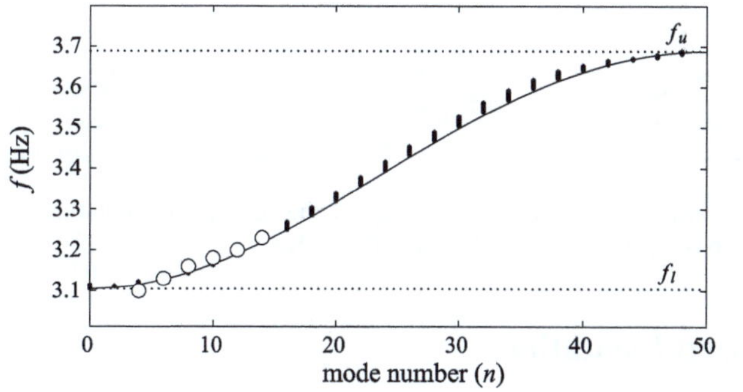

Fig. 2 Experimental eigenmodes (*open circles*) located on the theoretical dispersion curve. *Black circles* correspond to numerical results

where $\eta = d/L \in [0, 1]$ measures the strength of the coupling and n is the mode index number.

For the setup shown in Fig. 1 and used in our experiments, the parameters of the chain where $N = 39$ on the system shown on Fig. 1, $m = 5$ g, $L = 10$ cm, $d = 3$ cm, $a = 5$ cm, $R = 33$ cm (we note that the distances have a small variability along the chain, thus the experimental system is weakly inhomogeneous). For this set of parameters, the edges of the dispersion relation or cutoff frequencies, Eq. (5), occur at the frequencies $f = 3.11$ Hz and $f = 3.63$ Hz respectively (Fig. 2).

Because of the boundary conditions, in our closed system only even number of the mode index n are allowed. In Fig. 3 we show the first six eigenmodes excited in the system with increasing frequencies (see caption). The frequencies at which the different modes appear in the experiment follow the dispersion relation (5) with accuracy.

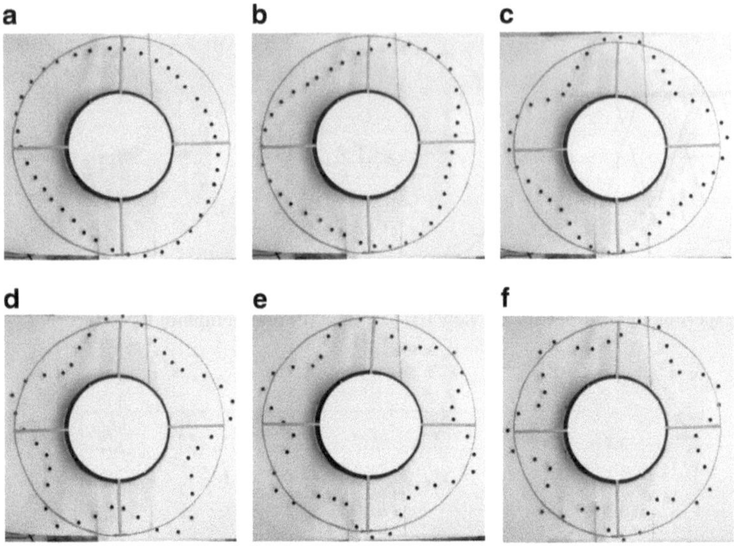

Fig. 3 Eigenmodes observed versus frequency of the parametric excitations: (**a**) 3.10 Hz, (**b**) 3.13 Hz, (**c**) 3.16 Hz, (**d**) 3.18 Hz, (**e**) 3.21 Hz, (**f**) 3.23 Hz

3 Localized Excitations

Driving the chain at frequencies slightly above or below the edges of the dispersion relation or phonon band, one can excite different types on localized modes in the chain. In this section we report the existence of four types of localized modes in the experimental ring chain described before. In a 1D linear (straight) chain, the motion of each oscillator is symmetric with respect to the angle [5]. For low amplitude oscillations, i.e. for linear modes, this is still a valid assumption for the ring-shaped chain. However, due to the curved geometry, this symmetry no more holds. The maximum amplitude is different in compression (motion towards the center of the ring) and expansion phases. Also the distance between neighbor oscillators shows such asymmetry, being larger for the expansion phase. This occurs because the masses approach or repel depending on the direction of motion. This effect is also observed in 3D (spherical) gas bubbles driven at large amplitude. In the following, we will ignore this effect. Note that the solutions given by Eq. (2) will not take into account this asymmetry, since curvature effects are not included there.

3.1 Lower Cutoff Modes

The NLS equation given by Eq. (4), predicts the existence of breathers at frequencies slightly below the lower cutoff. In Fig. 4, we show such localized solution, obtained

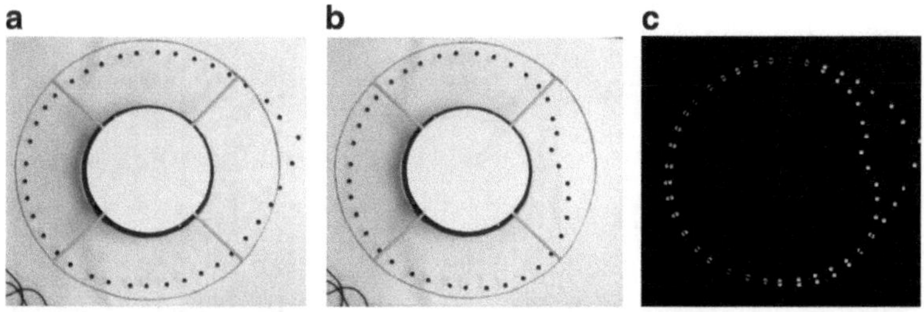

Fig. 4 Single breather as a example of localized solution (**a**): Maximum expansion phase; (**b**): Maximum compression phase; (**c**): Substraction between frames (**a**) and (**b**). Movie available at http://goo.gl/IK7RJ

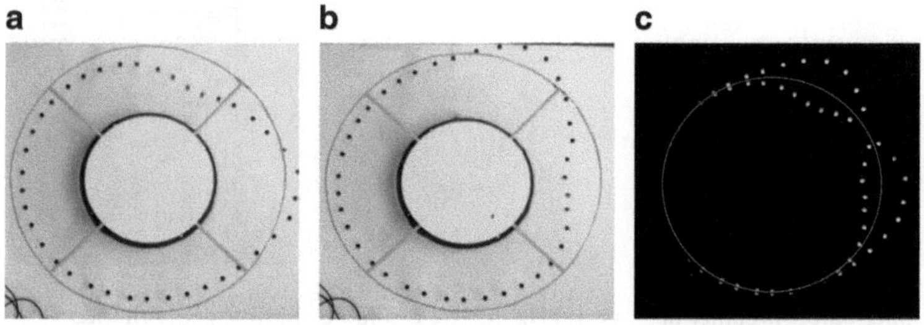

Fig. 5 A bound state of two breathers, at the same parameters as in Fig. 4. Movie available at http://goo.gl/nbPRP

by hitting a few oscillators when the chain is driven at $f = 3.03$ Hz. As in most of the paper, Fig. 4a, b show the chain distribution at the instant of maximum positive and negative peak amplitudes (expansion and compression phases, respectively), while figure (c) is the difference between both images, evidencing the breathing motion. Note the asymmetry between phases, as discussed above.

With the same parameters, one can easily excite also multiple-breather states. In Fig. 5 we show a bound state of two breathers, oscillating with opposite phases. Comparing with the single breather in Fig. 4, the bounded state presents distinctive features: the width of the individual breathers is smaller, their profiles (particularly in the compression phase) are smoothed, and the asymmetry in compression-expansion phases is enhanced.

We have also observed lower cutoff kinks, as those shown in Fig. 6. The kinks appear in pairs, and separate domains where a group of pendula oscillate with a given phase, shifted by π with respect to the neighbor domain. Differently from the nodes in the eigenmodes shown in Fig. 3, the distance between kinks (and the size of the domains) is arbitrary, and not fixed by the boundary conditions. We note that continuous theories based on NLS equations does not predict the existence of such lower cutoff kinks, even considering anharmonic coupling terms. This issue should be thus investigated further. Interestingly, kinks have still not been predicted to occur in gas bubbles.

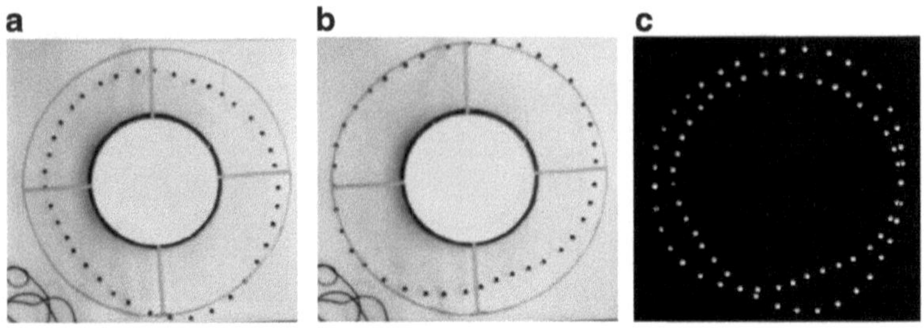

Fig. 6 Lower cutoff kink, obtained for a frequency $f = 3.08$ Hz

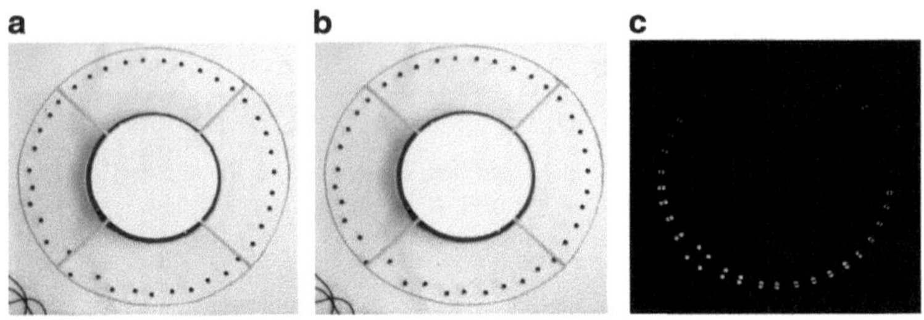

Fig. 7 Upper cutoff breather at $f = 4.0$ Hz. Movie available at http://goo.gl/h3Se5

These low frequency modes are more likely to be related to those observed in real gas bubbles, since oscillation of neighbor particles occurs in phase (as is always the case in a continuous system).

3.2 *Upper Cutoff Modes*

Both breathers and kinks exist for the chain at frequencies above the gap. This solutions do not find an equivalent in real bubbles (neighbor particles oscillate in anti-phase).

A remarkable fact is that the description of chain behavior as envelope wave solutions of the NLS equation [6] predicts that breather and kinks can not exist simultaneously in a given chain (they have incompatible existence domains). This is supported by the experimental observation in straight chains reported previously [9]. However, we have observed both structures (see figures) in our chain. This is another indication that the dynamics of our chain is not adequately described by continuous models of NLS type, and one must consider discreteness or curvature effects, which may enhance the stability domains. Also, the existence of weak randomness in the chain parameters may affect the stability of these structures. Both upper cutoff breathers and kinks are shown in Figs. 7 and 8 respectively.

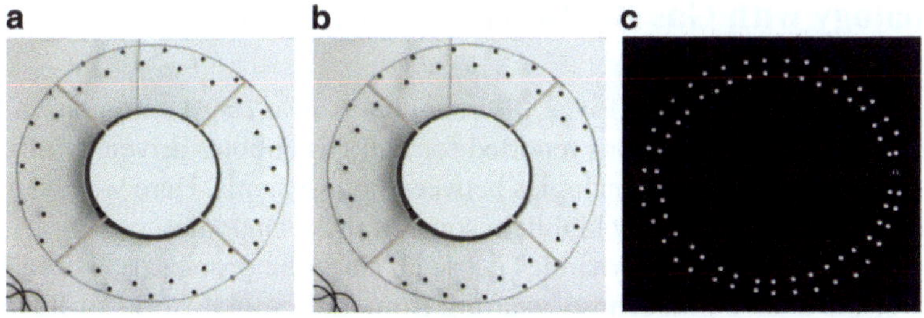

Fig. 8 Upper cutoff kink excited at $f = 4.1$ Hz. Movie available at http://goo.gl/hEOvy

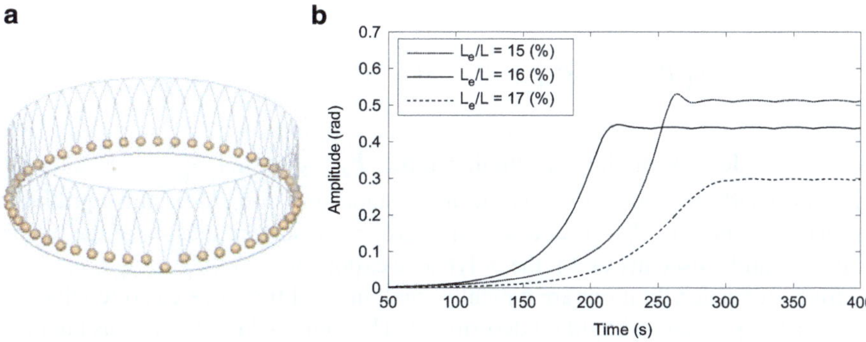

Fig. 9 (**a**) The ring chain with a point defect. (**b**) Time evolution of the breather amplitude, for different defect lengths

4 Defect-Induced Localization

The numerical study reveals that, in a perfectly homogeneous chain, localized solutions are difficult to excite, being very sensitive to the shape and amplitude of the initial condition. However, experimentally they appear much more easily, sometimes being enough to hit one oscillator to excite a breather or kink. Our experimental chain is far from ideal. There are no exactly equal oscillators, differing in the length, position of coupling knot, distance to the neighbors, etc. However it shows all the plethora of localized solutions described above, and predicted for the homogeneous chain. We have explored numerically the effect of two types of inhomogeneities in the chain: a localized defect (e.g. one pendulum with a length slightly different than the rest, Fig. 8a) and a random distribution of structural parameters. The effect of a single defect was discussed in [10], in the context of breather stabilization against space-time chaotic states. Applying a similar analysis to our ring chain, we observe and increase in the stabilization, according to our experimental observations. It occurs at particular values of the length of the defect: there exist an optimal length, at which the time of development of the breather is minimized, as shown in Fig. 8b). We also note that breathers tend to be localized at the defect position, a fact reported previously in the literature (Fig. 9).

5 Analogy with Gas Bubbles in an Ultrasound Field

The presented results about driven discrete chains in a ring configuration can be related to different observations reported for real gas bubbles driven by ultrasound fields, owing to the existing analogies between both systems. Here we review some basic results and subsequently highlight some of these analogies.

The literature on bubble dynamics is broad, since the problem has been investigated over the course of a century, starting from the seminal works of Rayleigh in 1912. Most of the efforts were devoted to describe the radial motion of a bubble under the action of an external pressure field, which is well described by the Rayleigh-Plesset equation for the time dependent radius $R(t)$

$$\rho\left(R\ddot{R}+\frac{3}{2}\dot{R}^2\right)=P_g-P_0-P_A(t)-\frac{2\sigma}{R}-4\mu\frac{\dot{R}}{R}, \tag{6}$$

with ρ the density of the surrounding fluid, $P_g=\left(P_0+\frac{2\sigma}{R_0}\right)\left(\frac{R_0}{R}\right)^{3\gamma}$, the gas pressure inside the bubble, P_0 is the hydrostatic pressure, R_0 is the equilibrium radius of the bubble, P_A the acoustic pressure, σ the surface tension of the bubble, μ the dynamic viscosity and γ the polytropic exponent.

Since acoustic vibrations are typically small in amplitude, one can often linearize the equations to get a simplified description. The radial vibration of a gas bubble in a fluid is well modeled as a simple harmonic oscillator whose stiffness and mass parameters relate to the properties of the gas and fluid. It is well known that the resonant frequency of the radial mode, known as the Minnaert frequency, is given by

$$f_0=\frac{1}{2\pi R_0}\sqrt{\frac{3\kappa p_0}{\rho_0}}. \tag{7}$$

For a microscopic bubble in water at standard pressure, this equation reduces to $f_0R_0=3.26\,\mathrm{m/s}$. Note that a similar scaling relation roughly holds for our macroscopic bubble.

The analysis of surface modes in bubbles implies also the angular dependence, and becomes more complex. Two different theoretical approaches have been used. One consists in assuming that the radius can be expanded in terms of spherical harmonics, as

$$r(t,\theta,\phi)=R(t)+\sum_{n=0}^{i}a_i(t)Y_i(\theta,\phi). \tag{8}$$

When $|a_n|\ll R$ linearization leads to the following set of equations

$$\ddot{a}_n+\gamma_n\dot{a}_n+\omega_n^2a_n=0 \tag{9}$$

where γ_n and ω_n are the damping and pulsation frequency of the different surface modes. In the stationary case $R(t) \rightarrow R_0$ they are given by

$$\begin{aligned} \gamma_n &= \frac{2(n+2)(2n+1)\mu}{\rho R^2} \\ \omega_n^2 &= \frac{(n-1)(n+1)(n+2)\sigma}{\rho R^3} \end{aligned} \tag{10}$$

Similar equations can be obtained, with different coefficients, for the macroscopic ring of pendula.

A different approach, which does not require linearization, has been followed in [11]. Starting from the Hamiltonian of the fluid system and using a multiple-scales expansion, an amplitude equation for the envelopes was derived.

Consider a solution in the form

$$r(t,\theta,\phi) = R_0 + \left(\frac{1}{2}\rho(\theta,t)e^{-i\omega t} + c.c.\right)\cos(n\phi) \tag{11}$$

then, it is shown that the amplitude obeys the equation

$$2i\omega\frac{\partial\rho}{\partial t} = \Delta\rho + c^2\frac{\partial^2\rho}{\partial x^2} + 2i\omega\delta\rho + \beta\rho^* + \alpha\rho\,|\rho|^2\,. \tag{12}$$

where $x = \cos(\theta)$ and

$$\begin{aligned} \Delta &= (\omega_1^2 - \omega^2) + \omega_1^2\left(\frac{5}{8n} + \frac{3x^2}{2}\right) \\ c^2 &= \frac{3\omega_1^2}{2n^2} \\ \alpha &= \frac{17}{32}\omega_1^2\left(\frac{n}{R_0}\right)^2 \end{aligned} \tag{13}$$

where $\omega_1^2 = (n-1)(n+1)(n+2)\sigma/(\rho R_0^3)$ and β and δ are parametric driving and losses introduced phenomenologically. This equation is structurally similar to the equation obtained for the chain of pendula, Eq. (4), but with different coefficients. The study predicts the existence of localized patterns on the bubble wall, as shown in[11], similar to those reported here for the macroscopic bubble pendula.

6 Conclusions

The spatio-temporal dynamics of a parametrically driven array of coupled oscillators forming a closed line or ring has been investigated. We demonstrate experimentally the existence of four types of localized solutions, in the form of kinks and breathers. These structures show particular features related to the asymmetry introduced by the curvature of the system. The effect of inhomogeneities in the chain is also discussed. This macroscopic system shows similarities with the acoustically driven gas bubble, where the parametric driving is exerted by a harmonic ultrasound field. We compared here both systems to show possible relations between the localized patterns that can be used novel features in gas bubble dynamics (e.g., the existence of kinks). However, formal techniques as the Similarity Analysis (SA) can be used to explore quantitatively the limits of this analogies. Work in this direction is in progress.

Acknowledgements The work was supported by Spanish Ministry of Science and Innovation and European Union FEDER through project FIS2011-29731-C02-02.

References

1. T. Leighton , *The Acoustic Bubble* (Academic, London, 1997)
2. A.I. Eller, L.A. Crum, Instability of the motion of a pulsating bubble in a sound field. J. Acoust. Soc. Am. **47**, 762 (1970)
3. A. Prosperetti, Viscous effects on perturbed spherical flows. Appl. Math. **84**, 339 (1977)
4. A.O. Maksimov, T. Leighton, Pattern formation on the surface of a bubble driven by an acoustic field. Proc. R. Soc. A **468**, 57 (2011)
5. B. Denardo, B. Galvin, A. Greenfield, A. Larraza, S. Putterman, W. Wright, Observations of localized structures in nonlinear lattices: domain walls and kinks. Phys. Rev. Lett. **68**, 1730 (1992)
6. W.-Z. Chen, Experimental observation of solitons in a 1D nonlinear lattice. Phys. Rev. B. **49**, 15063 (1994)
7. M. Versluis, D.E. Goertz, P. Palanchon, I.L. Heitman, S.M. van der Meer, B. Dollet, N. de Jong, D. Lohse, Microbubble shape oscillations excited through ultrasonic parametric driving. Phys. Rev. E **82** (2010)
8. O.M. Braun, Y.S. Kivshar, *The Frenkel-Kontorova model: Concepts, Methods and Applications* (Springer, Berlin/New York, 2004)
9. J. Cuevas, L.Q. English, P.G. Kevrekidis, M. Anderson, Discrete breathers in a forced-damped array of coupled pendula: modeling, computation, and experiment. Phys. Rev. Lett. 102, 224101 (2009)
10. N.V. Alexeeva, I.V. Barashenkov, G.P. Tsironis, Impurity-induced stabilization of solitons in arrays of parametrically driven nonlinear oscillators. Phys. Rev. Lett. **84**, 3053 (2000)
11. A.O. Maksimov, T.G. Leighton, P.R. Birkin, Self focusing of acoustically excited Faraday ripples on a bubble wall. Phys. Lett. A **372**, 3210 (2008)

Part IV
Quodons and Muscovite Mica

Theory of a Quodon Gas with Application to Precipitation Kinetics in Solids Under Irradiation

V. Dubinko and R. Shapovalov

Abstract Rate theory of the radiation-induced precipitation in solids is modified with account of non-equilibrium fluctuations driven by the "gas" of lattice solitons (a.k.a. "quodons") produced by irradiation. According to quantitative estimations, a steady-state density of the quodon gas under sufficiently intense irradiation can be as high as the density of phonon gas. The quodon gas may be a powerful driver of the chemical reaction rates under irradiation, the strength of which exponentially increases with irradiation flux and may be comparable with strength of the phonon gas that exponentially increases with temperature. The modified rate theory is applied to modelling of copper precipitation in FeCu binary alloys under electron irradiation. In contrast to the classical rate theory, which disagrees strongly with experimental data on all precipitation parameters, the modified rate theory describes quite well both the evolution of precipitates and the matrix concentration of copper measured by different methods.

1 Introduction

Radiation damage in crystals caused by energetic particles (gamma, electrons, neutrons, light and heavy ions, etc.) is traditionally characterized by the numbers of point defects, i.e. vacancies and self-interstitial atoms (a.k.a. Frenkel pairs) and their clusters produced in displacement events. Their subsequent evolution is governed by diffusion, which leads to segregation of point defects into vacancy and interstitial clusters, dislocation loops and voids, a.k.a. *extended defects*. The difference in the ability to absorb point defects by extended defects is one the main driving force of microstructural evolution under irradiation. A recovery from

V. Dubinko (✉) · R. Shapovalov
NSC Kharkov Institute of Physics and Technology, Kharkov 61108, Ukraine
e-mail: vdubinko@mail.ru

R. Carretero-González et al. (eds.), *Localized Excitations in Nonlinear Complex Systems*, Nonlinear Systems and Complexity 7, DOI 10.1007/978-3-319-02057-0_14,

radiation damage is traditionally thought to be driven by thermal fluctuations resulting in the evaporation of single vacancies or atoms from extended defects. These are examples of Schottky defects, defined as point defects ejected from an extended defect [1]. Another driver for radiation-induced microstructural evolution is based the forced atomic relocations resulting from nuclear collisions, a.k.a. *ballistic effects* [2] that have been taken into account for the explanation of the dissolution of precipitates under cascade damage. Later on it was recognized that the so-called "thermally activated" reactions may be strongly modified by irradiation resulting in the *radiation-induced production of Schottky defects* [1,3–6], which has essentially the same physical nature as the ballistic effects [12], but, in contrast to the latter, it operates under both cascade and non-cascade damage conditions, including sub-threshold electron irradiation that does not produce stable Frenkel pairs. The underlying mechanisms for these processes are based on the interaction of extended defects with unstable Frenkel pairs, focusing collisions (a.k.a. focusons) and with *lattice solitons*. The latter can be mobile, and these are referred to bellow as ***quodons***, *which are stable quasi-particles that propagate one-dimensionally and transfer energy along the close packed directions of the lattice*. Quodons may have more technological significance than focusons due to much longer propagation ranges expected from the nonlinear theory and demonstrated experimentally [6, 7]. Russell and Eilbeck [7] have presented evidence for the existence of quodons that propagate great distances in atomic-chain directions in crystals of muscovite, an insulating solid with a layered crystal structure. Specifically, when a crystal of muscovite was bombarded with alpha-particles ($E < 1\,\mathrm{keV}$) at a given point at room temperature, atoms were ejected from remote points on another face of the crystal, lying in atomic chain directions at more than 10^7 unit cells distance from the site of bombardment.

There is also evidence that quodons can occur in non-layered crystals of different classes, including insulators, semiconductors and metals. Some examples can be found in the radiation damage studies in silicon [8] and germanium [9] and in diffusion experiments in polycrystalline austenitic stainless steel [10] and single crystals of copper [11]. This points out to the necessity of the *modification of the chemical rate theory* with account of the quodon-induced energy deposition to the reaction area. Accordingly, the rate theory of microstructure evolution in solids has been modified with account of the production of Schottky defects at surfaces of extended defects due to their interaction with the radiation–induced lattice excitations [1,3–6]. The modified theory predictions include important phenomena, which have not been properly understood before, such as the irradiation creep [1,3], radiation-induced annealing of voids [1,6], saturation of the void growth under high dose irradiation [6], and the void lattice formation [1,5].

Until recently the evidence for the existence of lattice solitons provided by molecular-dynamics simulations was restricted mainly to one and two-dimensional networks of coupled nonlinear oscillators with simple "toy" potentials [12, 13], whereas reports on their observation in three-dimensional systems using "realistic" molecular–dynamics potentials were scarce and restricted to alkali halide crystals (see e.g. [14]). The lattice solitons found in these simulations always drop down from the optical band(s) into the phonon gap, and hence become unstable. Consequently, it has been assumed that the softening of atomic bonds with increasing vibrational amplitude is a general property of crystals, and therefore lattice solitons with frequencies above the top phonon frequency cannot occur. However, in their recent paper, Haas et al. [15] have provided a new insight on this problem by demonstrating that the anharmonicity of metals appears to be very different from that of insulators. The point is that the essential contribution to the screening of the atomic interactions in metals comes from free electrons at the Fermi surface. As a consequence, the ion-ion attractive force may acquire a nonmonotonic dependence on the atomic distance and may be enhanced resulting in an amplification of even anharmonicities for the resulting two-body potentials. This effect can counteract the underlying softening associated with the bare potentials with a moderate increase of vibrational amplitudes. As a result, in some metals, lattice solitons may exist with frequency *above the top of the phonon spectrum*. Using the known literature values of the pair potentials, Haas et al. have found that in *Ni* and *Nb* this condition is fulfilled. Their molecular-dynamics simulations of the nonlinear dynamics of *Ni* and *Nb* confirmed that high-frequency lattice solitons may exist in these metals, and their corresponding energies may be relatively small, starting from *threshold energy of* 0.2 *eV*, just above the phonon band. These results allow us to look at the modification of the rate theory based on the quodon dynamics from a different angle as compared to the one proposed in Refs. [1, 3–6]. Initially it has been assumed that the main difference between quodons and focusons is in their path lengths. Accordingly, it was postulated that (I) the quodon production rate was equal to the focuson production rate; and (II) similar to focusons, quodons could eject vacancies from extended defects provided that their energy exceeded the vacancy formation energy. This mechanism is close to the classical ballistic mechanism of the precipitate dissolution under cascade damage [2]. However, in view of the new results [15], one can assume that irradiation may produce quodons with energies almost as low as that of phonons, which are lower than typical focuson energies by orders of magnitude. This assumption has important consequences, as discussed in the first part of the paper, where we introduce and develop a new concept of quodon "gas" proposed recently in Ref. [16]. The concept is then used to modify a rate theory of the radiation-induced growth (or shrinkage) of precipitates of a new phase (extended defects) in solids. In the second part, the modified rate theory is applied to modelling of the nucleation and growth of copper precipitates (nano-sized clusters of copper atoms) in *FeCu* binary alloys under electron irradiation, which has been observed experimentally and characterized quantitatively by Mathon et al. [17].

2 Gas of Quodons and Its Effect on Reaction Rates in Solids

Based on the low threshold energies for the formation of lattice solitons, as compared to those for focusons, we shall state the following. First, the quodon generation rate may be expected to be much higher than that of focusons due to a higher cross-section for the low energy transfer from a projectile to the lattice. Second, a Schottky defect can not be produced in one collision event between an extended defect and a low energy quodon, which can deliver energy in portions small as compared to the Schottky defect formation barrier. So the ballistic mechanism of the Schottky defect production does not seem to be statistically relevant for a description of the effect caused by quodons. In the following sections we will develop a statistical approach to the modification of reaction rates by the quodon "gas".

2.1 Gas of Quodons

It is known that even in the case of displacement damage, only a small part of the energy of impinging particles is spent on generation of stable Frenkel pairs (that require 20–40 eV each) and their clusters (in energetic displacement cascades), while the major part of energy is dissipated into heat, or in other words, it is spent on generation of phonons.

The *first main assumption* in the present paper is that *quodons are the transient form of the heat generation under irradiation*, which means that they are constantly generated by irradiation, and subsequently lose energy by generating phonons. Let K_q be the average rate of quodon generation (per atomic site per second), which should be proportional to the flux of impinging particles, F_{irr}, and the energy deposition density by one particle, dE_{irr}/dx, and inversely proportional to the mean quodon energy, $\langle E_q \rangle$:

$$K_q(F_{irr}) = k_{eff}^q F_{irr} \left(\frac{dE_{irr}}{dx} \right) \frac{w}{\langle E_q \rangle}, \tag{1}$$

where w is the atomic volume, and k_{eff}^q is the quodon production efficiency that depends on material and irradiation conditions and can range from zero (no quodon generation) to unity (e.g. under sub-threshold irradiation that does not produce stable defects).

Then the mean density of quodon gas under steady-state irradiation (the number of quodons per unit volume) will be given simply by the product of their mean generation rate and the life-time, τ_q:

$$N_q\left(F_e, T_{irr}\right) = \frac{K_q\left(F_e, E_e\right)\tau_q}{w}, \quad \tau_q(T) = \frac{l_q(T)}{c_q}, \tag{2}$$

where c_q is the quodon propagation speed, which is assumed bellow to be close to the sound velocity, c_s, and $l_q(T)$ is the quodon propagation range before decay.

From a quantum theory of intrinsic localized modes [14] it is known that they decay by generating phonons. Similar to that, quodons are assumed lose energy in quodon-phonon collisions *by portions (or quanta)* ΔE_{qp}. Then the mean quodon propagation range may be written as

$$l_q(T, \langle E_q \rangle) = \frac{\langle l_{qp}(T) \rangle}{\epsilon_{pq}}, \quad \epsilon_{pq} \equiv \frac{\Delta E_{qp}}{\langle E_q \rangle}, \tag{3}$$

where $\langle l_{qp}(T) \rangle$ is the mean length of quodon free path between collisions with phonons, which is determined by a well known formula for a 1–D propagating particle in a medium with scattering centers of a given density and cross-section:

$$\langle l_{qp}(T) \rangle = \frac{1}{\pi R_{qp}^2 \langle N_p(T) \rangle}, \quad \langle N_p(T) \rangle = \frac{1}{w} \frac{1}{\exp\left(\dfrac{\hbar\omega_D}{k_B T}\right) - 1}, \tag{4}$$

where R_{qp} is the effective quodon-phonon cross-section radius, $\langle N_p(T) \rangle$ is the density of phonons having a high-frequency ω_D (a.k.a. Debye phonons), k_B is the Boltzmann constant and T is the absolute temperature. We assume here for simplicity that Debye phonons are the main contributors to the quodon decay due to the loss of energy in each collision between them represented by the energy coupling parameter $\Delta E_{qp} \approx \hbar\omega_D$.

These and other material parameters used in the calculations are listed in Table 1.

As can be seen, the quodon-phonon cross-section radius is extremely small ($\sim 10^{-12}$ m), and so the quodon propagation range can reach tens of centimeters (Fig. 1), which agrees with experimental data on tracks produced by quodons in mica muscovite [7]. This enormous ranges in real crystals that contain structural defects can be understood only assuming that quodons can both lose and gain energy in the scattering process with extended defects. So our second main assumption is that in the collision events between quodons and extended defects, quodons lose and gain energy with equal probability.

As can be seen from Table 1 and Fig. 1, the quodon generation rate under irradiation conditions [17] can exceed the displacement rate by 6 orders of magnitude, and the density of quodon gas becomes comparable to that of Debye phonons at the electron flux of 10^{24} m^{-2}s^{-1}, which corresponds to the displacement rate of $\sim$ 5×10^{-3}s^{-1} that is a typical value for radiation damage studies using electron beams. So we may conclude that under irradiation a crystal contains a mixture of "gases" of quasi-particles, namely, almost equilibrium phonons and strongly non-equilibrium quodons, the densities of which may be comparable. After irradiation is switched off, quodons transfer their energy to phonons and disappear over a short relaxation time $\sim\tau_q \approx 7 \times 10^{-5}$ s, as the crystal attains a thermal equilibrium state. In the following section, we consider the effect of quodon gas on chemical reaction rates under irradiation.

Table 1 Material (*Fe*) and irradiation parameters used in calculations

Parameter	Value
Atomic spacing, b, m	2.96×10^{-10}
Atomic volume of the host lattice, w	$0.5 \times b^3$
Sound velocity, c_s, m/s	3.82×10^3
Maximum phonon frequency in *Fe* and *Cu*, $\omega_D^{Fe,Cu}$, s^{-1}	$6 \times 10^{13}, 7 \times 10^{12}$
Quodon-phonon coupling, $\Delta E_{qp} \approx \hbar\omega_D^{Fe}$, eV	0.207
Quodon-phonon cross-section, R_{qp}, m	9.06×10^{-13}
Quodon-defect coupling, $V_q \approx \hbar\omega_D^{Cu}$, eV	0.029
Defect energy relaxation time, τ_0, s	8×10^{-8}
Mean quodon energy, $\langle E_q \rangle$, eV	1
Irradiation temperature [17], T_{irr}, K	563
Quodon life-time, τ_q (T_{irr}), s	7.184×10^{-5}
Electron flux [17], F_e, m^{-2}s^{-1}	4×10^{17}
Electron energy [17], E_e, MeV	2.5
Displacement energy [17], E_d, eV	30
Displacement rate [17], K_d (F_e, E_e), s^{-1}	2.0×10^{-9}
Quodon production efficiency, k_{eff}^q	1
Quodon production rate [17], K_q (F_e, E_e), s^{-1}	7.27×10^{-3}
Quodon propagation length, l_q (T_{irr}), m	0.23
Migration energy of vacancies, E_{vm}, eV	0.91
Pre-exponent factor, D_v^0, m^2/s	10^{-5}
Vacancy formation energy, E_{vf}^{th}, eV	2.0
Cu-vacancy migration energy, $\Delta E_{Cu,V}^m$, eV	0.89
Cu-vacancy binding energy, $E_{Cu,V}^b$, eV	0.6
Cu migration frequency factor, υ_{Cu}, s^{-1}	7×10^{-12}
Cu migration entropy factor, $\Delta S_{Cu,V}^m$	$2 \cdot k_B$
Cu dissolution energy, eV	0.586
Reaction activation volume, W_a	$10 \cdot w$

2.2 *Modification of Reaction Rates in Solids Under Irradiation*

The "gas" of phonons is responsible for the temperature effect on the reaction rates, $\dot{R}$, which is expressed by Arrhenius' law:

$$\dot{R} = \omega_p \exp\left(-\frac{E_a}{k_B T}\right), \tag{5}$$

where ω_p and E_a are the frequency factor and the activation barrier, respectively. The reaction activation barrier is determined by the maximum *free energy change* of the system imposed by reaction (a.k.a. the free energy barrier). *Phonons* represent thermal fluctuational forces acting on a system (e.g. point defect and its surrounding) and helping it to overcome the barrier. The strength of the phonon's "white noise" is proportional to the temperature and enters the denominator in the exponential argument.

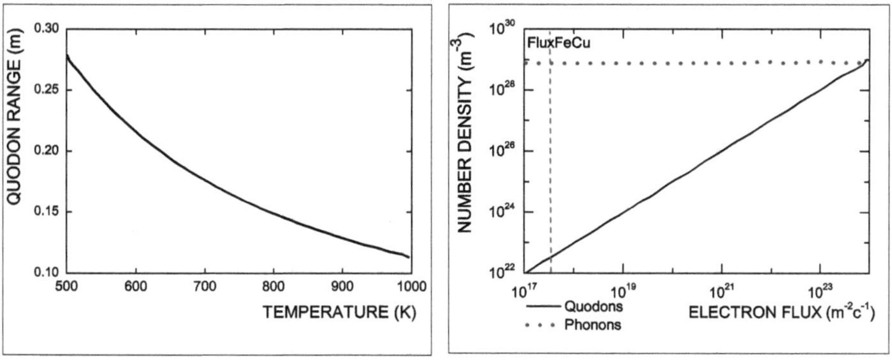

Fig. 1 The temperature dependence of the quodon propagation range $l_q(T)$ and the density of quodon gas vs. electron flux, $F_{irr} = F_e$ at the irradiation temperature 563 K. The *vertical dotted line* corresponds to irradiation flux [17]. Density of the gas of Debye phonons at 563 K is shown for comparison

On the other hand, unlike phonons, which are delocalized, quodons are strongly localized on several lattice sites, and so there is no reason to expect the effects on reaction rates of these two very different species to be similar. According to modelling in one-dimensional chains of coupled nonlinear oscillators, in the collision event between two quodons, they behave almost like particles, which undergo elastic and non-elastic collisions, accompanied with a change in momentum and energy. Based on that picture, we assume that in the collision event between a quodon and an extended defect, the latter can gain (or loose) some portion of energy at the place of collision, and so the resulting energy of the system including the extended defect will undergo stochastic deviations from its average value. To include these deviations in the reaction scheme let us recall the thermodynamic perturbation theory developed by Peierls for systems, the energy of which can not be defined precisely due to small (and hard to detect) deviations from the ground state [18]. Then the total *potential energy* of the system can be written as a sum

$$E = E_0 + V, \quad V << \min(k_B T, E_0), \tag{6}$$

where V is the stochastic deviation of the energy from the ground value E_0. The system free Helmholtz energy, Φ, is defined by the integral over the Gibbs ensemble:

$$\begin{aligned} \exp\left(-\frac{\Phi}{k_B T}\right) &= \int \exp\left(-\frac{E_0 + V}{k_B T}\right) d\Gamma \\ &\approx \int \exp\left(-\frac{E_0}{k_B T}\right) \left(1 - \frac{V}{k_B T} + \frac{V^2}{2(k_B T)^2}\right) d\Gamma, \end{aligned} \tag{7}$$

where $d\Gamma$ is the configuration volume in the phase space. Taking the logarithm of Eq. (7) and expanding into the Taylor's series again one gets the free energy in the following form:

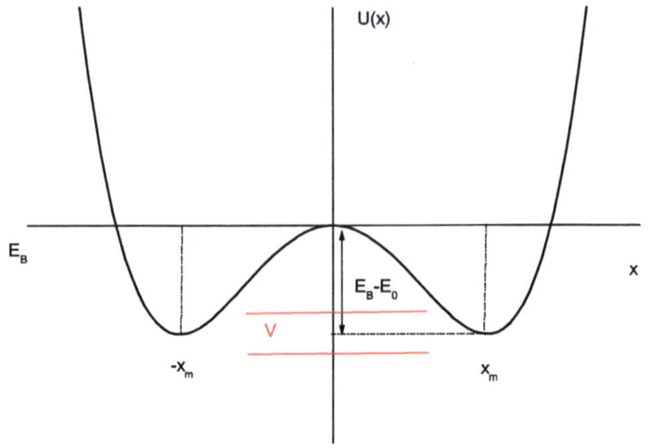

Fig. 2 Sketch of the double-well *potential landscape* with minima located at $\pm x_m$. These are stable states before and after reaction, separated by a potential "barrier" with the height changing randomly within the V band induced by quodons

$$\Phi = \Phi_0 + \int \left(V - \frac{V^2}{2k_B T} \right) \exp\left(-\frac{\Phi_0 - E_0}{k_B T} \right) d\Gamma + \frac{1}{2k_B T} \left[\int V \exp\left(-\frac{\Phi_0 - E_0}{k_B T} \right) d\Gamma \right]^2, \tag{8}$$

where Φ_0 is the ground free energy at $V = 0$. The integrals in Eq. (8) are the average values of the corresponding functions evaluated using the unperturbed Gibbs distribution function, and so Eq. (8) can be written as follows:

$$\Phi \approx \Phi_0 + \langle V \rangle - \frac{1}{2k_B T} \left[\langle V^2 \rangle - \langle V \rangle^2 \right], \tag{9}$$

where the brackets $\langle \;\; \rangle$ designate the integration over the Gibbs ensemble so that $\langle V \rangle$ is simply the average value of the perturbation energy (which vanishes in the case of random fluctuations of alternative signs) while the second term in Eq. (9) is *always negative* and it is proportional to the square of the energy deviation dispersion due to random fluctuations of the system energy.

Now this result can be applied to a system including the "reaction area" surrounding a point defect or some region near the surface of extended defect subjected to random collisions with quodons (Fig. 2). The *free energy* of such system will be decreased according to Eq. (9), and so decreased will be the maximum change of the system free energy required for the reaction to occur, i.e. the reaction activation barrier. This decrease has a purely *statistical nature* reflecting the increase of vibrational entropy of the system due to its collisions with quodons. Accordingly, the Arrhenius' law Eq. (5) can be rewritten with account of the energy exchange between quodons and the "reaction area":

$$\dot{R} = \omega_0 \exp\left(-\frac{E_a^q}{k_B T}\right), \quad E_a^q = \langle E_a \rangle - \Delta E_a, \quad E_a = \frac{\langle V_q^2 \rangle - \langle V_q \rangle^2}{2k_B T}, \tag{10}$$

where we have assumed that the average energy of the system will not change due to its collisions with quodons, which can give or take energy with equal probability, i.e. $\langle V_q \rangle = 0$.

This result means that under irradiation *all reaction barriers* in a crystal may be decreased by a value that depends on the statistics of the "gas" of quodons and their coupling with structural defects. It can be shown that at steady state the energy deviation dispersion in Eq. (10) can be expressed via the quodon generation rate K_q, the propagation range l_q and the mean portion of energy that is exchanged between a quodon and the reaction area in one collision event, V_q (for Cu precipitates $V_q \approx \hbar\omega_D^{Cu}$ – the Debye "cutoff" energy of phonons in Cu). To do so let us assume that the system can be found in two states: excited state with energy $\langle E_a \rangle \pm V_q$ due to collisions with quodons and in the ground state with energy $\langle E_a \rangle$. The relative fraction of time, in which the system is excited $\Delta t/t$ is given by the product of the frequency of collisions with quodons, ω_q, and the time it takes to relax to the ground state, τ_0. Then the square of the system energy averaged over large time t (which is equivalent to the averaging over the Gibbs ensemble) can be written as follows:

$$\begin{aligned} \langle V_q^2 \rangle &= \langle E_a \rangle^2 (t - 2\Delta t) + \left(\langle E_a \rangle + V_q\right)^2 \Delta t - \left(\langle E_a \rangle - V_q\right)^2 \Delta t \\ &= \langle E_a \rangle^2 \left(t - 2\omega_q \tau_0\right) + \left(\langle E_a \rangle + E_q\right)^2 \omega_q \tau_0 + \left(\langle E_a \rangle - V_q\right)^2 \omega_q \tau_0, \end{aligned} \tag{11}$$

whence it follows that the energy deviation dispersion in Eq. (10) is given by

$$\langle V_q^2 \rangle - \langle V_q \rangle^2 = 2V_q^2 \omega_q \tau_0, \tag{12}$$

The frequency of collisions between the reaction area of the radius R and the quodon gas of density N_q can be expressed via the quodon generation rate and propagation range:

$$\omega_q = N_q c_q 4\pi R^2 = \frac{K_q}{w} 4\pi R^2 l_q, \tag{13}$$

Since the quodon generation rate is proportional to the flux of impinging particles F_{irr} (see Eq. (1)), one can express the modified reaction rate as follows:

$$\dot{R} = \dot{R}_0 A\left(f_{irr}\right), \quad A\left(f_{irr}\right) = \exp\left(\frac{k_{mat} F_{irr}}{\left(k_B T\right)^2}\right), \quad k_{mat} \propto l_q \tau_0, \tag{14}$$

where $\dot{R}_0$ is the reaction rate in the ground state (no irradiation) and $A\left(f_{irr}\right)$ is the reaction amplification factor due to the interaction with quodon gas, and k_{mat} is the coefficient determined by quodon statistics and material parameters.

In the following section, the modified reaction rates will be applied to modelling of the second-phase precipitation under irradiation. This phenomenon is of both fundamental and technological importance, since it represents a phase transition of the first order in a strongly non-equilibrium system, i.e. a crystal under irradiation, which changes its service properties as a nuclear material.

3 Modeling of the Precipitation Kinetics Under Irradiation in the Modified Rate Theory

Service properties of pressure vessel steels (used in nuclear industry) are very sensitive to precipitation of copper (i.e. aggregation of copper impurities into nano-sized clusters of copper atoms – *precipitates*), since the precipitates acts as traps for gliding dislocations, which makes the material less ductile and more brittle. These phenomena are called the irradiation hardening and embrittlement of steels [17]. So the kinetics of precipitate nucleation and growth has been the focus of extensive experimental and theoretical studies [19–23]. However, the mechanisms of precipitation under irradiation are still not correctly understood. This is the reason we undertook a modelling of copper precipitation under electron irradiation in dilute (i.e. containing a low concentration of copper) *FeCu* binary alloys, where experimental trends have been rather comprehensively described by Mathon et al. [17]. One of the main advantages of the work [17] is that it describes not only evolution of copper precipitates but also the time dependence of the concentration of copper atoms in the matrix, $\overline{C}_{Cu}(t)$, both under thermal annealing at 773 K and under electron irradiation at 563 K. It appears that $\overline{C}_{Cu}$ at the end of the precipitation process, which corresponds to the copper solubility limit, is of the same order of magnitude in both cases. This result is in a marked disagreement with Arrhenius law that predicts the copper solubility limit at 563 K to be several orders of magnitude lower than observed.

In our view, this contradiction is a very principal one, since the solubility limit value practically determines the rate of the nucleation and growth, and a failure to evaluate it in the classical rate theory makes it impossible to describe correctly the precipitation kinetics under irradiation. To our knowledge, in all up to date models, the solubility limit under irradiation is assumed to be determined only by temperature, similar to the thermal case [20–27]. As will be demonstrated in the present work, this is a *misleading assumption*.

We present a detailed description of the rate theory scheme for the precipitation kinetics and describe the copper precipitation under thermal annealing and under irradiation in the framework of classical rate theory (Sect. 3.1). In Sects. 3.3–3.5, we modify the rate theory with account of non-equilibrium thermodynamics of "quodon gas" and apply it to modelling of copper precipitation in *FeCu* binary alloys under electron irradiation.

3.1 Classical Rate Theory of the Precipitation Kinetics

As is usual in the classical nucleation theory, *Cu*–precipitates are characterized by a distribution function of its sizes, f_n, that is the atomic fraction of precipitates consisting of n-atoms of copper. The time-evolution of the size-distribution function is described by the set of $N-1$ equations (the Becker-Doring *finite-difference* equation [19]), where N is the maximal size of the precipitates:

$$\frac{\partial f_n}{\partial t} = W_{n-1}^{+} f_{n-1} - W_n^{-} f_n - W_n^{+} f_n + W_{n+1}^{-} f_{n+1}, \quad n = 2, \ldots, N, \tag{15}$$

where W_n^+ and W_n^- are the *forward and backward transition rates*, or the probabilities to increase or decrease a number of atoms in a n-atomic precipitate by one per unite time (the reactions involving transitions $f_n \to f_{n\pm m}$, where $m > 1$ are neglected). So the precipitation is considered as a random walk in the size space (n) with a step of length $= 1$.

This equation can be transformed into a differential equation. Applying the Taylor's expansion in power series up to the second order one obtains the well known Fokker-Plank *differential* equation:

$$\frac{\partial f_n}{\partial t} = \frac{\partial}{\partial n}\left[\left(W_n^{-} - W_n^{+}\right) f_n\right] + \frac{\partial^2}{\partial n^2}\left[\frac{1}{2}\left(W_n^{-} + W_n^{+}\right) f_n\right], \tag{16}$$

In the adiabatic case (i.e. where the total number of solute atoms is constant), these equations can be integrated numerically provided that the forward and backward transition rates are known. In theory, precipitates are usually assumed to have a spherical form. In this case one has (see e.g. [21])

$$W_n^{+} = \frac{4\pi D}{w} R\overline{C}, \quad W_{n+1}^{-} = \frac{4\pi D}{w} R C_R, \tag{17}$$

where D is the monomer diffusivity, R is the precipitate radius, *overlineC* and C_R are the mean monomer concentrations in the bulk (the bar over C designates the average over a macroscopic volume) and the local concentration at the precipitate surface, respectively. The former can be found from the solute conservation low:

$$\overline{C} + \sum_{n=2}^{n=N} n f_n = \overline{Q}, \tag{18}$$

where $\overline{Q}$ is the total volume fraction of solute atoms, and the latter (C_R) is usually derived from thermodynamic considerations [17, 19]:

$$C_R = C_1^{eq} \exp\left(\frac{2\gamma w}{k_B T R}\right), \quad \gamma = 0.54 \frac{k_B T}{b^2}\left(\ln \frac{1}{C_1^{eq}} - 2\right), \tag{19}$$

Table 2 Diffusivity of copper in iron given by references listed in the last column

#	D^0_{Cu}, cm^2/s	E^m_{Cu}, eV	$D_{Cu}(773\,\text{K})$, cm^2/s	Ref
1	8.5	2.29	1.00×10^{-14}	This work
2	3.4	2.29	4.00×10^{-15}	[17]
3	7.2×10^{-2}	2.29	8.45×10^{-17}	[22]
4	6.3×10^{-1}	2.29	7.40×10^{-16}	[23]
5	7.08	2.53	2.56×10^{-16}	[20]

where C_1^{eq} is the equilibrium concentration of monomers at a free flat surface, γ is the precipitate surface energy according to Cahn-Hilliard theory [17].

Diffusion of copper in iron occurs by the vacancy mechanism, and so the copper diffusivity D_{Cu} is proportional to the mean concentration of vacancies in the bulk, $\overline{C}_V$:

$$D_{Cu} = \frac{\overline{C}_V}{C_V^{eq}} D^0_{Cu} \exp\left(-\frac{E^m_{Cu}}{k_B T}\right), \tag{20}$$

where E^m_{Cu} is the effective migration energy of copper, D^0_{Cu} is the pre-exponent factor, and C_V^{eq} is the equilibrium concentration of vacancies, which has the classical form (*in the case of thermal equilibrium only*), as well as the equilibrium concentration if copper atoms:

$$C_V^{eq} = C_V^0 \exp\left(-\frac{E_V^f}{k_B T}\right), \quad C_V^0 = \exp\left(\frac{\Delta S_V}{k_B}\right), \tag{21}$$

$$C_{Cu}^{eq} = C_{Cu}^0 \exp\left(-\frac{E_{Cu}^f}{k_B T}\right), \quad C_{Cu}^0 = \exp\left(\frac{\Delta S_{Cu}}{k_B}\right), \tag{22}$$

where E_V^f, E_{Cu}^f are the vacancy and copper formation energies, $\Delta S_{V,Cu}$ are the entropy factors. In this way, the rate theory is completed, and we present some results obtained for the copper precipitation under conditions of thermal annealing and under electron irradiation in dilute *FeCu* binary alloys (~1 at % *Cu*) in comparison with experimental data by Mathon et al. [17]. Under thermal annealing one has simply $\overline{C}_V = C_V^{eq}$, and so the precipitation rate is determined only by the copper diffusivity. We have tested several different diffusivities in order to achieve the best agreement with experimental data [17]. They are listed in Table 2.

The results for the mean precipitate radius and the concentration under thermal annealing at 773 K are presented in Figs. 3 and 4.

It can be seen from these figures that the best fit is achieved using the *Cu* diffusivity presented in the first line in the Table 2, which shows a very good agreement with experimental data in the most reliable region – at the late stage of precipitation, $t > 1{,}000$ s.

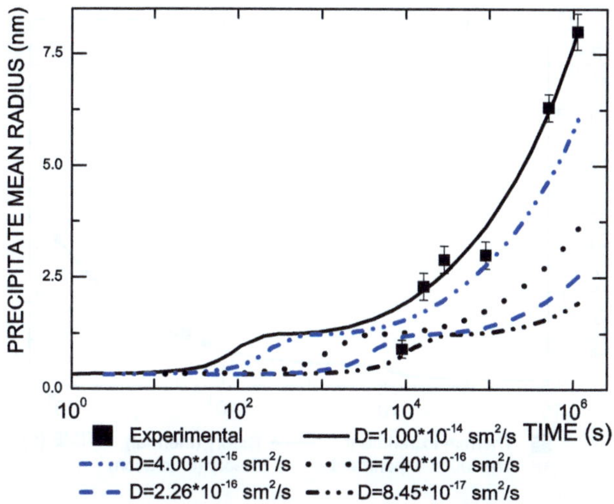

Fig. 3 Time evolution of the *Cu* precipitate mean radius under annealing at 773 K vs. experimental data [17]

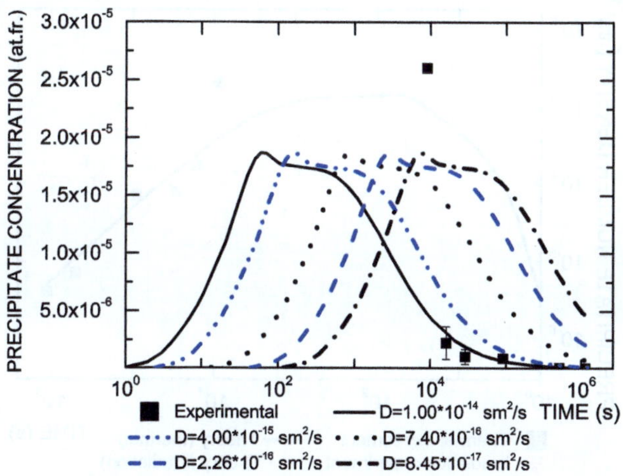

Fig. 4 Time evolution of the *Cu* precipitate concentration under annealing at 773 K vs. experimental data [17]

Under irradiation, the classical rate theory assumes that the solubility limit of copper *is given by Eq. (22) as well as under thermal annealing*, and the only effect of irradiation is the enhancement of copper diffusivity via radiation-induced increase of the mean vacancy concentration in the matrix, $\overline{C}_V >> C_V^{eq}$.

Irradiation conditions in the Ref. [17] are as follows: temperature, $T_{irr} = 563\,\mathrm{K}$, displacement rate $K = 2 \times 10^{-9}$ dpa/s, total irradiation time $t_{irr} = 7.75 \times 10^5$ s, which corresponds to the total irradiation dose of 1.55×10^{-3} displacements per atom (dpa).

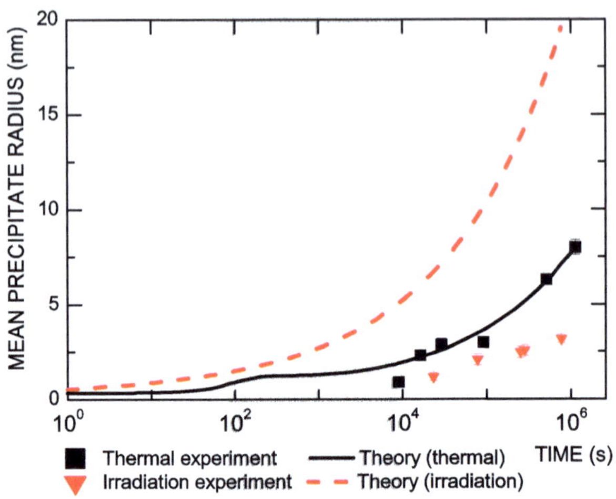

Fig. 5 Time evolution of the *Cu* precipitate mean radius under irradiation at 563 K vs. experimental data [17]

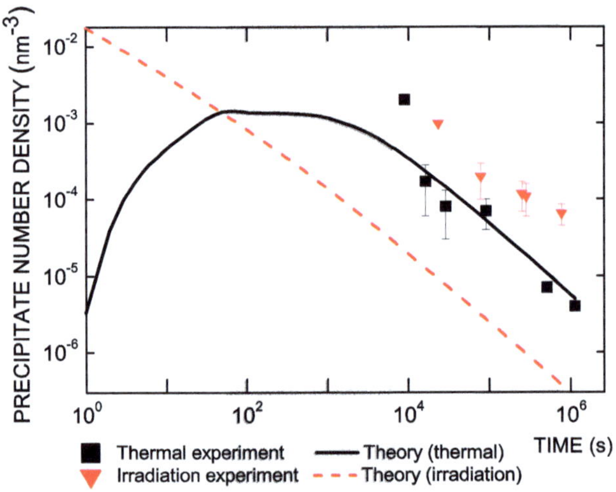

Fig. 6 Time evolution of the *Cu* precipitate concentration under irradiation at 563 K vs. experimental data [17]

It can be seen in Figs. 5 and 6, that the classical theory approach greatly overestimates the growth rate of the precipitate mean radius, and underestimates their concentration. As mentioned in the introduction, this discrepancy is a natural consequence of the main assumption of the classical theory on the solubility limit of copper being determined by Arrhenius formula (thermodynamic assumption). Since this assumption is used by all existing precipitation models, their results could not be validated by experimental data on the radiation-induced solubility limit, as

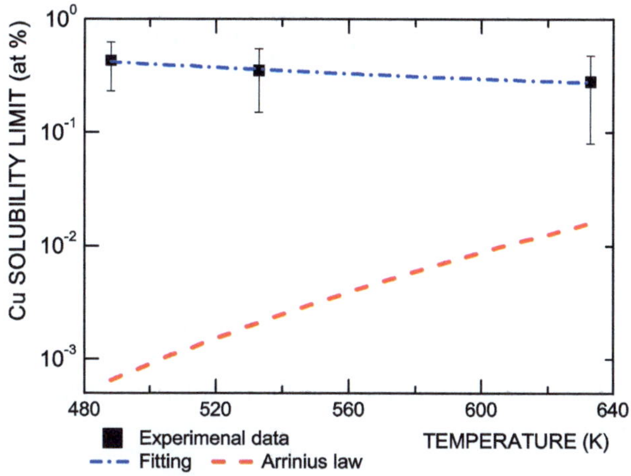

Fig. 7 Temperature dependence of the copper solubility limit under irradiation [17] vs. Arrhenius law

it becomes evident from comparison between theoretical and experimental values presented in Fig. 7. This discrepancy was actually admitted (as a puzzle) by the authors [17], and it will be shown to be a natural consequence of the radiation-induced increase of the solubility limit due to the interaction of precipitates with irradiation-induced quodons. Below we use the copper diffusivity data from the first line in the Table 1 as the best fit to experimental data for thermal annealing.

3.2 *Quodon-Induced Solubility Limit Change*

In this section, non-equilibrium thermodynamics of "quodon gas" produced by irradiation is taken into account in the evaluation of the solubility limit under irradiation.

Consider a free energy, $\Phi^{th}_{Cu}(n, c)$ of a system consisting of an extended defect (*Cu* precipitate) of n atoms and a diluted solid solution of point defects – *Cu* atoms in our case, with a concentration $c << 1$:

$$\Phi^{th}_{Cu}(n, c) = (N_0 - n)\left(\phi^m_{Cu} + k_B T \ln c\right) + 4\pi\gamma b^2 n^{2/3} + n\phi^p_{Cu} \tag{23}$$

where N_0 is the total number of point defects in the system, ϕ^m_{Cu} is the Gibbs potential one of that in the matrix, ϕ^p_{Cu} is the free energy of a defect in the precipitate, and γ the interfacial energy. Minimization of the free energy with respect to n variation, $\partial\Phi^{th}_{Cu}/\partial n = 0$, is known to result in the classical *thermal equilibrium* concentration of point defects, which depends exponentially on T and n:

$$C_{Cu}^{th}(T,n) = \exp\left(-\frac{\Delta\phi^{mp}}{k_B T}\right)\exp\left(\beta n^{-1/3}\right),$$
$$\Delta\phi^{mp} = \phi_{Cu}^{m} - \phi_{Cu}^{p} > 0, \qquad \beta = \frac{8\pi b^2\gamma}{3k_B T}, \tag{24}$$

and defines a thermodynamic solubility limit of point defects, $C_{Cu0}^{th}(T)$:

$$C_{Cu}^{th}(T, n \to \infty) \longrightarrow C_{Cu0}^{th}(T). \tag{25}$$

Here the difference $\Delta\phi^{mp}$ is the *minimum work* done by the system while transferring a point defect from a solution to the precipitate under constant volume and temperature, which is usually referred to as the *dissolution activation (free) energy*. It is defined as

$$\Delta\phi^{mp} = \Delta E^{mp} - T\Delta S^{mp}, \tag{26}$$

where ΔE^{mp} is the height of potential barrier (see Fig. 6), and ΔS^{mp} is the entropy change due to point defect transfer.

Under irradiation, free energy of the system is reduced due to fluctuations of the potential landscape near the interface region caused by the scattering of quodons at the interface (Fig. 2), which *reduces* the minimum work done by the system while transferring a point defect from a solid solution to the precipitate, as has been demonstrated in Sect. 2.2:

$$\Delta\phi^{mp} = \Delta\phi_{th}^{mp} - \Delta\phi_q(F_{irr}, T), \quad \Delta\phi_q(F_{irr}, T) = \frac{V_q^2\omega_q(F_{irr}, T)\tau_0}{k_B T}, \tag{27}$$

where $\Delta\phi_{th}^{mp}$ designates free energy of the system without irradiation and $F_{irr} > 0$ is a flux of impinging particles.

Accordingly, instead of thermodynamic expression (24) one has to consider a quodon-modified *dynamic equilibrium* concentration of point defects, which should depend on the irradiation flux that generates quodons:

$$C_{Cu}^{q}(F_{irr}, T, n) = C_{Cu}^{th}(T, n)\exp\left(\frac{\Delta\phi_q(F_{irr}, T)}{k_B T}\right), \tag{28}$$

which will give us the quodon-modified solubility limit, $C_{Cu0}^{q}(F_{irr}, T)$, as follows

$$C_{Cu0}^{q}(F_{irr}, T) = C_{Cu0}^{th}(T)\exp\frac{V_q^2\omega_q(F_{irr}, T)\tau_0}{(k_B T)^2}, \tag{29}$$

$$\omega_q(F_{irr}, T) = \frac{K_q\left(F_{irr} l_q(T)\right) W_a}{wb}, \tag{30}$$

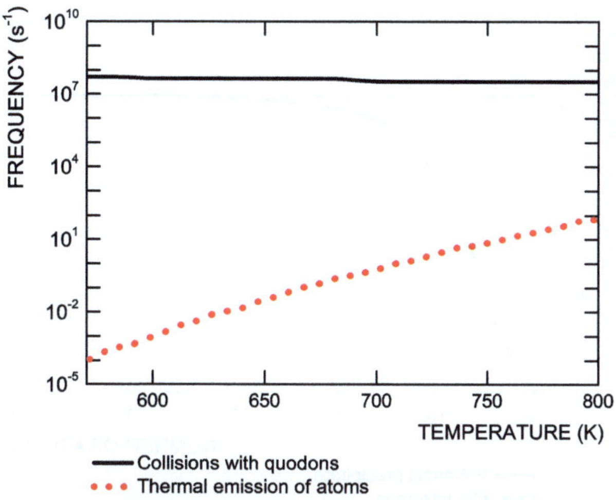

Fig. 8 Frequency of collisions between a precipitate and quodons vs. frequency of atom thermal emission

where $W_a = 4\pi R^2 b$ is the effective activation volume for the quodon-induced modification of the solubility barrier.

As shown in Fig. 8, the frequency of collisions between a precipitate and quodons is about $10^7\,\mathrm{s}^{-1}$, which greatly exceeds the frequency of thermal desorption of copper atoms into the matrix, which ranges from 6×10^{-5} to $22\,\mathrm{s}^{-1}$ in the temperature interval under investigation: $T = (563 \div 773\,\mathrm{K})$. This means that the interaction between quodon "gas" and structural defects (e.g. precipitates) can be viewed as that between an ideal molecular gas and Brownian particles. Statistical lows seem to be equally valid in both cases.

Naturally, the quodon scattering at structural defects should depend on the defect size, ranging from low values for point defects (almost "invisible" for quodons) to the values estimated above for large extended defects. This physical picture can be modeled simply by assuming that $\Delta E_q(F_e, T)$ depends on the precipitate size using the following *size factor fit*(n):

$$\Delta\phi_q(F_e, T, n) = \Delta\phi_q(F_e, T) \times \mathit{fit}(n), \quad \mathit{fit}(n) = 10^{-3} \times \left(1 + \frac{2 \times 10^3}{1 + \exp(5/n)}\right), \tag{31}$$

For example, at irradiation temperature $T = 563\,\mathrm{K}$, one gets in this way the barrier decrease $\Delta\phi_q(F_e, T, n) = 0.119\,\mathrm{eV}$ for a small precipitate, and $\Delta\phi_q(F_e, T, n) = 0.157\,\mathrm{eV}$ for a large precipitate, which coincides with that for a flat interfacial surface, $\Delta\phi_q(F_e, T, \infty)$, which determines the solubility limit.

Figure 9 shows dependence of the quodon-induced change of the dissolution energy on the number of atoms in a precipitate, given by Eq. (31) as compared to the value for a flat interface.

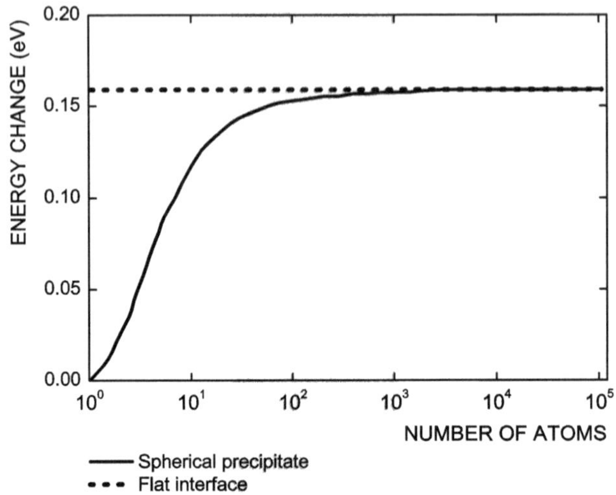

Fig. 9 Quodon-induced change of the dissolution free energy, $\Delta\phi_q$, vs. the number of atoms in a precipitate

3.3 *Quodon-Induced Interfacial Energy Change*

We will use an equation for the interfacial energy γ similar to that calculated by Mathon et al. [17] based on the Cahn-Hilliard theory adapted for taking into account of non-configurational entropy. In thermal case, the equation is as follows:

$$\gamma_0(T) = 0.54\frac{k_B T}{b^2}\left(\ln\frac{1}{C_{Cu0}^{th}(T)} - 2\right) = 0.54\frac{k_B T}{b^2}\left(\frac{E^f}{K_B T} - \frac{\Delta S^{non}}{k_B} - 2\right), \tag{32}$$

where $C_{Cu0}^{th}(T)$ is the *thermal equilibrium* solubility at a flat interface. Under irradiation, the equation should be modified to take into account the quodon interaction with the surface. Accordingly, for a flat interface one has

$$\begin{aligned}\gamma_0(F_{irr}, T) = \; & 0.54\frac{k_B T}{b^2}\left(\ln\frac{1}{C_{Cu0}^{q}(F_{irr}, T)} - 2\right) = \\ & 0.54\frac{k_B T}{b^2}\left(\frac{E^f - \Delta E_{q0}(F_{irr})}{K_B T} - \frac{\Delta S^{non}}{k_B} - 2\right),\end{aligned} \tag{33}$$

where $C_{Cu0}^{q}(F_{irr}, T)$ is the quodon-induced solubility limit.

In order to take into account size dependence of the quodon-precipitate scattering, we will modify Eq. (33) similar to Eq. (31), and introduce a size-dependent interfacial energy:

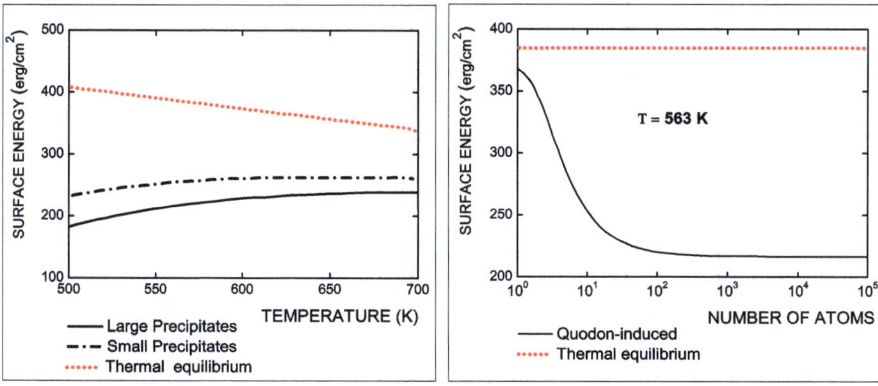

Fig. 10 Quodon-induced interface energy as a function of temperature and the number of atoms in the precipitate

$$\gamma_q(n) = 0.54\frac{k_B T}{b^2}\left(\frac{E^f - fit_\gamma(n)\Delta E_{q0}(F_{irr})}{K_B T} - \frac{\Delta S^{non}}{k_B} - 2\right), \tag{34}$$

where the size factor $fit_\gamma(n)$ is different from that in Eq. (31):

$$fit_\gamma(n) = 10^{-1} \times \left(1 + \frac{2 \times 10^1}{1 + \exp(5/n)}\right). \tag{35}$$

The resulting dependence of the interface energy on temperature and precipitate size is shown in Fig. 10. In this way one can evaluate the concentration of *Cu* in *dynamic equilibrium* with precipitates of different sizes at different temperatures and irradiation conditions, as demonstrated in Fig. 11.

One can see that our new concept is able to describe both high copper solubility limit (compared to the thermal value) in agreement with that measured in [17] (Fig 11a) and quite comparable "local equilibrium" concentrations near small copper clusters at thermal annealing at 773 K and under irradiation at 563 K (Fig. 11b), which actually determine the precipitate critical size and nucleation rate. So the diffusivities at these two experimental setups should be also comparable in order to get reasonable agreement with observed nucleation and growth rates.

3.4 Diffusivity of Cu Under Irradiation

Diffusivity of *Cu* is determined by the vacancy mechanism and it can be written as follows:

$$D_{Cu}^{irr} = 4b^2 \upsilon_{Cu} C_V^{irr} \exp\left(\frac{\Delta S_{Cu,V}^m}{k_B}\right) \exp\left(-\frac{\Delta E_{Cu,V}^m - E_{Cu,V}^b}{k_B T}\right), \tag{36}$$

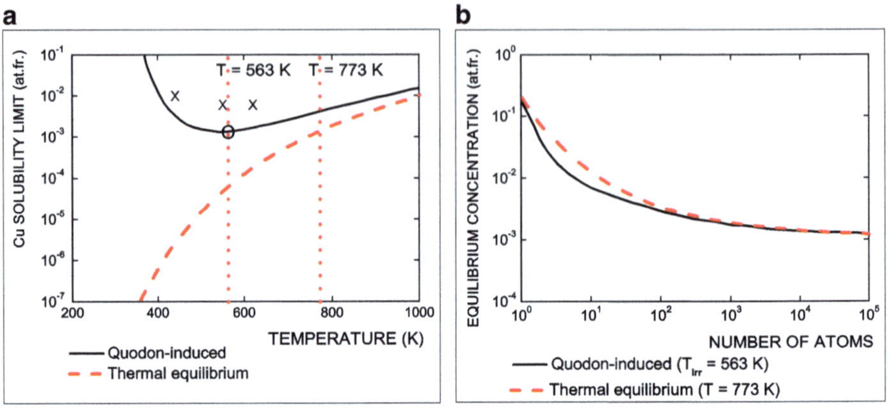

Fig. 11 **(a)** Thermal equilibrium and quodon-induced solubility limits of *Cu* in a *Fe* matrix; **(b)** equilibrium concentrations at precipitates of different sizes; **x** and **O** are the experimental data [17] for electron irradiation of *FeCu* alloy

where υ_{Cu} is the frequency factor, $\Delta S^m_{Cu,V}$ is the entropy factor, $\Delta E^m_{Cu,V}$ and $E^b_{Cu,V}$ are the *Cu*-vacancy migration and binding energies, respectively.

It appears that the result depends crucially on the mean vacancy concentration, C_V^{irr}, which is given by the balance of their production and annihilation in the bulk and at extended defects [1, 28]:

$$C_V^{irr}(F_e, E_e, T) = \frac{K_d(F_e, E_e)}{k_p^2 + k_d^2} + C_V^q(F_{irr}, T),$$
$$k_{ED}^2 \approx 4\pi N_p R_p + \rho_d, \tag{37}$$

where $K_d(F_e, E_e)$ is the displacement rate, k_{ED}^2 are the dislocation and precipitate sink strength for vacancies determined by the precipitate concentration N_p and mean radius R_p, and by the dislocation density, ρ_d; C_V^q is the mean dynamic equilibrium concentration of vacancies at all extended defects in the system, which (similar to the equilibrium concentration of *Cu*) depends both on temperature and irradiation flux [28], as shown in Fig. 12a.

Without irradiation, $C_V^q(0, T)$ corresponds to the thermal equilibrium concentration shown in Fig. 12a for comparison. Under irradiation it increases due to the radiation-induced vacancy emission from extended defects, but this effect is overshadowed by the vacancy production in the bulk and annihilation at extended defects. The main contribution to vacancy annihilation comes from precipitates, since their sink strength (measured experimentally in [17]) is higher than that of the dislocations by several orders of magnitude under these experimental conditions. Precipitates behave under irradiation similar to gas bubbles [28] adjusting the number of vacancies in them adiabatically to the number of copper atoms due to the positive feedback from the misfit stress on the bias of precipitates for absorption

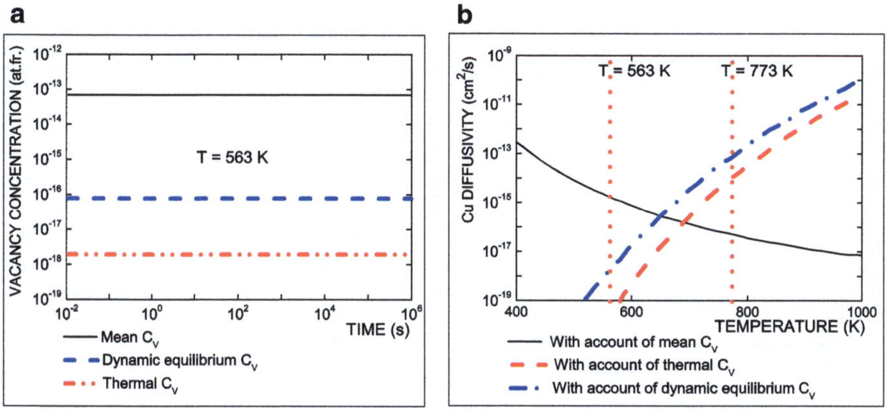

Fig. 12 Equilibrium (dynamic and thermal) and radiation-induced concentration of vacancies **(a)** and diffusivities of *Cu* in a *Fe* matrix **(b)** calculated for electron irradiation with account of different vacancy mechanisms [17]

of interstitial atoms. As a result, precipitates act as strong recombination centres for vacancies and interstitial atoms.

As shown in Fig. 12b, the *Cu* diffusivity under irradiation at 563 K is determined by the mean vacancy concentration and become comparable with that under thermal annealing at 773 K. The ratio of $D_{Cu}^{th}(773)/D_{Cu}^{irr}(563) \approx 6.9$, seems to be in excellent agreement with a factor of ~5 more rapid thermal evolution of precipitates at 773 K as compared to that under irradiation at 563 K observed by Mathon et al. [17].

In the following section we will test our predictions by modeling the evolution of *Cu* precipitates and the matrix concentration of *Cu* under electron irradiation *with and without* account of the radiation-induced quodons and compare the results with experimental data.

3.5 *Evolution of Cu Precipitates and the Matrix Concentration of Cu Under Electron Irradiation*

The evolution of *Cu* precipitates and the matrix concentration of *Cu* under electron irradiation have been modeled using parameters from Table 1. The results are compared with results of the classical model (without quodon-induced effects) and with experimental data [17] in Figs. 13 and 14.

One can see that the classical model disagrees strongly with experimental data on all precipitation parameters, and this discrepancy is especially pronounced for the matrix concentration of copper (Fig. 14). In contrast, the quodon-modified model describes quite well both the evolution of precipitates and the matrix concentration

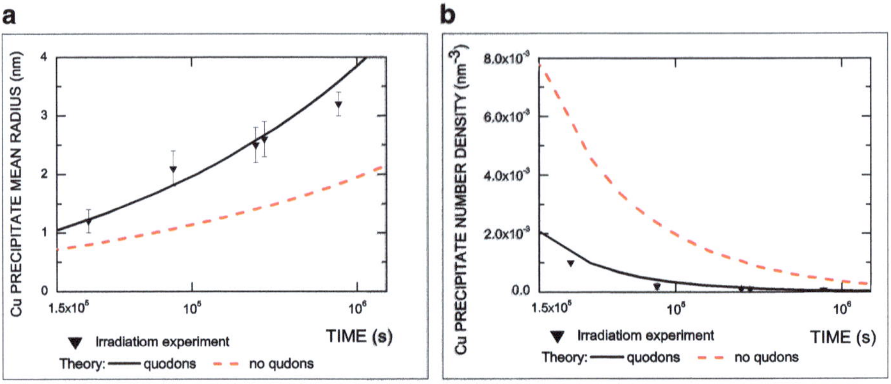

Fig. 13 Time evolution of the *Cu* precipitate: **(a)** – mean radius and **(b)** – concentration vs. experimental data [17]

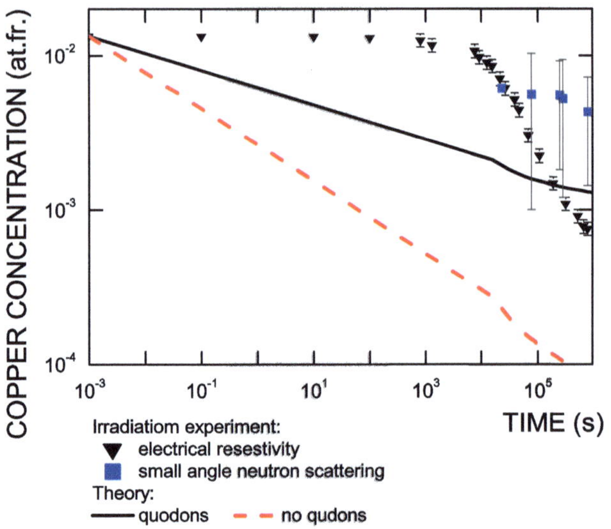

Fig. 14 Time evolution of the concentration of copper atoms under thermal annealing and irradiation. Experimental data by Mathon et al. [17] obtained by electrical resistivity method are shown by *triangles* together with results obtained by small angle neutron scattering shown by *squares*

of copper measured by different methods. The latter fact seems to be of particular importance, since it reflects the principal difference between the two concepts, namely, the "classical" and the new one, in relation to the mechanisms of production of Schottky defects. In the classical theory they are assumed to be emitted by extended defects exclusively due to thermal fluctuations, which, in our language, are driven by phonons. In the modified rate theory we have taken into account essentially *athermal fluctuation* mechanisms based on the interaction of extended defects with radiation-induced quodons. It should be noted that these mechanisms have been considered in the previous works [1, 3–6, 28] dealing with radiation-induced emission of another kind of Schottky defects, namely, vacancies. Similar to the solute atoms, the dynamic equilibrium concentration of vacancies has been evaluated (in the framework of "ballistic" concept) to be much higher than thermal equilibrium values. However, it is rather difficult technically to measure the dynamic equilibrium concentration of vacancies directly by electric resistivity measurements due to the lack of sufficient accuracy of this method. In contrast, the equilibrium concentration of solute atoms, such as *Cu*, can be measured directly at the final stage of the precipitate evolution, and this has been done by Mathon et al. [17], which makes their work particularly relevant for experimental discrimination between the classical and quodon-modified models.

4 Summary

From a methodological side, the proposed concept of the radiation-induced "gas" of quodons offers a new insight on the radiation-induced processes in solids. It appears that *quodons are the transient form of the heat generation under irradiation* that subsequently transfers energy to phonons. The quodon gas may be a powerful driver of the chemical reaction rates under irradiation, the strength of which exponentially increases with irradiation flux and may be comparable with strength of the phonon gas that exponentially increases with temperature. Phonons obey the lows of equilibrium thermodynamics, such as the minimization of a system free energy, and the latter could not be defined for a crystal under irradiation in the classical framework. The proposed method of the free energy modification by taking into account non-equilibrium fluctuations of the potential landscapes for chemical reactions offers a self-consistent way for description of the radiation-induced reactions and evolution of the microstructure in the nuclear materials.

A detailed description of the rate theory modelling of the precipitation kinetics has been presented and applied to describe the precipitation of copper in iron matrix under thermal annealing and under irradiation. The modelling has been performed both in the framework of a *classical* rate theory and the *modified* rate theory, which takes into account non-equilibrium fluctuations driven by the "gas" of radiation-induced quodons. The classical theory was shown to disagree strongly with experimental data on all precipitation parameters. In contrast, the quodon-modified theory describes quite well both the evolution of precipitates and the matrix concentration of copper measured by different methods in *FeCu* binary alloys under electron irradiation.

Acknowledgements This study has been supported by the STCU grants #5228; #5497, PNNL EED LDRD, and the U.S. Department of Energy, Office of Fusion Energy Sciences under contract DE-AC06-76RLO 1830.

Helpful discussions with Professors Vladimir V. Hizhnyakov and Juan F. R. Archilla are gratefully acknowledged.

References

1. V.I. Dubinko, Radiation damage and recovery of crystals: Frenkel vs. Schottky defect production, in *Nuclear Materials*, ed. by M.P. Hemsworth (Nova Science Publishers, Hauppauge, 2011), pp. 81–110
2. G. Martin, Phys. Rev. B **30**, 1424 (1984)
3. V.I. Dubinko, Radiat. Eff. Defects Solids **160**, 85 (2005)
4. V.I. Dubinko, V.F. Klepikov, J. Nucl. Mater. **362**, 146 (2007)
5. V.I. Dubinko, A.G. Guglya, S.E. Donnelly, Nucl. Inst. Methods Phys. Res. B **269**, 1634 (2011)
6. V.I. Dubinko, F.M. Russell, J. Nucl. Mater. **419**, 378 (2011)
7. F.M. Russell, J.C. Eilbeck, Europhys. Lett. **78**, 10004 (2007)
8. P. Sen, J. Akhtar, F.M. Russell, Europhys. Lett. **51**, 401 (2000)
9. F.D. Auret, S.M.M. Coelho, G. Myburg, P.J. Janse, W.E.M. van Rensburg, Thin Solid Films **518**, 2485 (2010)
10. G. Abrasonis, W. Moller, X.X. Ma, Phys. Rev. Lett. **96**, 065901 (2006)
11. L. Zhang, G. Tang, X. Ma, Phys. Lett. A **374**, 2137 (2010)
12. M.V. Ivanchenko, O.I. Kanakov, V.D. Shalfeev, S. Flach, Physica D **198**, 120 (2004)
13. J.F.R. Archilla, J. Cuevas, M.D. Alba, M. Naranjo, J.M. Trillo, J. Phys. Chem. B **110**, 24112 (2006)
14. V. Hizhnyakov, D. Nevedrov, A.J. Sievers, Physics B **316–317**, 132 (2002)
15. M. Haas, V. Hizhnyakov, A. Shelkan, M. Klopov, A.J. Sievers, Phys. Rev. B **84**, 144303 (2011)
16. V.I. Dubinko, A.V. Dubinko, Nucl. Instrum. Methods B (2012). http://dx.doi.org/10.1016/j.nimb.2012.10.014
17. M.H. Mathon, A. Barbu, F. Dunstetter, F. Maury, N. Lorenzelli, C.H. de Novion, J. Nucl. Mater. **245**, 224 (1997)
18. L.D. Landau, E.M. Lifshitz, *Statistical Physics, Part 1: (Course of Theoretical Physics, Volume 5)*, 3rd edn. (Reed Educational and Professional Publishing Ltd, 1980), Oxford, pp. 95–98
19. J.W.P. Schmelzer (ed.), *Nucleation Theory and Applications* (WILEY-VCH Verlag GmbH & Co. KGaA, Weinheim 2005)
20. S.I. Golubov, A. Serra, Yu.N. Osetsky, A.V. Barashev, J. Nucl. Mater. **277**, 113 (2000)
21. A.A. Turkin, A.S. Bakai, J. Nucl. Mater. **358**, 10 (2006)
22. A.V. Barashev, S.I. Golubov, D.J. Bacon, P.E.J. Flewitt, T.A. Lewis, Acta Materialia **52**, 877 (2004)
23. F. Christien, A. Barbu, J. Nucl. Mater. **324**, 90 (2004)
24. E. Meslin, A. Barbu, L. Boulanger, B. Radiguet, P. Pareige, K. Arakawa, C.C. Fu, J. Nucl. Mater. **382**, 190 (2008)
25. R. Monzen, M.L. Jenkins, A.P. Sutton, Philos. Mag. A **80:3**, 711 (2000)
26. Q.L. Wang, J.Z. Zhao, Mater. Sci. Eng. A **528**, 268 (2010)
27. S. Cammelli, C. Degueldre, G. Kuri, J. Bertsch, D. Lützenkirchen-Hecht, R. Frahm, J. Nucl. Mater. **385**, 319 (2009)
28. V.I. Dubinko, S. Hu, Yu Li, C.H. Henager Jr, R.J. Kurtz, Philos. Mag. doi:10.1080/14786435.2012.704425 (2012)

Energy Gain by Discrete Particle Non-linear Lattice Excitations

F.M. Russell

Abstract Evidence is given for the existence of discrete particle nonlinear lattice kink-pulses in crystals of the layered mineral muscovite. The kink-pulses, created in atomic cascades caused by nuclear scattering of high energy muons, propagate and are recorded in mono-atomic sheets of potassium in muscovite. The recording process involves exothermic precipitation of impurity atoms of iron on the paths of kinks. The lateral stability of multiple kink pulses was explored numerically in 2D arrays and by study of quasi-2D slices through molecular dynamics simulations of atomic cascades in non-layered single crystals leading to a description of atomic cascades in muscovite. Such cascades were identified in mica but their range was orders of magnitude larger than predicted. This discrepancy was resolved by showing that kink pulses gain energy from the exothermic precipitation of impurity when recording the paths of kink-pulses. It is suggested that in heavily irradiated metal crystals kink-pulses might persist over long ranges by gaining energy from interstitials created by irradiation. A test of this prediction would be if the range of so-called long-range-effects in irradiated metals increased with the magnitude of radiation damage. It is proposed that multiple kink-pulses can sweep forward crystal defects. The MD simulations gave no evidence for creation of breathers in cascades in non-layered crystals.

F.M. Russell (✉)
Heriot-Watt University, EH14 4AS Edinburgh, UK
e-mail: fr031c6637@ntlworld.com

R. Carretero-González et al. (eds.), *Localized Excitations in Nonlinear Complex Systems*, Nonlinear Systems and Complexity 7, DOI 10.1007/978-3-319-02057-0_15,

Main Abbreviations

- o MD: Molecular dynamics
- o IP: Interparticle interaction
- o KE: Kinetic energy
- o V1: Speed of sound
- o Vt: Lateral velocity
- o $\Delta\rho = \rho - 1$: Compression density
- o TEM: Transmission electron microscopy
- o KE: Kinetic energy
- o OS: On-site Potential
- o PE: Stored Potential Energy
- o Vl: Foward velocity
- o ρ: Density of particles
- o σ: Lattice unit

1 Introduction

Natural crystals of the mineral muscovite containing an impurity of iron can record transient disturbances of the lattice when in a meta-stable state during cooling after growth deep underground. These crystals have a layered structure characterised by mono-atomic sheets of potassium atoms. It was found that the tracks of charged particles, namely positrons from the decay of potassium isotope K^{40} nuclei within the crystals, could trigger locally the precipitation of the impurity thereby recording the path of the charged particles. The tracks of other high energy positively charged particles such as muons and protons also can be recorded. This precipitation is exothermic thereby releasing stored, or latent, energy in the immediate proximity of the lattice disturbance due to the moving particles. Most of the available latent energy is released after the disturbance has passed as the impurity continues to accrete on the initial nucleation precipitation sites thereby making the tracks visible. This recording process is very sensitive and offers a unique way to study transient disturbances of a crystal lattice. Hence, it is relevant to the study of how swift particles interact with crystals.

The impact of swift particles on a solid involves non-linear forces that can give rise to exotic lattice excitations such as intrinsic localized modes [1–3]. Here, swift means particles of mass m and energy E moving at supersonic speed relative to that in the solid, so that $2E/m > 10^8\,\text{m}^2/\text{s}^2$ or $E > 1\,\text{eV}$ for protons. In many cases of practical importance particles with orders of magnitude greater energy impinge on or penetrate in to a solid and create atomic cascades [4]. Inevitably, in the initial stages of development of a cascade the colliding particles move at highly supersonic speeds, causing the envelope enclosing the cascade to expand supersonically. This rapid initial dispersal of energy and momentum will create numerous discrete-particle lattice excitations in the solid that interact both with each other and with local thermal motions. These excitations can be stationary or mobile. Some of the mobile excitations, or pulses, eventually survive the chaos, overtake the cascade envelope and propagate away. These are not primarily created at the surface of the cascade. The type of excitation that makes up pulses depends on various factors. Such factors are the initial starting conditions during collisions of particles, the prevailing atomic motions due to the cascade including thermal motions, the lattice structure and the motions of atoms in the tails behind the moving pulses. This is further complicated by changes that occur in the pulses as they evolve in time and interact with lattice defects. Because of this complexity it is often not possible to give unambiguous descriptions for the different types of mobile discrete particle excitations that can exist but, usually, they fall into two types, namely, kinks and breathers [5]. Actual lattice disturbances would be expected to contain elements of both types plus some chaotic motions. Evidence for a particular kind of lattice excitation called a quodon, of remarkable stability and great range, was found in crystals of muscovite mica, which has a pronounced layered structure [6]. A quodon is well described as a subsonic optical mode breather. The properties of quodons were explored by modelling of the atomic chains and crystal layers in which they occurred [7]. These model and theoretical studies also showed that kinks could exist in muscovite but were laterally unstable resulting in very limited range [8]. The reality of quodons with exceptionally long ranges in crystals of muscovite was verified later in laboratory experiments [9].

However, a long standing problem was that whilst clear evidence of kinks was seen in videos of molecular dynamic (MD) studies of cascades in non-layered crystals there was little evidence for mobile breathers of any kind. The complexity of layered crystal structures deterred their study by MD methods. One possible explanation was that breathers could exist only in crystals that satisfied certain lattice structural conditions relating to layers. In a layered crystal a reasonable approximation allowed the equilibrium positions of atoms to be determined by an on-site potential well. Experiments with mechanical analogues of atomic chains, using gravity to give the on-site potential well, showed that the ability to support breathers decreased as the depth of the on-site potential well was reduced. This was achieved by tilting the analogue away from the vertical. Another possible explanation was that the magnitude of atomic motions of any breathers was much smaller than that of typical kinks, thereby being masked. Kinks would be more conspicuous than breathers in MD videos because, moving at supersonic speed, they rapidly reach the front or envelope of a cascade and then move forward into undisturbed parts of the lattice. Breathers move sub-sonically and so might be lost in the wake following kinks. Another factor is that since breathers move more slowly than kinks they might get caught up in reflections of kinks from computational cell boundaries in MD studies. A notable feature of breathers is their strong localisation in space on chains that intersect the initial impact site creating a cascade. This feature should assist in their identification in videos. To date clear evidence for them has not been seen but that does not exclude them.

This confusing state was further complicated by evidence from irradiation experiments on non-layered crystals, in particular metals, that showed some kind of mobile lattice excitation could propagate over distances that seemed incompatible with kinks but possible for breathers [10–12]. Since MD and other modelling methods give support for kinks in both layered and non-layered crystals a search was made for evidence of kinks in the fossil records in muscovite. It was found that previously unexplained patterns of precipitation in the shape of fans were consistent with a kink-based origin if the inevitable lateral spreading of kinks was compensated for by the kinks gaining energy from the crystal. It was already known that triggering the recording process that operated in muscovite crystals released energy stored in the lattice. Here attention is focused on the behaviour of discrete particles in kink pulses rather than treating them on a continuum basis. This is because of interest in the interaction of the pulses with individual atoms at crystal defects, such as interstitials, vacancies, dislocations, grain boundaries and voids.

2 Experimental Methods

The study of interactions of swift particles with solid matter has a long history starting with the discovery in 1896 of radioactivity and the particle-like nature of electrons but still there are unanswered questions. The main reason for this is that the interactions take place at the atomic level inside a solid where they are shielded from, or not available to, direct observation. Moreover, there are no techniques for observing directly the motions of individual atoms and particles within a solid. The nature and products of any interactions must be deduced from observable results of irradiation experiments. For example, these results might be changes in the electronic or other physical properties of the solid or creation of crystal defects observable by TEM, none of which relate to identifiable atoms. Further, all such effects are consequences of the transient events of irradiation. The inability to directly observe and study the motions of individual atoms in a solid is why numerical modelling of atomic motions in solids by molecular dynamics and other models is relied on so heavily in interpreting irradiation effects. But even these techniques have limitations. What is needed is a means to record permanently significant changes of positions of atoms in a solid and the ability to study those records quantitatively.

One attempt to achieve this was the use of stacks of photographic films by Hess in 1911 to study cosmic rays [13]. Swift particles induce electronic changes in grains of silver halides held as a colloidal suspension in a gelatine layer. The affected grains were developed chemically to metallic silver. The separated films then could be examined by microscopy. The minimum linear grain size of $\sim$0.1 μm set a lower limit on the volume within which the interaction occurred, of order 10^8 atoms. Moreover, the lattice orientation of the grains was unknown. Nevertheless, this technique allowed the tracks and flux of cosmic rays to be recorded and measured – and led to a Nobel Prize in 1936. To minimise this grain-size limitation in 1963 the technique of precipitation of silver from a glass containing silver oxide was explored [14]. After irradiation the glass was heated in hydrogen to precipitate silver at sites of gross damage to the lattice. Although the linear size limit was reduced to $\sim$5 nm the material was insensitive to all swift particles other than heavy fission fragments, which caused the technique to be abandoned. Its significance is that it pointed to the possibility of recording large displacements of atoms from their equilibrium positions by precipitation of an impurity within a solid. A more sensitive method was developed that involved single crystals of silver chloride. It depended on creating massive damage to the crystal by a cosmic ray particle followed by conversion of silver ions to metallic silver that migrated to the damaged regions. The presence of metallic silver rendered the damaged site visible [15]. A similar example of this process was found, also in 1963, to occur in crystals of the mineral muscovite; see below. Due to the difficulty of growing large crystals of muscovite in a laboratory it was necessary to rely on natural sources of swift particles to create nucleation sites. Fortunately, these were copious from both cosmic ray sources and from radioactivity in the muscovite crystals. These allowed identification of the tracks due to different swift particles [16]. The nuclear recoils from radioactive decays provided an in situ source of energetic lattice excitations.

Natural crystals of muscovite containing an impurity of iron can record, while still at high temperature, perturbations of its lattice caused by high energy charged particles passing through the crystal in the vicinity of sheets of potassium atoms. The recording process is exothermic and involves precipitation of the mobile iron impurity. In turn, this causes local chemical changes leading to formation of black magnetite surrounding the nucleation sites. The structural weakness of the lattice at the potassium sheets allows precipitation at adjacent sites to coalesce in to thin ribbons of magnetite, thereby delineating particle tracks. The nucleation sites can be caused by both electronic and nuclear scattering processes. The sensitivity of this recording process is remarkable, being able to record the tracks of minimum ionising leptons with $dE/dx \sim 10\,\text{keV/cm}$ [17]. A unique feature of this recording process is that it also can respond to large amplitude motions of neutral atoms about their equilibrium positions. The lower limit of kinetic energy of an atom to be recorded by this process is $\sim$1 eV. In responding to both charged and uncharged perturbations the nucleation sites are within one atom distance from the perturbation, giving ideal location of those sites. However, subsequent decoration by accretion of impurity at the nucleation sites increases the linear scale of the recording but this enables the tracks to be viewed by optical microscopy. The crystals grow deep underground at high temperature and can accept up to a maximum of $\sim$1.7 atomic % of Fe in to the lattice. This is equivalent to about one impurity atom of Fe per two unit cells but most crystals contain well below this limit. As new crystals slowly cool the concentration of impurity increases until it becomes supersaturated, when precipitation occurs at nucleation sites. Although the recording sensitivity increases with impurity content there is a compensating increase of impurity-based scattering that increases the rate of energy loss by swift particles, which shortens their range. In addition to this remarkable sensitivity and nearly ideal localisation of the nucleation sites muscovite crystals are optically transparent, thus revealing the fine ribbons of epitaxially grown black magnetite delineating the tracks. The pronounced layered structure also allows crystals to be cleaved in to very thin sheets. These two features enable events occurring within crystals to be studied easily and reconstructed in space.

Surprisingly, there is another recording process that can operate in muscovite that involves an excess of calcium instead of iron. This process responds to large amplitude motions of neutral atoms but not to the presence of either positive or negative free charges. The first process involves intrusive precipitation of iron as magnetite, Fe3O4. The second process involves a solid-state compositional phase change in which epitaxial ribbons of epidote, $Ca_2(Al,Fe)_3(SiO_4)_3(OH)$, replace muscovite [18]. This transition does not occur naturally in absence of intense lattice disturbance.

3 Layered Structure Effects

The layered structure of muscovite influences the passage of swift particles and propagation of kinks and breathers. The cell structure is shown diagrammatically in Fig. 1a, b [19].

It consists of a flat sheet of potassium atoms one atom thick in the (001)-plane sandwiched between identical slabs of linked silicate tetrahedra. The minimum spacing between the potassium atoms is unusually large at 0.53 nm. The potassium is weakly bound to the structure, which accounts for its easy cleavage in the (001)-plane. In essence, the potassium atoms can move relatively freely in the (001)-plane of the sheet about their equilibrium positions that are set by the rigid and surprisingly strong silicate slabs. They can also move in the direction perpendicular to the (001)-plane but less freely. It is fortunate that potassium has 0.01 % of the radioactive isotope K^{40} that emits electrons or positrons. Studies of the recorded tracks of the emitted leptons showed that the recording process operates in the vicinity of the potassium sheets, where there is space and structural flexibility to accommodate the precipitated iron impurity. Mobile impurity Fe ions initially are interstitial but they can fall in to potential wells created by large movements of potassium atoms. This precipitation process is exothermic [20].[1] For high energy, nearly relativistic, particles the release of energy is of little consequence. However, for slowly moving lattice excitations the release of energy can be critically important. This is because the insertion of an atom into a chain at which a kink or breather is located will modify the structure and energy of the excitation. This aspect is examined further in Sect. 7. The nuclear recoil accompanying lepton emission from K^{40} creates a non-linear lattice excitation that propagates along chains of potassium atoms. This quasi-particle was called a quodon. Analogue and numerical studies of the potassium chains and sheets showed that quodons were well described as a type of breather, namely, a mobile, longitudinal acoustic or optical-mode breather [21, 22]. The reality of energetic quodons was verified by an experiment in which they were created at one edge of a crystal and, after propagating $>10^7$ atoms along a chain at 300 K, ejected the last atom in the chain by elastic scattering [9]. The crystals used in this experiment contained very little Fe impurity, insufficient to record any lattice disturbances. As there was no precipitation of impurity the long range of quodons demonstrated in muscovite was not due to gain of energy from the crystal.

---

[1] An interesting introduction to solid state transitions is that by Polanyi [20].

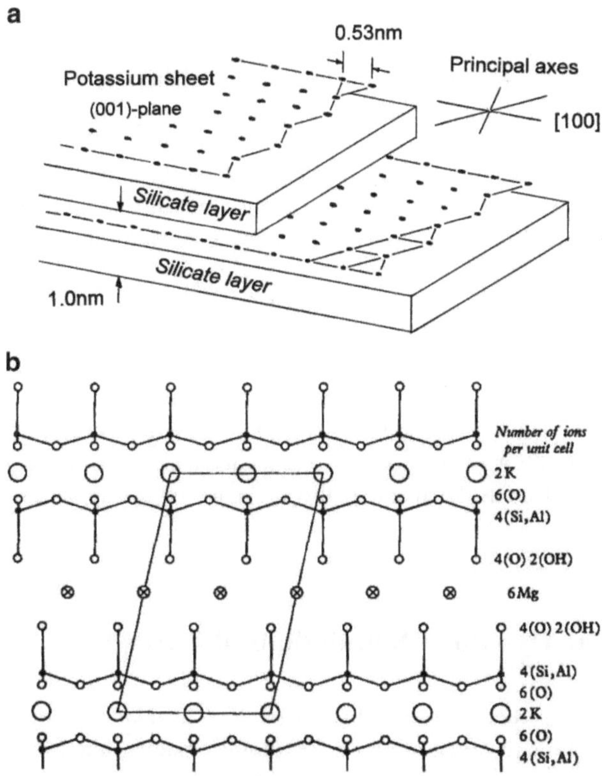

Fig. 1 (**a**) Schematic diagram of structure of muscovite mica crystal. Flat sheets of potassium atoms one atom thick in the (001)-plane are sandwiched between mirror silicate layers. The minimum spacing between potassium atoms is large at 0.53 nm and the perpendicular distance between potassium sheets is 1.0 nm. Due to the weak bonding of the potassium crystals of muscovite can be cleaved perfectly in the (001)-plane into thin transparent sheets. The weak bonding also allows impurity atoms, mainly of iron, to be precipitated as epitaxial ribbons between the silicate layers. (**b**) Arrangement of atoms in the unit cell. Muscovite has monoclinic symmetry and the most common stacking polymorph is 2M1 giving a unit cell with **a** 0.519, **b** 0.904, **c** 2.008 nm [19]

Numerical modelling of breathers shows that the envelope of a breather extends laterally only to the next immediately adjacent chains. The tracks of quodons created by recoil nuclei from the decay of K^{40} or by nuclear scattering in the potassium sheets show no evidence of spreading in to adjacent sheets. This can be understood empirically by the large perpendicular spacing of 1.04 nm between the potassium sheets in the 1M polymorph or 2.08 nm in the most common $2M_1$ polymorph of muscovite. This inhibits resonant coupling between atoms in different potassium sheets. There will be induced motions of atoms in the silicate layers but these will be small in amplitude partly because of the strong binding forces in those layers. To first order, the potassium sheets can be treated as isolated 2-D arrays positioned between fixed silicate layers. An alternative description of this isolation between sheets is as follows. The process of nuclear scattering will impart momentum p to a kink pulse or quodon. Each pulse will have an associated de Broglie wavelength given by h/p. For a kink pulse of energy of 1 eV in a potassium chain λ is $\sim$0.04 nm. As this is much smaller than the spacing between the potassium sheets a pulse should show classical behaviour with a low probability for coupling to adjacent sheets. Only at lower energies approaching thermal values will λ be sufficiently large for coupling between sheets to be appreciable. This argument suggests that kink pulses, as well as quodons, created in a nuclear scattering event in a potassium sheet should propagate predominately in that sheet.

Consider a lattice excitation composed of multiple kink pulses arranged in line abreast in adjacent chains in a sheet. If the rate of lateral spreading of a kink pulse is assumed constant then the average energy of kink pulses at the front of a multiple kink pulse should decrease inversely with range. In a non-layered crystal a kink pulse is expected to spread laterally in both directions orthogonal to the chain direction of propagation. This produces a conical envelope. In this case the average energy of kink pulses at the front should decrease inversely as the square of the range.

4 Molecular Dynamics Simulations of Cascades

A single kink pulse created in an undisturbed lattice by a kinetic nuclear collision or scattering has a limited range of about 100 atomic spacing due to lateral spreading [8]. In an atomic cascade many kink pulses will be created but the lattice where they are created will be greatly disturbed, with many atoms displaced from their initial equilibrium positions. This randomness of atomic positions restricts the amount of energy that can be given to a kink pulse without creating interstitials, usually less than 100 eV per atom. Hence, a high energy collision leading to an atomic cascade is expected to create many uncoordinated kink pulses that will interact with each other. The conditions for lateral stability of individual pulses will be compromised. This would lead to a lateral smearing of the moving fronts of the multiple kink pulses. Kink pulses will be created at different times and at different positions in a cascade and this will lead to interaction between pulses on nearby chains moving in the same low Miller index directions. For example, if a kink pulse on a chain is flanked on its sides by pulses that are slightly ahead or behind it then their interaction will lead to them lining up abreast as that is a minimum energy state. If this were not true then a plane wave front composed of multiple kink pulses would be unstable and disintegrate. As a consequence of these interactions it is expected that the moving fronts of multiple kink pulses emanating from a cascade would show evidence of self-assembly in forward motion but a smearing in radial angular distribution. The forward self-assembly should appear as flat fronts moving in low Miller directions. In a thin slice through a cascade these flat fronts would present as lines of kink pulses moving abreast.

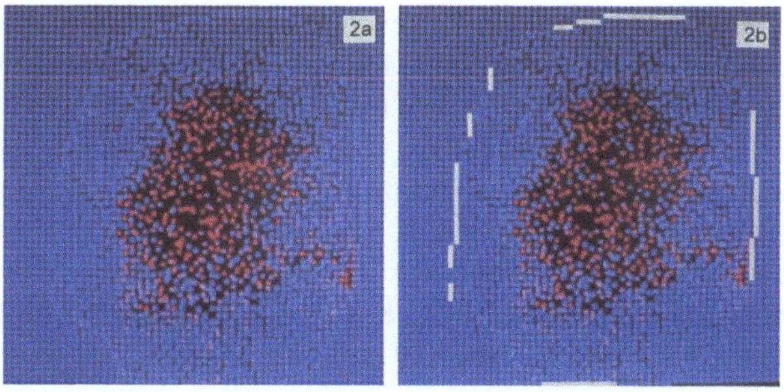

Fig. 2 (**a**) A single time frame from a video of an atomic cascade obtained by molecular dynamics. The cascade was created in a gold crystal by a gold nucleus moving with 10 keV energy. The elapsed time from the start was 1.0 ps. The frame is of a two atom thick slice through the cascade that intersects the initial impact. The instantaneous positions of moving multiple-kink pulses can be seen outside of the envelope containing atomic displacements [23]. (**b**) Some of the moving wave fronts of multiple kink-pulses are here highlighted by *white strips*. Evidence of self-assembly of the kinks to form plane wave fronts moving in chain directions is visible

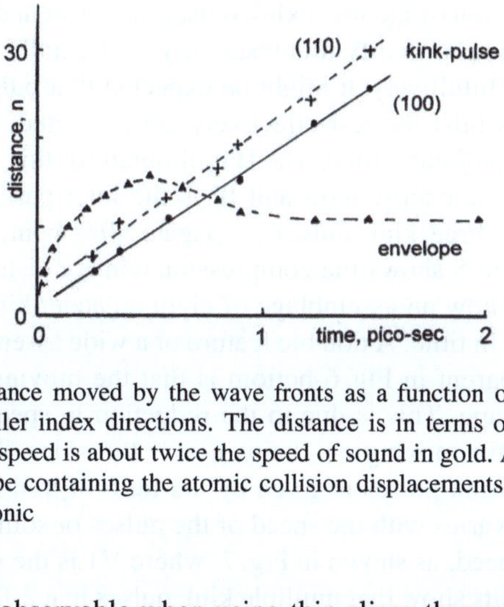

Fig. 3 Plots of the distance moved by the wave fronts as a function of time when moving in the (100) and (110) Miller index directions. The distance is in terms of the number of atomic spacings. The measured speed is about twice the speed of sound in gold. Also shown is the rate of expansion of the envelope containing the atomic collision displacements of the cascade. Initially this expansion is hypersonic

These effects are observable when using thin slices through a cascade to avoid obscuration by the bulk of the cascade. For example, Fig. 2a shows a single video frame taken at 1.0 ps after a 10 keV nuclear recoil in a gold crystal [23]. The slice is two atoms thick and intersects the initial recoil site. For identification purposes the positions of the kink pulse fronts are highlighted in Fig. 2b. The tendency to self-assembly in to planes is visible by the lateral growth of the fronts. At the edges of wave fronts moving in different Miller chain directions there is chaotic motion. The speed of these wave fronts can be found from successive video frames. Figure 3 shows that the speeds in the (100) and (110) chain directions are similar and nearly twice the speed of sound in gold in this particular case. Also shown is the speed of growth of the cascade envelope, showing that it expands initially at high supersonic speed but eventually collapses to leave a few permanent lattice defects. It is expected that the predicted behaviour of multiple kink pulses would be more evident in MD simulations of lower energy cascades than that above, as the origin of the pulses then would be more compact. The momentum vector of the initial particle collision then should influence more strongly the direction of propagation of a multiple kink pulse. The above videos show that multiple kink pulses propagate away from a cascade in low Miller chain directions. To see what kind of recorded pattern they might create in crystals of muscovite it is necessary to examine further how multiple kink pulses should behave in a 2-dimensional sheet of atoms.

5 Numerical Modelling in 2D Arrays

As stated above, a moving kink pulse on an isolated chain is stable but becomes unstable by lateral spreading when embedded in a 2-dimensional sheet, giving an effective range of $\sim$100 atoms before the atoms in the moving front become thermalised. The addition of identical kink pulses on adjacent chains, such that they are in line abreast, should modify this instability by partially screening the pulses in the middle chains. Intuitively, it might be expected that pulses on the outermost chains would decay whilst the rest effectively are protected. This effect has been examined in a 2-D triangular lattice. The Hamiltonian for this system $H = KE + IP$ where KE is the kinetic energy term and IP is the inter-particle interaction term. Figure 4 shows how three kink pulses propagate after being started in phase in adjacent chains. Figure 5 shows the compression density distribution in mid-flight [24]. Figure 6 shows how an assemblage of eight adjacent kink pulses, initially in line abreast, develops in time. A notable feature of a wide assemblage of kink pulses that begins to be apparent in Fig. 6-bottom is that the moving front progressively develops a curved shape. This is due to the reduction in speed of a kink pulse as it loses energy to kinks moving on chains nearer the sides. The lateral spreading of an assemblage of kink pulses is given by the ratio of transverse to longitudinal velocities. This ratio varies with the speed of the pulses or solitons, decreasing with increasing forward speed, as shown in Fig. 7, where V1 is the speed of sound in the 2-D sheet. These results show that multiple kink pulses in a 2-D sheet, when created nearly simultaneously to be in phase and in line abreast, should propagate further than an isolated kink pulse and should progressively develop a curved front as the forward moving front spreads laterally.

Assume a given starting energy for a kink pulse and that, to first order, by spreading laterally to form several kink pulses in adjacent chains their energy becomes uniformly distributed across the moving front. An effective range can be defined as that at which the energy per kink pulse drops to where they become ineffective in interacting with crystal defects, say, below 0.1 eV. If the speed of kink pulses is approximately twice the speed of sound, as found from the MD data, then the predicted minimum spreading angle should be about 3°. In a 2-D sheet with starting energy of 50 eV the effective range is then $500/\tan(3^\circ)$ or $\sim 10^4$ atomic

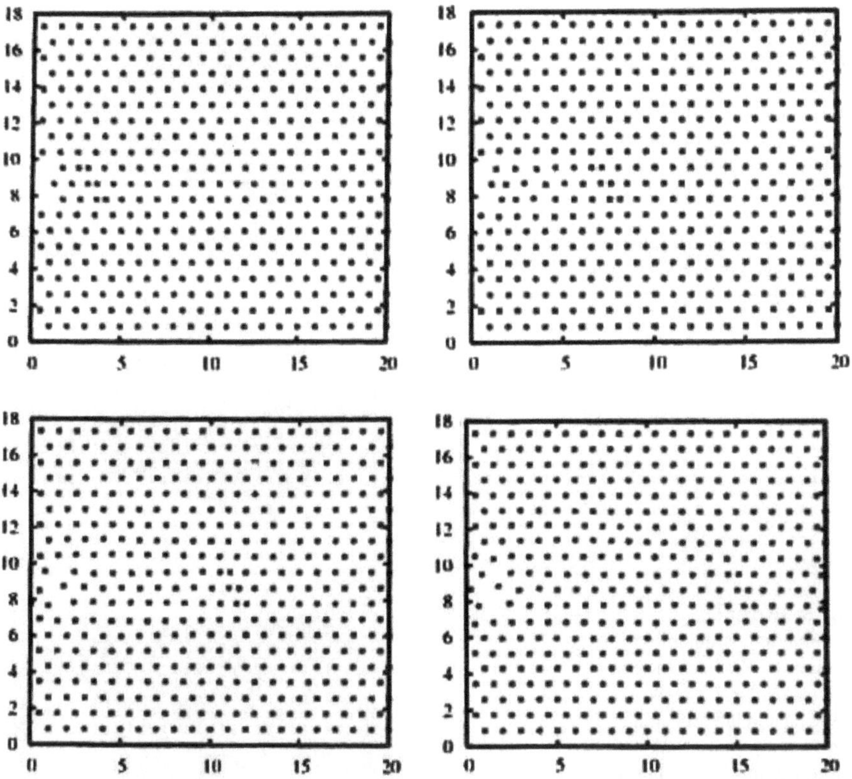

Fig. 4 Plots of numerical simulations of three kink-pulses moving in a triangular 2D array. See text

spacing. This starting energy is realistic because a K^{40} decay gives a nuclear recoil of 48 eV, which often produces only one quodon or kink depending on the initial conditions. For muscovite the range in or near a chain direction in the potassium sheets should then be $\sim$5 μm. The cut-off energy for triggering precipitation is $\sim$1 eV so the longest recorded trace of a kink pulse would be $\sim$0.5 μm. As this is comparable with the variable width of decoration with magnetite on the recorded tracks of swift particles in muscovite it was assumed for many years that the tracks of kink pulses could not be differentiated from random variations in the extent of decoration and so would not be observable.

For a non-layered 3-D crystal, with lateral spreading within a cone, the expected range would be $\sim$430 atomic spacing for a 50 eV starting energy and 0.1 eV cut-off. For the case of gold as used in the above MD study the range is $\sim$120 nm. Even if the starting energy of kink pulses could be as high as 5 keV – and did not cause a cascade – then the range would be only $\sim$1.2 μm. Based on the assumption that kink pulses progressively loose energy by lateral spreading it was clear that they

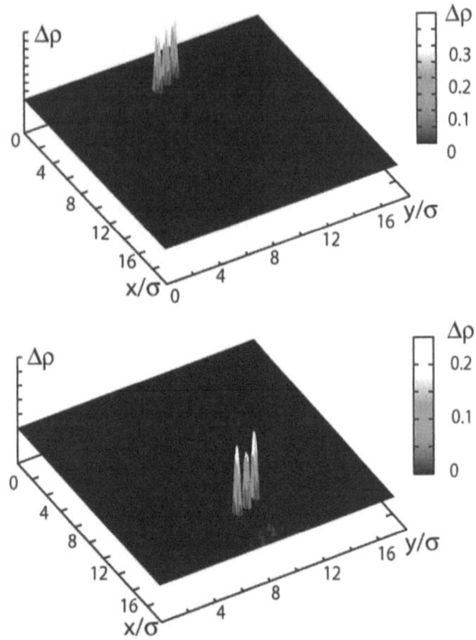

Fig. 5 Evolution of numerical simulations of three kink-pulses moving in a triangular 2D array. Initial moment is strong and it leads to decrease of lateral radiation and increase of life-time and free length. $\Delta\rho = \rho - 1$ is the compression density, where ρ is the density of particles, σ is the lattice unit

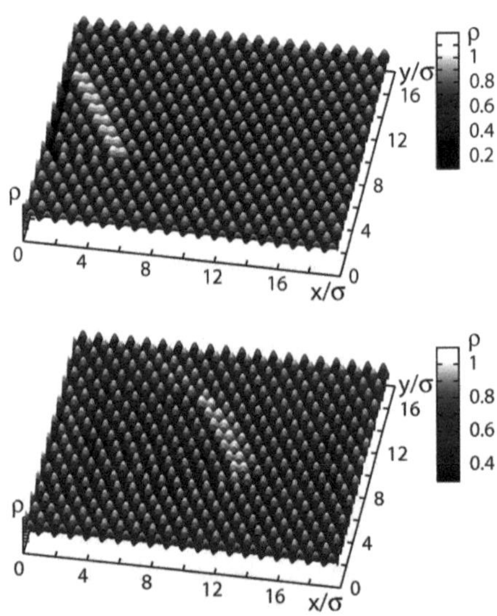

Fig. 6 Evolution of numerical simulations of eight kink-pulses moving in a triangular 2D array. They have relatively small momentum. Since kink solitons with higher energy move faster the loss of energy at the sides of the moving front causes the front to become curved. ρ is the density of particles, σ is the lattice unit

could not be the cause of reported long-range-effects in irradiated materials, which can exceed 1.5 mm, without augmentation of energy in some way [12].

The so-called long-range-effect, also known as anomalous penetration, has a long history. It refers to various physical processes or effects that are observed to occur in materials under irradiation at distances that greatly exceed the penetration depth of the bombarding particles after allowing for channelling. There are numerous such effects. Some examples are changes in void size, micro-hardness, density of crystal

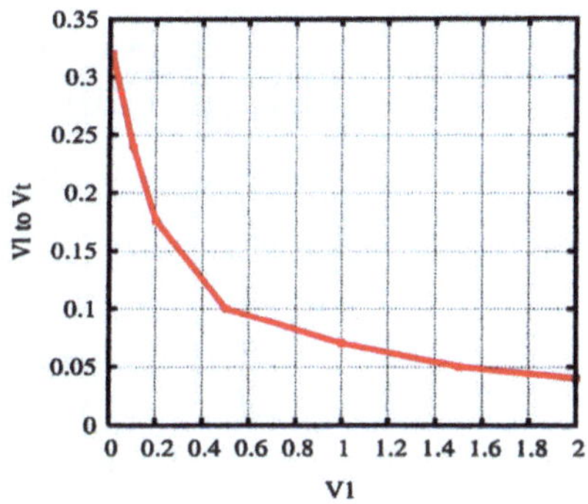

Fig. 7 Plot of the ratio of lateral spreading velocity to forward velocity of a kink front. The faster the front moves the less is the spreading. V1 is the speed of sound

defects and migration of atomic species at interfaces [25–28]. At first sight, it is not obvious that kink pulses could account for such effects, because of their short range. However, a combination of two effects associated with kink pulses might provide a solution whereby defects are pushed forward. This is examined in Sect. 10 below.

6 Structure of Non-linear Intrinsic Localised Modes

For purposes of comparison the structures of a quodon and a kink pulse are illustrated in Fig. 8a, b by their creation in a chain following an impulse. The figure shows the positions of particles in a chain with regular spacing as a function of time. In both cases the particles are subject to non-linear long range inter-particle forces plus at both ends of the chain there are fixed particles to prevent expansion. In the case of quodons an on-site potential defines the equilibrium positions. In the kink pulse case there are no on-site potentials. The particles are constrained to move only along the chain axis. In neither case does the initial impulse add an additional particle to the chain. A kink pulse forms immediately whereas a quodon evolves more slowly. A kink pulse results in a net displacement of the particles which is cancelled out upon reflection at the chain end. In a quodon adjacent particles move in opposite directions in optical mode. Typically, this results in about three pairs of particles that are alternately closer together or further apart than their equilibrium spacing. For a quodon propagating in a crystal any impurity ions are inhibited from entering a chain when the atoms are close together but may enter during their subsequent moving apart. The oscillation that gives the highest probability for insertion of an impurity moves at the phase velocity within the quodon envelope, which moves at the group velocity. An insertion interferes with the internal oscillations. It can either add or subtract energy from the quodon depending where it occurs but averages to zero.

The instantaneous positions of atoms or particles in kink pulses in the above MD video frame of a 3-D lattice and in the 2-D triangular lattice have been measured and are shown in Fig. 9a, b, respectively. The ratio of their displacements from

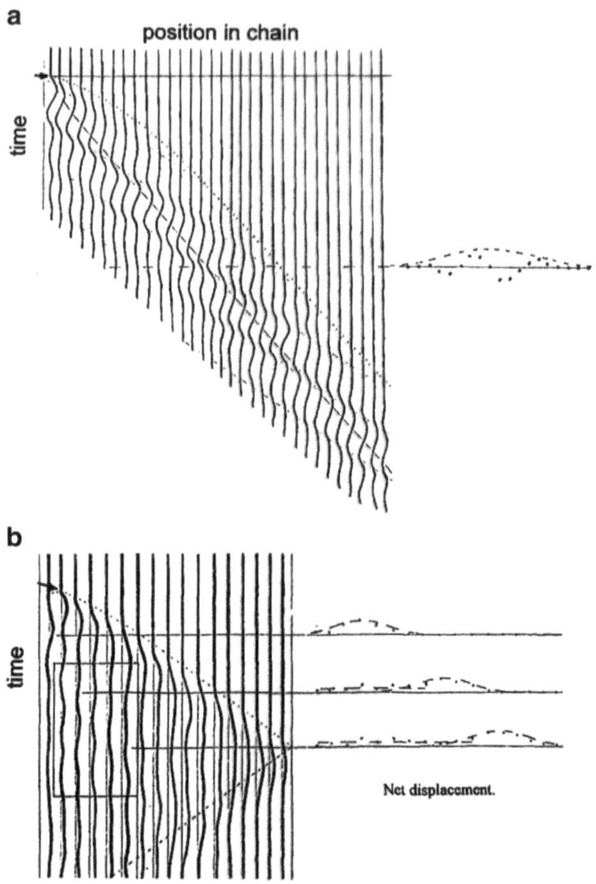

Fig. 8 Time plots of motions of particles in a chain of interacting particles following an impulse to the first particle. (**a**) In this case there are on-site potentials that define equilibrium positions of the particles. An impulse creates a quodon that shows oscillatory internal motions of longitudinal optical mode and moves sub-sonically. (**b**) In this case there are no on-site potentials except at the ends of the chain to contain the particles. An impulse creates a kink pulse that leaves particles with a finite displacement from their equilibrium positions. The kink moves supersonically. This case applies to typical metal crystal structures

equilibrium positions relative to the equilibrium spacing are shown for two arbitrary but comparable amplitudes of kinks. Such pulses can be considered to be a fast-kink followed by a slow anti-kink. In practice, the eventual positions of atoms in a chain following the creation of a kink pulse will depend on how that pulse was formed. In particular, it will be different depending on whether or not the initial impulse, due to a swift particle, resulted in the addition of an atom to the chain. This also will affect the final positions of atoms in chains lying adjacent to those travelled by kink pulses. The profile in Fig. 9a does not follow in detail the simple kink profile shown in Fig. 9b. This is not due to measurement error and is examined next. It is instructive

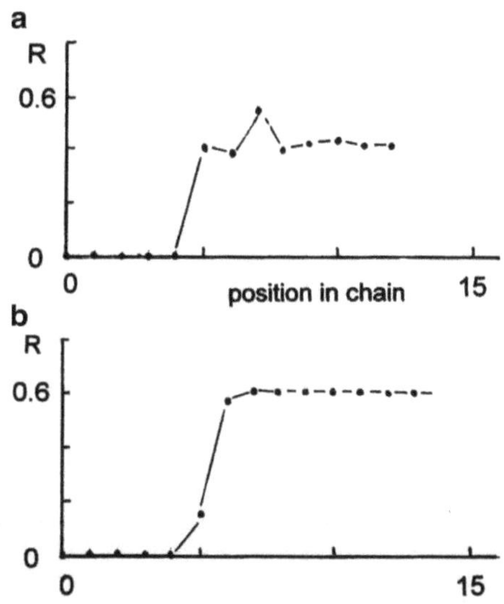

Fig. 9 Plots of the longitudinal motions of particles in the vicinity of the moving kink-pulse, given as the ratio of displacements relative to the undisturbed particle spacing. Plot (**a**) is from the MD video for a gold crystal and (**b**) is from the data in Fig. 2a for a 2D array. In both cases the displacement kink is rapid and spans at most two particles

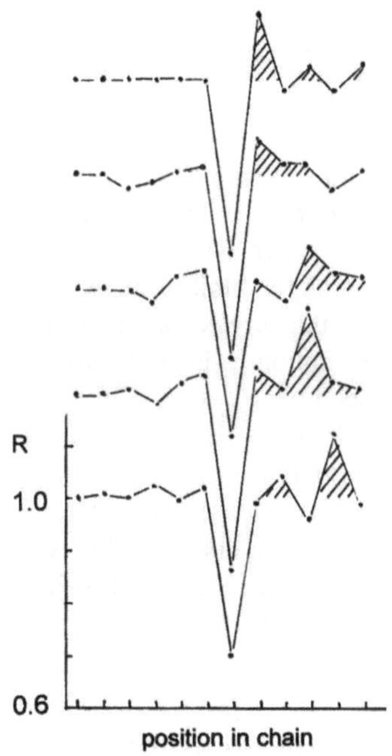

Fig. 10 Plots of the separation of adjacent atoms in different kink pulses at different positions along a chain. In each case the kink is moving from *right* to *left*. The plots show that the largest separation between adjacent atoms is not always directly behind the strong compressive motion in the kink but is usually within the first four pairs

to plot how the distance between adjacent atoms in a chain varies along the chain, as shown in Fig. 10 for several different chains, thereby giving a sampling in time. There is a large decrease in spacing at the kink front followed by a modest oscillation that decays along the chain. The separations are largest in the region behind the kink front, typically for the first or the third following separations. Unfortunately, there

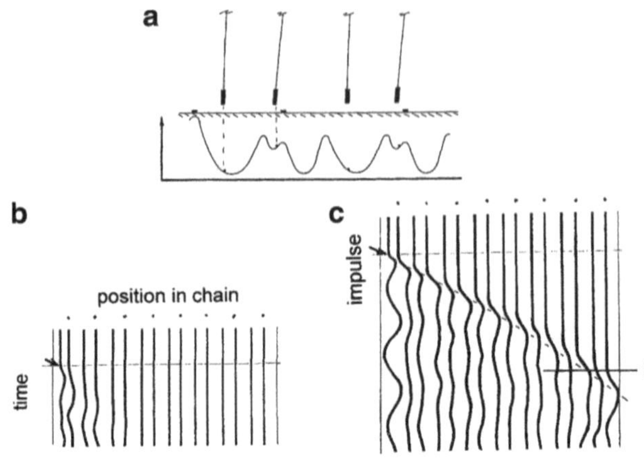

Fig. 11 Time plots of motions of particles in a chain of interacting particles following an impulse to the first particle. Alternate particles, indicated by a *dot* at the *top* of alternate traces, are held initially in a higher energy level, as shown in diagram (**a**). In plot (**b**) the impulse is insufficiently strong to release the particle from the higher energy level and results in a breather that propagates at sub-sonic speed. In contrast, in plot (**c**) a stronger impulse allows the particles in the higher levels to release their energy, resulting in the formation of a strong kink-soliton moving at supersonic speed. A stationary breather also is created in the initial impact

is no simple mathematical expression describing this wave form. These results are, of course, for an undisturbed lattice in the absence of any nearby interstitials. The plot shown in Fig. 11 also shows that the largest separation is not always directly behind the kink front, as can be seen by examining the line marked 1. Although the spacing between atoms following a kink front show clear evidence of optical mode oscillations it is unlikely they constitute a breather because they move at supersonic speed.

In principle, there is no obvious upper limit to how much energy can be given to a kink pulse that is confined to a single chain. In reality, thermal motions in a solid will encourage lateral spreading. However, if there are many identical kink pulses in adjacent chains and they are in phase abreast then the lateral spreading of those far from the sides will be impeded. This is important for long range propagation of kink-pulses.

7 Energy Gain by Kink Pulses

It is reasonable to assume that in the recording process impurity atoms migrate to a lower energy position in the lattice that is not accessible in absence of lattice disturbances caused by kinks or breathers. If this was not true then the recording process could be reversed by heating of the crystal, which has been shown by experiment not to occur. Since the mobile impurity is initially interstitial the lower energy position is most probably when the impurity inserts in to a chain.

A study of the muscovite lattice using numerical methods showed that a secondary potential well of about 2 eV exists with centre located equidistant from three closest potassium atoms. The depth of this well increased when the next closest atom to this well moved towards the well, such as when a kink-pulse propagates. It is possible that this secondary well plays an important part in the highly localised initial precipitation that occurs in the recording process by providing a resting place for the displaced potassium atom. The large atomic spacing of atoms in the sheets of potassium allows impurity atoms of iron to migrate. The presence of impurity atoms will locally distort the equilibrium positions of potassium atoms. When atoms in a kink pulse in a potassium sheet are closest together the change in local crystal potentials inhibits an interstitial impurity atom from inserting in to the chain. The hatched regions in Fig. 10 indicate where an impurity atom has the highest probability for insertion in to a lower energy well in a chain. If an insertion occurs then the displaced potassium atom is prevented from returning to its original position. In effect, there will be an additional force pushing the kink forward. Thus it seems inevitable that kink pulses should gain energy from defects in a crystal involving interstitials. Hence, in muscovite containing sufficient Fe impurity to be precipitated kink pulses should gain energy from the crystal via the recording process.

The way a kink pulse can gain energy stored in a lattice is illustrated in Fig. 11a–c. It relates to a model of a chain of interacting particles with the Hamiltonian H = KE + IP + OS + PE where the terms relate to kinetic energy, inter-particle interactions, on-site potentials and stored potential energy, respectively [29]. The potential energy term PE introduces some double-wells in the system, as indicated in Fig. 11a. In the case shown, a double-well alternates with a single-well along the chain. In Fig. 11b an excitation produced by the initial impulse to the chain is too weak to release the stored energy and propagates sub-sonically as a breather or quodon. In sharp contrast, the stronger impulse in Fig. 11c releases stored energy and results in a supersonic kink pulse. In principle, therefore, the range of kink pulses in muscovite should increase as the available stored energy increases. Of course, a simple 1-D model only crudely simulates the effect of off-chain interstitials.

The preceding results from numerical and analogue studies give some insight in to the behaviour of atoms in the tail behind the advancing front of a multiple kink pulse. Both show that there is a partial dragging of atoms along a chain but this does not constitute a full knock-on as occurs in focused collision sequences. However, due to the artificial initial starting conditions in both of these approaches, and the influence of boundaries, it is probable that this partial displacement fades in time, especially at elevated temperatures. Moreover, near a cascade the surrounding lattice initially is distorted by the swelling caused by creation of multiple vacancies, most of which eventually disappear.

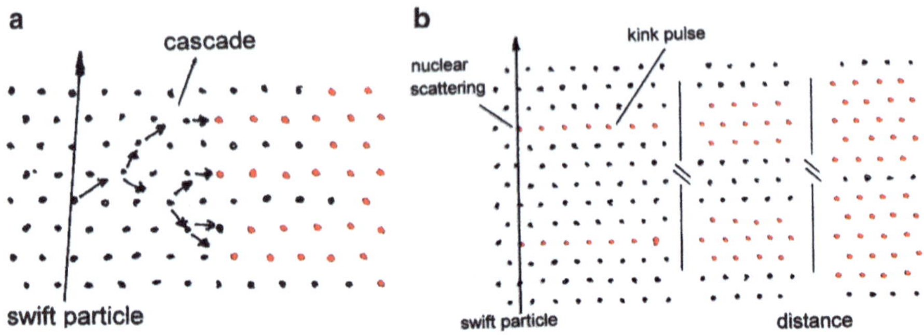

Fig. 12 (**a**) Plot of a single nuclear scattering event creating a cascade that then creates multiple kink-pulses. (**b**) Schematic diagram of nearly simultaneous nuclear scattering events creating multiple kink-pulses that spread laterally in time as they propagate. The speed of the swift particles is usually orders of magnitude greater than the speed of the kink-pulses

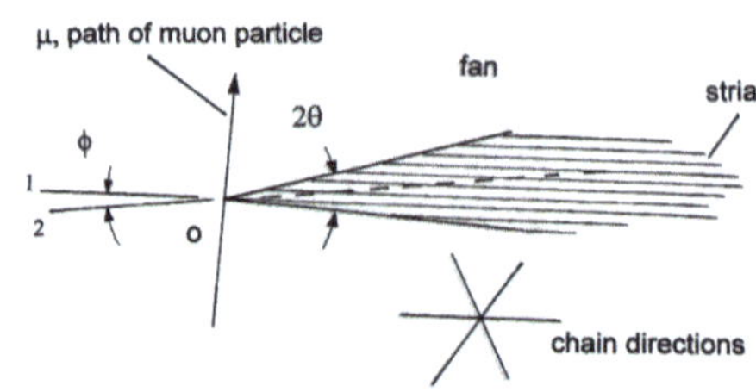

Fig. 13 Diagram showing the expected shape of the disturbance to the lattice due to a nuclear scattering event of a relativistic muon μ that generates a cascade leading to multiple kink-pulses. As the kink-pulses spread sideways with lateral spreading angle is they form a fan-shaped pattern with opening angle of 2θ The centre-line of the fan lies at the angle ϕ relative to a chain direction. The striae within the fan all lie exactly in an atomic chain direction, indicated by the line ⟨1⟩. The angle between the centre line of the fan indicated by line ⟨2⟩ and the chain direction is ϕ

8 Creation of Coordinated Kink Pulses

In principle, a swift particle passing through a crystal could suffer multiple successive nuclear scatterings. This is the basis of channelling. If the particle is nearly relativistic then kink pulses from successive scatterings would be created almost instantaneously, as the speed of kink pulses is many orders of magnitude slower than that of light. Although kink pulses are unlikely to be created in every adjacent chain usually they would all be moving in the same chain direction. Lateral spreading would then smear out missing pulses. If these coordinated multiple kink pulses were created in a sheet in a layered crystal then the result would be similar to that shown in Fig. 12a, b. This process of creation of coordinated multiple kink pulses would be expected also in the early stages in development of atomic cascades when the colliding atoms are moving at high supersonic speeds. Hence, coordinated pulses should result from development of a cascade in the 2-D sheet of potassium atoms in muscovite. In this case the mean direction of propagation of the slowly laterally expanding multiple kink pulse, that is, the direction of the centre line of the enclosing envelope, would be determined by the momentum vector initiating the cascade. This is illustrated in Fig. 13 where θ is the lateral spreading angle and ϕ is the angle between the centre line of the expanding pulse and a chain direction.

9 Evidence for Multiple Kink Pulses from Nuclear Scattering

Since evidence of multiple, laterally extended, kink/anti-kink pulses are visible in MD studies of atomic cascades in metals and can be numerically modelled in 2-dimensional arrays representing the potassium sheets in muscovite, a search was made for evidence of them in muscovite crystals. There are two main ways for high energy charged particles to lose energy when passing through matter. Firstly, by electronic processes, such as ionisation and production of delta rays. Secondly, by Coulomb-based elastic nuclear scattering. The only other significant ways for loss of energy are by bremsstrahlung and inelastic scattering, both of which have small cross-sections and so are unlikely to occur. Conditions conducive for the creation of coordinated multiple kink pulses, as explained above, are most likely to occur when a high energy charged particle undergoes channelling in passing through a crystal. The resulting pattern of a slowly spreading pulse recorded by the precipitation process should resemble a fan with a small opening angle 2θ.

A photomicrograph of a thin sheet of mica containing several fossil tracks of muons are shown in Fig. 14 the tracks lying in random directions in the (001)-plane of the crystal. That is, they are experiencing planar channelling. It is seen that numerous fan-shaped appendages are intimately associated with the tracks. The fans have a similar general shape. They are consistent with the expected appearance of multiple kink pulses resulting from elastic nuclear scattering events. The probability that these fans arose from some effect unrelated to the high energy particles is negligible for several reasons. Firstly, the fans occur only in the same layers as the charged particle tracks in the plane of recording. Secondly, the contiguous relationship between the fans and the tracks means the causative agent for the fans must have originated on a track and propagated away. Thirdly, fans do not occur at random away from charged particle tracks. This fact rules out the possibility that fans result from build-up of argon from radioactive decay of potassium that collected at random voids and distorted the lattice. Lastly, neither dislocations nor other know types of crystal defects could produce these fans. The angular distribution of fans in the mica sample of Fig. 14 is shown in Fig. 15, showing that ϕ clusters round a chain direction. It follows from Sect. 4 that the spreading angle θ should always be greater than approximately 2°. This critical test is borne out by the measured angles shown in Fig. 16.

Fig. 14 Negative photomicrograph of a sheet of muscovite mica showing straight tracks of high energy muons. Many fan-shaped appendages arise from the muon tracks, with lengths of order 1 mm, as indicated by the scale bar of 1 mm length. These are consistent with multiple-kink-pulses created by coordinated nuclear scattering events as the muons undergo channelling. The fans are aligned approximately with atomic chains in the principal crystal directions. The tracks and fans are made visible by the precipitation of an impurity of iron from solid solution in the mica at nucleation sites caused by disturbances to the lattice. Size of the specimen: 22 × 13 mm

There is, however, one major difficulty in identifying these fans with the expected properties of coordinated multiple kink pulses. It is the large discrepancy between the range of the fans of $\sim$1 mm measured in the crystal sample and the expected range, from Sect. 4, for kink pulses of about 1 μm. However, this discrepancy can be resolved if kink pulses can gain energy from the crystal via the recording process. After a crystal of muscovite has cooled sufficiently for the recording process to start then a significant disturbance anywhere in the crystal will initiate precipitation. This uniformity of properties is due to the slow rate of cooling of crystals embedded in surrounding rock and the fast diffusion rate of impurity within a crystal at high temperature. That is, the latent energy of precipitation is essentially uniformly distributed in the crystal. Hence, it does not matter how many kink pulses are propagating in any particular region as each one will initiate precipitation locally. Given that kink pulses slowly lose energy by lateral spreading, if they gain energy from the latent energy in the crystal then there should be a proportional increase in their range. The latent energy is proportional to the amount of iron precipitated, which can be determined. It is known that the epitaxially grown ribbons of precipitated material have approximately constant thickness, so the amount of precipitated material is proportional to its area A_p in a volume V of material. Fans are not observed for A_p/V less than about 0.3 cm^{-1}, corresponding to about one impurity atom of iron per 200 unit cells of mica. This might be due to insufficient release of energy to sustain the kink pulses or failure of the recording process or both. Several examples of fans in crystals containing different concentrations of Fe impurity are shown in Fig. 17. The relation between measured range of these fans and the energy released in the recording process A_p/V is shown in the log/log plot of Fig. 18. The slope of the line indicates a linear relationship, that is, the range of

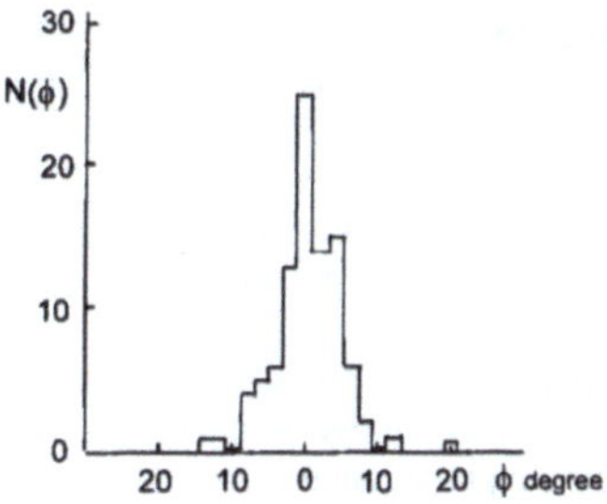

Fig. 15 Plot of the angular distribution of the centre line of fans relative to a chain direction, showing that they cluster about chain directions

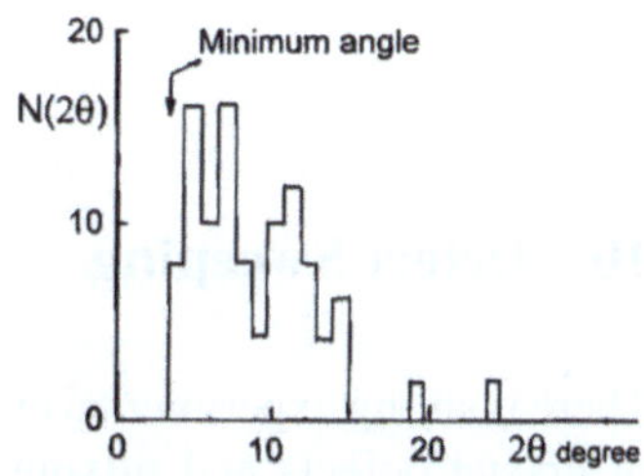

Fig. 16 Plot of the full opening angle 2θ of fans. This shows that there is a minimum lateral spreading angle of about 2°, as expected from the lateral spreading of multiple kink-pulses

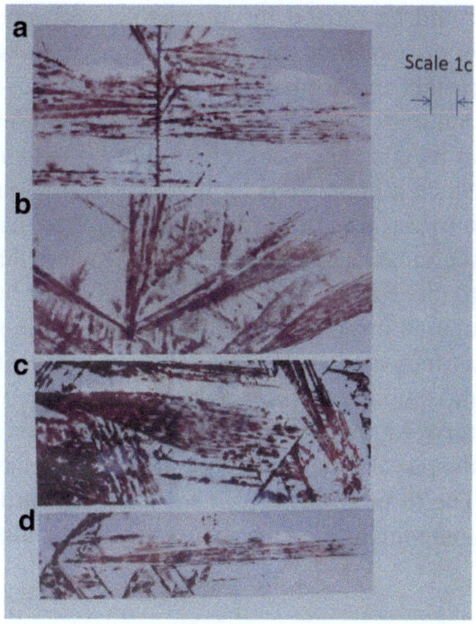

Fig. 17 Examples of near full size multiple kink-pulses in mica, showing typical fan-shape with an inner structure of multiple striae. In example (**a**) multiple fans are created by passage of a swift heavy ion in the (001)-plane in mica. The track of the swift ion is clearly visible. Similar multiple creation of fans is seen in Fig. 14, where muons are moving in the (001)-plane. In examples (**b**)–(**d**) the swift particles are moving at an angle to the (001)-plane so their tracks are not visible

the fans is proportional to the amount of energy released by the recording process. This result, together with the similarity of the properties of the fans with those expected of coordinated multiple kink pulses, supports the hypothesis that kink pulses can gain energy from that stored in a crystal.

Fig. 18 Log-log plot of the length L of fans as a function of the parameter (A_p/V), which is proportional to the concentration of iron impurity in the crystal. The length of a fan is approximately proportional to the amount of energy released by the precipitated iron

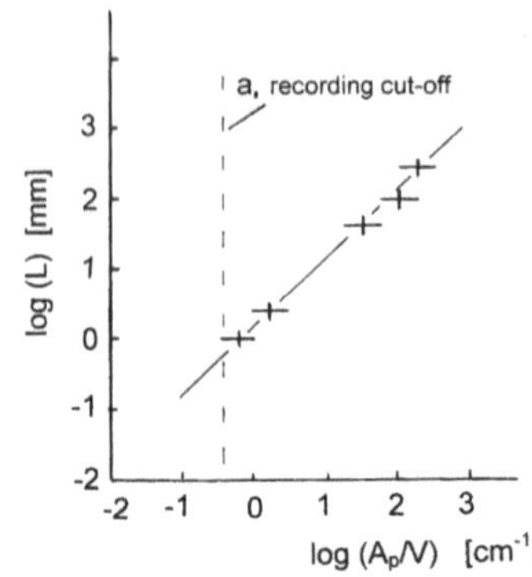

10 Defect Sweeping

There is strong experimental evidence, in metals and semiconductors, for movement of crystal defects and mixing of atomic species at interfaces far removed from primary sites of irradiation with swift particles [12]. It has been shown that vacancies could be moved by interaction with energetic breathers [30]. However, as there is no evidence for the creation of stable mobile breathers in single crystals of metals an alternative source of compact mobile lattice excitations must exist. The demonstration by MD of copious creation of mobile kink pulses in atomic cascades in metals suggests kinks might be responsible for the reported long-range-effects. Since a breather is highly localised about a single chain it follows that the initial movement of impurity atoms involved in recording its path must also be localised to the same chain. It has been shown here that the most probable place for the impurity to move to is the rarefaction region just behind a kink pulse. It is highly improbable that an insertion could take place in the strong compaction region of a kink pulse, as indicated in Fig. 10. The approach of a multiple kink pulse wave front to an interstitial will have only two possible outcomes. The interstitial could be pushed forward – analogous to channelling – or it could retain its interstitial position with the possibility of insertion into the rarefaction region behind the compaction region. The former route produces a net forward migration of crystal defects and the later route might increase the energy in the kink pulse.

11 Inner Structure of Fans

All fans show evidence of an inner structure in the form of striations. These are more clearly seen in the longer fans shown in Fig. 17a–d. The fans shown at b, c and d contain at least 19, 20, and 190 striae, respectively. The angular orientation of the striae in the (001)-plane shows that they lie exactly in chain directions. That is, the maxima of the disturbances to the lattice causing the striae move along atomic chains. The simplest interpretation of these striae is that they arise from localisation of energy in the multiple kink pulses making up the moving front. Within a given fan the spacing between adjacent striae is uniform and constant but the spacing varies from crystal to crystal. Inspection of the fans shows that striae begin at the side boundaries of the laterally spreading multiple kink pulses. Although individual striae are similar to the tracks of quodons they are inconsistent with that origin for two reasons. Firstly, they do not show evidence of scattering at lattice defects creating secondary quodons. Secondly, the curved shape of the moving front is inconsistent with independently moving quodons. A more likely explanation of striae is a lateral instability in the moving front, such as a Benjamin-Feir instability. Similar effects are reported in coupled waveguides [31]. Whatever the correct explanation, the striae show that there is some degree of self-assembly of kink pulses that prevents rapid dispersal of the striae. The minimum scale or degree of localisation of the moving pulse or excitation that forms a striation is not known but an upper limit is set by measurements. The minimum observed width of striae is 0.8 μm and their minimum separation between centres is 1.5 μm. They could be highly localised like a quodon but decoration by accretion of impurity atoms masks the actual size of the envelope of the pulse. Until the cause and structure of striae are better known there must remain the possibility that they occur only in quasi-two-dimensional lattices or layered crystals. However, since kinks require no special lattice structural conditions it is likely that atomic motion causing striae are ubiquitous. The existence of striae is further evidence for the fans having a discrete particle origin as opposed to some kind of solid-state crystal defect origin.

The localisation of energy on certain chains will increase the probability for extracting more energy from that stored in the crystal. This is illustrated in the analogue results shown in Fig. 11, where a weak pulse is unable to extract stored energy but a strong pulse can. If striae also occur in non-layered crystals, as seems probable, then they would form rod-like channels showing enhanced annealing of defects. In effect, this localisation of energy in a multiple kink pulse could allow it to burrow through the bulk of a crystal. This might provide a plausible explanation of the so-called long range effect reported in irradiated crystals.

12 Discussion

In the demonstration of ejection of atoms from the rear face of a crystal of muscovite by quodons, after they had propagated $>10^7$ atoms along chains, care was taken to use a crystal of very low iron content. Although the potassium in muscovite is radioactive only about one in 10^5 atoms in a chain would have decayed before the crystal was used. Hence, it is very unlikely that there would have been any significant energy gained by quodons in the ejection experiment. There would have been about two orders of magnitude fewer decays before the recording process ceased, so the impact of energy stored from decays on the propagation of fans would have been negligible. The ejection experiment confirmed their localisation in space and stability against thermal motion. If quodons could pick-up energy then it would be expected that their tracks would extend indefinitely as long as iron was available. Quodons can be scattered by crystal defects to create secondary quodons of lower energy, which increases their probability for further scattering at defects. It was found that the length of these secondary quodons tracks became progressively shorter at each scattering. This showed that quodons do not on average gain energy otherwise all of their tracks would be long. This still allows quodons or breathers to be active in connection with annealing radiation damage but they would obtain no net gain of energy [32–34]. In principle, much of the studies of how breathers interact with crystal defects should apply to kink pulses, too.

The localisation of energy on certain chains in a striation will increase the probability for extracting more energy from that stored in the crystal. This effect was illustrated in the analogue result shown in Fig. 11, where a weak pulse is unable to extract stored energy but a strong pulse could. It is also a common feature of nonlinear systems. If striae also occur in non-layered crystals, as seems probable, then they would form rod-like channels showing enhanced annealing of defects. In effect, this localisation of energy in a multiple kink pulse could allow it to burrow through the bulk of a crystal. This might provide a plausible explanation of the so-called 'long range effect' reported in irradiated crystals. Without gaining energy from the lattice multiple kink pulses have only a very limited range. To achieve long flight paths commensurate with 'long-range-effects' it is necessary for energy to be stored in the material, so such effects should not appear immediately but only after energy has been stored in the material. This prediction should be testable by experiment. Of particular significance in this connection, the ability to pick up energy would help in combating any loss of energy due to coupling to free or conduction band electrons. Hence, kink pulses are more likely to survive in metals than quodons. The great range of quodons has been demonstrated only in muscovite which is an excellent insulator.

On the basis that irradiation of a crystal with energetic particles can create atomic cascades, which can then create semi-permanent defects such as interstitials, there will also be created multiple kink pulses that could pick-up energy from, or anneal, those defects. There will be a balance between these competing processes of creating and annealing defects. It is anticipated that there would de-stabilising interactions between kink pulses if they are created too close together on a given chain. Hence, it is expected that reducing the energy of the incoming particles causing irradiation would lead to a shift in the above competing processes towards annealing. That is, for enhanced annealing of defects after irradiation at a given energy the energy of the irradiating particles should be lowered. The finding that compact lattice excitations such as multiple kink pulses can gain energy from that stored in a lattice invites application to other areas. For example, it raises the question of whether an intense multiple kink pulse, or shock-wave, could lead to a runaway release of stored energy arising from radiation damage. It is not so much the magnitude of the energy released that might be of concern but the possibility that defects could be quickly swept to grain boundaries, thereby cause weakness and susceptibility to stress-fracture. Since multiple kink pulses show some aspects of self-assembly it is possible that they could survive passage across dislocations without scattering. It is less clear what might happen at a grain boundary but it is possible that some transmission might occur.

Preliminary studies of transient effects caused by cascades in silicon, which does not possess straight atomic chains, show marked differences compared to those from crystals with such chains. Intuitively, the branching of the bonds would be expected to spread localised excitations laterally. This topic needs further study by MD methods at relatively low impact energies by giving certain atoms specified initial velocities. An experimental search for ejection of atoms by inelastic scattering of mobile lattice excitations able to penetrate 2.1 mm of a silicon single crystal bombarded with 800 keV argon ions was made but no evidence was found [35].

The question of possible existence of breathers in certain metals has been examined in the case where electronic screening of interaction between adjacent atoms is significant [36]. Evidence from numerical simulations suggest breathers might occur in Ni and Nb given certain starting conditions for the excitation. However, they seem to be stationary breathers and so of restricted relevance to annealing radiation damage.

13 Conclusion

Evidence for relatively stable multiple kink pulses created by nuclear scattering events has been found in crystals of muscovite. They propagate at supersonic speed along atomic chains. Although the experimental evidence for their existence relates to a layered crystal structure, numerical modelling of them indicates that they should exist also in non-layered crystals. It has been shown that kink pulses can gain energy from potential energy previously stored in a crystal. This enables kink pulses to propagate great distances of the order of 10^8 atomic spacing in muscovite containing a high concentration of impurity. A notable feature of multiple kink pulses is the spontaneous localisation of energy in their wave-front but the degree of lateral localisation is not yet known to better than 0.8 μm. It is proposed that multiple kink pulses can sweep forward crystal point defects. This process might account for experimental observations of long range effects. The possibility is raised of a runaway release of energy stored in a crystal by passage of an intense shock-wave composed of many multiple kink pulses. Of course, the release of stored energy by propagation of a compressive pulse is well known, for example, by a detonation wave front in chemical explosives. Here, this process is examined in relation to discrete particle events with application to annealing of radiation damage in nuclear power materials. It is predicted that such effects occur only after build-up of stored energy in crystal defects. If confirmed, it would indicate that kink pulses were the primary excitation involved. The question of whether or not breathers can exist in non-layered crystals and in metals remains unanswered but no evidence for them has been seen in MD studies of cascades in metals.

Acknowledgements It is a pleasure to acknowledge the essential help given by A P Chetverikov in connection with the lateral instability of solitons in 2D arrays. I wish to thank K. Nordlund for supplying data on MD studies of cascades and J. F. R. Archilla and J. C. Eilbeck for very helpful discussions. Finally, I also wish to thank J. F. R. Archilla for essential help and guidance in related technical matters.

References

1. A.M. Kosevich, A.S. Kovalev, Self–localization of vibrations in a one-dimensional anharmonic chain. Sov. Phys. JETP **67**, 1793 (1974)
2. A.J. Sievers, S. Takeno, Intrinsic localized modes in anharmonic crystals. Phys. Rev. Lett. **61**, 970 (1988)
3. V.M. Burlakov, S.A. Kisilev, V.I. Rupasov, Localized vibrations of homogeneous anharmonic chains. Phys. Lett. A **147**, 130–134 (1990)
4. J. Gibson, A. Goland, M. Milgram, G. Vineyard, Dynamics of radiation damage. Phys. Rev. **120**(4), 1229 (1960)
5. Q. Dou, J. Cuevas, J.C. Eilbeck, F.M. Russell, Breathers and kinks in a simulated crystal experiment. Discret. Contin. Dyn. Syst. Ser.-S **4**, **5**, 1107–1118 (2011)
6. F.M. Russell, D.R. Collins, Lattice-solitons in radiation damage. Nucl. Instrum. Methods Phys. Res. B **105**, 30–34 (1995)
7. J.L. Marín, J.C. Eilbeck, F.M. Russell, Localized moving breathers in a 2D hexagonal lattice. Phys. Lett. A **248**, 225–229 (1998)
8. Yu.V. Martynenko, P.G. Moskovkin, Solitons in radiation physics of crystals. Radiat. Eff. Defects Solids **117**, 321–328 (1991)
9. F.M. Russell, J.C. Eilbeck, Evidence for moving breathers in a layered crystal insulator at 300 K. Europhys. Lett. **78**, 10004 (2007)
10. P. Sen, J. Akhtar, F.M. Russell, MeV ion-induced movement of lattice disoder in single crystalline silicon. Europhys. Lett. **51**, 401 (2000)
11. G. Abrasonis, W. Möller, X.X. Ma, Anomalous ion accelerated bulk diffusion of interstitial nitrogen. Phys. Rev. Lett. **96**, 065901 (2006)
12. L. Zhang, G. Tang, X. Ma, F.M. Russell, X. Cao, B. Wang, P. Zhang, Long-range effect of ion irradiation on Cu surface segregation in a CuNi system. Phys. Lett. A **375**, 1976–1979 (2011)
13. V.F. Hess, Über beobachtungen der durchdringenden strahlung bei sieben freiballonfahrten. Phys. Z. **13**, 1084–1091 (1912)
14. R.L. Fleischer, P.B. Price, Charged particle tracks in glass. J. Appl. Phys. **34**, 2903–2904 (1963)
15. C.B. Childs, L.M. Slifkin, Delineating of tracks of heavy cosmic rays and nuclear processes within large crystals of silver chloride. Rev. Sci. Instrum. **34**, 101–104 (1963)
16. F. Russell, Identification and selection criteria for charged lepton tracks in mica. Nucl. Tracks Radiat. Meas. **15**, 41–44 (1988)
17. F.M. Russell, Decorated track recording mechanisms in muscovite mica. Nucl. Tracks Radiat. Meas. **19**, 109–113 (1991)
18. J.W. Steeds, F.M. Russell, W.J. Vine, Formation of epidote fossil positron tracks in mica. Optik **92**, 149–154 (1993)
19. W.A. Deer, R.A. Howie, J. Zussman, *Rock-Forming Minerals*, vol. 3 (Longmans, London, 1965)
20. J.C. Polanyi, in *The Chemical Bond, Structure and Dynamics*, ed. by A.H. Zewail (Academic, Boston, 1992), pp. 149–175. ISBN:0-12-779620-7
21. F.M. Russell, D.R. Collins, Lattice-solitons and non-linear phenomena in track formation. Radiat. Meas. **25**, 67–70 (1995)

22. F.M. Russell, Y. Zolotaryuk, J.C. Eilbeck, T. Dauxois, Moving breathers in a chain of magnetic pendulums. Phys. Rev. B **55**, 6304 (1997)
23. K. Nordlund, A classical molecular dynamics computer simulation of a collision cascade (2008), http://en.wikipedia.org/wiki/File:10kevau_au.gif. Accessed 26 Dec 2012
24. A.P. Chetverikov, W. Ebling, M.G. Velarde, Soliton-like excitations and solectrons in two-dimensional nonlinear lattices. Eur. Phys. J. B **80**, 137–145 (2011)
25. D.I. Tetelbaum, E.V. Kuril'chik, N.D. Latisheva, Long-range effect at low-dose ion and electron irradiation of metals. Nucl. Instrum. Methods Phys. Res. B **127/128**, 153–156 (1997)
26. D.I. Tetelbaum, A.Y. Azov, E.V. Kuril'chik, V.Y. Bayankin, The long range influence of ion and photon irradiation on the mechanical properties and on the composition of the permalloy-79. Vacuum **70**, 169–173 (2003)
27. D.I. Tetelbaum, E.V. Kuril'chik, Y.A. Mendeleva, Long range effect under low intensity irradiation of solids. J. Surf. Investig. X-ray Synchrotron Neutron Tech. **5**(2), 249–258 (2009)
28. V.L. Levshunova, G.P. Pokhil, D.I. Tetelbaum, Oscillations of the charge in oxide at silicon surface as an origin of the process leading to the long range effects. J. Surf. Investig. X-ray Synchrotron Neutron Tech. **3**(2), 276–278 (2010)
29. F.M. Russell, J.C. Eilbeck, Persistent mobile lattice excitations in a crystalline insulator. Discret. Contin. Dyn. Syst. Ser.-S **4**, **5**, 1267–1285 (2011)
30. J. Cuevas, C. Katerji, J.F.R. Archilla, J.C. Eilbeck, F.M. Russell, Influence of moving breathers on vacancies migration. Phys. Lett. A **315**, 364–371 (2003)
31. D.N. Christodoulides, R.I. Joseph, Discrete self-focusing in nonlinear arrays of coupled waveguides. Opt. Lett. **13**(9), 784 (1988)
32. V.I. Dubinko, F.M. Russell, Radiation damage and recovery due to the interaction of crystal defects with anharmonic lattice excitations. J. Nucl. Mater. **419**(1), 378–385 (2011)
33. V.I. Dubinko, Radiation damage and recovery of crystals: Frenkel vs. Schottky defect production, in *Nuclear Materials*, ed. by M.P. Hemsworth (Nova Science Publishers, Hauppauge, 2011), pp. 81–110
34. V.I. Dubinko, P.A. Selyshchev, J.F.R. Archilla, Reaction-rate theory with account of the crystal anharmonicity. Phys. Rev. E **83**(4), 13 (2011)
35. D. Schlößer, K. Kroneberger, M. Schosnig, F.M. Russell, K.O. Groeneveld, Search for solitons in solids. Radiat. Meas. **23**, 209–213 (1994)
36. M. Hass, V. Hizhnyakov, A. Shelkan, M. Klopov, A.J. Sievers, Prediction of high-frequency intrinsic localised modes in Ni and Nb. Phys. Rev. B **84**, 144303 (1988)

Supersonic Kinks in Coulomb Lattices

Juan F.R. Archilla, Yuriy A. Kosevich, Noé Jiménez, Víctor J. Sánchez-Morcillo, and Luís M. García-Raffi

Abstract There exist in nature examples of lattices of elements for which the interaction is repulsive, the elements are kept in place because different reasons, as border conditions, geometry (e.g., circular) and, certainly, the interaction with other elements in the system, which provides an external potential. A prime example are layered silicates as mica muscovite, where the potassium ions form a two dimensional lattice between silicate layers. We propose an extremely simplified model of this layer in order to isolate the properties of a repulsive lattice and study them. We find that they are extremely well suited for the propagation of supersonic kinks and multikinks. Theoretically, they may have as much energy and travel as fast as desired. This striking results suggest that the properties of repulsive lattices may be related with some yet not fully explained direct and indirect observations of lattice excitations in muscovite.

J.F.R. Archilla (✉)
Grupo de Física No Lineal, Departamento de Física Aplicada I, ETSI Informática, Universidad de Sevilla, Avda. Reina Mercedes, s/n. 41012-Sevilla, Spain
e-mail: archilla@us.es

Y.A. Kosevich
Semenov Institute of Chemical Physics, Russian Academy of Sciences, ul. Kosygina 4, 119991 Moscow, Russia

N. Jiménez · V.J. Sánchez-Morcillo
Instituto de Investigación para la Gestión, Integrada de las Zonas Costeras, Universidad Politécnica de Valencia, c/ Paranimfo no. 1, 46730 Grao de Gandia, Spain
e-mail: victorsm@upv.es

L.M. García-Raffi
Instituto de Matemática Pura y Aplicada, Universidad Politécnica de Valencia, Camino de Vera s/n, 46022 Valencia, Spain

R. Carretero-González et al. (eds.), *Localized Excitations in Nonlinear Complex Systems*, Nonlinear Systems and Complexity 7, DOI 10.1007/978-3-319-02057-0_16,

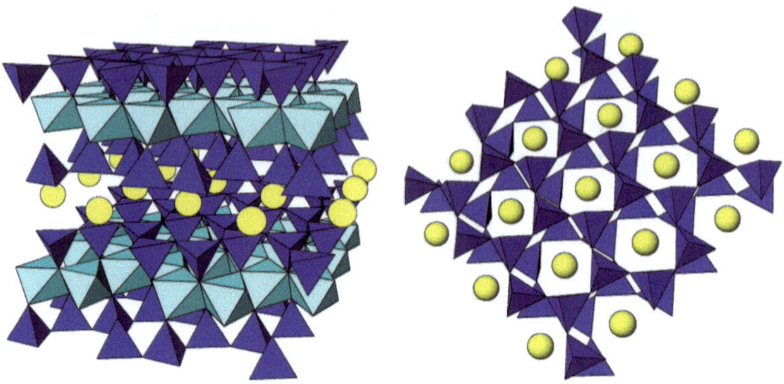

Fig. 1 Crystal unit cell of muscovite ICSD 34406. The *circles* represent the potassium ions forming the interlayer sheet. (a = 5.19 Å; b = 9.02 Å; c = 20.0 Å; β=95.7°)

1 Introduction

Layered silicates are made up from two basic building blocks: a sheet of edge-sharing [SiO_4] units, the tetrahedral sheet, and another one of edge-sharing [MO_6] units, the octahedral sheet. Muscovite is a mica in which a charged layer results from the isomorphic substitution of silicon by aluminium in the tetrahedral sheet [1]. The structure of muscovite is depicted at the left of Fig. 1. The figure at the right shows an illustration of the interlayer space of muscovite. It can be observed that both surfaces of the upper and lower tetrahedral sheets are formed by the basal oxygen atoms from the [SiO_4] tetrahedra, which form a rough hexagonal honeycomb structure. The interlayer balancing sheet is therefore sandwiched in between, potassium occupying the dimples left at the center of each pair of hexagonal cells. Each potassium ion is surrounded by six other potassium ions in the interlayer sheet with a ditrigonal symmetry.

Essentially, in both clay minerals and mica, [SiO_4] tetrahedra are linked into infinite two-dimensional networks by sharing three oxygens.

Mica muscovite has a very special characteristic, its recording capability. Tracks from particles as muons can be distinguished in the complex decoration, but the only explanation to most of the tracks is the presence of localized excitations, called quodons [2]. They propagate in the cation layer without mass transport along the lattice directions. Another interesting effect probably related with the existence of fast travelling localized perturbations was reported in Ref. [3]: experiments show that an alpha particle hitting one boundary of a muscovite layer provokes the ejection of an atom at the other side of the layer, several millimeters far from the excitation.

What makes the two dimensional, hexagonal lattice that form the cation layer so special? Transversal breathers related with the kinetics [4–6] have been found in the cation layer, they have small energies below a few eV, move slowly and radiating, therefore they are not good candidates for quodons. We advance the hypothesis that one reason could be that the ion lattice is made out of repulsive cations, K^+ ions in muscovite. As a first step to explore the implications of the hypothesis we study a one-dimensional model of potassium ions coupled through Coulomb potentials, and observe that supersonic kinks [7] with a very different range of energies can propagate long distances riding a sea of phonons and rebounding with others.

It is certainly probable that a straightforward 2D generalization of the one dimensional model will not support one dimensional kinks. On the other hand, if the tracks in mica correspond to lattice excitations as we believe, they are quasi one-dimensional. Why they do not spread laterally is a key question that we are not prepared to answer at this stage but which is one of our most important midterm objectives. For that we will most likely have to introduce more characteristics of real mica and our goal is to find exactly which ones. But we think that understanding first what is happening in one dimension is a necessary first step as have been done so often and successfully in solid state physics.

Certainly, repulsive Coulomb lattices are not limited to the potassium sheet in mica. Electrons form a Wigner crystal [8] which have been extensively studied in several dimensions. See for example Ref. [9] on one dimensional crystal and Ref. [10] for the first experimental 2D realization on a liquid-helium surface. Macroscopic repulsive Coulomb systems can be constructed in one dimension with magnets located with opposite poles in front of each other, although in this case friction is also present. This system has already been constructed by the authors and research on it is planned.

The outline of this article is as follows. In Sect. 2 we describe the model and its justification, in Sect. 3 we obtain the dispersion curves for the linearized system and compare them with the data for mica muscovite. Section 4 describes the tail analysis and their implication for possible travelling localized solutions. In Sect. 5 the velocities of the kinks are obtained theoretically, while Sect. 6 describes the results of numerical simulations and compare with the theory. After some comments about the effect of long range interaction the article finishes with the conclusions.

2 Model Equations

We simplify the model of the potassium ions inside the layer of muscovite making the following assumptions: (a) We consider the motions only along a lattice direction where the K^+ ions have the shortest distance $a = 5.19\,\text{Å}$, which becomes the lattice unit; (b) We substitute the electric field created by the silicate layers by an uniform field, which is taken into account only conceptually; (c) We consider only Coulomb interaction and only between nearest neighbors. This is justified because the negative charge above and below the K^+ ions is rapidly screened by the surrounding negative charge. (d) The effect of the surrounding negative charge is to keep the ions in place. This is incorporated into the system through fixed particles

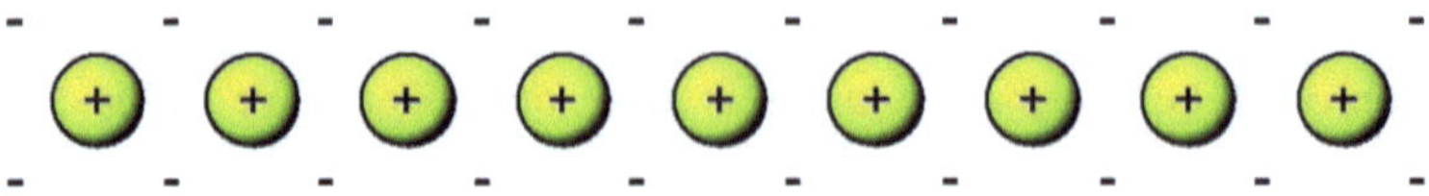

Fig. 2 Sketch of the Coulomb's chain. The *yellow spheres* represent the K^+ ions surrounded by the negative charge of the silicate layers above and below

at the ends, so as the interatomic distance is the one observed in real muscovite. In Fig. 2 we show a sketch of the model.

Note, that however important the periodic potential in real mica might be, the existence of the two-dimensional K^+ does not need it. It comes about for the need to compensate the excess negative charge of the silicate layers. The latter determines also the surface charge density of the K^+ layer, and, thus, the interatomic spacing. Repulsion of the ions determines a closed packed hexagonal structure.

With these assumptions the equation of motion of a K^+ ion is given by:

$$m\ddot{x}_n = -\frac{K_c e^2}{(x_{n+1} - x_n)^2} + \frac{K_C e^2}{(x_n - x_{n-1})^2}. \tag{1}$$

These coupled equations can be made dimensionless, by choosing the unit of distance, $a = 5.19\,\text{Å}$, time, $\tau = \sqrt{m_K a^3/K_C e^3} = 0.1989\,\text{ps} \simeq 0.2\,\text{ps}$, mass, $m_{K+} = 39.1\,\text{amu}$, and also defining the displacement with respect to the equilibrium position u_n, measured in lattice units as $x_n = an + au_n$. We get:

$$\ddot{u}_n = -\frac{1}{(1 + u_{n+1} - u_n)^2} + \frac{1}{(1 + u_n - u_{n-1})^2}. \tag{2}$$

The system of N atoms is kept in place (the chain is stable at equilibrium) by setting two extra fixed particles at positions n = 0 and n = N + 1. Therefore, x_0 and x_{N+1} are always zero. The equations for the particles n = 1 and n = N, become

$$\begin{aligned}\ddot{u}_1 &= -\frac{1}{(1 + u_2 - u_1)^2} + \frac{1}{(1 + u_1)^2}\\ \ddot{u}_N &= -\frac{1}{(1 - u_N)^2} + \frac{1}{(1 + u_N - u_{N-1})^2}\end{aligned} \tag{3}$$

These boundary conditions have been chosen for convenience, because with periodic boundary conditions two kinks are produced travelling with opposite directions and eventually colliding. In this article we are not trying to model border effects and far away from the borders the properties of phonons and kinks with different boundary condition are virtually identical in the simulations. As our physical model consists of a large number of atoms, periodic conditions are more convenient for the theory, but fixed ends are preferably for the numerical calculations. Somehow, fixed end are more realistic to model the interaction of a swift particle with the lattice border, but certainly the boundary potential has to be carefully chosen, which is also one midterm objective of the authors' research.

3 Phonons

For small amplitudes, the power-law potentials can be expanded in series, using that $1/(1+y)^2 \simeq 1 - 2y + 3y^2 - 4\,y^3 \ldots$. Retaining cubic terms and smaller, Eq. (2) is reduced to

$$\begin{aligned}\ddot{u}_n &= 2(u_{n+1} + u_{n-1} - 2u_n) - 3(u_{n+1} - u_n)^2 + 3(u_n - u_{n-1})^2 \\ &\quad + 4(u_{n+1} - u_n)^3 - 4(u_n - u_{n-1})^3 + \ldots . \end{aligned} \tag{4}$$

which is the $\alpha - \beta$ FPU equation. Neglecting nonlinear terms, it reduces to the discrete linear wave equation,

$$\ddot{u}_n = \mathrm{c}^2(u_{n+1} + u_{n-1} - 2u_n) \tag{5}$$

where $\mathrm{c} = \sqrt{2}$ will be shown to be the speed of sound. Note that, although it can be scaled out, we keep here the value of c in the equations. This makes easier the comparison with other results in the literature (different scalings are often used), and mainly with the standard scaling corresponding to $\mathrm{c} = 1$. Also note that this linearized equation is *attractive*, i.e., if the distance $u_{n+1} - u_n$ increases the force on the particle u_n is positive. The reason for it is the decreasing nature of the Coulomb force: when the distance with the particle in front increases the repulsion weakens and the repulsion from the particle behind becomes larger pushing the particle n forward.

A phonon corresponds to small amplitude particle displacements of the form $u_n = \exp(\mathrm{i}(qn - \omega t))$, with q and ω being the phonon wave number and frequency, respectively. By substitution in Eq. (5) we obtain the phonon dispersion equation:

$$\omega = \omega_M \sin(q/2) \quad \text{with} \quad \omega_M = 2\mathrm{c}, \quad \text{and} \quad q = \frac{2\pi m}{N}, \qquad m = 0, 1, \ldots N\,. \tag{6}$$

$\omega_M = 2\mathrm{c}$ is the maximum frequency, corresponding to the mode $q = \pi$. The phase velocity $V_{ph} = \omega/q$ and the group velocity $V_g = \partial\omega/\partial q$ are given by

$$V_{ph} = \mathrm{c}\frac{\sin(q/2)}{q/2} \quad , \quad V_g = \mathrm{c}\cos(q/2) \tag{7}$$

The speed of sound c is recovered in the long wavelength limit, i.e., $\lim_{q\to 0} V_{ph} = \lim_{q\to 0} V_g = \mathrm{c}$, as anticipated. This is also the maximum of both the phase and group velocities. The latter becomes zero for the boundary mode $q = \pi$.

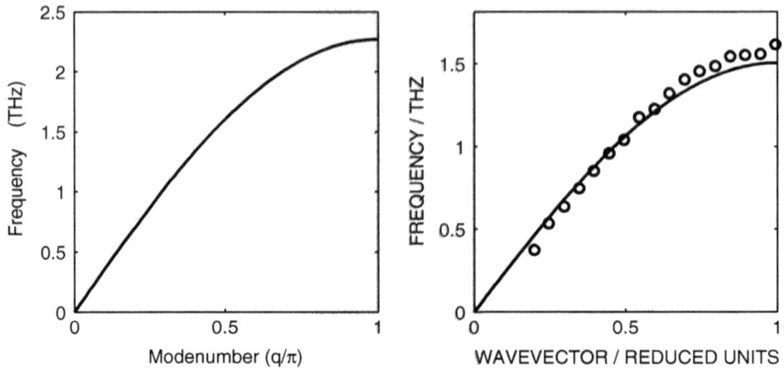

Fig. 3 Phonon dispersion curves obtained in the Coulomb's chain (*right*) and reproduced from Ref. [11] (*left*)

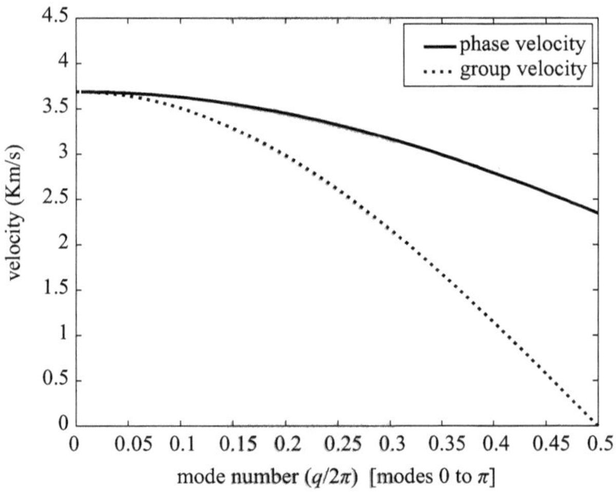

Fig. 4 Phase and group velocities in the Coulomb's chain, both give at $q \to 0$ the sound velocity, which can be compared with the values of 3.4–3.7 km/s obtained in Ref. [14]

The above are standard calculations, we include them for completeness but specially to compare the phonon properties in our model with actual experimental measurements in real samples. In Fig. 3 we plot the dispersion curve as obtained for our model in physical units (left), and compare it with data obtained from neutron spectroscopy and molecular dynamics in Ref. [11] (right). We can see that there is a general agreement in the shape and values. Similar curves have been also obtained in Refs. [12, 13].

Figure 4 shows the phase and group velocities in physical units. Note that the predicted speed of sound is similar to that found in Ref. [14] of 3.4–3.7 km/s.

Since we are using a very simplified model, it may result surprising that it gives values similar to much more complex models and experimental data. The reason is that in our systems there are many real things: the atomic mass, the Coulomb interaction, and some external potentials that keep the chain at equilibrium at the real interatomic distance. This also supports our assumption that at least for small perturbations the dominant interaction is the Coulombian one.

4 Localized Travelling Solutions

In this section we consider different types of localized solutions supported by the Coulomb lattice. Although these solutions are large enough to be intrinsically nonlinear, they also have tails of small amplitude which abide the linear equation. Tails cannot exist by themselves but depends on the existence of the nonlinear solution, however they also exist and its study gives information about the *parent* excitation. We obtain first the tail properties and later we consider the fully nonlinear solutions, kinks and solitons. Tails are obtained with the linear system, therefore their properties are common to any system with the same linearization. However, they only give the conditions that the tails have to fulfill. The properties of existence of the nonlinear excitation, its velocity, amplitude and other characteristics have to be worked out with the whole nonlinear system.

4.1 Tail Analysis

Consider a nonlinear localized moving excitation, travelling with velocity V. We also assume that its amplitude decays exponentially far away enough from the center. Tails are well described by the expression

$$u_n = \exp(-\xi(n - Vt))\exp(\mathrm{i}(qn - \omega t)), \tag{8}$$

which is valid for $n > Vt$ if $\xi > 0$ corresponding to the front tail of the structure and for $n < Vt$ if $\xi < 0$ corresponding to the back tail. We call $|\xi|$ the localization parameter and its inverse $\Lambda = 1/|\xi|$ the decay length. The limit case $\xi = 0$ corresponds to extended phonons. Since the tail amplitude is by definition small, it satisfies the linear equation (5). Using that $u_{n+1} = \exp(-\xi + \mathrm{i}q)u_n$, $u_{n-1} = \exp(\xi - \mathrm{i}q)u_n$ and $\dot{u}_n = (\xi v - \mathrm{i}\omega)u_n$, substitution of u_n yields the following two equations for the real and imaginary part of Eq. (5):

$$\xi^2 V^2 - \omega^2 = 2\mathrm{c}^2(\cosh(\xi)\cos(q) - 1) \quad ; \quad \xi V\omega = \mathrm{c}^2\sinh(\xi)\sin(q). \tag{9}$$

As there are two equation with four unknowns, we can express two of them as ω and V as functions of the other two ξ and q. First, the frequency ω is given by:

$$\omega = \cosh(\xi/2)\omega_M \sin(q/2) = \cosh(\xi/2)\omega_{\text{phonons}}, \tag{10}$$

with $\omega_M = 2c$. Note that Eq. (10) reduces to the phonon dispersion equation when there is no localization, i.e. $\xi \to 0$.

The velocity of the tail is given by:

$$V = \frac{\sinh(\xi/2)}{\xi/2} c \cos(q/2) = \frac{\sinh(\xi/2)}{\xi/2} V_{\text{g,phonons}} \,. \tag{11}$$

Note that both $\sinh(\xi/2)/(\xi/2)$ and $\cosh(\xi/2)$ are larger than the unity and also monotonically increasing functions in $|\xi|$, which means that the tails oscillate and move faster than the phonons. Except for the degenerate case $\xi = 0$, which corresponds to no localization and no tail, i.e., to phonons.

4.2 Oscillating Tails

Breathers are localized oscillating solutions with a frequency outside the phonon band, but there are neither stationary nor moving breathers in this system, a fact known for repulsive FPU systems [15]. However, the tail analysis can be applied to oscillating tails following or preceding a soliton or kink, which do exist in this system.

Travelling tails correspond to localized, vibrating, travelling excitations, that is ξ, V and ω are all different from zero. From Eq. (10) we conclude that the tail frequency ω increases with the localization ξ. However, being inside the phonon band, i.e., $0 < \omega < \omega_M = 2c$, the interaction with the phonons would produce the tail instability. The mode $q = \pi$ would be the most stable, as its frequency is always larger than ω_M, but its velocity is zero. However, increasing the localization ξ we can increase both the frequency and the velocity. However, values of ξ larger than a few units are not realistic, neither values of the frequency of many times the maximum phonon frequency. So a kind of balance is needed. One notable particular case, introduced in Ref. [7], allows analytical estimations (see next section). It corresponds to solutions with $q = 2\pi/3$, known as the *magic number*. For this case we obtain

$$\omega = \sqrt{3}c \cosh(\xi/2) \quad ; \quad V = \frac{c \sinh(\xi/2)}{\xi} \,. \tag{12}$$

Some estimations can be done. If $\omega > 2c$ then $\xi > \xi_0 = 1.0986$, for $\xi = \xi_0$ each atom amplitude is exactly a third smaller than the previous one. For $\xi = \xi_0$, the oscillating tail velocity is $V \approx c/3$ or one third of the sound velocity. We need values of about $\xi = 6.5$ to approach the sound velocity, but at such strong localization value the frequency is 11 times the maximum phonon frequency and a particle has an amplitude one thousand smaller than the previous one. Since such conditions are unrealistic, we can, in principle, consider that oscillating tails

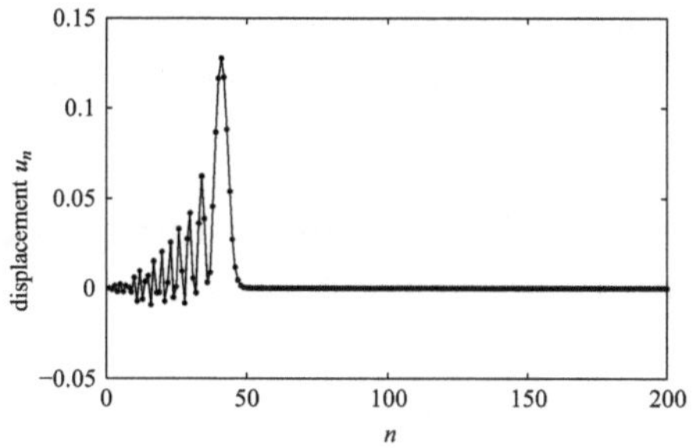

Fig. 5 Profile of the transitory state of a soliton, the subsonic oscillatory back tail will be left behind the supersonic soliton. The front tail becomes so steep that, in practice, it becomes an abrupt end of the perturbation. Therefore, the soliton will transform to a kink and the back tail without the support of the soliton will be dispersed in phonons. Dimensionless units

are subsonic. As we will see below, our simulations show that the most common behavior is an oscillating tail that is left behind the supersonic soliton bringing about a kink.

4.3 Supersonic Solitons and Kinks

Solitons are localized, travelling non oscillatory excitations, that is with $V \neq 0$ and $\omega = 0$. From Eq. (10) it follows that $\sin(q/2) = 0$ or $q = 0$ (solitons are in-phase modes). Substituting into Eq. (11) we obtain the soliton velocity as a function of the localization ξ:

$$V_{\text{sol}} = \frac{\sinh(\xi/2)}{\xi/2} \text{c} \,. \tag{13}$$

Two important conclusions arise: (a) Solitons do not have an internal structure, as they come from the mode $q = 0$ all the atoms are in–phase, and, the minimum value of the velocity of a soliton is c the sound velocity, which is also the maximum phonon velocity. Therefore, the solitons, differently from oscillating tails, are supersonic.

As commented above it is possible to have different tail solutions for the front and back. For example, in principle, it is possible a front soliton, followed by a simple translation $x_n = a$, that would be a kink which leaves behind the atoms displaced one lattice unit, or the back tail may be an oscillatory tail, travelling at subsonic speed, being left behind by the supersonic soliton at the front as shown in Fig. 5. The front tail needs to be so steep to move in front of the soliton that it in practice means that the front of the soliton finishes abruptly, and the soliton becomes a kink. Of course, if these properties hold requires the study of the full system with the full potentials, which will be done in the next sections, but it is also essential to know the behavior of the system at low amplitudes, because part of the system correspond to that description when there is a localized nonlinear solution.

5 Kink Characteristics Within the Rotating Wave Approximation: Magic Wave Number

The previous tail analysis was derived from the linear equation (5). Is thus generic for many systems of coupled oscillators, although the specific values in physical units are specific for our system, as the phonon frequencies and speeds. We can obtain properties of the full nonlinear system using the Rotating Wave Approximation (RWA) [16], which consists in an approximation of the Fourier series up to the first harmonic.

It is convenient to define a new variable, the relative displacement or strain of the chain, as $v_n = u_n - u_{n-1}$. From Eq. (3), we can easily obtain the evolution equation for the strain as

$$\ddot{v}_n = 2F_n - F_{n+1} - F_{n-1}; \quad \text{with} \quad F_n = \frac{1}{(1+v_n)^2}. \tag{14}$$

Kinks are characterized by a very steep profile, with only a few particles moving at a given time. A solution presenting this behavior is given (in the strain representation) by a half-wavelength structure with the *magic* wave number $q = 2\pi/3$. Following Ref. [7], we propose the following ansatz, describing a kink travelling to the right

$$v_n = -\frac{A}{2}(1 + \cos(qn - \omega t)) \quad \text{if} \quad -\pi < qn - \omega t < \pi \tag{15}$$

and $v_n = 0$ otherwise. Equation (15) represents a solution where three particles (the kink core) are in motion, while the others remain at rest. Note that this ansatz is only an approximation, although very good for intermediate to large amplitudes as it will be shown below.

When we substitute (15) in the equation of motion (14), nonlinearity generates multiple frequencies. The RWA approximation states that nonlinear contributions higher than second harmonics can be neglected. This implies that the force terms can be approximated by their first-order Fourier expansions, as $F_n = a_0 + a_1 \cos(qn - \omega t)$. On the other hand, the left high side of Eq. (14) gives, after substitution of Eq. (15), $\ddot{v}_n = (A/2)\omega^2 \cos(qn - \omega t)$ which is equal to the first harmonic term of the right hand side. The zero-harmonic cancels out due to the dependency of the rhs on F_n. The first harmonic of F_n is given by:

$$a_1 = \frac{1}{\pi} \int_{-\pi}^{\pi} \frac{\cos(\theta) d\theta}{(1 - \frac{A}{2} - \frac{A}{2}\cos(\theta))^2}. \tag{16}$$

For $A < 1$ we obtain

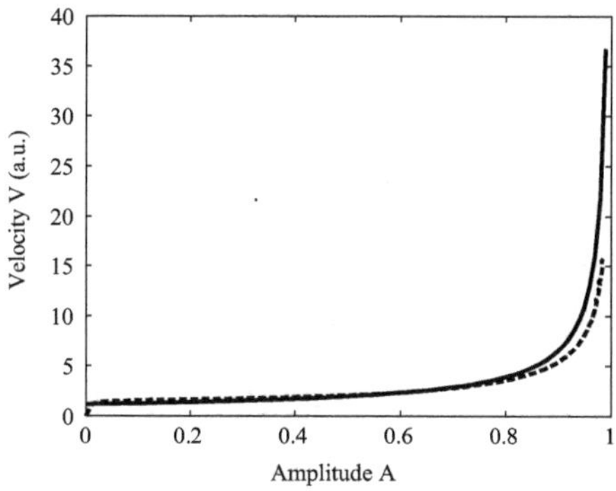

Fig. 6 Comparison of the numerical (*dashed line*) and theoretical (*continuous line*) velocities of kinks within the RWA, for the magic wave number $q = 2\pi/3$. Dimensionless units

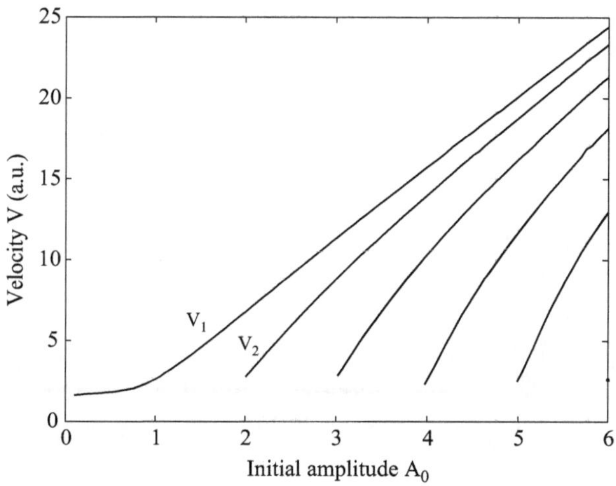

Fig. 7 Velocity and number of kinks as a function of the initial excitation amplitude A_0. Dimensionless units

$$a_1 = \frac{A}{(1-A)^{3/2}}. \tag{17}$$

Therefore, substituting $\ddot{v}_n$ and $F_n = a_0 + a_1 \cos(qn - \omega t)$ in Eq. (14), is straightforward to obtain the frequency and velocity, as functions of the amplitude A

$$\omega = \frac{1}{(1-A)^{3/4}}\,\omega_M \sin(q/2); \quad V = \frac{\omega}{q} = \frac{1}{(1-A)^{3/4}}\,\mathrm{c}\,\frac{\sin(q/2)}{q/2} \tag{18}$$

These magnitudes are equal to the corresponding ones for phonons, multiplied by a factor $1/(1-A)^{3/4}$, which of course tend to the unity when A is small, but that diverges when A approaches to unity. We can see the kink velocities in Fig. 6.

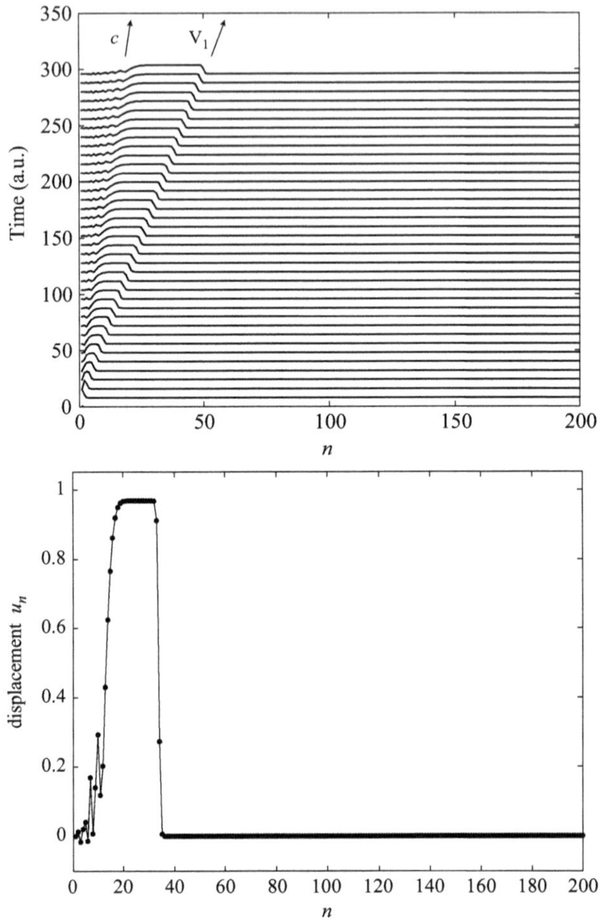

Fig. 8 *Top*: Profile and evolution of a single kink for $A_0 = 1.05$. *Bottom*: Detail of the displacements at $t = 200$

6 Numerical Results

We have done a systematic numerical study to determine the kink properties. The method to excite kinks is described in Ref. [7]. It consists in perturbing sinusoidally the first particle in the chain during half a period, with an amplitude A_0, so that the initial and final positions are both zero. Increasing the amplitude of the perturbation A_0 the velocity of the kink increases, and eventually, two, three, and any number of kinks are produced, travelling one after the other. The dependency of the velocity and number of kinks with the initial amplitude is shown in Fig. 7.

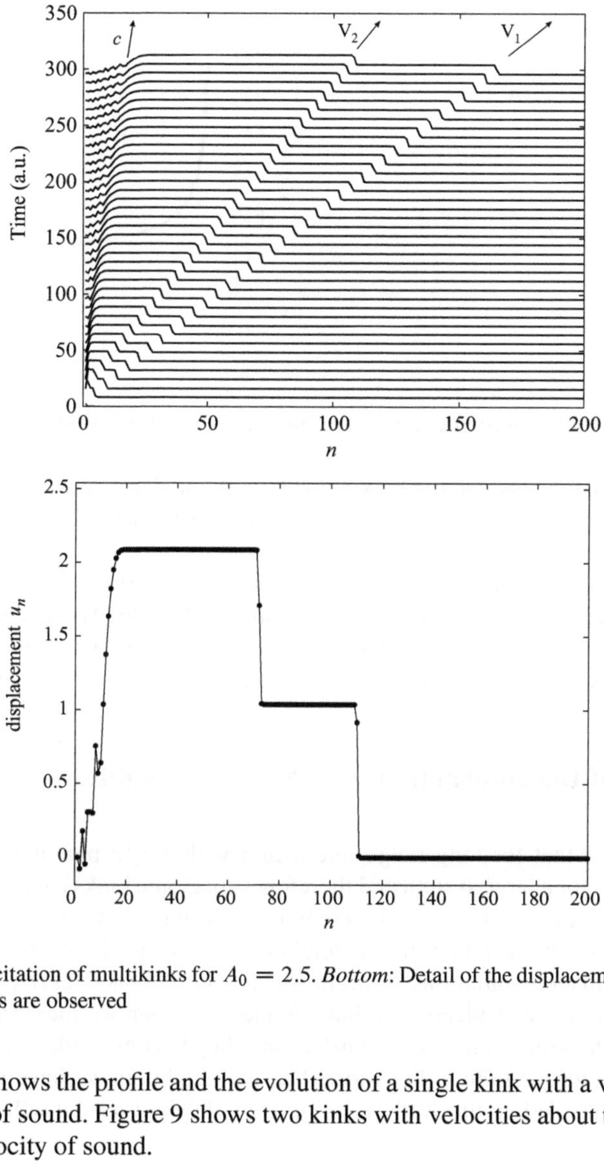

Fig. 9 *Top*: excitation of multikinks for $A_0 = 2.5$. *Bottom*: Detail of the displacements at $t = 200$ where two kinks are observed

Figure 8 shows the profile and the evolution of a single kink with a velocity twice the velocity of sound. Figure 9 shows two kinks with velocities about three and five times the velocity of sound.

The kinks rebound at the borders and when they encounter each other. They are composed of particles moving in the direction of the kink and particles travelling in opposite directions cannot cross because that would need an infinite potential energy.

The velocity of the kinks V increases with their amplitude A, tending to infinity with an asymptote at $A = 1$. Note that $A = 1$ would mean that the distance between particles $d_n = 1 + v_n \to 0$ which means an infinite energy. We have already

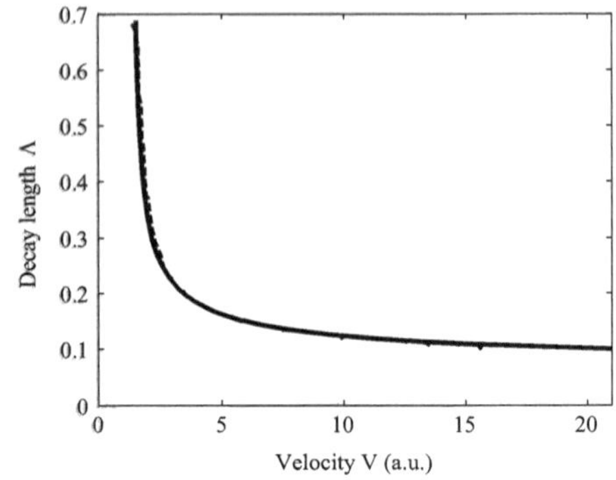

Fig. 10 Comparison of the numerical (*dashed line*) and theoretical (*continuous line*) decay lengths as a function of the velocity

commented the good agreement with the numerical results with the theory shown in Fig. 6.

The relationship between the kink velocity with the decay length $\Lambda = 1/\xi$ given in Eq. (11) is shown in Fig. 10 compared with the numerical results, showing a perfect agreement.

In physical units, kinks can have any velocity and energy, for example we have found velocities up to 27 km/s or eight times the sound velocity and energies from a few eV to hundreds of them. Whether these kinks exist and with which energies in mica muscovite remains and open question, but they seem a very good candidate.

7 Effect of the Interaction with Several Neighbors

It is well know that for long-range interaction with a $1/r$ potential the speed of sound diverges for a large system and therefore supersonic kinks cannot exist. Sound propagation is restored due to the screening of Coulomb's law by the electrons in a solid. It is not clear how many neighbors is reasonable to take into account and we have made simulations in order to find out if kinks survive when longer range interaction is considered. We have found that when we increase the number of neighbors the kinks velocity diminishes and they become wider, with more than three particles forming them. For more than ten neighbors we have not been able to produce kinks. When the study is completed, detailed results will be published elsewhere.

8 Conclusions

In minerals like mica muscovite there exist layers of repulsive ions all with the same charge sign forming a two-dimensional lattice. Muscovite shows the striking property of having been able to record the tracks of energetic particles as positrons and muons, but also shows tracks of what seem lattice localized excitations called quodons. Also an experiment seems not to have any other explanation. Motivated by the fact that an essential, but disregarded, characteristic of the K^+ lattice is that all the particles repel each other, we study a minimal system made out of repulsive particles. We have found both theoretically and numerically that solitons in a transitory state and supersonic kinks and multikinks can travel very fast and thus are an efficient means of transporting localized energy along the lattice. The study of more realistic extensions of this model are underway and will be reported elsewhere.

Acknowledgements JFRA acknowledges financial support from the project FIS2008-04848. The work of VSM and LMGR was also supported by projects FIS2011-29731-C02-02 and MAT2009-09438 respectively. All projects are granted by the Spanish Ministerio de Ciencia e Innovación. All authors acknowledges Mike Russell for continuous discussions.

References

1. L.A. Pérez-Maqueda, F. Franco, M.A. Avilés, J. Poyato, J.L. Pérez-Maqueda, Effect of sonication on particle-size distribution in natural muscovite and biotite. Clays Clay Miner. **51**, 701–708 (2003)
2. K. Kroneberger, M. Schosnig, F.M. Russell, K.O. Groeneveld, Search for solitons in solids. Radiat. Meas. **23**, 209–213 (1994)
3. F.M. Russell, J.C. Eilbeck, Evidence for moving breathers in a layered crystal insulator at 300 K. Europhys. Lett. **78**, 10004 (2007)
4. J.F.R. Archilla, J. Cuevas, M.D. Alba, M. Naranjo, J.M. Trillo, Discrete breathers for understanding reconstructive mineral processes at low temperatures. J. Phys. Chem. B **110**(47), 24112–24120 (2006)
5. J.F.R. Archilla, J. Cuevas, F.R. Romero, Effect of breather existence on reconstructive transformations in mica muscovite. AIP Conf. Proc. **982**(1), 788–791 (2008)
6. V.I. Dubinko, P.A. Selyshchev, J.F.R. Archilla, Reaction-rate theory with account of the crystal anharmonicity. Phys. Rev. E **83**(4), 13 (2011)
7. Yu.A. Kosevich, R. Khomeriki, S. Ruffo, Supersonic discrete kink-solitons and sinusoidal patterns with *magic* wave number in anharmonic lattices. Europhys. Lett. **66**, 21–27 (2004)
8. E. Wigner, On the interaction of electrons in metals. Phys. Rev. Lett. **46**, 1002–1011 (1934)
9. K.A. Matveev, A.V. Andreev, M. Pustilnik, Equilibration of a one-dimensional Wigner crystal. Phys. Rev. Lett. **105**, 046401 (2010)
10. C.C. Grimes, G. Adams, Evidence for a liquid-to-crystal phase transition in a classical, two-dimensional sheet of electrons. Phys. Rev. Lett. **42**, 795–798 (1979)
11. D.R. Collins, W.G. Stirling, C.R.A. Catlow, G. Rowbotham, Determination of acoustic phonon dispersion curves in layer silicates by inelastic neutron scattering and computer simulation techniques. Phys. Chem. Miner. **19**, 520–527 (1993)
12. N. Wada, W.A. Kamitakahara, Inelastic neutron and Raman scattering studies of muscovite and vermiculite layered silicates. Phys. Rev. B **43**, 2391 (1991)
13. S.L. Chaplot et al., Inelastic neutron scattering and lattice dynamics of minerals. Eur. J. Miner. **14**, 291 (2002)
14. G. Brudeylins, D. Schmicker, Elastic and inelastic helium atom scattering at a cleaved mica sheet. Surf. Sci. **333**, 237–242 (1995)
15. G. James, P. Kevrekidis, J. Cuevas, Breathers in oscillator chains with hertzian interactions. Physica D **251**, 34–39 (2013)
16. J.B. Page, Asymptotic solutions for localized vibrational modes in strongly anharmonic periodic systems. Phys. Rev. B **41**, 7835 (1990)

Part V
Magnetic Metamaterials

Travelling Waves in Nonlinear Magnetic Metamaterials

J. Diblík, M. Fečkan, M. Pospíšil, V.M. Rothos, and H. Susanto

Abstract In this article, a model of one-dimensional metamaterial formed by a discrete array of nonlinear resonators is considered. The existence and uniqueness results of periodic and asymptotic travelling waves of the system are presented. The existence and the stability of asymptotic waves are also computed and discussed numerically.

J. Diblík
Department of Mathematics, Faculty of Electrical Engineering and Communication, Brno University of Technology, Technická 3058/10, 616 00 Brno, Czech Republic
e-mail: diblik@feec.vutbr.cz

M. Fečkan (✉)
Department of Mathematical Analysis and Numerical Mathematics, Comenius University, Mlynská dolina, 842 48 Bratislava, Slovakia

Mathematical Institute of Slovak Academy of Sciences, Štefánikova 49, 814 73 Bratislava, Slovakia
e-mail: Michal.Feckan@fmph.uniba.sk

M. Pospíšil
Centre for Research and Utilization of Renewable Energy, Faculty of Electrical Engineering and Communication, Brno University of Technology, Technická 3058/10, 616 00 Brno, Czech Republic
e-mail: pospisilm@feec.vutbr.cz

V.M. Rothos
Department of Mathematics, Physics and Computational Sciences, Mathematics Division, Faculty of Engineering, Aristotle University of Thessaloniki, Thessaloniki GR54124, Greece
e-mail: rothos@auth.gr

H. Susanto
School of Mathematical Sciences, University of Nottingham, University Park, Nottingham NG7 2RD, UK
e-mail: Hadi.Susanto@nottingham.ac.uk

R. Carretero-González et al. (eds.), *Localized Excitations in Nonlinear Complex Systems*, Nonlinear Systems and Complexity 7, DOI 10.1007/978-3-319-02057-0_17,

1 Introduction

Metamaterials are artificial materials that are engineered to have properties that may not be found in nature. The modification is achieved by composing inhomogeneities, i.e. structural rather than chemical, to create desirable effective behaviour. Primarily the study is on engineering the refractive index of a material. Since the proposal [21], a new paradigm in electromagnetism has emerged due to these new types of artificial composites, including cloaking devices [22] (see also [18] for a review of recent results for electromagnetic manipulation enabled by metamaterials). The canonical constituent components, or meta-atoms from which metamaterials are fashioned are the split ring resonator, which consists of an inductive metallic ring with a gap to provide capacitance, or its U-shaped modifications.

Previous studies of metamaterials were focused on the linear properties of the medium during wave propagation, in which case magnetic permeability and material permittivity are non-dependent on the intensity of the electromagnetic field. When nonlinearity is present in metamaterials due to either employing a nonlinear host medium [20] or by engineering the elements of a metamaterial with a nonlinear component [13], nontrivial properties and behaviours can be present, i.e. dynamic tunable systems and active artificial media, such as materials that are compressible by magnetic field [15] and devices that can switch from a normal flat mirror to a focusing/defocusing one without changing shape [24] (see also [23] for a review).

The propagation of electromagnetic waves in metamaterials has been discussed and modelled by several types of nonlinear equations, such as a nonlinear Klein-Gordon equation [19], coupled short-pulse equation [28], higher-order nonlinear Schrödinger equations [27] and coupled Klein-Gordon equations [14,17]. The latter equations are of the form

$$\begin{aligned} \frac{dq_n}{dt} &= i_n, \qquad n \in \mathbb{Z}, \\ \frac{d}{dt}(\lambda i_{n-1} - i_n + \lambda i_{n+1}) &= \gamma i_n - f(t) + \varphi(q_n), \end{aligned} \tag{1}$$

with loss coefficient γ, coupling parameter λ, external forcing f and nonlinear function φ. The coupled equations model the dynamics of the capacitor charge of split-ring resonators or U-type elements in magnetic metamaterials [14, 17]. Due to the nonlinearity, localization of electromagnetic waves is possible. For Eq. (1), localized excitations in the form of discrete breathers and domain walls were observed. In this article, we investigate the existence of travelling periodic solutions of system (1) with periodic forcing, and the bifurcation of periodic solutions and asymptotic waves of (1) with small γ, λ and f. We consider in particular the case when the external force also varies in space.

The present paper is organized as follows. In Sect. 2 we are looking for a periodic travelling wave when its amplitude is limited by the magnitude of the forcing. In Sect. 3, we suppose that an unperturbed system (see (21)) has a periodic solution, and we use the subharmonic Melnikov bifurcation method to find conditions under which this periodic solution persists. Homoclinic Melnikov bifurcation method is used in Sect. 4. We obtain large solutions in Sects. 3 and 4 under small perturbations, while in Sect. 2 we obtain a small solution under small perturbation. The main difference is the resonance property, i.e. Sect. 2 is dealing with non-resonances (see condition (9)), while in Sect. 3 there is a resonance assumption (H2). In Sect. 5 we solve the governing equations numerically, in particular for the asymptotic waves. Comparisons with the analytical results are presented where we obtain good agreement.

2 Existence Results on Periodic Solutions

In this section, we study the existence of periodic solutions of system (1) forced by a travelling wave field when the amplitude of periodic solutions is proportional to the amplitude of forcing. Thus, we consider the equivalent equation

$$\lambda \ddot{q}_{n+1} - \ddot{q}_n + \lambda \ddot{q}_{n-1} = \gamma \dot{q}_n + \varphi(q_n) + f \cos(\omega t + pn), \tag{2}$$

where as above $\gamma \geq 0$ is the dissipative loss of the medium, $\lambda \in \mathbb{R}$ is the coupling constant between nearest neighbor resonators, $\omega > 0$ is the external driving frequency, $f \neq 0$ is the amplitude of the external force, $p \neq 0$ is the wavenumber of the travelling wave field and φ is the nonlinearity of the magnetic material, which is normally assumed to be of Kerr-type [17]. Therefore, we assume that $\varphi(z)$ is an odd analytic function in its variable z with radius of convergence $\rho > 0$ such that

$$\mathrm{D}\varphi(0) = 0. \tag{3}$$

Seeking for waves travelling in the same direction as the external drive, we take $q_n(t) = U(z)$, $z = \omega t + pn$, in (2) with $U(z + \pi) = -U(z)$ to obtain

$$\omega^2(\lambda U''(z+p) - U''(z) + \lambda U''(z-p)) = \gamma \omega U'(z) + \varphi(U(z)) + f \cos z. \tag{4}$$

Note that $p \in \mathbb{R}\backslash\{0\}$. Considering the case $p = -\pi$, one will obtain alternating charges between the nearest-neighbor resonators as $f \cos(\omega t + pn) = (-1)^n f \cos \omega t$.

We take Banach spaces

$$W := \left\{ U \in C(\mathbb{R},\mathbb{R}) \mid U(z) = \sum_{k\in\mathbb{Z}} d_k \, \mathrm{e}^{k\imath z}, \sum_{k\in\mathbb{Z}} |d_k| < \infty \right\},$$

$$X := \left\{ U \in C(\mathbb{R},\mathbb{R}) \mid U(z) = \sum_{k\in\mathbb{Z}} c_k \, \mathrm{e}^{(2k+1)\imath z}, \sum_{k\in\mathbb{Z}} |c_k| < \infty \right\},$$

$$Y := \left\{ U \in C^1(\mathbb{R},\mathbb{R}) \mid U(z) = \sum_{k\in\mathbb{Z}} c_k \, \mathrm{e}^{(2k+1)\imath z}, \sum_{k\in\mathbb{Z}} |2k+1||c_k| < \infty \right\},$$

$$Z := \left\{ U \in C^2(\mathbb{R},\mathbb{R}) \mid U(z) = \sum_{k\in\mathbb{Z}} c_k \, \mathrm{e}^{(2k+1)\imath z}, \sum_{k\in\mathbb{Z}} (2k+1)^2|c_k| < \infty \right\}$$

with the norms

$$\|U\| := \sum_{k\in\mathbb{Z}} |d_k|, \quad \|U\| := \sum_{k\in\mathbb{Z}} |c_k|,$$

$$\|U\|_1 := \sum_{k\in\mathbb{Z}} |2k+1||c_k|, \quad \|U\|_2 := \sum_{k\in\mathbb{Z}} (2k+1)^2|c_k|,$$

respectively. It is easy to verify that $Z \circlearrowleft Y \circlearrowleft X$ are compact embeddings, $X \subset W$ and $\|U\| \le \|U\|_1 \; \forall U \in Y$, $\|U\|_1 \le \|U\|_2 \; \forall U \in Z$.

The following lemma is clear.

Lemma 1. *If $U_1, U_2 \in W$ then $U_1U_2 \in W$ and $\|U_1U_2\| \le \|U_1\|\|U_2\|$. For each $k \in \mathbb{N}$, if $U_1, U_2, \ldots, U_{2k+1} \in X$ then $U_1U_2\ldots U_{2k+1} \in X$.*

By setting

$$\mathcal{K}U := \omega^2(\lambda U''(z+p) - U''(z) + \lambda U''(z-p)) - \gamma\omega U'(z),$$
$$\mathcal{F}(U, f) := \varphi(U) + f \cos z,$$

Eq. (4) has the form

$$\mathcal{K}U = \mathcal{F}(U, f).$$

Denote $B(\rho) := \{U \in Y \mid \|U\|_1 < \rho\}$.

We have the next result.

Lemma 2. *Function $\mathcal{F} : B(\rho) \times \mathbb{R} \to X$ fulfils*

$$\|\mathcal{F}(U, f)\| \le \sum_{k=3}^{\infty} \frac{|\mathrm{D}^k\varphi(0)|}{k!} \|U\|_1^k + |f|, \tag{5}$$

$$\|\mathcal{F}(U_1, f) - \mathcal{F}(U_2, f)\| \le \|U_1 - U_2\|_1 \sum_{k=3}^{\infty} \frac{|\mathrm{D}^k \varphi(0)|}{k!} \sum_{j=0}^{k-1} \|U\|_1^j \|U_2\|_1^{k-j-1}, \quad (6)$$

$$\|\mathcal{F}(U, f_1) - \mathcal{F}(U, f_2)\| \le |f_1 - f_2| \quad (7)$$

for any $U, U_1, U_2 \in B(\rho) \subset Y$ *and* $f, f_1, f_2 \in \mathbb{R}$.

Proof. Since $\cos z = \frac{\mathrm{e}^{\imath z} + \mathrm{e}^{-\imath z}}{2}$, we get $\|\cos z\| = 1$ and (7) easily follows. By applying Lemma 1 and estimating the Taylor series

$$\|\varphi(U)\| \le \sum_{k=3}^{\infty} \frac{|\mathrm{D}^k \varphi(0)|}{k!} \|U\|_1^k$$

we obtain (5). Using similar estimation for $\|\varphi(U_1) - \varphi(U_2)\|$ and the identity

$$a^k - b^k = (a - b)(a^{k-1} + a^{k-2} b + \cdots + b^{k-1})$$

for any $a, b \in \mathbb{C}$, we arrive at (6).

Now, if $U \in Z$ with $U(z) = \sum_{k \in \mathbb{Z}} c_k\, \mathrm{e}^{(2k+1)\imath z}$ then

$$\mathcal{K}U(z) = \sum_{k \in \mathbb{Z}} \left[-\omega^2 (2k+1)^2 \,(2\lambda \cos(2k+1)p - 1) - \imath \gamma \omega (2k+1) \right] c_k\, \mathrm{e}^{(2k+1)\imath z} \quad (8)$$

and so $\mathcal{K} \in L(Z, X)$ with

$$\|\mathcal{K}\|_{L(Z,X)} \le \omega^2 (1 + 2|\lambda|) + \gamma \omega.$$

If

$$\Theta := \inf_{k \in \mathbb{Z}} \sqrt{\omega^4 (2k+1)^2 \,(2\lambda \cos(2k+1)p - 1)^2 + \gamma^2 \omega^2} > 0 \quad (9)$$

is a constant depending on γ, λ, ω and p, then we also have $\mathcal{K}^{-1} \in L(X, Z) \subset L(X, Y) \subset L(X)$. So $\mathcal{K}^{-1} : X \to Y$ is compact such that

$$\|\mathcal{K}^{-1}\|_{L(X,Y)} \le \frac{1}{\Theta}. \quad (10)$$

Now we can prove the following existence results on (4) when all parameters except f are fixed.

Theorem 1. *Assume* (9) *along with*

$$|f| < |f_l| \quad (11)$$

for f_l satisfying

$$A(r) := \Theta r - \sum_{k=3}^{\infty} \frac{|\mathrm{D}^k \varphi(0)|}{k!} r^k = |f_l|, \tag{12}$$

$$\mathrm{D}A(r) = 0 \tag{13}$$

for some $r \in (0, \rho)$. Then Eq. (4) *has a unique solution $U(f) \in \overline{B(\rho_f)}$ in a closed ball where $\rho_f < \rho$ is the smallest positive root of $A(r) = |f|$. Moreover, $U(f)$ can be approximated by an iteration process. Finally, it holds*

$$\|U(f_1) - U(f_2)\|_1 \leq \frac{|f_1 - f_2|}{\Theta - \sum_{k=3}^{\infty} \frac{|\mathrm{D}^k \varphi(0)|}{(k-1)!} \rho_{\max\{|f_1|,|f_2|\}}^{k-1}} \tag{14}$$

for any $f_1, f_2 \in \mathbb{R}$ satisfying (11).

Proof. We rewrite (4) as a parameterized fixed point problem

$$U = \mathcal{R}(U, f) := \mathcal{K}^{-1} \mathcal{F}(U, f)$$

in $B(\rho) \subset Y$. We already know that $\mathcal{R} : B(\rho) \times \mathbb{R} \to Y$ is compact, continuous and by (5), (10) such that

$$\|\mathcal{R}(U, f)\|_1 \leq \frac{1}{\Theta} \left(\sum_{k=3}^{\infty} \frac{|\mathrm{D}^k \varphi(0)|}{k!} \|U\|_1^k + |f| \right).$$

Next, if there is $0 < \rho_f < \rho$ such that

$$A(\rho_f) = |f|, \tag{15}$$

then $\mathcal{R}(\cdot, f)$ maps $\overline{B(\rho_f)}$ into itself. So it remains to study (15). In order to find the largest f_l for which (15) has a solution $\rho_{f_l} > 0$, we need to solve $A(r) = |f_l|$ together with (13) for $r \in (0, \rho)$. This implies (11). Note that

$$\pm \mathrm{D}A(r) = \pm \Theta \mp \sum_{k=3}^{\infty} \frac{|\mathrm{D}^k \varphi(0)|}{(k-1)!} r^{k-1} > \mathrm{D}A(\rho_{f_l}) = 0$$

for $r \in (0, \rho)$, $\pm r < \pm \rho_{f_l}$, and $\lim_{r \to \rho^-} \mathrm{D}A(r) = -\infty$ (see Sects. 7.21, 7.22 and 7.31 of [26]). Hence ρ_{f_l} is uniquely determined by (13). Moreover, continuity of $A(r)$ with $A(0) = 0$, $A(\rho_{f_l}) = |f_l|$ yield that $0 < \rho_f < \rho_{f_l}$ whenever $0 < |f| < |f_l|$, i.e. $\mathrm{D}A(\rho_f) > 0$. So assuming (9), (11) and by (3), we know that (15) has a positive solution $\rho_f < \rho$. We take the smallest one. So $\mathcal{R}(\cdot, f)$ maps $\overline{B(\rho_f)}$ into itself and, moreover, by (6), (10)

$$\|\mathcal{R}(U_1, f) - \mathcal{R}(U_2, f)\|_1 \le \frac{\|U_1 - U_2\|_1}{\Theta} \sum_{k=3}^{\infty} \frac{|\mathrm{D}^k \varphi(0)|}{(k-1)!} \rho_f^{k-1}$$

for any $U_1, U_2 \in \overline{B(\rho_f)}$. Hence, $\mathcal{R}(\cdot, f)$ is a contraction on $\overline{B(\rho_f)}$ with a contraction constant

$$\frac{1}{\Theta} \sum_{k=3}^{\infty} \frac{|\mathrm{D}^k \varphi(0)|}{(k-1)!} \rho_f^{k-1} = \frac{\Theta - \mathrm{D}A(\rho_f)}{\Theta} < \frac{\Theta - \mathrm{D}A(\rho_{f_l})}{\Theta} = 1.$$

The proof of the existence and uniqueness is finished by the Banach fixed point theorem [2]. Next, let $f_1, f_2 \in \mathbb{R}$ satisfy (11), then $U(f_i) \in \overline{B(\rho_{f_i})} \subset \overline{B(\rho_{f_3})}$ for $i = 1, 2$ and $f_3 := \max\{|f_1|, |f_2|\}$. Note f_3 satisfies (11). By (6), (7) and (10), we derive

$$\begin{aligned} \|U(f_1) - U(f_2)\|_1 &= \|\mathcal{R}(U(f_1), f_1) - \mathcal{R}(U(f_2), f_2)\|_1 \\ \le \|\mathcal{R}(U(f_1), f_1) - \mathcal{R}(U(f_2), f_1)\|_1 &+ \|\mathcal{R}(U(f_2), f_1) - \mathcal{R}(U(f_2), f_2)\|_1 \\ \le \frac{\|U(f_1) - U(f_2)\|_1}{\Theta} &\sum_{k=3}^{\infty} \frac{|\mathrm{D}^k \varphi(0)|}{(k-1)!} \rho_{f_3}^{k-1} + \frac{|f_1 - f_2|}{\Theta} \end{aligned}$$

which implies (14).

Theorem 2. *Let* (9) *be fulfilled and f_l satisfies* (12), (13) *for some $r \in (0, \rho)$. Then Eq.* (4) *has a solution $U(f_l) \in \overline{B(\rho_{f_l})}$ in a closed ball where $\rho_{f_l} < \rho$ is the smallest positive root of $A(r) = |f_l|$.*

Proof. We already know that $\mathcal{R}(\cdot, f_l)$ maps $\overline{B(\rho_{f_l})}$ into itself, and $\mathcal{R}(\cdot, f_l)$ is compact. The Schauder fixed point theorem implies the result.

Remark 1. In the following special cases, (9) holds and we can replace Θ with the corresponding Θ_i in the above considerations:

1. If $\gamma > 0$ then $\Theta \ge \Theta_1 := \gamma\omega > 0$.
2. If $|\lambda| < 1/2$ then

$$\Theta \ge \inf_{k \in \mathbb{Z}} \omega^2 |2k+1| (1 - 2\lambda \cos(2k+1)p) \ge \omega^2 (1 - 2|\lambda|) =: \Theta_2 > 0.$$

3. If $|\lambda| > 1/2$, $\mu := \arccos \frac{1}{2\lambda}$, we can apply the identity

$$\cos x - \cos y = -2 \sin \frac{x+y}{2} \sin \frac{x-y}{2}$$

to obtain

$$2\lambda \cos(2k+1)p - 1 = -4\lambda \sin \frac{(2k+1)p + \mu}{2} \sin \frac{(2k+1)p - \mu}{2}.$$

Next, if $\nu := \frac{\mu}{p} \in 2\mathbb{Z}$ and p has the form $p = \frac{2p_1+1}{p_2}\pi$ with $2p_1 + 1$ and p_2 relatively prime integers (their only common divisor is 1), then we can write

$$|2\lambda \cos(2k+1)p - 1| = 4|\lambda||\sin K_+(k)||\sin K_-(k)| \tag{16}$$

where $K_\pm(k) := \frac{(2k+1\pm\nu)(2p_1+1)\pi}{2p_2}$. So it holds

$$K_\pm(k) = \frac{(\text{odd number})\pi}{\text{even number}} \notin \pi\mathbb{Z}$$

for each $k \in \mathbb{Z}$. Moreover, note that $\sin K_\pm(k+p_2) = -\sin K_\pm(k)\ \forall k \in \mathbb{Z}$. Therefore, it is sufficient to take in (9) inf for k from a set of only $|p_2|$ subsequent integers, i.e.

$$\begin{aligned} \Theta &\geq \inf_{k\in\mathbb{Z}} \omega^2|2k+1||2\lambda\cos(2k+1)p-1| \\ &\geq 4\omega^2|\lambda| \min_{k=1,\dots,|p_2|} |\sin K_+(k)||\sin K_-(k)| =: \Theta_3 > 0. \end{aligned} \tag{17}$$

4. If $|\lambda| > 1/2$ and $\nu := \frac{\mu}{p} \in \mathbb{R}\backslash\mathbb{Q}$ for $\mu := \arccos\frac{1}{2\lambda}$, $p = \frac{p_1}{p_2}\pi$ with relatively prime $p_1, p_2 \in \mathbb{Z}\backslash\{0\}$, then we get equality (16) where

$$K_\pm(k) := \frac{(2k+1\pm\nu)p_1\pi}{2p_2} \notin \pi\mathbb{Z} \tag{18}$$

for each $k \in \mathbb{Z}$. Furthermore, $\sin K_\pm(k+p_2) = (-1)^{p_1}\sin K_\pm(k)\ \forall k \in \mathbb{Z}$. Hence, (17) holds with $K_\pm(k)$ given by (18).

Remark 2. If (3) does not hold, i.e., $\mathrm{D}\varphi(0) \neq 0$ then we include $-\mathrm{D}\varphi(0)U(z)$ into $\mathcal{K}$ to get:

$$\tilde{\mathcal{K}}U := \omega^2(\lambda U''(z+p) - U''(z) + \lambda U''(z-p)) - \gamma\omega U'(z) - \mathrm{D}\varphi(0)U(z).$$

So if $U(z) = \sum_{k\in\mathbb{Z}} c_k\, \mathrm{e}^{(2k+1)\imath z}$ then

$$\tilde{\mathcal{K}}U(z) = \sum_{k\in\mathbb{Z}} \left[-\omega^2(2k+1)^2\,(2\lambda\cos(2k+1)p-1) - \imath\gamma\omega(2k+1) - \mathrm{D}\varphi(0)\right] c_k\, \mathrm{e}^{(2k+1)\imath z}.$$

Hence by supposing

$$\tilde{\Theta} := \inf_{k\in\mathbb{Z}} \sqrt{\left(\omega^2(2k+1)(2\lambda\cos(2k+1)p-1) - \frac{\mathrm{D}\varphi(0)}{2k+1}\right)^2 + \gamma^2\omega^2} > 0$$

we get that $\tilde{\mathcal{K}}^{-1} : X \to Y$ is compact such that

$$\|\tilde{\mathcal{K}}^{-1}\|_{L(X,Y)} \leq \frac{1}{\tilde{\Theta}}.$$

So we can follow our method.

3 Bifurcation Results for Periodic Travelling Waves

In this section, we consider (1) with small γ, λ and f. So we consider the system

$$\begin{aligned} \frac{dq_n}{dt} &= i_n, \quad n \in \mathbb{Z} \\ \frac{d}{dt}\left(\varepsilon\lambda i_{n-1} - i_n + \varepsilon\lambda i_{n+1}\right) &= \varepsilon\gamma i_n - \varepsilon h(\omega t + pn) + \varphi(q_n), \end{aligned} \tag{19}$$

for C^2-smooth and 2π-periodic h, $\varphi \in C^2(\mathbb{R}, \mathbb{R})$ and $\omega > 0$, $p \neq 0$, and $\varepsilon \neq 0$ is a small parameter. Equation (19) implies

$$(\varepsilon\lambda\ddot{q}_{n-1} - \ddot{q}_n + \varepsilon\lambda\ddot{q}_{n+1}) = \varepsilon\gamma\dot{q}_n - \varepsilon h(\omega t + pn) + \varphi(q_n). \tag{20}$$

Putting $q_n(t) = U(\omega t + pn)$ for $U \in C^2(\mathbb{R}, \mathbb{R})$ in (20), we get

$$\omega^2 U''(z) + \varphi(U(z)) - \varepsilon\lambda\omega^2\left(U''(z-p) + U''(z+p)\right) + \varepsilon\gamma\omega U'(z) - \varepsilon h(z) = 0. \tag{21}$$

Now, we suppose

(H1) $U''(z) + \varphi(U(z)) = 0$ has a $\overline{T}$-periodic solution U_0.

Remark 3. Since $U_0(-z + c_0)$ also solves $U''(z) + \varphi(U(z)) = 0$ and there is $z_0 \in \mathbb{R}$ such that $U_0'(z_0) = 0$, we may suppose that $U_0(0) = 0$ and then $U_0(z) = U_0(-z)$.

Then

$$U_\omega(z) := U_0(z/\omega)$$

satisfies $\omega^2 U_\omega''(z) + \varphi(U_\omega(z)) = 0$. Note U_ω is $T_\omega := \overline{T}\omega$-periodic and even. We assume the resonance condition

(H2) $T_\omega = 2\pi\frac{u}{v}$ for $u, v \in \mathbb{N}$.

Now, we follow the standard subharmonic Melnikov method [2, 3, 12] to (21) based on the Lyapunov-Schmidt method, but for reader's convenience we present more details. First, we take Banach spaces

$$X := \{U \in C^2(\mathbb{R}, \mathbb{R}) \mid U(z+T) = U(z)\ \forall z \in \mathbb{R}\},$$
$$Y := \{U \in C(\mathbb{R}, \mathbb{R}) \mid U(z+T) = U(z)\ \forall z \in \mathbb{R}\}$$

with the usual maximum norms $\|\cdot\|_2$ and $\|\cdot\|_0$, respectively, where $T := 2\pi u = \overline{T}\omega v$. Then we split

$$Y = Y_1 \oplus Y_2, \quad X = X_1 \oplus X_2$$

with

$$Y_1 := \text{span}[U'_\omega], \quad Y_2 := \left\{U \in Y \mid \int_0^T U(z)U'_\omega(z)dz = 0\right\},$$
$$X_1 := Y_1, \quad X_2 := \left\{U \in X \mid \int_0^T U(z)U'_\omega(z)dz = 0\right\}.$$

Next, we take the projections $P : Y \to Y_1$ and $Q = \mathbb{I} - P : Y \to Y_2$ defined as

$$PU := \frac{\int_0^T U(s)U'_\omega(s)ds}{\int_0^T U'^2_\omega(s)ds} \times U_\omega.$$

Now, we split

$$U(z+\alpha) = U_\omega(z) + V(z), \quad V \in X_2 \tag{22}$$

in (21) to get

$$\begin{aligned}\omega^2 V''(z) + \varphi'(U_\omega(z))V(z) = \varepsilon\lambda\omega^2\left(U''_\omega(z-p) + U''_\omega(z+p)\right) - \varepsilon\gamma\omega U'_\omega(z) + \varepsilon h(z+\alpha)\\ +\varphi(U_\omega(z)) + \varphi'(U_\omega(z))V(z) - \varphi(U_\omega(z) + V(z))\\ +\varepsilon\lambda\omega^2\left(V''(z-p) + V''(z+p)\right) - \varepsilon\gamma\omega V'(z).\end{aligned} \tag{23}$$

Now, we use the following well-known result [3, 12].

Lemma 3. *Equation $\omega^2 V''(z) + \varphi'(U_\omega(z))V(z) = \tilde{Y}(z) \in Y$ has a solution $V \in X$ if and only if $\tilde{Y} \in Y_2$. This solution is unique requiring $V \in X_2$. Moreover, there is a constant $C_\omega > 0$ such that $\|V\|_2 \le C_\omega \|\tilde{Y}\|_0$.*

Lemma 3 allows us to apply the Lyapunov-Schmidt procedure to (23) as follows

$$\omega^2 V''(z) + \varphi'(U_\omega(z))V(z) = QH(z) \tag{24}$$

and

$$PH(z) = 0 \tag{25}$$

for

$$\begin{aligned}H(z) := \varepsilon\lambda\omega^2\left(U_\omega''(z-p)+U_\omega''(z+p)\right)-\varepsilon\gamma\omega U_\omega'(z)+\varepsilon h(z+\alpha)\\ +\varphi(U_\omega(z))+\varphi'(U_\omega(z))V(z)-\varphi(U_\omega(z)+V(z))\\ +\varepsilon\lambda\omega^2\left(V''(z-p)+V''(z+p)\right)-\varepsilon\gamma\omega V'(z).\end{aligned}$$

Using

$$H(z) := O(\varepsilon)+O\left(\|V^2\|_2\right)$$

together with Lemma 3 and the implicit function theorem [2], we can uniquely solve (24) to get $V(z) = V_\varepsilon(z) = O(\varepsilon)$. Putting V_ε in (25) we get

$$\begin{aligned}\int_0^T \Big\{&\lambda\omega^2\left(U_\omega''(z-p)+U_\omega''(z+p)\right)-\gamma\omega U_\omega'(z)+h(z+\alpha)\\ &+\frac{\varphi(U_\omega(z))+\varphi'(U_\omega(z))V_\varepsilon(z)-\varphi(U_\omega(z)+V_\varepsilon(z))}{\varepsilon}\\ &+\lambda\omega^2\left(V_\varepsilon''(z-p)+V_\varepsilon''(z+p)\right)-\gamma\omega V_\varepsilon'(z)\Big\}U_\omega'(z)dz = 0\end{aligned}\tag{26}$$

for $\varepsilon \neq 0$ small. Since

$$\begin{aligned}\frac{\varphi(U_\omega(z))+\varphi'(U_\omega(z))V_\varepsilon(z)-\varphi(U_\omega(z)+V_\varepsilon(z))}{\varepsilon} = O\left(\frac{\|V_\varepsilon\|_2^2}{\varepsilon}\right) = O(\varepsilon),\\ \lambda\omega^2\left(V_\varepsilon''(z-p)+V_\varepsilon''(z+p)\right)-\gamma\omega V_\varepsilon'(z) = O(\varepsilon),\end{aligned}$$

(26) is equivalent to

$$M^{u/v}(\alpha)+O(\varepsilon) = 0$$

for

$$\begin{aligned}M^{u/v}(\alpha) :&= \int_0^T \left(\lambda\omega^2\left(U_\omega''(z-p)+U_\omega''(z+p)\right)-\gamma\omega U_\omega'(z)+h(z+\alpha)\right)U_\omega'(z)dz\\ &= \int_0^T \left(-\gamma\omega U_\omega'(z)+h(z+\alpha)\right)U_\omega'(z)dz,\end{aligned}\tag{27}$$

since U_ω is even. So the small coupling parameter $\varepsilon\lambda$ has no influence in the first order Melnikov function.

Summarizing, we obtain the following result.

Theorem 3. *Suppose (H1) and (H2). If there is a simple zero α_0 of a Melnikov function (27), i.e. $M^{u/v}(\alpha_0) = 0$ and $D_\alpha M^{u/v}(\alpha_0) \neq 0$, then there is a $\delta > 0$ such that for any $0 \neq \varepsilon \in (-\delta, \delta)$ there is a unique $2\pi u$-periodic solution $U(z)$ of (21) with*

$$U(z) = U_0\left(\frac{z-\alpha_0}{\omega}\right) + O(\varepsilon).$$

Proof. The proof follows immediately from the implicit function theorem applied to (26).

Note

$$\begin{aligned} M^{u/v}(\alpha) &= \int_0^T \left(-\gamma\omega U_\omega'(z) + h(z+\alpha)\right) U_\omega'(z)dz \\ &= \frac{1}{\omega}\int_0^{\overline{T}\omega v} \left(-\gamma U_0'(z/\omega) + h(z+\alpha)\right) U_0'(z/\omega)dz \\ &= \int_0^{\overline{T}v} \left(-\gamma U_0'(z) + h(\omega z+\alpha)\right) U_0'(z)dz \\ &= -v\gamma\int_0^{\overline{T}} U_0'(z)^2dz + \int_0^{\overline{T}}\sum_{i=0}^{v-1} h(\omega\overline{T}i + \omega z + \alpha)U_0'(z)dz. \end{aligned}$$

In what follows, we study for simplicity the case with $v = 1$, so then (H2) becomes

(H3) $\omega = \frac{2\pi u}{\overline{T}}$ for $u \in \mathbb{N}$.

Hence

$$\begin{aligned} M^u(\alpha) &= -\gamma\int_0^{\overline{T}} U_0'(z)^2dz + \int_0^{\overline{T}} h\left(\frac{2\pi u}{\overline{T}}z + \alpha\right) U_0'(z)dz \\ &= -\gamma\int_0^{\overline{T}} U_0'(z)^2dz - \frac{2\pi u}{\overline{T}}\int_0^{\overline{T}} h'\left(\frac{2\pi u}{\overline{T}}z + \alpha\right) U_0(z)dz. \end{aligned}$$

Example 1. To illustrate the theory, we consider $\varphi(U) = U + U^3$ and then equation from (H1) is the Duffing equation

$$U''(z) + U(z) + U^3(z) = 0 \tag{28}$$

possessing a family of periodic solutions

$$U_{0,a}(z) = a\,\mathrm{cn}\left(\sqrt{1+a^2}z\right)$$

for $a > 0$ with periods $\overline{T} = \overline{T}(a) = \frac{4K(k)}{\sqrt{1+a^2}}$, $k = \frac{a}{\sqrt{2+2a^2}}$. Note $U_{0,a}(0) = a$ and $U_{0,a}'(0) = 0$. Here cn is the Jacobi elliptic function, $K(k)$ is the complete elliptic function of the first kind and k is the elliptic modulus [16]. Moreover, we have

$$\overline{T}'(a) = \frac{8(E(k) - K(k)) - 4a^2K(k)}{a\sqrt{1+a^2}\,(2+a^2)} < 0,$$

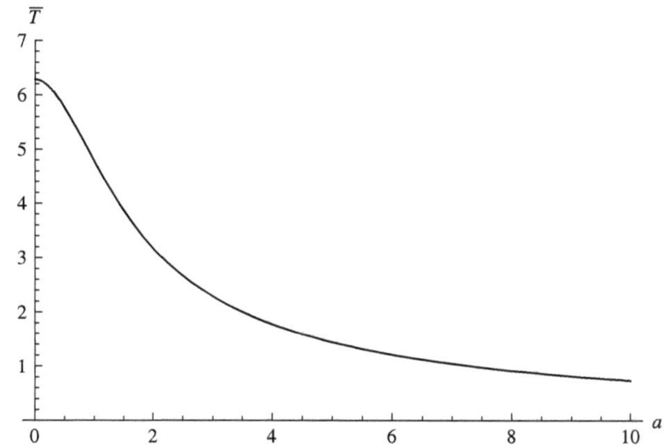

Fig. 1 The graph of function $\overline{T}(a)$

since $E(k) \leq K(k)$, where $E(k)$ is the complete elliptic function of the second kind. So $\overline{T}(a)$ is decreasing from $\overline{T}(0) = 2\pi$ to 0 (see Fig. 1). Then condition (H3) holds for any $\omega > 1$.

Now, we compute

$$\begin{aligned}
&\int_0^{\overline{T}(a)} U_{0,a}'(z)^2 dz = a^2\left(1+a^2\right)\int_0^{\overline{T}(a)} \mathrm{dn}^2\left(\sqrt{1+a^2}z\right)\mathrm{sn}^2\left(\sqrt{1+a^2}z\right)dz \\
&= a^2\sqrt{1+a^2}\int_0^{4K(k)} \mathrm{dn}^2 z\,\mathrm{sn}^2 z dz = 2\frac{a^2}{k^2}\sqrt{1+a^2}\int_0^{2K(k)} \mathrm{dn}^2 z(1-\mathrm{dn}^2 z)dz \\
&= 2(1+a^2)^{3/2}\int_0^{2K(k)} \left(\mathrm{dn}^2 z - \mathrm{dn}^4 z\right)dz = \frac{2}{3}\sqrt{1+a^2}\left((2+a^2)K(k) - 2E(k)\right)
\end{aligned}$$

by using (2.1.10), (2.2.11–13), (2.5.3), Sect. 3.5 and (3.10.14) of [16]. Here dn and sn are the other Jacobi elliptic functions.

Now, we take $h(z) = \cos z$ and evaluate

$$\begin{aligned}
&\frac{2\pi u}{\overline{T}(a)}\int_0^{\overline{T}(a)} h'\left(\frac{2\pi u}{\overline{T}(a)}z+\alpha\right)U_{0,a}(z)dz \\
&= -\frac{2\pi a u}{\overline{T}(a)}\int_0^{\overline{T}(a)} \sin\left(\frac{2\pi u}{\overline{T}(a)}z+\alpha\right)\mathrm{cn}\left(\sqrt{1+a^2}z\right)dz \\
&= -\frac{\pi a u}{2K(k)}\int_0^{4K(k)} \sin\left(\frac{\pi u}{2K(k)}z+\alpha\right)\mathrm{cn}\, z dz \\
&= -\frac{\pi a u}{2K(k)}\left(\sin\alpha\int_{-2K(k)}^{2K(k)} \cos\left(\frac{\pi u}{2K(k)}z\right)\mathrm{cn}\, z dz + \cos\alpha\int_{-2K(k)}^{2K(k)} \sin\left(\frac{\pi u}{2K(k)}z\right)\mathrm{cn}\, z dz\right) \\
&= -\frac{\pi a u}{2K(k)}\sin\alpha\int_0^{4K(k)} \cos\left(\frac{\pi u}{2K(k)}z\right)\mathrm{cn}\, z dz \\
&= \frac{(-1)^u - 1}{2}\frac{\pi^2 a u}{kK(k)}\sin\alpha\,\mathrm{sech}\left(\frac{\pi K(\sqrt{1-k^2})u}{2K(k)}\right)
\end{aligned}$$

since $\cos z$ and $\operatorname{cn} z$ are even and $\sin z$ is odd, and the Fourier expansion of cn (see (8.7.7) and (2.6.5) of [16]) are used as well. Summarizing, the Melnikov function is now

$$
\begin{aligned}
M^u(\alpha) &= -\gamma\frac{2}{3}\sqrt{1+a^2}\left((2+a^2)K(k)-2E(k)\right)\\
&-\frac{(-1)^u-1}{2}\frac{\pi^2 au}{kK(k)}\sin\alpha\,\operatorname{sech}\left(\frac{\pi K(\sqrt{1-k^2})u}{2K(k)}\right), \quad k=\frac{a}{\sqrt{2+2a^2}}.
\end{aligned}
\tag{29}
$$

Clearly, we need $u \in \mathbb{N}$ to be odd. Then to have a simple zero of $M^u(\alpha)$, we need

$$
\gamma\frac{\sqrt{2}K(k)\left((2+a^2)K(k)-2E(k)\right)}{3\pi^2 u}\cosh\left(\frac{\pi K(\sqrt{1-k^2})u}{2K(k)}\right)<1, \quad k=\frac{a}{\sqrt{2+2a^2}}.
$$

Setting

$$
\Lambda(a,u) := \frac{\sqrt{2}K(k)\left((2+a^2)K(k)-2E(k)\right)}{3\pi^2 u}\cosh\left(\frac{\pi K(\sqrt{1-k^2})u}{2K(k)}\right)<1,
$$

$k=\frac{a}{\sqrt{2+2a^2}}$, we see that

$$
\gamma < \frac{1}{\Lambda(a,u)}
\tag{30}
$$

gives the magnitude for the damping in order to apply Theorem 3. Note $\Lambda(a,u)>0$ for any $a>0$, $u>0$, $\Lambda(0,u)=0$ and $\Lambda(a,u)\to\infty$ as $a\to\infty$.

4 Bifurcation Results for Asymptotic Travelling Waves

In this section, we first consider, instead of (H1), the following one

(C1) $\varphi(0)=0$, $\varphi'(0)<0$ and $U''(z)+\varphi(U(z))=0$ has an asymptotic solution $\Gamma \in C^2(\mathbb{R},\mathbb{R})$ such that $\lim_{|z|\to\infty}\Gamma(z)=0$ and $\lim_{|z|\to\infty}\Gamma'(z)=0$.

Remark 4. We may again suppose like in Remark 3 that Γ is even.

Then

$$
\Gamma_\omega(z) := \Gamma(z/\omega)
$$

satisfies $\omega^2\Gamma_\omega''(z)+\varphi(\Gamma_\omega(z))=0$, $\lim_{|z|\to\infty}\Gamma_\omega(z)=0$ and Γ_ω is even. Next, when we take $U_\omega=0$ in Sect. 3, then $X_1=\{0\}$ and $Y_1=\{0\}$. So there is no bifurcation Eq. (25). This implies the following result.

Theorem 4. *Suppose (C1). Then there is a $\delta > 0$ such that for any $\varepsilon \in (-\delta, \delta)$ there is a unique 2π-periodic solution $W_\varepsilon(z)$ of (21) with $W_\varepsilon(z) = O(\varepsilon)$.*

Remark 5. A proof of Theorem 4 follows also directly from Sect. 2. Under (C1), the equation $\omega^2 U''(z) + \varphi(U(z)) = 0$ has a hyperbolic/nonresonant equilibrium $U_\omega(z) = 0$. The linearization at U_ω is $\omega^2 U''(z) + \varphi'(0)U(z) = 0$, which has no nonzero periodic solutions. So for any periodic perturbation, i.e. also for (21), there is a small periodic solution. This is a statement of Theorem 4.

Now, we follow the standard homoclinic Melnikov method [3, 12] to (21) based on the Lyapunov-Schmidt method, but for reader's convenience we again present some details. First, we take Banach spaces

$$X := \left\{U \in C^2(\mathbb{R}, \mathbb{R}) \mid \|U\|_{2,\infty} = \|U\|_\infty + \|U'\|_\infty + \|U''\|_\infty < \infty\right\},$$
$$Y := \{U \in C(\mathbb{R}, \mathbb{R}) \mid \|U\|_\infty < \infty\}$$

with the usual supremum norm $\|U\|_\infty = \sup_{z\in\mathbb{R}} |U(z)|$. Then we split

$$Y = Y_1 \oplus Y_2, \quad X = X_1 \oplus X_2$$

with

$$Y_1 := \operatorname{span}[\Gamma_\omega'], \quad Y_2 := \left\{U \in Y \mid \int_{-\infty}^{\infty} U(z)\Gamma_\omega'(z)dz = 0\right\},$$
$$X_1 := Y_1, \quad X_2 := \left\{U \in Y \mid \int_{-\infty}^{\infty} U(z)\Gamma_\omega'(z)dz = 0\right\}.$$

Next, we take the projections $P : Y \to Y_1$ and $Q = \mathbb{I} - P : Y \to Y_2$ defined as

$$PU := \frac{\int_{-\infty}^{\infty} U(s)\Gamma_\omega'(s)ds}{\int_{-\infty}^{\infty} \Gamma_\omega'^2(s)ds} \times \Gamma_\omega.$$

Now, we use the following well-known analogy of Lemma 3 [3, 12].

Lemma 4. *Equation $\omega^2 V''(z) + \varphi'(U_\omega(z))V(z) = \tilde{Y}(z) \in Y$ has a solution $V \in X$ if and only if $\tilde{Y} \in Y_2$. This solution is unique requiring $V \in X_2$. Moreover, there is a constant $c_\omega > 0$ such that $\|V\|_{2,\infty} \le c_\omega \|\tilde{Y}\|_\infty$.*

Consequently, we can follow the approach of Sect. 3 as in [3, 12] to derive the following result.

Theorem 5. *Suppose (C1). If there is a simple zero β_0 of the Melnikov function*

$$M(\beta) := \int_{-\infty}^{\infty} \left(-\gamma\Gamma'(z) + h(\omega z + \beta)\right) \Gamma'(z)dz. \tag{31}$$

Then there is a $\theta > 0$ *such that for any* $0 \neq \varepsilon \in (-\theta, \theta)$ *there is a unique bounded solution* $U(z)$ *of (21) on* $\mathbb{R}$ *with*

$$U(z) = \Gamma\left(\frac{z-\beta_0}{\omega}\right) + O(\varepsilon).$$

Next, putting $U(z) = W_\varepsilon(z) + W(z)$ in (21), where $W_\varepsilon(z)$ is a small 2π-periodic solution from Theorem 4, we get

$$\begin{aligned} &\omega^2 W''(z) + \varphi(W_\varepsilon(z) + W(z)) - \varphi(W_\varepsilon(z)) \\ &-\varepsilon\lambda\omega^2 \left(W''(z-p) + W''(z+p)\right) + \varepsilon\gamma\omega W'(z) = 0. \end{aligned} \tag{32}$$

Now, we take Banach spaces

$$\begin{aligned} X_\varsigma &:= \left\{U \in C^2(\mathbb{R}, \mathbb{R}) \mid \|U\|_{2,\varsigma} = \|U\|_\varsigma + \|U'\|_\varsigma + \|U''\|_\varsigma < \infty\right\}, \\ Y_\varsigma &:= \left\{U \in C(\mathbb{R}, \mathbb{R}) \mid \|U\|_\varsigma < \infty\right\} \end{aligned}$$

with the weighted supremum norm $\|U\|_\varsigma = \sup_{z\in\mathbb{R}} |U(z)|\, e^{\varsigma|z|}$ for a sufficiently small $\varsigma > 0$. Then we can repeat the above arguments to solve (32) in X_ς under assumptions of Theorem 5. Indeed, Lemma 4 holds in X_ς and Y_ς [11]. Denoting $\Upsilon_\omega := \frac{d}{d\varepsilon} W_0$, (32) has the form

$$\begin{aligned} 0 &= \omega^2 W''(z) + \varphi(W(z)) + \varphi(W_\varepsilon(z) + W(z)) - \varphi(W_\varepsilon(z)) - \varphi(W(z)) \\ &\quad -\varepsilon\lambda\omega^2 \left(W''(z-p) + W''(z+p)\right) + \varepsilon\gamma\omega W'(z) \\ &= \omega^2 W''(z) + \varphi(W(z)) + \varepsilon\left(\varphi'(W(z)) - \varphi'(0)\right)\Upsilon_\omega(z) + O(\varepsilon^2) \\ &\quad -\varepsilon\lambda\omega^2 \left(W''(z-p) + W''(z+p)\right) + \varepsilon\gamma\omega W'(z). \end{aligned} \tag{33}$$

Note $U_\omega \in X_\varsigma$ and $\omega^2\Upsilon''_\omega(z) + \varphi'(0)\Upsilon_\omega(z) - h(z) = 0$. Equation (33) has a similar form as (21), so we set $W(z+\beta) = U_\omega(z) + W_1(z)$ as in (22) and apply the above approach to get the Melnikov function

$$\begin{aligned} M(\beta) &:= \int_{-\infty}^{\infty} \Big(\lambda\omega^2 \left(U''_\omega(z-p) + U''_\omega(z+p)\right) - \gamma\omega U'_\omega(z) \\ &\qquad - \left(\varphi'(U_\omega(z)) - \varphi'(0)\right)\Upsilon_\omega(z+\beta)\Big) U'_\omega(z) dz \\ &= \int_{-\infty}^{\infty} \left(-\gamma\omega U'_\omega(z) - \varphi'(U_\omega(z))\Upsilon_\omega(z+\beta) - \omega^2\Upsilon''_\omega(z+\beta) + h(z+\beta)\right) U'_\omega(z) dz, \end{aligned} \tag{34}$$

since U_ω is even. By using the integration by parts, we derive

$$\begin{aligned} &\int_{-\infty}^{\infty} \left(\varphi'(U_\omega(z))\Upsilon_\omega(z+\beta) + \omega^2\Upsilon''_\omega(z+\beta)\right) U'_\omega(z) dz \\ &= \int_{-\infty}^{\infty} \left(\frac{d}{dz}\left[\varphi(U_\omega(z))\right] \Upsilon_\omega(z+\beta) + \omega^2\Upsilon''_\omega(z+\beta) U'_\omega(z)\right) dz \\ &= \int_{-\infty}^{\infty} \left(-\varphi(U_\omega(z))\Upsilon'_\omega(z+\beta) - \omega^2\Upsilon'_\omega(z+\beta) U''_\omega(z)\right) dz = 0. \end{aligned}$$

Hence (34) has the form

$$\begin{aligned} M(\beta) &= \int_{-\infty}^{\infty} \left(-\gamma\omega U_\omega'(z) + h(z+\beta)\right) U_\omega'(z)dz \\ &= \frac{1}{\omega}\int_{-\infty}^{\infty} \left(-\gamma\Gamma'(z/\omega) + h(z+\alpha)\right) \Gamma'(z/\omega)dz \\ &= \int_{-\infty}^{\infty} \left(-\gamma\Gamma'(z) + h(\omega z+\alpha)\right) \Gamma'(z)dz, \end{aligned} \tag{35}$$

which is just (31). Consequently, under the assumptions of Theorem 5, there is a $\theta_1 > 0$ such that for any $0 \neq \varepsilon \in (-\theta_1, \theta_1)$ there is a unique solution $W(z) \in Y_\varsigma$ of (32) with

$$W(z) = \Gamma\left(\frac{z-\beta_0}{\omega}\right) + O(\varepsilon)$$

in Y_ς. Then $U_1(z) := W_\varepsilon(z) + W(z) = \Gamma\left(\frac{z-\beta_0}{\omega}\right) + O(\varepsilon)$ is a bounded solution of (21) on $\mathbb{R}$ with the same property as $U(z)$ in Theorem 5. The uniqueness of this solution gives $U(z) = U_1(z)$. Consequently, the solution predicted in Theorem 5 exponentially tends to the small periodic solution.

Example 2. To illustrate the theory, we consider $\varphi(U) = -U + U^3$ and then equation from (C1) is the Duffing equation

$$U''(z) - U(z) + U^3(z) = 0 \tag{36}$$

possessing a homoclinic solution

$$\Gamma(z) = \sqrt{2}\,\text{sech}\, z.$$

Again $h(z) = \cos z$. Then the Melnikov function (31) is now [12, p. 191]

$$\begin{aligned} M(\beta) &:= -\sqrt{2}\int_{-\infty}^{\infty} \left(\sqrt{2}\gamma\,\text{sech}\, z\tanh z + \cos(\omega z+\beta)\right)\text{sech}\, z\tanh z dz \\ &= -\frac{4\gamma}{3} + \sqrt{2}\omega\pi\,\text{sech}\left(\frac{\omega\pi}{2}\right)\sin\beta. \end{aligned}$$

If

$$\gamma < \frac{3\sqrt{2}}{4}\omega\pi\,\text{sech}\left(\frac{\omega\pi}{2}\right), \tag{37}$$

then $M(\beta)$ has a simple zero and so Theorem 5 can be applied.

Next, we consider the following condition.

(C2) $\varphi(\pm 1) = 0$, $\varphi'(\pm 1) < 0$ and $U''(z) + \varphi(U(z)) = 0$ has an asymptotic solution $\Gamma \in C^2(\mathbb{R}, \mathbb{R})$ such that $\lim_{z\to\pm\infty} \Gamma(z) = \pm 1$ and $\lim_{|z|\to\infty} \Gamma'(z) = 0$.

Remark 6. Now $U'(z) > 0$ on $\mathbb{R}$. Next $\Gamma_-(z) := \Gamma(-z)$ is a solution satisfying $\lim_{z\to\pm\infty} \Gamma_-(z) = \mp 1$ and $\lim_{|z|\to\infty} \Gamma_-'(z) = 0$. So Γ and Γ_- create a heteroclinic cycle. Next, we can suppose that $\Gamma(0) = 0$. If φ is odd then Γ is odd. So Γ' is even and Γ'' is odd.

Similarly, we get 2π-periodic solutions $W_{\varepsilon,\pm}(z) = \pm 1 + O(\varepsilon)$ of (21), for any $\varepsilon \neq 0$ small. Now, we can repeat the above approach to derive the Melnikov function

$$M(\beta) := \int_{-\infty}^{\infty} \left(\lambda\left(\Gamma''\left(z-\frac{p}{\omega}\right)+\Gamma''\left(z+\frac{p}{\omega}\right)\right)-\gamma\Gamma'(z)+h(\omega z+\beta)\right)\Gamma'(z)dz. \tag{38}$$

Remark 7. By Remark 6 we know that if φ is odd then Γ' is even and Γ'' is odd. Then (38) possesses the form of (31).

Under assumptions of Theorem 5 with (38), we get a bounded solution $U(z)$ of (21) with the same properties and exponentially tending to the small periodic solutions $W_{\varepsilon,\pm}$. We note that to show these exponential attractions, we consider (32) for $W_{\varepsilon,\pm}$ on the Banach spaces

$$X_\varsigma^\pm := \left\{U \in C^2(\mathbb{R}_\pm, \mathbb{R}) \mid \|U\|_{2,\varsigma} = \|U\|_\varsigma + \|U'\|_\varsigma + \|U''\|_\varsigma < \infty\right\},$$
$$Y_\varsigma^\pm := \left\{U \in C(\mathbb{R}_\pm, \mathbb{R}) \mid \|U\|_\varsigma < \infty\right\},$$

respectively.

Example 3. To illustrate the theory, we consider $\varphi(U) = U - U^3$ and then equation from (C2) is the Duffing equation

$$U''(z) + U(z) - U^3(z) = 0 \tag{39}$$

possessing a heteroclinic solution

$$\Gamma(z) = \tanh(z/\sqrt{2}).$$

Again $h(z) = \cos z$. Clearly φ is odd, so by Remark 7, the Melnikov function (38) has the form (31), so it is given by [3]

$$\begin{aligned} M(\beta) &:= \frac{1}{2}\int_{-\infty}^{\infty}\left(-\gamma\,\text{sech}^2\left(\frac{z}{\sqrt{2}}\right) + \sqrt{2}\cos(\omega z+\beta)\right)\text{sech}^2\left(\frac{z}{\sqrt{2}}\right)dz \\ &= \frac{\sqrt{2}}{3}\left(-2\gamma + 3\omega\pi\cos\beta\,\text{csch}\left(\frac{\omega\pi}{\sqrt{2}}\right)\right). \end{aligned}$$

If

$$\gamma < \frac{3}{2}\omega\pi \operatorname{csch}\left(\frac{\omega\pi}{\sqrt{2}}\right),$$

then $M(\beta)$ has a simple zero and so Theorem 5 can be applied.

Remark 8. Under either (C1) or (C2) there is an accumulation of periodic travelling waves of (21) on the asymptotic travelling waves predicted in Theorem 5 with periods tending to infinity. This follows from similar arguments as in [12, p. 197]. So we have the so called blue sky catastrophe [6].

Remark 9. By following [5, 6], method of Sects. 3 and 4 can be applied to (19) to show the bifurcation of periodic and asymptotic breathers. Related results are given in [4, 7–10].

Finally, we construct a non-odd φ satisfying (C2). We look for φ in the form $\varphi(U) = U(1-U^2)\left(a_1U^3 + a_2U^2 + a_3U + 1\right)$ for some $a_i \in \mathbb{R}$, $i = 1, 2, 3$. Note $\varphi'(0) = 1$. Then

$$\Phi(U) := \int \varphi(U)\, dU = \frac{U^2}{2} + \frac{a_3U^3}{3} - \frac{U^4}{4} + \frac{a_2U^4}{4} + \frac{a_1U^5}{5} - \frac{a_3U^5}{5} - \frac{a_2U^6}{6} - \frac{a_1U^7}{7}.$$

The condition $\Phi(-1) = \Phi(1)$ implies $3a_1 + 7a_3 = 0$, so we take $a_1 = -7a$, $a_3 = 3a$ and $a_2 = b$. Then

$$\varphi(U) = U(1-U^2)\left(1 + bU^2 + aU\left(3 - 7U^2\right)\right) \tag{40}$$

and

$$\Phi(U) = \frac{1}{12}U^2\left(6 + 3(-1+b)U^2 - 2bU^4 + 12aU\left(-1+U^2\right)^2\right).$$

Set

$$\mathcal{M} = \left\{(a,b) \in \mathbb{R}^2 \mid \min_{U\in[-1,1]} \left(1 + bU^2 + aU\left(3 - 7U^2\right)\right) > 0\right\}.$$

Clearly, if $(a,b) \in \mathcal{M}$ then φ given by (40) satisfies (C2). Next, since $\max_{U\in[0,1]} \left|U\left(3 - 7U^2\right)\right| = 4$. Then for any $b \geq 0$ and $|a| < 1/4$ it holds

$$1 + bU^2 + aU\left(3 - 7U^2\right) \geq 1 - 4|a| > 0.$$

Hence $\mathcal{M}$ is nonempty. Of course, further computations must be done numerically.

5 Numerical Results

To illustrate the theoretical results obtained in the previous sections, we have solved the governing equation (4) (cf. (21)) numerically. The advance-delay equation is solved using a pseudo-spectral method. We express the solution U in a Fourier series

$$U(z) = \sum_{j=1}^{J} \left[A_j \cos\left((j-1)\tilde{k}z\right) + B_j \sin\left(j\tilde{k}z\right)\right], \tag{41}$$

where $\tilde{k} = 2\pi/L$ and $-L/2 < z < L/2$. The Fourier coefficients A_j and B_j are then found by requiring the series to satisfy (4) at several collocation points. Hence, $2J$ collocation points are required, which are chosen with uniform grid points. The stability of a solution obtained from (4) is then determined numerically by evolving it through the original equation (1). The governing equation is integrated using a Runge-Kutta method of order four with periodic boundary conditions. In the following, we only illustrate the results of Sect. 4 on the existence of asymptotic travelling waves. In particular, we consider the nonlinearity discussed in Example 2. The nonlinearity studied in Example 3 will be mentioned briefly. We have considered periodic wave solutions as well, but they are not presented here as calculations can be done rather straightforwardly using the numerical method.

It is important to note that the physically relevant range for the coupling parameter λ is $|\lambda| < 1/2$ for the following reason. When we consider (1) in $\ell^2 \times \ell^2$, then we need to solve the equation

$$\lambda x_{n-1} - x_n + \lambda x_{n-1} = y_n \quad n \in \mathbb{Z}$$

in ℓ^2, i.e. the linear equation $D(\lambda)x = y$, $x, y \in \ell^2$. The spectrum of the corresponding left-hand side linear self-adjoint operator $D(\lambda)$ is $\sigma(D(\lambda)) = [-2|\lambda| - 1, 2|\lambda| - 1]$ (see [25, p. 20]). When $|\lambda| < 1/2$, then $\mathrm{dist}\{0, \sigma(D(\lambda))\} \geq 1 - 2|\lambda| > 0$. This implies that $\|D(\lambda)^{-1}\| \leq \frac{1}{1-2|\lambda|}$. On the other hand, if $|\lambda| \geq 1/2$ then $0 \in \sigma(D(\lambda)) = 0$, and (1) is problematic. The situation is different on $\ell^\infty \times \ell^\infty$, since we can consider periodic boundary conditions $i_j = i_{n+j}$, $q_j = q_{n+j}$, $j \in \mathbb{Z}$. So we truncate D and solve a linear equation

$$\lambda x_2 - x_1 + \lambda x_n = y_1\,,$$

$$\vdots$$

$$\lambda x_1 - x_n + \lambda x_{n-1} = y_n\,.$$

The spectrum of the corresponding left-hand side linear symmetric matrix $D_n(\lambda)$ is $\sigma(D_n(\lambda)) = \{2\lambda \cos\frac{2\pi}{n} j - 1 \mid j = 0, \cdots, n-1\}$. So the above linear equation is $D_n(\lambda)x = y$. When $|\lambda| < 1/2$, then $\mathrm{dist}\{0, \sigma(D_n(\lambda))\} \geq 1 - 2|\lambda| > 0$ and

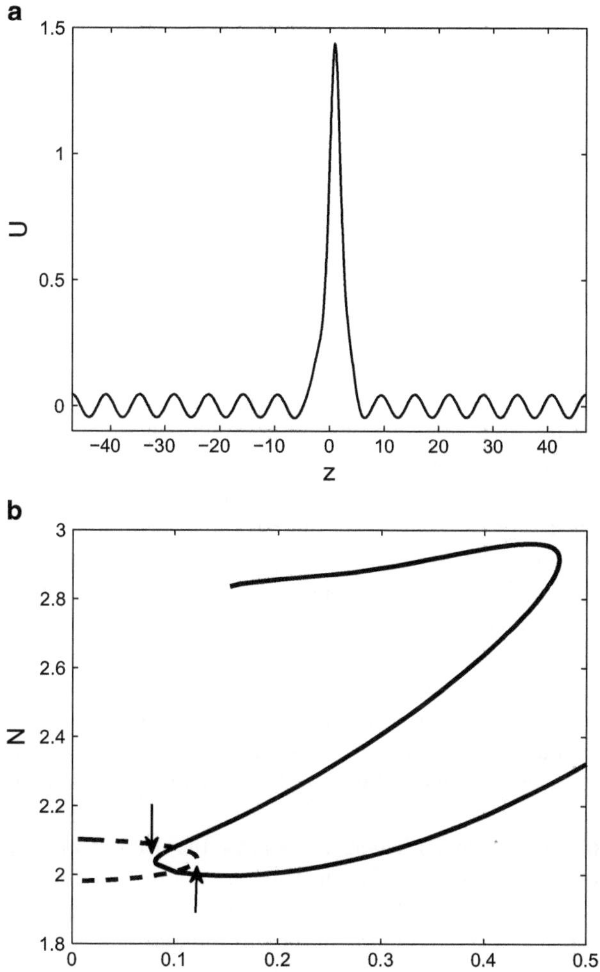

Fig. 2 (**a**) An asymptotic travelling wave for a Duffing nonlinearity $\varphi = -U + U^3$. (**b**) Continuations of the solution in (**a**) for varying γ (*dashed*) and f (*solid*). On the vertical axis is the solution norm (see the text)

$\|D_n(\lambda)^{-1}\| \leq \frac{1}{1-2|\lambda|}$ independently of n. Moreover, $\sigma(D_n(\lambda)) \subset (-\infty, 0)$. On the other hand, if $|\lambda| > 1/2$ then $\sigma(D_n(\lambda)) \cap (0, \infty) \neq \emptyset$. Therefore, in this case one will observe a blow-up at finite time. Nevertheless, $D_n(\lambda)$ could still be invertible, which explains why periodic travelling waves could also exist for $|\lambda| > 1/2$ (see Remark 1).

For the nonlinearity discussed in Example 2, i.e. $\varphi = -U + U^3$, shown in Fig. 2a is the profile of an asymptotic wave for $\lambda = \gamma = f = 0.1$, $\omega = 1$ and $p = \pi$. The numerical solution is computed using $L = 30\pi$ and $J = 100$. We have used larger values of J as well, but we did not see any significant quantitative difference. One can observe that the single-hump profile in Fig. 2a is accompanied by periodic waves as suggested by Theorem 4. The asymptotic wave is also found in the existence region given by Theorem 5 (see also Eqs. (35) and (37)).

Regarding the existence of the asymptotic wave, we have performed numerical continuations of the wave in the panel by varying one parameter using a pseudo-arclength method. Shown in Fig. 2b is the continuation of the figure as the damping parameter γ or the driving amplitude f varies that is shown in dashed and solid line, respectively. The vertical axis is the norm of the solution

$$N = \sqrt{\int_{-L/2}^{L/2} |U|^2 \, dx}.$$

Note that there are two intersections between the dashed and the solid lines. The wave in Fig. 2a corresponds to the upper intersection.

For fixing $f = 0.1$ and varying γ, we obtained numerically that there is a saddle-node bifurcation at $\gamma \approx 0.122$. In Fig. 2b, the bifurcation is indicated by the right arrow. From the Melnikov function (37), the approximate boundary is calculated as $\gamma \approx 0.133$, which is quantitatively close to the numerical value. As for varying the driving amplitude f while fixing $\gamma = 0.1$, we also observed saddle-node bifurcations. Upon decreasing f, there is a saddle-node bifurcation between the middle branch, corresponding to single-hump asymptotic waves, and the lower branch that corresponds to periodic waves. The bifurcation, which is indicated by the left arrow, occurs at $f \approx 0.082$. From Theorem 5 (cf. Eq. (37)), the existence for asymptotic waves is bounded by $f \approx 0.075$, which agrees quite well with the numeric.

From the solution in Fig. 2a, when we increase the driving amplitude f, we also observe another saddle-node bifurcation. The upper branch in the panel corresponds to asymptotic waves with double-hump. We did not study the multi-hump waves further because it is beyond the scope of the present paper.

After analysing the existence, next we consider the stability of the asymptotic waves. For the profile shown in Fig. 2a, depicted in Fig. 3 is the typical time-dynamics of the wave obtained from integrating the governing equation (1). One can observe that the travelling wave is strongly unstable. The hump could only travel for one site before the background becomes excited and destroys the localised profile. The instability is naturally expected due to the fact that the zero solution (when there is no drive) forming the background of the asymptotic wave is unstable, i.e. it is a saddle point. For that reason, we believe that all the branches in Fig. 2b correspond to unstable solutions. For the same reason, the asymptotic waves in Example 3 would also be unstable as the background, i.e. $U = 1$ when there is no drive, is also a saddle point. Moreover, differently from the zero background in Example 2, the instability in this case will create an unbounded blow-up (see Ref. [1] for a related problem).

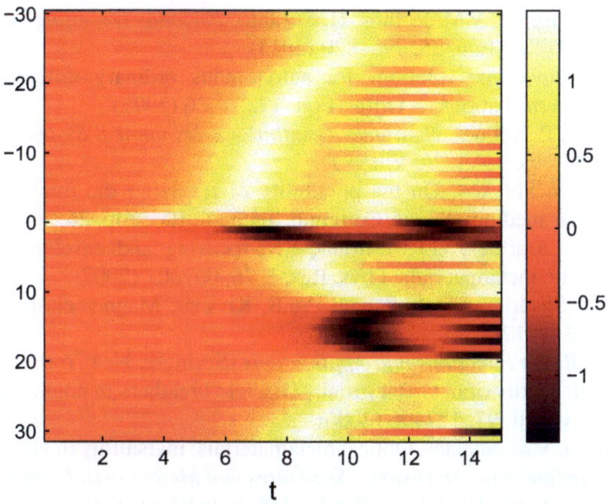

Fig. 3 Time dynamics of the wave shown in Fig. 2a. Plotted is the top view of $q_n(t)$

Acknowledgements We thank the reviewers for valuable comments and suggestions which improved our paper. J.D. is supported by the Grant GAČR P201/11/0768. M.F. is partially supported by grants VEGA-MS 1/0507/11, VEGA-SAV 2/0029/13 and APVV-0134-10. M.P. is supported by the project No. CZ.1.07/2.3.00/30.0005 funded by European Regional Development Fund. V.R. and H.S. acknowledge a partial support from the London Mathematical Society through a visitor grant. This research (V.R.) has been co-financed by the European Union (European Social Fund – ESF) and Greek national funds through the Operational Program "Education and Lifelong Learning" of the National Strategic Reference Framework (NSRF) – Research Funding Program: THALES – Investing in knowledge society through the European Social Fund. V.R. also acknowledges support from AUTH-Research Council Grant.

References

1. V. Achilleos, A. Álvarez, J. Cuevas, D.J. Frantzeskakis, N.I. Karachalios, P.G. Kevrekidis, B. Sánchez-Rey, *Escape Dynamics in the Discrete Repulsive φ^4 – Model*. arXiv:1206.1363
2. M.S. Berger, *Nonlinearity and Functional Analysis* (Academic, New York, 1977)
3. C. Chicone, *Ordinary Differential Equations with Applications*. Texts in Applied Mathematics, vol. 34 (Springer, New York, 2006)
4. J. Diblík, M. Fečkan, M. Pospíšil, Forced Fermi-Pasta-Ulam lattice maps. Miskolc Math. Notes **14**, 63–78 (2013)
5. M. Fečkan, *Topological Degree Approach to Bifurcation Problems* (Springer, Berlin, 2008)
6. M. Fečkan, *Bifurcation and Chaos in Discontinuous and Continuous Systems* (Springer, Berlin, 2011)
7. M. Fečkan, V.M. Rothos, Travelling waves in Hamiltonian systems on 2D lattices with nearest neighbour interactions. Nonlinearity **20**, 319–341 (2007)
8. M. Fečkan, V.M. Rothos, Travelling waves of discrete nonlinear Schrödinger equations with nonlocal interactions. Appl. Anal. **89**, 1387–1411 (2010)
9. M. Fečkan, V.M. Rothos, Travelling waves of forced discrete nonlinear Schrödinger equations. Discret. Contin. Dyn. Syst.: S **4**, 1129–1145 (2011)

10. M. Fečkan, M. Pospíšil, V.M. Rothos, H. Susanto, *Periodic Travelling Waves of Forced FPU Lattices*. J. Dyn. Differ. Equ. **25**, 795–820 (2013)
11. J. Gruendler, Homoclinic solutions for autonomous ordinary differential equations with nonautonomous perturbations. J. Differ. Equ. **122**, 1–26 (1995)
12. J. Guckenheimer, P. Holmes, *Nonlinear Oscillations, Dynamical Systems, and Bifurcations of Vector Fields* (Springer, New York, 1983)
13. V.A. Kalinin, V.V. Shtykov, On the possibility of reversing the front of radio waves in an artificial nonlinear medium. Sov. J. Commun. Technol. Electron. **36**, 96–102 (1991)
14. I. Kourakis, N. Lazarides, G.P. Tsironis, Self-focusing and envelope pulse generation in nonlinear magnetic metamaterials. Phys. Rev. E **75**, 067601 (2007)
15. M. Lapine, I.V. Shadrivov, D.A. Powell, Yu.S. Kivshar, Magnetoelastic metamaterials. Nat. Mater. **11**, 30–33 (2012)
16. D.F. Lawden, *Elliptic Functions and Applications* (Springer, New York, 1989)
17. N. Lazarides, M. Eleftheriou, G.P. Tsironis, Discrete breathers in nonlinear magnetic metamaterials. Phys. Rev. Lett. **97**, 157406 (2006)
18. N.M. Litchinitser, V.M. Shalaev, Optical metamaterials: invisibility in visible and nonlinearities in reverse, in *Nonlinearities in Periodic Structures and Metamaterials*, ed. by C. Denz, S. Flach, Yu.S. Kivshar (Springer, Heidelberg/Dordrecht/London/New York, 2010), pp. 217–240
19. S. Longhi, Gap solitons in metamaterials. Waves in Random and Complex Media **15**, 119–126 (2005)
20. J.B. Pendry, A.J. Holden, D.J. Robbins, W.J. Stewart, Magnetism from conductors and enhanced nonlinear phenomena. IEEE Trans. Microw. Theory Tech. **47**, 2075–2084 (1999)
21. J.B. Pendry, D. Schurig, D.R. Smith, Controlling electromagnetic fields. Science **312**, 1780–1782 (2006)
22. D. Schurig, J.J. Mock, B.J. Justice, S.A. Cummer, J.B. Pendry, A.F. Starr, D.R. Smith, Metamaterial electromagnetic cloak at microwave frequencies. Science **314**, 977–980 (2006)
23. I.V. Shadrivov, Nonlinear metamaterials, in *Nonlinearities in Periodic Structures and Metamaterials*, ed. by C. Denz, S. Flach, Yu.S. Kivshar (Springer, Heidelberg/Dordrecht/London/New York, 2010), pp. 241–257
24. I.V. Shadrivov, P.V. Kapitanova, S.I. Maslovski, Yu.S. Kivshar, Metamaterials controlled with light. Phys. Rev. Lett. **109**, 083902 (2012)
25. G. Teschl, *Jacobi Operators and Completely Integrable Nonlinear Lattices*. Mathematical Surveys and Monographs, vol. 72 (American Mathematical Society, Providence, 2000)
26. E.C. Titchmarsh, *Theory of Functions*, 2nd edn. (Oxford University Press, London, 1939)
27. N.L. Tsitsas, N. Rompotis, I. Kourakis, P.G. Kevrekidis, D.J. Frantzeskakis, Higher-order effects and ultrashort solitons in left-handed metamaterials. Phys. Rev. E **79**, 037601 (2009)
28. N.L. Tsitsas, T.P. Horikis, Y. Shen, P.G. Kevrekidis, N. Whitaker, D.J. Frantzeskakis, Short pulse equations and localized structures in frequency band gaps of nonlinear metamaterials. Phys. Lett. A **374**, 1384–1388 (2010)

Stability of Discrete Breathers in Magnetic Metamaterials

Dmitry Pelinovsky and Vassilis Rothos

Abstract We consider the discrete Klein–Gordon equation for magnetic metamaterials derived by Lazarides, Eleftheriou, and Tsironis Phys Rev Lett 97:157406, 2006). We obtain a general criterion for spectral stability of multi-site breathers for a small coupling constant. We show how this criterion differs from the one derived in the standard discrete Klein–Gordon equation (Koukouloyannis and Kevrekidis, Nonlinearity 22:2269–2285, 2009; Pelinovsky and Sakovich, Nonlinearity 25:3423–3451, 2012).

1 Introduction

We address space-localized and time-periodic breathers in the discrete Klein–Gordon equation describing magnetic metamaterials which consist of periodic arrays of split-ring resonators [4, 7]:

$$\ddot{q}_n + V'(q_n) = \epsilon(\ddot{q}_{n+1} + \ddot{q}_{n-1}), \quad n \in \mathbb{Z}, \tag{1}$$

where $t \in \mathbb{R}$ is the evolution time, $q_n(t) \in \mathbb{R}$ is the normalized charge stored in the capacitor of the n-th split-ring resonator, $V : \mathbb{R} \to \mathbb{R}$ is a smooth on-site potential for the voltage across the slit of the n-th resonator, and $\epsilon \in \mathbb{R}$ is the coupling constant from the mutual inductance. In particular, the voltage $u = f(q) = V'(q)$

D. Pelinovsky (✉)
Department of Mathematics and Statistics, McMaster University, Hamilton, ON L8S 4K1, Canada
e-mail: dmpeli@math.mcmaster.ca

V. Rothos
Faculty of Engineering, School of Mathematics, Physics and Computer Sciences, Aristotle University, GR54124 Thessaloniki, Greece
e-mail: rothos@auth.gr

R. Carretero-González et al. (eds.), *Localized Excitations in Nonlinear Complex Systems*, Nonlinear Systems and Complexity 7, DOI 10.1007/978-3-319-02057-0_18,

is found by inverting the charge-voltage dependence near small charge:

$$q = u + \alpha u^3 \quad \Rightarrow \quad u = f(q) = q - \alpha q^3 + \mathcal{O}(q^5) \quad \text{as} \quad q \to 0, \tag{2}$$

where α is the parameter for the self-focusing ($\alpha > 0$) or self-defocusing ($\alpha < 0$) nonlinearity. These parameter values correspond to the soft and hard potentials V respectively, for sufficiently small values of q. Note that V is an even function of q.

Discrete breathers in both one-dimensional and two-dimensional lattices were approximated numerically in the limit of small coupling constant ϵ [4,7]. Excitations of discrete breathers near the edge of a one-dimensional lattice created by a truncated array of nonlinear split-ring resonators were considered numerically in [8].

It is the purpose of this paper to consider spectral stability of multi-site discrete breathers in the limit of small coupling constant ϵ. This limit is referred usually as the *anti-continuum* limit and it has been considered before in the context of spectral stability of discrete breathers in the standard discrete Klein–Gordon equation [1,3,6,10] and in the discrete nonlinear Schrödinger equation [11,13]. Recent works [12, 14, 16] were devoted to the derivation of the most general stability criterion for multi-site breathers in Klein–Gordon lattices. Our main result shows that the stability criterion for multi-site breathers in the discrete Klein–Gordon equation (1) differs from the one derived in the standard discrete Klein–Gordon equation [6, 12].

The paper is organized as follows. We formulate the discrete Klein–Gordon equation (1) as an evolution problem in Sect. 2. The existence and continuation results for multi-site discrete breathers in the limit of small coupling constant ϵ are reviewed in Sect. 3. Spectral stability of multi-site breathers for small coupling constants is considered in Sect. 4. Section 5 discusses application of the stability criterion to the multi-site breathers in magnetic metamaterials.

2 Formalism

In what follows, we shall use bold-faced notations for vectors in discrete space $l^p(\mathbb{Z})$ defined by their norms

$$\|\mathbf{q}\|_{l^p} := \left(\sum_{n \in \mathbb{Z}} |q_n|^p \right)^{1/p}, \quad p \geq 1.$$

Components of $\mathbf{q}$ are denoted by q_n for $n \in \mathbb{Z}$. These components can be functions of t, in which case they can be considered either in the space $C^2(0,T)$ of twice continuously differentiable functions on $(0,T)$ or in the L^2-based Sobolev space $H^s_{\text{per}}(0,T)$ of T-periodic functions equipped with the norm,

$$\|f\|_{H^s_{\mathrm{per}}} := \left(\sum_{m\in\mathbb{Z}}(1+m^2)^s|c_m|^2\right)^{1/2}, \quad s\geq 0,$$

where the coefficients $\{c_m\}_{m\in\mathbb{Z}}$ define the Fourier series of a T-periodic function f,

$$f(t) = \sum_{m\in\mathbb{Z}} c_m \exp\left(\frac{2\pi imt}{T}\right), \quad t\in[0,T].$$

To start analysis, we set up the discrete Klein–Gordon equation (1) as an evolution problem in t in the phase space $C^2([0,T), l^2(\mathbb{Z}))$, where $T > 0$ is the maximal existence time (which may be infinite). Let us consider the bounded operator

$$M(\epsilon) = I - \epsilon(\sigma_+ + \sigma_-) : l^2(\mathbb{Z}) \to l^2(\mathbb{Z}),$$

where the shift operators $\sigma_\pm : l^2(\mathbb{Z}) \to l^2(\mathbb{Z})$ are defined by

$$(\sigma_\pm \mathbf{q})_n = q_{n\pm 1}, \quad n\in\mathbb{Z}. \tag{3}$$

For any $\epsilon \in \left(-\frac{1}{2},\frac{1}{2}\right)$, the operator $M(\epsilon) : l^2(\mathbb{Z}) \to l^2(\mathbb{Z})$ is diagonally dominant and hence invertible and the inverse operator $M^{-1}(\epsilon) : l^2(\mathbb{Z}) \to l^2(\mathbb{Z})$ is bounded. Moreover, the operator $M^{-1}(\epsilon)$ is analytic at $\epsilon = 0$ and admits the Taylor series,

$$M^{-1}(\epsilon) = I + \sum_{k=1}^{\infty} \epsilon^k(\sigma_+ + \sigma_-)^k, \quad \epsilon\in\left(-\frac{1}{2},\frac{1}{2}\right). \tag{4}$$

The discrete Klein–Gordon equation (1) can be written in the operator form as follows:

$$M(\epsilon)\frac{d^2\mathbf{q}}{dt^2} + \mathbf{f}(\mathbf{q}) = \mathbf{0}, \tag{5}$$

where $(\mathbf{f}(\mathbf{q}))_n = V'(q_n)$. Inverting $M(\epsilon)$ for any $\epsilon \in \left(-\frac{1}{2},\frac{1}{2}\right)$, we obtain the evolution form of the discrete Klein–Gordon equation (5):

$$\frac{d^2\mathbf{q}}{dt^2} + M^{-1}(\epsilon)\mathbf{f}(\mathbf{q}) = \mathbf{0}. \tag{6}$$

With this formulation, we prove the first result on local existence of solutions of the Cauchy problem associated with the evolution equation (6).

Proposition 1. *Let $V \in C^2(\mathbb{R})$ and $\mathbf{q}_0, \mathbf{q}_1 \in l^2(\mathbb{Z})$. For any $\epsilon \in \left(-\frac{1}{2},\frac{1}{2}\right)$, there exist $T > 0$ and a unique local solution of the evolution problem (6) in the phase space $\mathbf{q} \in C^2([0,T), l^2(\mathbb{Z}))$ such that $\mathbf{q}(0) = \mathbf{q}_0$ and $\dot{\mathbf{q}}(0) = \mathbf{q}_1$.*

Proof. Because $V \in C^2(\mathbb{R})$ and $l^2(\mathbb{Z})$ is a Banach algebra with respect to pointwise multiplication, the map $\mathbf{f}(\mathbf{q}) : l^2(\mathbb{Z}) \to l^2(\mathbb{Z})$ is C^1. For any $\epsilon \in \left(-\frac{1}{2}, \frac{1}{2}\right)$, there exists $C(\epsilon) > 0$ such that $\|M^{-1}(\epsilon)\|_{l^2 \to l^2} \leq C(\epsilon)$. Therefore, the vector field $M^{-1}(\epsilon)\mathbf{f}(\mathbf{q})$ is a bounded C^1 map from $l^2(\mathbb{Z})$ to $l^2(\mathbb{Z})$, hence, it is locally Lipschitz. The result of Proposition 1 follows from the standard existence theory of second-order evolution equations in Banach spaces [2, Chap. 2]. □

Remark 1. For the particular function V defined by (2), we note that the assumption $V \in C^2(\mathbb{R})$ is satisfied for any $\alpha > 0$ (in which case, $V \in C^\infty(\mathbb{R})$), because $1 + 3\alpha u^2 > 0$ for all $u \in \mathbb{R}$ and $f : \mathbb{R} \to \mathbb{R}$ is one-to-one and onto. However, for $\alpha < 0$, the function f is one-to-one and onto on $(-Q_0, Q_0)$ with the range in $(-U_0, U_0)$, where

$$Q_0 = \frac{2}{\sqrt{27|\alpha|}}, \quad U_0 = \frac{1}{\sqrt{3|\alpha|}}. \tag{7}$$

Therefore, in this case, we only have $V \in C^2(-Q_0, Q_0)$ (in fact, $V \in C^\infty(-Q_0, Q_0)$), so a unique local solution of the evolution problem (6) exists if $\mathbf{q}_0 \in l^2(\mathbb{Z})$ satisfies further restriction: $(\mathbf{q}_0)_n \in (-Q_0, Q_0)$ for all $n \in \mathbb{Z}$.

Remark 2. We shall only consider the Klein–Gordon lattice (1) with the nonlinear potential (2) for small values of q_n.

3 Existence of Multi-site Discrete Breathers

We consider space-localized and time-periodic breathers of the discrete Klein–Gordon equation (1) in the space $\mathbf{q} \in H^2_{\text{per}}((0,T), l^2(\mathbb{Z}))$, where $T > 0$ represents the fundamental period.

Remark 3. Note that the space $H^2_{\text{per}}((0,T), l^2(\mathbb{Z}))$ for discrete breathers is actually weaker than the space $C^2([0,T), l^2(\mathbb{Z}))$, for which the existence of a unique local solution is established in Proposition 1, but Sobolev's embedding of $H^2_{\text{per}}(0,T)$ to $C_{\text{per}}(0,T)$ and the bootstrapping arguments from the evolution equation (6) show that if $\mathbf{q} \in H^2_{\text{per}}((0,T), l^2(\mathbb{Z}))$, then $\mathbf{q} \in C^2_{\text{per}}((0,T), l^2(\mathbb{Z}))$ (the opposite is true immediately).

Accounting for symmetries, we shall work in the restriction of $H^2_{\text{per}}(0,T)$ to the space of even T-periodic functions,

$$H^2_e(0,T) = \left\{ f \in H^2_{\text{per}}(0,T) : \quad f(-t) = f(t), \quad t \in \mathbb{R} \right\}.$$

We shall also assume everywhere that the nonlinear potential V is an even function of q, which agrees with the potential defined by (2). This assumption is not very restrictive and is used to simplify the technical computations.

At $\epsilon = 0$, we have many possible configurations of multi-site breathers,

$$\mathbf{Q}^{(0)}(t) = \sum_{k \in S} \sigma_k \varphi(t) \mathbf{e}_k, \tag{8}$$

where $\mathbf{e}_k$ is the unit vector in $l^2(\mathbb{Z})$ associated with the site $k \in \mathbb{Z}$, $S \subset \mathbb{Z}$ is a finite set of excited sites of the lattice, $\sigma_k \in \{+1, -1\}$ encodes the phase factor of the k-th oscillator, and $\varphi \in H_e^2(0, T)$ is an even solution of the nonlinear oscillator equation at the energy level E,

$$\ddot{\varphi} + V'(\varphi) = 0 \quad \Rightarrow \quad E = \frac{1}{2}\dot{\varphi}^2 + V(\varphi). \tag{9}$$

Remark 4. Note that if φ is a solution of (9), then $-\varphi$ is also a solution of (9) because V' is an odd function of φ. This motivates the notations in (8) due to the technical simplification that V is even. If V is of general type, we would need to modify the representation formula (8) and the subsequent analysis.

The unique even solution $\varphi(t)$ satisfies the initial condition,

$$\varphi(0) = a, \quad \dot{\varphi}(0) = 0, \tag{10}$$

where a is the smallest positive root of $V(a) = E$ for a fixed value of E. Period of oscillations T is uniquely defined by the energy level E, according to the following formula:

$$T = \sqrt{2} \int_{-a}^{a} \frac{d\varphi}{\sqrt{E - V(\varphi)}}. \tag{11}$$

Remark 5. All nonlinear oscillators at the excited sites of $S \subset \mathbb{Z}$ in the configuration (8) have the same period T. Two oscillators at the j-th and k-th sites are said to be in-phase if $\sigma_j \sigma_k = 1$ and anti-phase if $\sigma_j \sigma_k = -1$.

Persistence of the limiting configuration (8) as a space-localized and time-periodic breather of the discrete Klein–Gordon equation (1) for small values of ϵ is established by MacKay and Aubry [9]. Using this theory, we prove the next result on the existence and continuation of the multi-site discrete breathers.

Proposition 2. *Fix the period T and the solution $\varphi \in H_e^2(0, T)$ of the nonlinear oscillator equation (9) with an even function $V \in C^\infty(\mathbb{R})$ such that $V''(0) = 1$. Assume that $T \neq 2\pi n$, $n \in \mathbb{N}$ and $T'(E) \neq 0$. Define $\mathbf{Q}^{(0)}$ by the representation (8) with fixed finite $S \subset \mathbb{Z}$ and $\{\sigma_k\}_{k \in S}$. There are $\epsilon_0 \in \left(0, \frac{1}{2}\right)$ and $C > 0$ such that for all $\epsilon \in (-\epsilon_0, \epsilon_0)$, there exists a unique solution $\mathbf{Q}^{(\epsilon)} \in H_e^2((0, T), l^2(\mathbb{Z}))$ of the discrete Klein–Gordon equation (1) satisfying*

$$\|\mathbf{Q}^{(\epsilon)} - \mathbf{Q}^{(0)}\|_{H^2_{\mathrm{per}}((0,T),l^2(\mathbb{Z}))} \leq C|\epsilon|. \tag{12}$$

Moreover, the map $(-\epsilon_0, \epsilon_0) \ni \epsilon \mapsto \mathbf{Q}^{(\epsilon)} \in H_e^2((0,T), l^2(\mathbb{Z}))$ *is* C^∞.

Proof. We shall write $M^{-1}(\epsilon) = I + \epsilon K(\epsilon)$, where $K(\epsilon) : l^2(\mathbb{Z}) \to l^2(\mathbb{Z})$ is a bounded analytic operator for all $\epsilon \in \left(-\frac{1}{2}, \frac{1}{2}\right)$. Then, the discrete Klein–Gordon equation (1) for T-periodic solutions $\mathbf{Q}$ can be rewritten in the perturbed form:

$$\frac{d^2\mathbf{Q}}{dt^2} + \mathbf{f}(\mathbf{Q}) = -\epsilon K(\epsilon)\mathbf{f}(\mathbf{Q}), \tag{13}$$

where $(\mathbf{f}(\mathbf{Q}))_n = V'(Q_n)$.

Substituting $\mathbf{Q} = \mathbf{Q}^{(0)} + \mathbf{W}$, where $\mathbf{Q}^{(0)}$ is given by (8), we obtain the coupled system of differential-difference equations in the form:

$$L_e W_n = -\epsilon \left(K(\epsilon)\mathbf{f}(\mathbf{Q}^{(0)})\right)_n + (\mathbf{N}(\mathbf{W}, \epsilon))_n, \quad n \in S \tag{14}$$

and

$$L_0 W_n = -\epsilon \left(K(\epsilon)\mathbf{f}(\mathbf{Q}^{(0)})\right)_n + (\mathbf{N}(\mathbf{W}, \epsilon))_n, \quad n \in \mathbb{Z}\backslash S, \tag{15}$$

where the linear operators are

$$\begin{aligned} L_e &= \partial_t^2 + V''(\varphi(t)) : \quad H_e^2(0,T) \to L_e^2(0,T), \\ L_0 &= \partial_t^2 + 1 : \qquad\qquad H_e^2(0,T) \to L_e^2(0,T) \end{aligned}$$

and the nonlinear vector field is

$$\begin{aligned} (\mathbf{N}(\mathbf{W}, \epsilon))_n &= -\epsilon \left(K(\epsilon)(\mathbf{f}(\mathbf{Q}^{(0)} + \mathbf{W}) - \mathbf{f}(\mathbf{Q}^{(0)}))\right)_n \\ &\quad + V'(Q_n^{(0)}) + V''(Q_n^{(0)})W_n - V'(Q_n^{(0)} + W_n). \end{aligned}$$

We have used here that V is even and $V''(0) = 1$.

Under the condition $T'(E) \neq 0$, the operator L_e is invertible, because the only eigenvector $\dot{\varphi}$ of $L = \partial_t^2 + V''(\varphi(t)) : H^2_{\text{per}}(0,T) \to L^2_{\text{per}}(0,T)$ is odd in t. Similarly, operator L_0 is invertible if $T \neq 2\pi n$, $n \in \mathbb{N}$.

Thanks to Banach algebra of $H_e^2((0,T), l^2(\mathbb{Z}))$ and the assumption $V \in C^\infty(\mathbb{R})$, the map $\mathbf{N}(\mathbf{Q}, \epsilon) : H_e^2((0,T), l^2(\mathbb{Z})) \times \mathbb{R} \to H_e^2((0,T), l^2(\mathbb{Z}))$ is C^∞, hence it is locally Lipschitz. Thanks to the invertibility of the linearized operators L_e and L_0 on $L_e^2(0,T)$, the result of the theorem follows from the Implicit Function Theorem and the map $\epsilon \mapsto \mathbf{Q}^{(\epsilon)}$ is C^∞ for small ϵ (Theorem 4.E in [17]). □

Remark 6. Although persistence of other breather configurations, where oscillators are neither in-phase nor anti-phase, cannot be apriori excluded, we restrict our studies to the breather configurations covered by Proposition 2.

4 Stability of Multi-site Breathers

Let $\mathbf{Q} \in H_e^2((0,T), l^2(\mathbb{Z}))$ be a multi-site breather in Proposition 2. To study the spectral stability of multi-site breathers, we substitute the decomposition $\mathbf{q}(t) = \mathbf{Q}(t)+\mathbf{w}(t)$ to the discrete Klein–Gordon equation (1), neglect quadratic and higher-order terms in $\mathbf{w}$, and obtain the linearized discrete Klein–Gordon equation,

$$\ddot{w}_n + V''(Q_n)w_n = \epsilon\left(\ddot{w}_{n+1} + \ddot{w}_{n-1}\right), \quad n \in \mathbb{Z}. \tag{16}$$

Using the abstract evolution form (6) and the decomposition $M^{-1}(\epsilon) = I + \epsilon K(\epsilon)$, we can rewrite the linearized equations (16) in the equivalent form:

$$\frac{d^2\mathbf{w}}{dt^2} + \mathbf{f}'(\mathbf{Q})\mathbf{w} = -\epsilon K(\epsilon)\mathbf{f}'(\mathbf{Q})\mathbf{w}, \tag{17}$$

where $\mathbf{f}'(\mathbf{Q})$ is the diagonal operator with entries $V''(Q_n)$, $n \in \mathbb{Z}$.

Because $\mathbf{Q}(t+T) = \mathbf{Q}(t)$, an infinite-dimensional analogue of the Floquet theorem applies and the Floquet monodromy matrix $\mathcal{M}$ is defined by $\mathbf{w}(T) = \mathcal{M}\mathbf{w}(0)$. We say that the breather is stable if all eigenvalues of $\mathcal{M}$, called Floquet multipliers, are located on the unit circle and it is unstable if there is at least one Floquet multiplier outside the unit disk. Because the linearized system (16) is reversible, Floquet multipliers come in pairs μ_1 and μ_2 with $\mu_1\mu_2 = 1$.

To consider Floquet multipliers, we can introduce the characteristic exponent λ in the decomposition $\mathbf{w}(t) = \mathbf{W}(t)e^{\lambda t}$. If $\mu = e^{\lambda T}$ is the Floquet multiplier of the monodromy operator $\mathcal{M}$, then $\mathbf{W} \in H^2_{\text{per}}((0,T), l^2(\mathbb{Z}))$ is a solution of the eigenvalue problem,

$$\frac{d^2\mathbf{W}}{dt^2} + 2\lambda\frac{d\mathbf{W}}{dt} + \lambda^2\mathbf{W} + \mathbf{f}'(\mathbf{Q})\mathbf{W} = -\epsilon K(\epsilon)\mathbf{f}'(\mathbf{Q})\mathbf{W}. \tag{18}$$

In particular, Floquet multiplier $\mu = 1$ corresponds to the characteristic exponent $\lambda = 0$. The generalized eigenvector $\mathbf{Z}_0 \in H^2_{\text{per}}((0,T), l^2(\mathbb{Z}))$ of the eigenvalue problem (18) for $\lambda = 0$ solves the inhomogeneous problem,

$$\frac{d^2\mathbf{Z}_0}{dt^2} + \mathbf{f}'(\mathbf{Q})\mathbf{Z}_0 = -\epsilon K(\epsilon)\mathbf{f}'(\mathbf{Q})\mathbf{Z}_0 - 2\frac{d\mathbf{W}_0}{dt}, \tag{19}$$

where $\mathbf{W}_0$ is the eigenvector of (18) for $\lambda = 0$. To normalize $\mathbf{Z}_0$ uniquely, we add a constraint that $\mathbf{Z}_0$ is orthogonal to $\mathbf{W}_0$ with respect to the natural inner product

$$\langle \mathbf{W}_0, \mathbf{Z}_0 \rangle_{L^2_{\text{per}}((0,T),l^2(\mathbb{Z}))} := \int_0^T \sum_{n\in\mathbb{Z}} (\bar{\mathbf{Z}}_0)_n(t)(\mathbf{W}_0)_n(t)dt.$$

At $\epsilon = 0$, the eigenvector $\mathbf{W}_0$ of the eigenvalue problem (18) for $\lambda = 0$ is spanned by the linear combination of N fundamental solutions,

$$\mathbf{W}^{(0)}(t) = \sum_{k \in S} c_k \dot{\varphi}(t) \mathbf{e}_k, \tag{20}$$

where N is the number of sites in the set S. The generalized eigenvector $\mathbf{Z}_0$ is spanned by the linear combination of N solutions,

$$\mathbf{Z}^{(0)}(t) = -\sum_{k \in S} c_k v(t) \mathbf{e}_k, \quad v := 2L_e^{-1}\ddot{\varphi}, \tag{21}$$

where $L_e = \partial_t^2 + V''(\varphi(t)) : H_e^2(0,T) \to L_e^2(0,T)$ is invertible and $\ddot{\varphi} \in L_e^2(0,T)$. Note that $\langle \dot{\varphi}, v \rangle_{L^2_{\rm per}(0,T)} = 0$ because $\dot{\varphi}$ is odd and v is even in t.

We proceed now with perturbation expansions for particular configurations S of the limiting breather (8). Perturbation expansions are different depending if the set S has no holes (the excited oscillators are adjacent) or includes some holes (oscillators at rest are located between excited oscillators).

4.1 Adjacent Excited Oscillators

We consider here the set $S = \{1, 2, \ldots, N\}$ of N adjacent sites with excited oscillators at $\epsilon = 0$. By Proposition 2, the breather solution $\mathbf{Q}^{(\epsilon)}$ can be expanded into the power series

$$\mathbf{Q}^{(\epsilon)} = \mathbf{Q}^{(0)} + \sum_{m=1}^{\infty} \epsilon^m \mathbf{Q}^{(m)}, \tag{22}$$

where $\mathbf{Q}^{(0)}(t) = \sum_{k=1}^{N} \sigma_k \varphi(t) \mathbf{e}_k$ and the correction terms are computed recursively from the system of linear inhomogeneous equations. In particular, for the first-order correction term, we write the linear inhomogeneous problem explicitly as follows:

$$\left(\frac{d^2}{dt^2} + V''(Q_n^{(0)}) \right) Q_n^{(1)} = -\left((\sigma_+ + \sigma_-) \mathbf{f}(\mathbf{Q}^{(0)}) \right)_n, \quad n \in \mathbb{Z}. \tag{23}$$

where again $(\mathbf{f}(\mathbf{Q}))_n = V'(Q_n)$.

Let φ be an even T-periodic solution of the nonlinear oscillator equation (9) subject to the initial conditions (10). Let ψ and ϕ be even T-periodic solutions of the linear inhomogeneous equations

$$\ddot{\psi} + V''(\varphi)\psi = V'(\varphi), \tag{24}$$

and

$$\ddot{\phi} + \phi = V'(\varphi). \tag{25}$$

Note that the unique even solutions exist for the linear equations (24) and (25) under the conditions $T'(E) \neq 0$ and $T \neq 2\pi m$, $m \in \mathbb{N}$ because of invertibility of operators L_e and L_0 defined in the proof of Proposition 2. By using solutions ψ and ϕ, we can write the first-order correction term $\mathbf{Q}^{(1)}$ explicitly as follows:

$$\mathbf{Q}^{(1)}(t) = -\sigma_1\phi(t)\mathbf{e}_0 - \sum_{k=1}^{N}(\sigma_{k-1} + \sigma_{k+1})\psi(t)\mathbf{e}_k - \sigma_N\phi(t)\mathbf{e}_{N+1}, \tag{26}$$

where we have used the convention: $\sigma_0 = \sigma_{N+1} = 0$. The following theorem represents the main result of the perturbation computations.

Theorem 1. *Under assumptions of Proposition 2, let $\mathbf{Q}^{(0)} = \sum_{k=1}^{N} \sigma_k \varphi \mathbf{e}_k$ yield a solution $\mathbf{Q}^{(\epsilon)} \in H_e^2((0,T), l^2(\mathbb{Z}))$ of the discrete Klein–Gordon equation (1) for small $\epsilon > 0$. Then the eigenvalue problem (18) for small $\epsilon > 0$ has $2N$ small eigenvalues,*

$$\lambda = \epsilon^{1/2}\Lambda + \mathcal{O}(\epsilon),$$

where Λ is an eigenvalue of the matrix eigenvalue problem

$$\frac{T^2(E)}{T'(E)M_1}\Lambda^2\mathbf{c} = \mathcal{S}\mathbf{c}, \quad \mathbf{c} \in \mathbb{C}^N. \tag{27}$$

Here M_1 is a positive numerical coefficient given by

$$M_1 = \int_0^T \ddot{\varphi}^2 dt > 0$$

and the $N \times N$ matrix $\mathcal{S}$ is given by

$$\mathcal{S} = \begin{bmatrix} -\sigma_1\sigma_2 & 1 & 0 & \dots & 0 & 0 \\ 1 & -\sigma_2(\sigma_1+\sigma_3) & 1 & \dots & 0 & 0 \\ 0 & 1 & -\sigma_3(\sigma_2+\sigma_4) & \dots & 0 & 0 \\ \vdots & \vdots & \vdots & \ddots & \vdots & \vdots \\ 0 & 0 & 0 & \dots & -\sigma_{M-1}(\sigma_{M-2}+\sigma_M) & 1 \\ 0 & 0 & 0 & \dots & 0 & -\sigma_M\sigma_{M-1} \end{bmatrix}.$$

Proof. At $\epsilon = 0$, the eigenvalue problem (18) admits eigenvalue $\lambda = 0$ of geometric multiplicity N and algebraic multiplicity $2N$, which is isolated from the rest of the spectrum. Perturbation theory in ϵ applies thanks to the smoothness of $\mathbf{Q}^{(\epsilon)}$ in ϵ and

V' in u. Perturbation expansions (so-called Puiseux series, see Chap. 2 in [5]) are smooth in powers of $\epsilon^{1/2}$ thanks to the Jordan block decomposition at $\epsilon = 0$.

We need to find out how the eigenvalue $\lambda = 0$ of algebraic multiplicity $2N$ split for small $\epsilon > 0$. Therefore, we are looking for the eigenvectors of the eigenvalue problem (18) in the subspace associated with the eigenvalue $\lambda = 0$ using the substitution $\lambda = \epsilon^{1/2}\tilde{\lambda}$ and the decomposition

$$\mathbf{W} = \mathbf{W}^{(0)} + \epsilon^{1/2}\tilde{\lambda}\mathbf{W}^{(1)} + \epsilon\tilde{\mathbf{W}},$$

where

$$\mathbf{W}^{(0)} = \sum_{k=1}^{N} c_k \dot{\varphi} \mathbf{e}_k$$

and

$$\mathbf{W}^{(1)} = -\sum_{k=1}^{N} c_k v(t) \mathbf{e}_k, \quad v := 2L_e^{-1}\ddot{\varphi}.$$

The error term $\tilde{\mathbf{W}}$ satisfies a residual equation, which we only write on the active sites $k \in S$ in the perturbation form:

$$\begin{aligned}\ddot{\tilde{W}}_k + V''(\varphi)\tilde{W}_k &= -\left((\sigma_+ + \sigma_-)\mathbf{f}'(\mathbf{Q}^{(0)})\mathbf{W}^{(0)}\right)_k - c_k\sigma_k V'''(\varphi)Q_k^{(1)}\dot{\varphi} \\ &\quad + \tilde{\lambda}^2 c_k(2\dot{v} - \dot{\varphi}) + \mathcal{O}(\epsilon^{1/2}),\end{aligned} \tag{28}$$

where again $\mathbf{f}'(\mathbf{Q})$ is the diagonal operator with entries $V''(Q_n)$, $n \in \mathbb{Z}$.

Expanding $\tilde{\lambda} = \Lambda + \mathcal{O}(\epsilon^{1/2})$, projecting the system of linear inhomogeneous equations (28) to $\dot{\varphi} \in H^2_{\text{per}}(0,T)$, the kernel of $L = \partial_t^2 + V''(\varphi) : H^2_{\text{per}}(0,T) \to L^2_{\text{per}}(0,T)$, and truncating at the leading order, we obtain the system of difference equations for $k \in S$:

$$\begin{aligned}\Lambda^2 c_k \int_0^T \left(\dot{\varphi}^2 + 2v\ddot{\varphi}\right) dt &= -(c_{k+1} + c_{k-1}) \int_0^T V''(\varphi)\dot{\varphi}^2 dt \\ &\quad + \sigma_k(\sigma_{k+1} + \sigma_{k-1})c_k \int_0^T V'''(\varphi)\dot{\varphi}^2\psi dt,\end{aligned} \tag{29}$$

where the explicit expression (26) has been used, as well as the convention: $c_0 = c_{N+1} = 0$.

It is proved with the standard computation [12] that

$$\int_0^T \left(\dot{\varphi}^2 + 2v\ddot{\varphi}\right) dt = -\frac{T^2(E)}{T'(E)}. \tag{30}$$

On the other hand, differentiating the linear inhomogeneous equation (24) and projecting it to $\dot{\varphi}$, we infer that

$$\int_0^T V'''(\varphi)\dot{\varphi}^2\psi dt = \int_0^T V''(\varphi)\dot{\varphi}^2 dt, \tag{31}$$

Using now the equation

$$\dddot{\varphi} + V''(\varphi)\dot{\varphi} = 0, \tag{32}$$

we finally obtain

$$\int_0^T V''(\varphi)\dot{\varphi}^2 dt = -\int_0^T \dot{\varphi}\dddot{\varphi} dt = \int_0^T \ddot{\varphi}^2 dt = M_1.$$

Combining all together, the system of difference equations (29) yields the matrix eigenvalue problem (27), which defines $2N$ small eigenvalues that bifurcate from $\lambda = 0$ for small $\epsilon > 0$. □

We recall the result obtained by Sandstede in Lemma 5.4 and Appendix C [15], which we reproduce here without a proof.

Proposition 3. *Let n_0 be the number of negative elements in $\{\sigma_j\sigma_{j+1}\}_{j=1}^{N-1}$. Matrix $\mathcal{S}$ in Theorem 1 has exactly n_0 positive and $N-1-n_0$ negative eigenvalues counting their multiplicities, in addition to the simple zero eigenvalue.*

Remark 7. Because $M_1 > 0$, the matrix eigenvalue problem (27) differs from the similar reduction for the standard Klein–Gordon equation in [12] by the sign change in front of the matrix $\mathcal{S}$. In particular, if $T'(E) < 0$, the matrix eigenvalue problem (27) has n_0 pairs of purely imaginary eigenvalues Λ and $N - 1 - n_0$ pairs of purely real eigenvalues Λ counting their multiplicities, in addition to the double zero eigenvalue. If $T'(E) > 0$, the conclusion changes to the opposite.

4.2 Oscillators at Rest Between Excited Oscillators

We consider here the set $S = \{1, 3, \ldots, 2N - 1\}$ of N sites with excited oscillators separated by exactly one oscillator at rest at $\epsilon = 0$. By using the power series expansions (22) in Proposition 2 with $\mathbf{Q}^{(0)}(t) = \sum_{k=1}^{N} \sigma_{2k-1}\varphi(t)\mathbf{e}_{2k-1}$, we compute a different explicit solution of the linear inhomogeneous equation (23):

$$\mathbf{Q}^{(1)}(t) = -\sum_{k=0}^{N}(\sigma_{2k-1} + \sigma_{2k+1})\phi(t)\mathbf{e}_{2k}, \tag{33}$$

where ϕ is an even T-periodic solution of the linear inhomogeneous equation (25) and we have used the convention: $\sigma_{-1} = \sigma_{2N+1} = 0$.

To find the second-order correction term, we write the linear inhomogeneous problem explicitly as follows:

$$\left(\frac{d^2}{dt^2} + V''(Q_n^{(0)})\right) Q_n^{(2)} = -\left((\sigma_+ + \sigma_-)^2 \mathbf{f}(\mathbf{Q}^{(0)})\right)_n - \left((\sigma_+ + \sigma_-)\mathbf{f}'(\mathbf{Q}^{(0)})\mathbf{Q}^{(1)}\right)_n,$$

where we have used the fact that $V'''(Q_n^{(0)})(Q_n^{(1)})^2 = 0$ for all $n \in \mathbb{Z}$.

Let θ and ζ be even T-periodic solutions of the linear inhomogeneous equation

$$\ddot{\theta} + V''(\varphi)\theta = \phi, \quad t \in \mathbb{R} \tag{34}$$

and

$$\ddot{\zeta} + \zeta = \phi, \quad t \in \mathbb{R}, \tag{35}$$

which exist and are unique under the conditions that $T'(E) \neq 0$ and $T \neq 2\pi m$, $m \in \mathbb{N}$. By using these solutions, we can write the second-order correction term $\mathbf{Q}^{(2)}$ explicitly as follows:

$$\begin{aligned}\mathbf{Q}^{(2)}(t) = &-\sum_{k=1}^{N}(\sigma_{2k-3} + 2\sigma_{2k-1} + \sigma_{2k+1})(\psi(t) - \theta(t))\mathbf{e}_{2k-1} \\ &-\sigma_1(\phi(t) - \zeta(t))\mathbf{e}_{-1} - \sigma_{2N-1}(\phi(t) - \zeta(t))\mathbf{e}_{2N+1}.\end{aligned} \tag{36}$$

The following theorem represents the main result of the perturbation computations.

Theorem 2. *Under assumptions of Proposition 2, let* $\mathbf{Q}^{(0)} = \sum_{k=1}^{N} \sigma_{2k-1}\varphi\mathbf{e}_{2k-1}$ *yield a solution* $\mathbf{Q}^{(\epsilon)} \in H_e^2((0,T), l^2(\mathbb{Z}))$ *of the discrete Klein–Gordon equation (1) for small* $\epsilon > 0$. *Then the eigenvalue problem (18) for small* $\epsilon > 0$ *has* $2N$ *small eigenvalues,*

$$\lambda = \epsilon\Lambda + \mathcal{O}(\epsilon^2),$$

where Λ *is an eigenvalue of the matrix eigenvalue problem*

$$\frac{T^2(E)}{T'(E)M_2}\Lambda^2\mathbf{c} = \mathcal{S}\mathbf{c}, \quad \mathbf{c} \in \mathbb{C}^N, \tag{37}$$

associated with the same matrix $\mathcal{S}$ *as in Theorem 1 and a different numerical coefficient* M_2 *given by*

$$M_2 = -\int_0^T \ddot{\varphi}\ddot{\phi}\,dt.$$

Proof. Similarly to the proof of Theorem 1, we are looking for the eigenvectors of the eigenvalue problem (18) in the subspace associated with the eigenvalue $\lambda = 0$ using the substitution $\lambda = \epsilon\tilde{\lambda}$ and the decomposition

$$\mathbf{W} = \mathbf{W}^{(0)} + \epsilon\tilde{\lambda}\mathbf{W}^{(1)} + \epsilon\mathbf{W}^{(2)} + \epsilon^2\tilde{\mathbf{W}},$$

where

$$\mathbf{W}^{(0)} = \sum_{k=1}^{N} c_{2k-1}\dot{\varphi}\mathbf{e}_{2k-1}, \tag{38}$$

$$\mathbf{W}^{(1)} = -\sum_{k=1}^{N} c_{2k-1}v\mathbf{e}_{2k-1}, \tag{39}$$

$$\mathbf{W}^{(2)} = -\sum_{k=0}^{N} (c_{2k-1} + c_{2k+1})\dot{\phi}\mathbf{e}_{2k}, \tag{40}$$

subject to the convention: $c_{-1} = c_{2N+1} = 0$. The error term $\tilde{\mathbf{W}}$ satisfies a residual equation, which we only write on the active sites $2k-1 \in S$ in the perturbation form:

$$\begin{aligned}\ddot{\tilde{W}}_{2k-1} + V''(\varphi)\tilde{W}_{2k-1} =& -\left((\sigma_+ + \sigma_-)^2\mathbf{f}'(\mathbf{Q}^{(0)})\mathbf{W}^{(0)}\right)_{2k-1}\\ &-\left((\sigma_+ + \sigma_-)\mathbf{f}'(\mathbf{Q}^{(0)})\mathbf{W}^{(2)}\right)_{2k-1}\\ &-c_{2k-1}\sigma_{2k-1}V'''(\varphi)Q_k^{(2)}\dot{\varphi} + \tilde{\lambda}^2 c_{2k-1}(2\dot{v} - \dot{\varphi})\\ &+\mathcal{O}(\epsilon),\end{aligned} \tag{41}$$

where we have used properties of the explicit solutions (33), (36), and (38)–(40).

Expanding $\tilde{\lambda} = \Lambda + \mathcal{O}(\epsilon)$, projecting the system of linear inhomogeneous equations (41) to $\dot{\varphi} \in H^2_{\text{per}}(0,T)$, and truncating at the leading order, we obtain the system of difference equations for $2k-1 \in S$:

$$\begin{aligned}\Lambda^2 c_{2k-1}\int_0^T \left(\dot{\varphi}^2 + 2v\ddot{\varphi}\right)dt =& -(c_{2k+1} + 2c_{2k-1} + c_{2k-3})\int_0^T (V''(\varphi)\dot{\varphi}^2 - \dot{\varphi}\dot{\phi})dt\\ &+\sigma_{2k-1}(\sigma_{2k+1} + 2\sigma_{2k-1} + \sigma_{2k-3})c_{2k-1}\int_0^T V'''(\varphi)\dot{\varphi}^2(\psi - \theta)dt.\end{aligned} \tag{42}$$

Differentiating the linear inhomogeneous equation (34) and projecting it to $\dot{\varphi}$, we infer that

$$\int_0^T V'''(\varphi)\dot{\varphi}^2\theta dt = \int_0^T \dot{\varphi}\dot{\phi}dt. \tag{43}$$

Combining (30), (31), and (43), the system of difference equations (42) yields the matrix eigenvalue problem (37) with

$$M_2 = \int_0^T (V''(\varphi)\dot{\varphi}^2 - \dot{\varphi}\dot{\phi})dt = \int_0^T \ddot{\varphi}(\ddot{\varphi} + \phi)dt$$
$$= \int_0^T \ddot{\varphi}(\phi - V'(\varphi))dt = -\int_0^T \ddot{\varphi}\ddot{\phi}dt,$$

where we have used Eqs. (25) and (32), as well as integration by parts. The matrix eigenvalue problem (37) defines $2N$ small eigenvalues that bifurcate from $\lambda = 0$ for small $\epsilon > 0$. □

Remark 8. The matrix eigenvalue problem (37) differs from the similar reduction for the standard Klein–Gordon equation in [12] by the sign change in front of the matrix $\mathcal{S}$ and by the replacement of the quantity M_2 with the quantity

$$M_2^{[12]} = \int_0^T \dot{\varphi}\dot{\phi}dt,$$

where ϕ is now a solution of the linear inhomogeneous equation $\ddot{\phi} + \phi = \varphi$ instead of Eq. (25).

5 Discussion

We consider the example of the discrete Klein–Gordon equation (1) related to the potential (2). Because V is even, the even solution $\varphi \in H_e^2(0,T)$ satisfies the symmetry

$$\varphi\left(\frac{T}{2} - t\right) = -\varphi(t), \quad t \in \mathbb{R}, \tag{44}$$

so that it can be expanded into the Fourier cosine series,

$$\varphi(t) = \sum_{n \in \mathbb{N}_{\text{odd}}} c_n(T) \cos\left(\frac{2\pi n t}{T}\right), \tag{45}$$

with zero coefficients c_n for all even n. Because $V'(\varphi) = -\ddot{\varphi}$, a solution of the linear inhomogeneous equation (25) can also be found in the form of the Fourier cosine series:

$$\phi(t) = \sum_{n \in \mathbb{N}_{\text{odd}}} \frac{4\pi^2 n^2 c_n(T)}{T^2 - 4\pi^2 n^2} \cos\left(\frac{2\pi n t}{T}\right). \tag{46}$$

Table 1 Stable multi-site breathers in the hard and soft potentials

	Theorem 1	Theorem 2
$T'(E) < 0$ $0 < T < 2\pi$	Anti-phase	Anti-phase
$T'(E) > 0$ $2\pi < T < 6\pi$	In-phase	$2\pi < T < T_*$ anti-phase $T_* < T < 6\pi$ in-phase

Using Parseval's equality, we compute the numerical coefficient M_2 in Theorem 2 in the Fourier series form:

$$M_2 = -\int_0^T \ddot{\varphi}\ddot{\phi}dt = \sum_{n\in\mathbb{N}_{\text{odd}}} \frac{(4\pi^2n^2)^3|c_n(T)|^2}{T^3(4\pi^2n^2 - T^2)}. \tag{47}$$

Consider now the dependence $T(E)$ defined by the integral formula (11). Because $V''(0) = 1$, we have $T(E) \to 2\pi$ as $E \to 0$. For small values of E, the cubic term in the expansion (2) shows that the case $\alpha < 0$ gives a hard potential with $T'(E) < 0$, whereas the case $\alpha > 0$ gives a soft potential with $T'(E) > 0$ for small E.

If $T'(E) < 0$ and $T(E) < 2\pi$, then $M_2 > 0$. Also recall that $M_1 > 0$. In this case, Proposition 3 implies that the only stable configuration of the multi-site breathers in Theorems 1 and 2 is the one with all alternating $\{\sigma_k\}_{k=1}^N$ or $\{\sigma_{2k-1}\}_{k=1}^N$ (anti-phase breathers). This conclusion is recorded in the first line of Table 1.

If $T'(E) > 0$ and $T(E) > 2\pi$, then the situation is different between Theorems 1 and 2. Because $M_1 > 0$, the only stable configuration of the multi-site breathers in Theorem 1 is the one with all equal $\{\sigma_k\}_{k=1}^M$ (in-phase breathers).

On the other hand, the quantity M_2 changes sign in the interval $T(E)$ between two resonances at 2π and 6π, because the first negative term in the series (47) dominates if $T(E)$ is close to 2π whereas the second positive term dominates if $T(E)$ is close to 6π. Therefore, there exists a period $T_* \in (2\pi, 6\pi)$ such that $M_2 < 0$ for $T \in (2\pi, T_*)$ and $M_2 > 0$ for $T \in (T_*, 6\pi)$. Stable configurations of discrete solitons for $T'(E) > 0$ and $2\pi < T < 6\pi$ are recorded in the second line of Table 1.

We can now compare these analytical results with numerical simulations of one-dimensional discrete breathers in [4, 7]. Figure 1 in [7] and Fig. 6 in [4] show stable propagation of the so-called fundamental breather ($N = 1$ in Theorem 1) for $\alpha > 0$. Profiles of stable breathers are also shown in Fig. 4 for $\alpha > 0$ and Fig. 5 for $\alpha < 0$ [4]. The stable fundamental breather corresponds to the sign-definite (positive) amplitudes for $\alpha > 0$ and sign-alternating amplitudes for $\alpha < 0$, which is in agreement with the results of Table 1. The two-site twisted (sign-alternating) mode ($N = 2$ in Theorem 2) is reported to be stable both for $\alpha > 0$ and $\alpha < 0$, which is also in agreement with Table 1 for $2\pi < T < T_*$.

We note that the results of Table 1 apply only to the small-amplitude discrete breathers in the nonlinear potential (2) because the sign of $T'(E)$ may change for large amplitudes. In particular, the potential may become hard for large amplitudes in the case $\alpha > 0$ because $f(q) \sim q^{1/3}$ as $q \to \infty$. Similarly, the potential may become soft for large amplitudes in the case $\alpha < 0$ because $f(q)$ only exists for $q \in (-Q_0, Q_0)$, where Q_0 is given by (7).

We do not also know if any discrete breather in the nonlinear potential given by (2) can have the period close to the resonant value 6π, to observe additional phenomena such as pitchfork bifurcations of single-site and multi-site breathers [12]. These open questions will await further detailed numerical studies of the discrete Klein–Gordon equation (1).

Acknowledgements The research D.P. is supported by the NSERC Discovery grant. The research of V.R. has been co-financed by the European Union (European Social Fund – ESF) and Greek national funds through the Operational Program "Education and Lifelong Learning" of the National Strategic Reference Framework (NSRF) – Research Funding Program: THALES – Investing in knowledge society through the European Social Fund.

References

1. J.F.R. Archilla, J. Cuevas, B. Sänchez-Rey, A. Alvarez, Demonstration of the stability or instability of multibreathers at low coupling. Physica D **180**, 235–255 (2003)
2. P. Cherrier, A. Milani, *Linear and Quasi-linear Evolution Equations in Hilbert Spaces* (AMS, Providence, 2012)
3. J. Cuevas, V. Koukouloyannis, P.G. Kevrekidis, J.F.R. Archilla, Multibreather and vortex breather stability in Klein–Gordon lattices: equivalence between two different approaches. Int. J. Bifurc. Chaos **21**, 2161–2177 (2011)
4. M. Eleftheriou, N. Lazarides, G.P. Tsironis, Magnetoinductive breathers in metamaterials. Phys. Rev. E **77**, 036608 (13 pages) (2008)
5. T. Kato, *Perturbation Theory for Linear Operators* (Springer, Berlin, 1995)
6. V. Koukouloyannis, P.G. Kevrekidis, On the stability of multibreathers in Klein–Gordon chains. Nonlinearity **22**, 2269–2285 (2009)
7. N. Lazarides, M. Eleftheriou, G.P. Tsironis, Discrete breathers in nonlinear magnetic metamaterials. Phys. Rev. Lett. **97**, 157406 (4 pages) (2006)
8. N. Lazarides, G.P. Tsironis, Yu.S. Kivshar, Surface breathers in discrete magnetic metamaterials. Phys. Rev. E **77**, 065601(R) (4 pages) (2008)
9. R.S. MacKay, S. Aubry, Proof of existence of breathers for time-reversible or Hamiltonian networks of weakly coupled oscillators. Nonlinearity **7**, 1623–1643 (1994)
10. A.M. Morgante, M. Johansson, G. Kopidakis, S. Aubry, Standing wave instabilities in a chain of nonlinear coupled oscillators. Physica D **162**, 53–94 (2002)
11. D.E. Pelinovsky, A. Sakovich, Internal modes of discrete solitons near the anti-continuum limit of the dNLS equation. Physica D **240**, 265–281 (2011)
12. D.E. Pelinovsky, A. Sakovich, Multi-site breathers in Klein–Gordon lattices: stability, resonances, and bifurcations. Nonlinearity **25**, 3423–3451 (2012)
13. D.E. Pelinovsky, P.G. Kevrekidis, D.J. Frantzeskakis, Stability of discrete solitons in nonlinear Schrödinger lattices. Physica D **212**, 1–19 (2005)
14. Z. Rapti, Multi-breather stability in discrete Klein–Gordon equations: beyond nearest neighbor interactions. The paper was published in Phys. Lett. A **377**, 1543–1553 (2013)
15. B. Sandstede, Stability of multiple-pulse solutions. Trans. Am. Math. Soc. **350**, 429–472 (1998)
16. K. Yoshimura, Stability of discrete breathers in nonlinear Klein–Gordon type lattices with pure anharmonic couplings. J. Math. Phys. **53**, 102701 (20 pages) (2012)
17. E. Zeidler, *Applied Functional Analysis. Main Principles and Their Applications*. Applied Mathematical Sciences, vol. 109 (Springer, New York, 1995)

Part VI
Other Nonlinear Models

Checking the Validity of Truncating the Cumulant Hierarchy Description of a Small System

Manuel Morillo, José Gómez-Ordóñez, and José M. Casado

Abstract We analyze the behavior of the first few cumulant moments in an array with a small number of coupled identical particles. Desai and Zwanzig (J Stat Phys 19(1):1 1978) studied noisy arrays of nonlinear units with global coupling and derived an infinite hierarchy of differential equations for the cumulant moments. They focused on the behavior of infinite size systems using a strategy based on truncating the hierarchy. In this work we explore the reliability of such an approach to describe systems with a small number of elements. We carry out an extensive numerical analysis of the truncated hierarchy as well as numerical simulations of the full set of Langevin equations governing the dynamics. We find that the results provided by the truncated hierarchy for finite systems are at variance with those of the Langevin simulations for large regions of parameter space. The truncation of the hierarchy leads to a dependence on initial conditions and to the coexistence of states which are not consistent with the theoretical expectations based on the multidimensional linear Fokker-Planck equation for finite arrays.

1 Introduction

The description of nonlinear stochastic systems can hardly be carried out without approximations due to the interplay of noise and nonlinearity. In some problems, the stationary distribution for the relevant variables is available in analytical form, but in general very little information can be obtained without approximations. A convenient way of describing the system dynamics is in term of cumulant moments satisfying an infinite set of coupled ordinary differential equations [1]. For all practical purposes, this infinite hierarchy needs to be truncated in order to

M. Morillo (✉) · J. Gómez-Ordóñez · J.M. Casado
Facultad de Física, Área de Física Teórica, Universidad de Sevilla, Apartado de Correos 1065, 41080 Sevilla, Spain
e-mail: morillo@us.es

R. Carretero-González et al. (eds.), *Localized Excitations in Nonlinear Complex Systems*, Nonlinear Systems and Complexity 7, DOI 10.1007/978-3-319-02057-0_19,

obtain a finite set of closed equations. A very much used approximation consists in truncating the infinite hierarchy at the Gaussian level by neglecting cumulants of order three and higher. For a stochastic variable $Y(t)$, Marcinkiewicz [2] indicated that the characteristic function $\Gamma(\lambda) = \langle \exp(i\lambda Y(t)) \rangle$ can be expressed as $\Gamma(\lambda) = \exp(P(\lambda, t))$ where $P(\lambda, t)$ is a polynomial of first or second degree in λ. Consequently, as pointed out by Hänggi and Talkner [3], the truncation of the hierarchy at levels higher than two is problematic, although, as these authors emphasize, it is not an empty concept. In their own words "it only means that the neglect of cumulants beyond a given order cannot be justified a priori".

In this work we consider the statistical mechanical description of a stochastic array containing a small finite number N of coupled identical elements. The dynamics will be given by a set of N coupled nonlinear stochastic differential equations for the degrees of freedom characterizing the elements of the array. Equivalently, we can describe the system in terms of an N-dimensional joint probability distribution. We will assume that this joint probability distribution satisfies an N-dimensional linear Fokker-Planck equation (FPE).

Obtaining dynamical information from the FPE is plagued with difficulties due to the nonlinear character of the dynamics. As mentioned above, a way of dealing with them is to consider the cumulant moments satisfying an infinite hierarchy of ordinary differential equations. In the case of the arrays studied in this work, Desai and Zwanzig [4] derived from the FPE such an infinite hierarchy. Two types of cumulants appear: diagonal cumulant moments associated to a single degree of freedom, $M_i(t)$, and cross cumulant moments, $\mu_{ijk\ldots}(t)$ involving two or more variables. For an infinite system, the off-diagonal cumulant moments are negligible if they were zero at the initial preparation (see [4]). In other words, off-diagonal cumulant moments are not generated by dynamical evolution in the case of infinitely large systems. But even then, for all practical purposes, the hierarchy has to be approximated by truncating it at a certain level. The results of truncating at different levels the hierarchy of diagonal cumulant moments in the $N \to \infty$ limit was already discussed by Desai and Zwanzig. As also indicated in [3], the cumulant truncation is a succesful strategy in that limit.

By contrast with the previous work of Desai and Zwanzig, the present work focuses on systems with a small number N of elements. In this case, both the diagonal and off-diagonal cumulant moments have to be taken into account. It is then an open question whether truncating the infinite hierarchy is an adequate approximation strategy. It is not possible to solve analytically the truncated set of equations at any level. So we will rely on a numerical treatment of the set. Our goal is to elucidate whether truncation of the infinite hierarchy of cumulant moments at different levels provides a reliable approximation for small finite arrays. To this end, we will compare the results for the first few cumulant moments obtained from the truncated hierarchy with those obtained by numerically solving the set of the N coupled Langevin equations. Besides the numerical simulations of the Langevin dynamics, the results of the truncated hierarchy will also be contrasted with the exact results expected from the stationary solution of the N-dimensional Fokker-Planck equation, as well as the global stability H-theorem. We will see that the simulation results agree with the exact known information, while the truncation strategy might lead to results incompatible with the exact results.

2 The Model

We consider a global coupling model that can be viewed as a set of nonlinear "oscillators", each of them described by a "coordinate" x_i. The dynamics of the system is given by the set of coupled Langevin equations (in dimensionless form)

$$\dot{x}_i = x_i - x_i^3 + \frac{\theta}{N} \sum_{j=1}^{N} (x_j - x_i) + \xi_i(t), \tag{1}$$

where θ is the strength of the global coupling term. This parameter will be taken to be either positive or negative. The terms $\xi_i(t),\ i = 1, 2, \ldots N$ represent N uncorrelated Gaussian white noises with zero averages and $\langle \xi_i(t_1) \xi_j(t_2) \rangle = 2D\delta_{ij}\delta(t_1 - t_2)$. This model was introduced by Kometani and Shimizu [5] as a model for muscle contractions, and it was later on analyzed by Desai and Zwanzig [4] from a Statistical Mechanics perspective. The model describes N degrees of freedom each of them globally coupled to all the other ones. Each degree of freedom has an intrinsic nonlinear dynamics. The nonlinearity and the presence of the noise terms render the behavior of the system far from trivial.

An alternative formulation of the dynamics is in terms of the linear Fokker-Planck equation for the joint probability density $f_N(x_1, x_2, \ldots, x_N, t)$,

$$\frac{\partial f_N}{\partial t} = \sum_i \frac{\partial}{\partial x_i} \left(\frac{\partial U}{\partial x_i} f_N \right) + D \sum_i \frac{\partial^2 f_N}{\partial x_i^2}, \tag{2}$$

where U is the potential energy relief,

$$U = \sum_{i=1}^{N} V(x_i) + \frac{\theta}{4N} \sum_{i=1}^{N} \sum_{j=1}^{N} (x_j - x_i)^2, \tag{3}$$

with the single particle potential

$$V(x_i) = \frac{x_i^4}{4} - \frac{x_i^2}{2}. \tag{4}$$

The term $V(x_i)$ describes a symmetrical potential with two wells of equal depths separated by a hump at $x_i = 0$. The interaction part of the full potential modifies it in such a way that for $\theta > 1$ the two wells blend into a single minimum at $x_i = 0$. For $\theta < 1$, the two wells exist, but their locations and the barrier height depend on θ. Note that for $\theta > 0$, the interaction energy contribution to the full potential favors that any pair x_i and x_j should have the same sign (both either positive or negative), while for $\theta < 0$, the opposite happens and the interaction tends to favor configurations with positive and negative values of the variable.

The only explicit solution of Eq. (2) is its long time stationary one, given by

$$f_N^{st}(x_1, x_2, \ldots, x_N) = \frac{1}{Z} \exp\left(-\frac{U}{D}\right) \tag{5}$$

where Z is a normalization function. This solution is independent of the initial condition, and the system will necessarily relax to it, although the time it takes to do it might be, depending on the system parameters, extremely large. It should be noted that we keep the finite number of particles N fixed, while we take the long time limit. Had we have taken the limit $N \to \infty$ first, as done in the infinite size limit studies, and afterwards the long time limit, the result would differ.

As mentioned above, Desai and Zwanzig [4] derived an infinite hierarchy of ordinary differential equations for the cumulant moments. Neglecting the cumulant moments of order five and higher one gets a set of 11 equations for the first four order (diagonal and off-diagonal) cumulant moments (see the Appendix in [4]). For asymptotically large systems ($N \to \infty$), Desai and Zwanzig found that, for some regions of parameter space, zero is the only stationary value for the first cumulant moment. For other regions of parameter values, there are two stable nonzero stationary values, while the zero value becomes unstable. The regions are separated by a transition line whose shape depends on whether $\theta > 1$ or $0 < \theta < 1$. As we will see in the next section, the truncated hierarchy of cumulants for small systems might yield several stable stationary moments depending upon the initial preparation, as obtained within the $N \to \infty$ limit. Nonetheless, those results are incompatible with the unicity of the solution of a multidimensional linear FPE and with the independence from initial condition of the long-time results, required by the H-theorem.

3 Numerical Simulations

We have carried out numerical simulations of the whole set of Langevin equations, Eq. (1). Using the procedure detailed in Ref. [6], we have integrated the Langevin equations for a large number of noise realizations (typically 5,000 realizations). Averaging over them, we estimate the first two cumulant moments of a single variable by

$$M_1(t) = \langle x(t) \rangle \approx \frac{1}{\mathcal{N}} \sum_{\alpha=1}^{\mathcal{N}} x^{(\alpha)}(t), \tag{6}$$

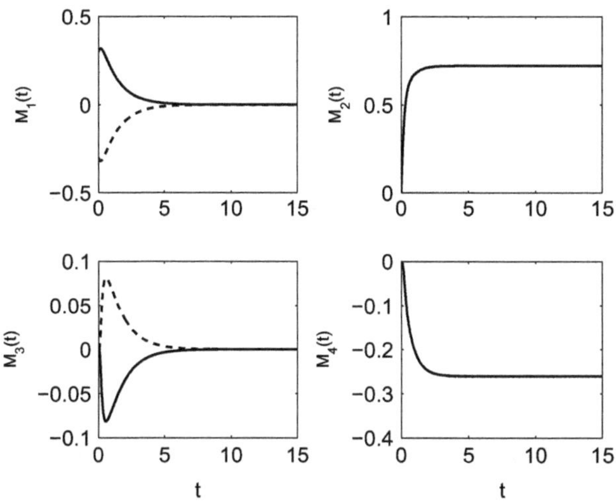

Fig. 1 Behavior of the first four diagonal cumulants obtained from the truncated set of 11 equations for the cumulants for $N = 5$, $D = 1.33$, $\theta = 2$ and two different initial conditions: $M_1(0) = 0.3$ (*solid lines*); $M_1(0) = -0.3$ (*dashed lines*)

and

$$M_2(t) = \langle x^2(t) \rangle - M_1^2(t) \approx \frac{1}{\mathcal{N}} \sum_{\alpha=1}^{\mathcal{N}} (x^{(\alpha)})^2(t) - M_1^2(t), \tag{7}$$

where $\mathcal{N}$ indicates the total number of trajectories and $x^{(\alpha)}(t)$ indicates the numerically obtained single particle trajectory in the α noise realization.

Let us first investigate what happens for parameter values such that, in the infinite size limit, Desai and Zwanzig obtained a single stable stationary first moment. In Fig. 1 we depict the behavior of the first four diagonal cumulants obtained from the truncated hierarchy of 11 equations for the cumulants for $N = 5$, $D = 1.33$, $\theta = 2$ and two sets of initial conditions: $M_1(0) = 0.3$ (solid lines) and $M_1(0) = -0.3$ (dashed lines). From the numerical simulations of Langevin equations for $N = 5$, $D = 1.33$, $\theta = 2$ and the same two sets of initial conditions we get the results depicted in Fig. 2. The long-time limit results are independent of the initial preparation and, except for the long time value of M_4, the truncated hierarchy yields a time evolution of the diagonal moments very much in agreement with those obtained with the Langevin simulations. We have numerically analyzed other set of parameter values and initial conditions and the numerical findings lead us to conclude that, for parameter values such that $M_1^{st} = 0$ is the only stable stationary solution, the set of 11 equations represents a reliable approximation to the correct behavior for systems of even very modest sizes.

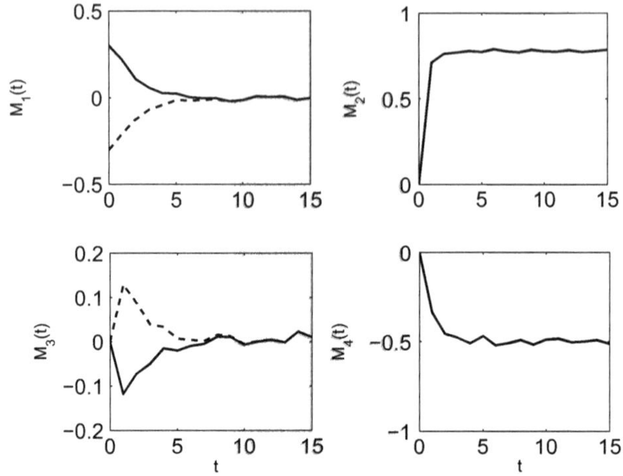

Fig. 2 Behavior of the first four diagonal cumulants obtained from the simulations of Langevin equations for the cumulants for $N = 5$, $D = 1.33$, $\theta = 2$ and two sets of initial conditions: $M_1(0) = 0.3$ (*solid lines*); $M_1(0) = -0.3$ (*dashed lines*). $M_2(t)$ and $M_4(t)$ for the two sets of initial conditions are indistinguishable with the graph resolution

Let us next consider a set of values for the parameters θ and D such that, for infinite systems, Desai and Zwanzig [4] found that truncation of the hierarchy of cumulant moments leads, in the $N \to \infty$ limit, to two coexisting stable stationary nonzero values for the first moment while $M_1^{st} = 0$ is unstable. In the case of small finite systems, the numerical solution of the truncated hierarchy still leads to three stationary values but $M_1^{st} = 0$ might be stable for a range of initial conditions. This range depends on the system size, the parameter values and the level of truncation. In Fig. 3, we depict the results for M_1^{st} obtained from the long time solution of the hierarchy of cumulant equations truncated at the fourth order level (black dots) and at the second order level (red squares) with $M_1(0) \neq 0$ and all the other cumulant set to zero. For all values of N considered, we have used $D = 0.33$ and $\theta = 2$. With the fourth order truncation, the long time solution $M_1^{st} = 0$ is stable for $N < 10$ regardless of the initial condition. For $N > 10$, the stability of the zero solution depends on the initial value $M_1(0)$. With the second order truncation, the solution $M_1^{st} = 0$ is stable for $N < 4$, while it becomes unstable for some range of initial conditions as $N > 4$. Thus, there is an influence of the level of truncation in the stability diagram of the zero stationary solution.

In Fig. 4 we depict the results for the time evolution of the first four diagonal cumulants obtained from the numerical solution of the truncated hierarchy of equations for $N = 15$, $D = 0.33$, $\theta = 2$ for two sets of initial conditions ($M_1(0) = \pm 0.3$).

In Fig. 5 we depict the behavior of the first four diagonal cumulants as obtained from numerical simulations of the Langevin equations for a system with $N = 15$, $D = 0.33$ and $\theta = 2$ with the two sets of initial conditions. The truncated

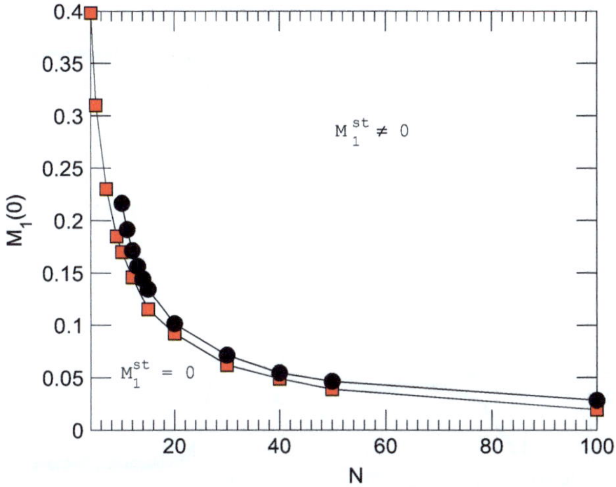

Fig. 3 The (*black*) *dots* and the (*red*) *squares* represent the limiting initial values of $M_1(0)$ vs. N separating the region of stability of the long time zero solution of the hierarchy of equations for the cumulants truncated at the fourth order or the second order levels respectively. Note that for $N < 10$, $M_1^{st} = 0$ for the fourth order truncation regardless of the initial preparation, while for the second order truncation, this fact is true for $N < 4$. The initial conditions are $M_1(0) \neq 0$ and all the other cumulant are taken to be zero initially. The lines are a guide to the eye. System parameters: $D = 0.33$ and $\theta = 2$

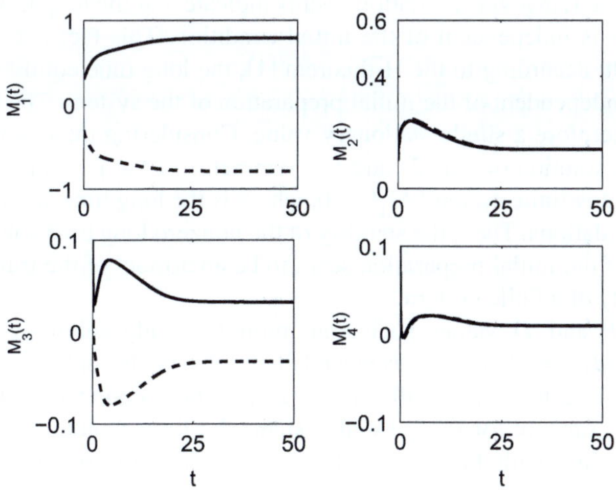

Fig. 4 Behavior of the first four diagonal cumulants obtained from the truncated hierarchy of equations for the cumulants for $N = 15$, $D = 0.33$, $\theta = 2$ and two sets of initial conditions, $M_1(0) = 0.3$ (*solid lines*); $M_1(0) = -0.3$ (*dashed lines*)

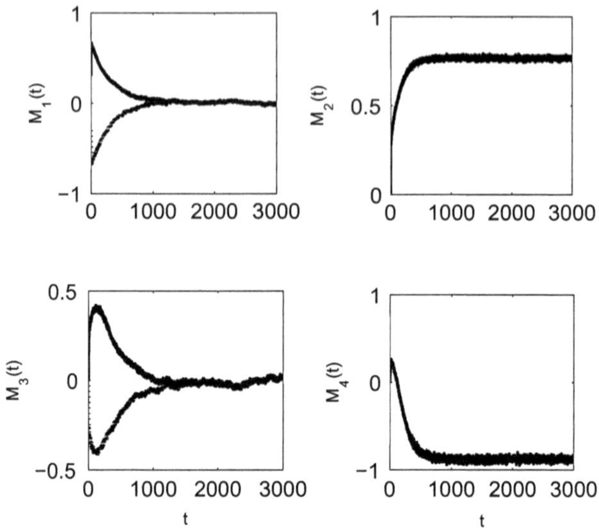

Fig. 5 Behavior of the first four diagonal cumulants obtained from the simulations of Langevin equations for the cumulants for $N = 15$, $D = 0.33$, $\theta = 2$ and two sets of initial conditions

set of cumulant equations leads to two nonzero first moment stationary values. The discrepancies between the results in Figs. 4 and 5 for the same parameter values are evident. The Langevin simulation results indicate that the long time behavior of the cumulants is independent of the initial condition. This fact is consistent with an exact result: according to the H-theorem [1], the long time equilibrium solution of Eq. (2) is independent of the initial preparation of the system. The first moment, $M_1(t)$, has therefore a single stationary value. Considering the canonical form of the long time solution of Eq. (2) and the symmetry of the potential in Eq. (3) we have that, for any finite system, $M_1^{st} = 0$, which is the long time value obtained via Langevin simulations. Then, the stability of the nonzero long time solutions and the dependence of the initial preparation seem to be an artifact of the truncation rather than a property of a finite system.

Even for θ and D values such that multiple steady solutions are possible, the solution $M_1^{st} = 0$ of the truncated hierarchy is the only stable stationary solution for N sufficiently small ($N < 10$ for the parameter values in Fig. 3). It is worth to compare for these small systems the time evolution of the first few cumulants obtained with the truncated hierarchy with the one provided by Langevin simulations. In Fig. 6 we depict the results of the evolution of the first four diagonal cumulants obtained with the numerical simulations of Eq. (1) for $N = 3$, as well as from the truncated hierarchy of equations for $\theta = 2$, $D = 0.33$. We see that, as expected, for this small system the truncated hierarchy leads to $M_1^{st} = 0$, in agreement with the results obtained from the Langevin simulations. In contrast, the second and fourth order cumulants steady values obtained from the truncated

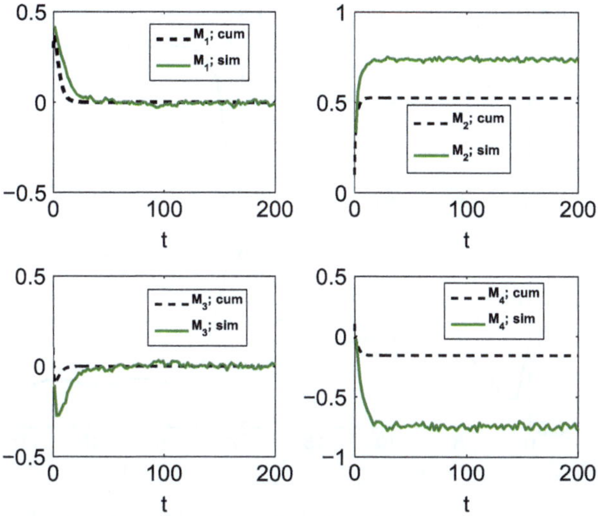

Fig. 6 Behavior of the first four diagonal cumulants estimated from the numerical simulations of Eq. (1) for $N = 3$ (*solid green lines*), as well as from the truncated hierarchy of equations for the cumulant moments (*black dashed lines*). Other parameter values $D = 0.33$ and $\theta = 2$

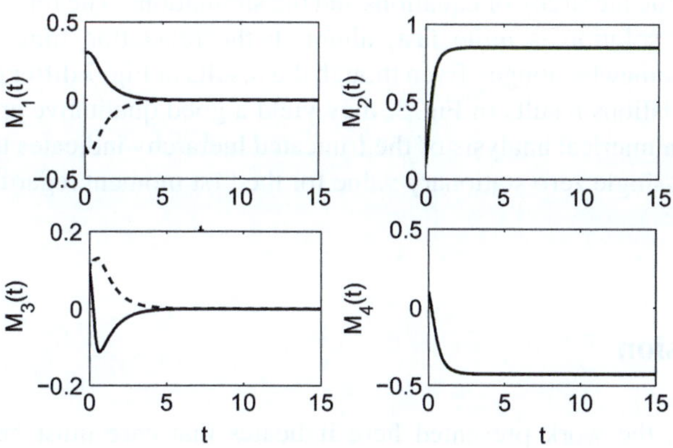

Fig. 7 Behavior of the first four diagonal cumulants of the truncated hierarchy of equations for $N = 10$, $D = 0.5$, $\theta = -0.5$ for two sets of initial conditions

hierarchy and from the Langevin simulations show large quantitative differences, indicating the limitations of the truncated hierarchy.

We now turn our attention to the model with a negative global coupling parameter $\theta < 0$. Figures 7 and 8 show the results obtained, respectively, with the truncated hierarchy of equations for the cumulant moments, and with the full simulation of the Langevin equations for systems with $N = 10$, $D = 0.5$ and $\theta = -0.5$. We see that the long time limit is independent of the initial conditions. There are

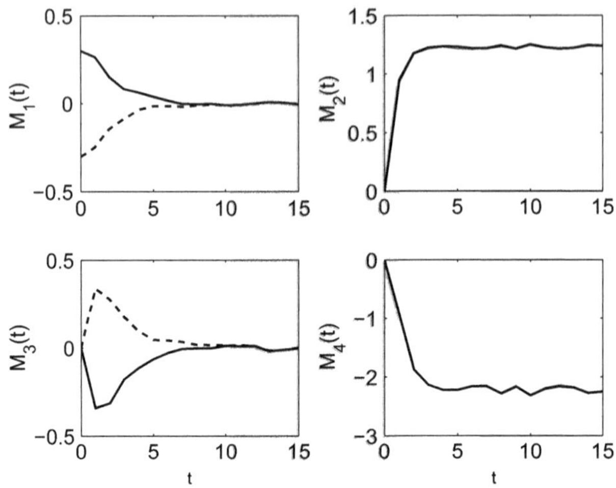

Fig. 8 Behavior of the first four diagonal cumulants obtained from numerical simulations of the Langevin equations for $N = 10$, $D = 0.5$, $\theta = -0.5$ for two sets of initial conditions

differences in the long time value for the second and fourth cumulant moments obtained with the hierarchy of equations and the simulations. The relaxation towards the stationary solution is quite fast, although the relaxation time with the full simulation is somewhat longer. Even though the results in Fig. 7 differ quantitatively from the simulations results in Fig. 8, they yield a good qualitative approximation. An extensive numerical analysis of the truncated hierarchy indicates that for $\theta < 0$ there is only a single zero stationary value for the first moment regardless the value of N.

4 Conclusion

In conclusion, the work presented here indicates that care must be taken when using an approximation to the dynamical behavior in a chain of interacting identical objects, based on truncating the infinite hierarchy of cumulant moments. Even for very large systems, if the parameter values considered are such that the truncated hierarchy leads to two stable coexisting solutions, the approximation is not correct. The results of the numerical simulations of the Langevin equations and the exact properties of the Fokker-Planck equation for finite systems of any size indicate that the coexistence of two stationary solutions is an artifact of the truncation. On the other hand, when the truncated hierarchy has a single stationary stable solution, it provides a reliable approximation of the system dynamics even for systems of very modest size. Although our conclusions are based on the study of a particular model, we think that they are qualitatively relevant for other cases. Langevin dynamics with additive noise are good general representations of dynamical systems in contact with a thermal environment. Actually, Langevin dynamics are more realistic than their deterministic limits often used to study dissipative nonlinear systems, where dissipation is included but all fluctuations are neglected. Global interactions are also quite general. The quartic nonlinearity we study here is a particular case. But having a different nonlinearity should not change our qualitative conclusions. The nonlinearity considered includes already the nontrivial effects associated to the dynamical coupling between noise and nonlinearity.

Acknowledgements We acknowledge the support of the Ministerio de Ciencia e Innovaci6n of Spain (FIS2008-04120).

References

1. H. Risken, *The Fokker-Planck Equation* (Springer, Berlin/Heidelberg/New York, 1984)
2. J. Marcinkiewicz, Math. Z. **19**, 612 (1938)
3. P. Hänggi, P. Talkner, J. Stat. Phys. **22**, 65 (1980)
4. R.C. Desai, R. Zwanzig, J. Stat. Phys. **19**, 1 (1978)
5. K. Kometani, H. Shimizu, J. Stat. Phys. **13**, 473 (1975)
6. J. Casado-Pascual, C. Denk, J. G6mez-Ord6ñez, M. Morillo, P. Hänggi, Phys. Rev. E **67**, 036109 (2003)

Shape-Invariant Solitons in Nematic Liquid Crystals: The Influence of Noise

Milivoj R. Belić, Milan S. Petrović, Aleksandra I. Strinić, and Najdan B. Aleksić

Abstract In a numerical study we investigate the influence of noise on the shape-invariant solitons in nematic liquid crystals. We use the modified Petviashvili's method for finding eigenvalues and eigenfunctions of the evolution partial differential equations, to determine the shape-invariant solitons in a realistic scalar three-dimensional model that includes the highly nonlocal nature of uniaxial nematic liquid crystals. We check the stability of such solutions by propagating them for long distances, without or within the presence of white noise. Without noise, we find them stable. In the presence of noise (added to the medium), we find them breathing, which renders shape-invariant solitons difficult to observe. After prolonged propagation, the noise leads to the dissipation of solitons.

1 Introduction

1.1 Fundamental Solitons

Fundamental optical spatial solitons are laser beams that propagate in nonlinear media without changing their transverse profiles [1]. Such shape-invariant solutions commonly appear in $(1+1)$-dimensional $[(1+1)\text{D}]$ nonlinear systems, especially the ones based on the nonlinear Schrödinger equation (NLSE). In that and other nonlinear evolution partial differential equations (PDEs), the inverse scattering

M.R. Belić (✉)
Texas A&M University at Qatar, P. O. Box 23874, Doha, Qatar
e-mail: milivoj.belic@qatar.tamu.edu

M.S. Petrović
Institute of Physics, P. O. Box 57, 11001, Belgrade, Serbia

A.I. Strinić · N.B. Aleksić
Institute of Physics, University of Belgrade, P. O. Box 68, 11001, Belgrade, Serbia

R. Carretero-González et al. (eds.), *Localized Excitations in Nonlinear Complex Systems*, Nonlinear Systems and Complexity 7, DOI 10.1007/978-3-319-02057-0_20,

theory, formulated to treat such equations, guaranties their existence [2]. However, the situation is more complex in the multidimensional and multicomponent systems. No complete inverse scattering theory is formulated in more than one dimension and even when the localized solutions are found, no rigorous procedure for guaranteeing their stability is established. In fact, wave instability and collapse of solutions are overriding concerns in multidimensional nonlinear systems [3]. Additional compounding difficulties arise in the multicomponent vector models or in the scalar nonlocal models, in which the medium response is driven by the optical field itself. Such are the models describing the generation of solitary waves—nematicons— in nematic liquid crystals (NLCs).

1.2 Nonlocality

Nonlocality is an important feature of many nonlinear media. Generically, it refers to a situation where the medium response at a point depends not only on the excitation at that point, but also on the excitation in the region around the point. A highly nonlocal situation arises in a nonlocal nonlinear medium in which the characteristic size of the medium response is much wider than the size of the excitation itself [4,5]. In nematic liquid crystals, both experiments [6,7] and theoretical calculations [8,9] demonstrated that the nonlinearity is highly nonlocal. Often, a highly nonlocal situation in systems described by the NLSE leads to the appearance of a parabolic potential in the equation [10]. Soliton profile calculations in such media have been presented in a number of papers [11–13]. Interestingly, even a high degree of nonlocality does not guarantee the existence of stable higher-order soliton structures [14, 15]. Orientational nonlinearity in NLCs is highly nonlocal, but the nonlinear response is not perfectly parabolic, implying that if one launches a Gaussian beam into the cell, it is only possible to observe breathing solitons [5, 16–18].

1.3 Status of the Field

A puzzling element in the status of the field is that even though everybody agrees that the fundamental shape-preserving solitons do exist in highly nonlocal nematic liquid crystals, practically nobody cared to present them explicitly. Experimental accounts often mention steady nematicons, but careful inspection of published figures reveals self-focusing oscillations. This is expected, as the beams launched into the crystals usually are Gaussian. An exact (shape-invariant) nematicon cannot be generated experimentally; only numerically. Experimental setups feature a few mm long cells only, which cannot capture slow (if any) convergence to a steady profile. In addition, as it will be demonstrated below, noise in NLCs exerts a negative influence on the stability of nematicons, by making the medium lossy.

In some publications the soliton profiles have been calculated using semi-analytical models [4, 19–21]. For the more general vectorial model, in which the order parameter of liquid crystal is not constant, steady elliptical soliton profiles have been found numerically [22]. To determine such profiles, the authors of this noticeable paper have demonstrated that it is necessary to include all three components of the optical electric field. In that case, the largest is the component parallel to the bias field, followed by the perpendicular component and in the end is the longitudinal component, as the smallest.

1.4 Contents of the Chapter

In this chapter we investigate numerically laser beam propagation in a highly nonlocal medium, utilizing widely accepted scalar model of uniaxial nematic liquid crystals. To find exact fundamental soliton solutions in a model that is multidimensional but not vectorial, we use an iterative numerical eigenvalue technique. We discuss the influence of boundary conditions on the shape and power of these solutions, and analyze soliton and Gaussian propagation using two different propagation methods. We check their stability in propagation and demonstrate that any small change in the input shape, as well as in the medium, leads to the soliton breathing. Adding white noise to the medium causes the solitons to breathe and eventually dissipate in the medium. Consequently, we question the real physical observability of such shape-invariant solitons in realistic media.

A note of caution is necessary here. We do not question the observability of nematicons in general—only the shape-invariant ones. Breathing nematicons are commonly observed. Furthermore, we do not invalidate any of the theories used to describe nematicons or experiments which depict them—each is valid under approximations utilized and in a region of physical (experimental) reality. However, beyond that—and particularly in the presence of noise—shape-invariant solitons cease to exist.

The text is divided into five sections. Section 2 deals with the model adopted for studying fundamental optical solitons in nematic liquid crystals. Section 3 presents the numerical procedure and the shape-invariant solutions obtained. Section 4 is concerned with the influence of noise on the propagation of localized solitary beams in a strongly nonlocal medium. Section 5 brings conclusions.

2 The Model

2.1 Scalar Nonlocal Model of Nematic Liquid Crystals

We adopt the well-known nonlocal nonlinear scalar model of nematic liquid crystals which provides good agreement with experiments [7]. The liquid crystal cell of interest is sketched in Fig. 1. The optical beam polarized along the x axis propagates

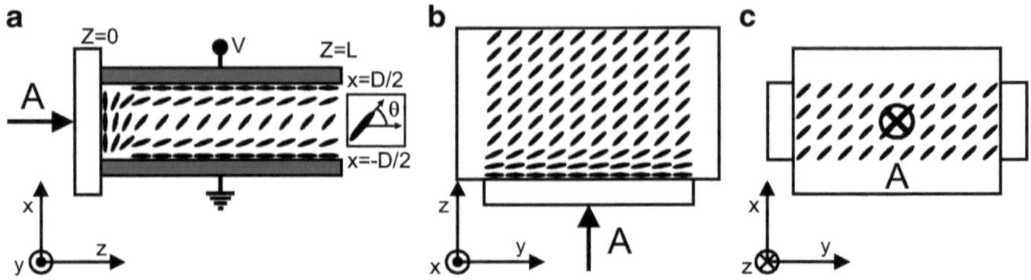

Fig. 1 Liquid-crystal cell model adopted: (**a**) *Side view*, (**b**) *top view*, (**c**) *front view*. *A* represents the launching beam

in the z direction, while the NLC molecules can rotate in the x–z plane. The total orientation of molecules with respect to the z axis is denoted as θ, whereas the orientation induced by the static electric field only is denoted by θ_0 (the pre-tilt angle). The bias field points in the x direction and is uniform in the z direction; hence the pre-tilt angle is uniform along the z axis as well. The quantity $\hat{\theta} = \theta - \theta_0$ corresponds to the optically induced molecular reorientation. The system is described by two PDEs, one for the electric field of the beam and the other for the reorientation angle $\hat{\theta}$ of the director field of NLC molecules.

2.2 Model Equations

The system of equations of interest consists of the Schrödinger-like equation for the propagation of the optical field A and the diffusion-like equation for the molecular orientation angle θ [7, 8]. After the rescaling $x/x_0 \to x$, $y/x_0 \to y$, $z/L_D \to z$, where x_0 is the transverse scaling length and $L_D = kx_0^2$ is the corresponding diffraction length, the following model equations in the computational domain are obtained [16]:

$$2i\frac{\partial A}{\partial z} + \Delta_{x,y}A + \alpha[\sin^2\theta - \sin^2\theta_0]A = 0, \tag{1}$$

$$\tau\frac{\partial \theta}{\partial t} = 2\Delta_{x,y}\theta + [\beta + \alpha|A|^2]\sin(2\theta), \tag{2}$$

where $\Delta_{x,y}$ is the transverse Laplacian. We introduced the notation: $\alpha = k_0^2 x_0^2 \Delta\varepsilon^{OPT}$, $\beta = \varepsilon_0 x_0^2 \Delta\varepsilon^{DC}|E^{DC}|^2/K$, and $\tau = 2\gamma x_0^2/K$. We also scaled the optical field intensity $\varepsilon_0|A|^2/2Kk_0^2 \to |A|^2$. The wave numbers in the medium and vacuum are k and k_0, respectively. The amplitude of the static bias electric field is $E^{DC} = V/D$, where V is the applied bias voltage and D is the cell thickness. $\Delta\varepsilon^{OPT}$ and $\Delta\varepsilon^{DC}$ are the optical and static permittivity anisotropies of the NLC molecules, respectively. K is Frank's elastic constant and γ the viscous coefficient.

We adopt the simplified one-constant approximation, in which the elastic constants for splay, twist, and bend deformation are all set equal to K.

It is worthwhile noting the mixed nature of PDEs: One is the spatial paraxial wave equation for the complex field A, the other the temporal diffusion equation for the scalar field θ. Normally, we are looking for localized solutions that decay fast in the transverse directions. Gaussian beams are used as inputs, except when the propagation of shape-invariant fundamental solitons is considered. Thus, the localized paraxial field A propagates along the z axis, adjusting at all times to the slowly-varying material field θ, which in turn is influenced by A through the optically-induced change in the index of refraction. Hard boundary conditions (BCs) on the molecular orientation at the liquid crystal cell faces in the x direction are assumed: $\theta(x = -D/2, y) = \theta(x = D/2, y) = 2^o$ [23], while in the y direction different BCs are assumed, depending on the situation. The cell is usually much wider in the y direction, allowing for some freedom in the choice of BCs. Different conditions affect the solutions differently, as it will be seen below.

3 Computations

3.1 Numerical Procedure

Our numerical procedure is quite complex. The procedure for the propagation equation is the split-step beam propagation method based on the fast Fourier transform (FFT). The diffusion equation for the optically induced molecular reorientation is treated using the successive over-relaxation (SOR) method. However, before the treatment of Eqs. (1) and (2) can commence, the pre-tilt angle θ_0 must be determined. It is found from Eq. (2) in the absence of the optical field and in the steady state:

$$2\Delta_{x,y}\theta_0 + \beta \sin(2\theta_0) = 0. \tag{3}$$

This equation represents an elliptic boundary value problem. It must be solved separately, before the propagation of the optical field is considered. Note that the pre-tilt angle $\theta_0(x, y)$ depends on the transverse variables only, being invariant along the z axis. Furthermore, owing to the experimental boundary conditions, θ_0 depends on x only. We do not use the condition $\theta_0 = \pi/4$, imposed elsewhere [5].

In our computations we use data corresponding to typical experimental conditions [8, 16, 23]. In all the simulations the following data are kept constant: $L_D = 75\,\mu\text{m}$, $x_0 = 2\,\mu\text{m}$, $\lambda = 514\,\text{nm}$, $n_0 = 1.53$, $K = 12 \cdot 10^{-12}$ N, $D = 75\,\mu\text{m}$, and $V = 1$ V. In the case of fundamental beam propagation, with single-peak on-axis intensity, we observe slow convergence of beam amplitude A and θ to their steady-state values; therefore, we confine our attention to the steady state only.

3.2 Modified Petviashvili Procedure

The localized eigen-solutions are determined from the system of Eqs. (1) and (2) using the modified Petviashvili's iteration method [18,24–26]. Equation (1) suggests the existence of a fundamental soliton in the form $A = a(x, y)e^{i\mu z}$, where μ is the propagation constant. The real-valued function $a(x, y)$ satisfies the equation

$$-\Delta a + (2\mu + \mathbf{P})a = \mathbf{Q}, \tag{4}$$

where $\mathbf{P} = \alpha \sin^2(\theta_0)$ and $\mathbf{Q} = \alpha \sin^2(\theta)a$. After Fourier transforming this equation, we get:

$$\overline{a} = \frac{1}{|\mathbf{k}|^2 + 2\mu}(\overline{\mathbf{Q}} - \overline{\mathbf{P}a}), \tag{5}$$

where overbar denotes Fourier transform. Straightforward iteration of Eq. (5) does not converge in general, so the stabilizing factors have to be introduced [25, 26]. There are two such factors, involving integrals over Fourier transforms in Eq. (5). They are of the form $\mathbf{a} = \int [(|\mathbf{k}|^2 + 2\mu)\overline{a} + \overline{\mathbf{P}a}]\overline{a}^* d\mathbf{k}$ and $\mathbf{b} = \int \overline{\mathbf{Q}}\overline{a}^* d\mathbf{k}$, and the iteration equation becomes:

$$\overline{a}_{m+1} = \frac{1}{|\mathbf{k}|^2 + 2\mu}\left[\left(\frac{\mathbf{a}_m}{\mathbf{b}_m}\right)^{3/2}\overline{\mathbf{Q}}_m - \left(\frac{\mathbf{a}_m}{\mathbf{b}_m}\right)^{1/2}\overline{\mathbf{P}a_m}\right]. \tag{6}$$

When iterated this equation behaves more friendly than Eq. (5). Clearly, when $\mathbf{a}_m$ converges to $\mathbf{b}_m$, convergence in Eq. (5) is achieved as well. In this manner, Eq. (2) is treated using the successive overrelaxation (SOR) method until convergence, for each iteration step given by Eq. (6). Our method is suitable for finding both fundamental and higher order soliton solutions.

Converged fundamental soliton solutions are presented in Fig. 2 for two different boundary conditions. They are stable in propagation, as it will be discussed below. The shape and the power of the fundamental shape-invariant solutions depend on the BCs applied. Zero boundary conditions ($\hat{\theta} = 0$ on all boundaries) correspond to the Dirichlet BCs. Periodic boundary conditions correspond to the mixed conditions—Dirichlet along the y axis and Neumann along the x axis.

Typically, the cell is much wider in the y direction than in the x direction; at the edges of the transverse region of interest the derivatives in the y direction are much smaller than the derivatives in the x direction. In addition, the two transverse directions are not equivalent—the external electric field in the x direction breaks the symmetry. It affects the molecular dipole field strongly. This allows us to consider different BCs in the y direction. Of the two, the solution with the periodic boundary conditions is more appropriate to the geometry of the problem; furthermore, it is more acceptable on physical grounds. The fundamental soliton so obtained requires less beam power for the same value of the propagation constant and identical

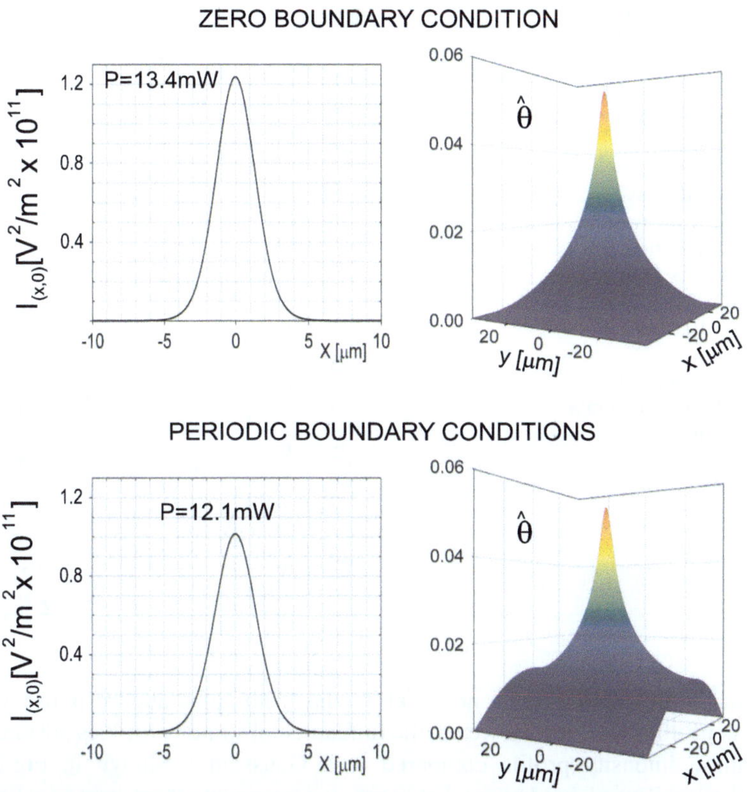

Fig. 2 Fundamental soliton solutions for $\mu = 5L_D^{-1}$. Intensity profiles and reorientation angle distributions are shown for two different BCs. Other parameters are the same

other parameters. Under identical conditions (but for the boundary conditions) a solution requiring less power should be more favorable. The often used $\theta_0 = \pi/4$ approximation corresponds to the zero boundary conditions.

3.3 *The Stability of Fundamental Solitons*

Checking the stability of spatial solitons involves propagating a perturbed soliton through the medium. If the perturbation dies away with propagation, then presumably the soliton is stable. An early account on the propagation of perturbed solitons in NLSE is provided in Ref. [27]; however, that paper deals with the 1D NLSE in the temporal domain. We need an analysis of 2D spatial solitons propagating in a noisy nonlocal medium.

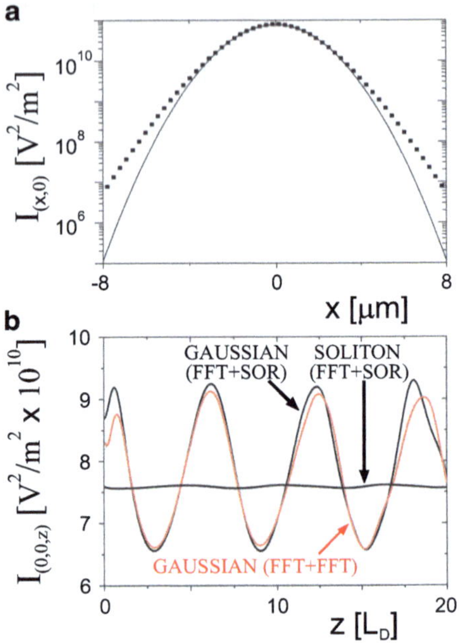

Fig. 3 (**a**) Fundamental soliton intensity profile obtained by the eigenvalue method (*black dots*), fitted with a Gaussian. Parameters: $P = 10.6\,\mathrm{mW}$, $\mu = 3.84 L_D^{-1}$; zero BCs. (**b**) Soliton and Gaussian propagation using two different propagation methods, FFT + SOR and FFT + FFT. The soliton power $P = 10.6\,\mathrm{mW}$; Gaussian power $P = 10.6\,\mathrm{mW}$ for FFT + SOR, $P = 10.1\,\mathrm{mW}$ for FFT + FFT

Spatial solitons in highly nonlocal media with quadratic response possess Gaussian profiles [4,5,10]. However, the fundamental soliton profile is not Gaussian. The soliton intensity profile compared to a Gaussian is shown in Fig. 3a. To check the stability of fundamental solitons, we propagate them numerically; peak intensities as functions of the propagation distance are presented in Fig. 3b. We find the fundamental shape-invariant soliton stable in propagation. Also included in Fig. 3b is a case presenting the propagation of a Gaussian with similar parameters, but obtained using two different numerical methods. In both methods a split-step beam propagation procedure based on the fast Fourier transform (FFT) is used for the propagation of the optical field. In the first method the diffusion equation for the optically induced molecular reorientation is treated using the SOR method; this procedure is referred to as the FFT + SOR. In the second method the diffusion equation is treated using the split-step procedure again—this is the FFT + FFT procedure. One can see that the methods provide similar results; however the first method is more accurate.

The problem with the FFT + FFT procedure is that it treats an array of transversely periodic cells. Since the molecular reorientation is wide, it tends to slightly spill over into the adjacent cells, i.e. back onto itself, adding perturbation to the optical field. This is not an overriding problem in the propagation of a Gaussian, as it only leads to a slightly amplified oscillation of the breathing solution. However, it makes a huge difference in the propagation of the fundamental soliton—it makes it impossible for the field to keep its shape-invariant input profile.

4 The Influence of Noise

4.1 *The Breathing Problem*

Any small change in the orientation angle θ pushes the system a bit from the self-organized equilibrium and forces it to oscillate about the shape-invariant soliton. In a highly nonlocal system, the potential is broad and parabolic, making it difficult for the narrow localized solution to radiate and relax to the fundamental soliton. It just keeps oscillating, forming a long-lived breathing soliton. The FFT procedure artificially amplifies that breathing. Therefore, the FFT procedure in the treatment of θ should be discarded.

Even the solution obtained by the SOR method slightly oscillates, with the same period as the Gaussian, if an insufficient numerical resolution is imposed on the Petviashvili procedure; however, the amplitude of that oscillation can be reduced to an arbitrarily small level provided higher accuracy of the eigenfunction profiles is requested. Then, the soliton propagates without any change for arbitrary long. In Fig. 3b we show a case where the oscillation of the amplitude is still perceptible. This brings us to an important point.

4.2 *Propagation in the Presence of Noise*

When one considers the propagation of a Gaussian beam using the two propagation methods, the results are close. The propagation of a Gaussian invariably leads to breathing beams, regardless of the method of integration. However, when the fundamental soliton is propagated through the medium in which a small amplitude noise is added to the pre-tilt angle θ_0, a breathing solution is also obtained. In addition, the presence of noise brings profound new features to the propagation of solitons.

We introduce noise in our simulations by adding randomly distributed white noise to θ_0 at each propagation step. Adding more noise leads to larger and more irregular oscillations, although for awhile one can discern a simple sinusoidal breather in the background, with the same period (Fig. 4). Such a situation is physically plausible: the existence of noise or random fluctuations in the director field of NLCs is a well established fact [28]. A similar induced oscillation phenomenon happens as well when a small intensity noise is added to the fundamental profile, but θ_0 kept unchanged. The perturbed fundamental soliton also oscillates during propagation in a noisy medium. However, as mentioned, this induced breather is not stable.

The oscillating phenomenon in the presence of noise is confirmed in our computations and is not difficult to understand. In a highly nonlocal nonlinear medium any additional energy from noise, no matter how small, cannot be easily radiated away and the solution has no way to relax to the fundamental soliton. Therefore, it

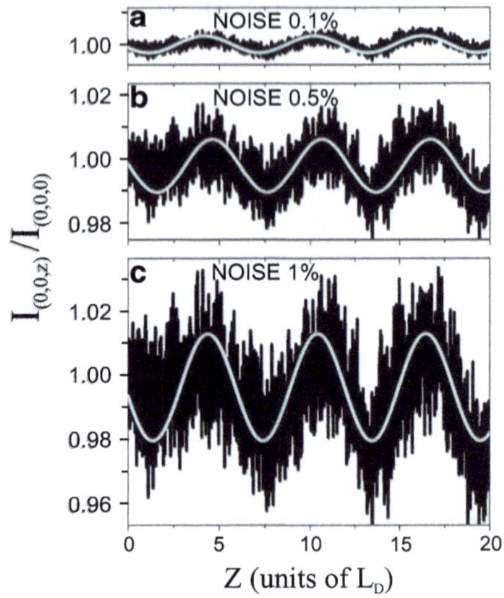

Fig. 4 Propagation of the fundamental soliton in a noisy medium. On-axis peak intensity is shown as a function of the propagation distance. An amount of (**a**) 0.1 %, (**b**) 0.5 %, and (**c**) 1 % randomly changing *white* noise is added to θ_0. *Blue* sinusoidal fit is to guide eye. Parameters: $\mu = 4L_D^{-1}$, zero BCs

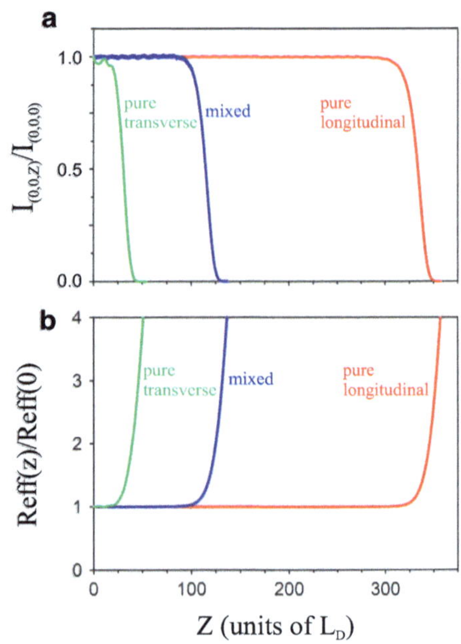

Fig. 5 Propagation of the fundamental soliton in a noisy medium for longer propagation distances and for the three models of noise: pure transverse, pure longitudinal, and mixed. (**a**) On-axis peak intensity; (**b**) effective beam width $R_{eff} = \sqrt{\int |A(x,y)|^2(x^2+y^2)dxdy/P}$. An amount of 0.1 % randomly changing *white* noise is added to the pre-tilt angle, as in Fig. 4a. Here $\mu = 2L_D^{-1}$

keeps oscillating about the shape-invariant soliton, forming a quasi-stable breathing solution. We say quasi-stable, because upon increasing the propagation distance, the soliton starts to disperse. An increase in the level of noise speeds up the dispersal. An example of such behavior is depicted in Fig. 5. It is not at all unexpected.

4.3 Destructive Influence of Noise

Figure 5 presents the peak intensity of a breathing soliton in propagation (Fig. 5a) and its effective width (Fig. 5b). Three noisy scenarios are employed in Fig. 5: pure transverse noise, pure longitudinal noise, and a mixed transverse-longitudinal noise. The transverse noise acts in the transverse plane and is uniform in the z direction; the longitudinal noise acts along the z direction and is uniform in the transverse plane. The mixed noise acts in the whole space—it is delta-correlated white noise added to θ_0, acting at each computational point. The mixed example in Fig. 5 is essentially the case (a) from Fig. 4, continued in propagation after 20 L_D. It is stable in propagation up to about 100 L_D and then starts to dissipate—its peak intensity decays and its width broadens. The power stays practically constant—it decays very slowly over the whole propagation region.

It is seen that the soliton dissipates in any scenario. The transverse noise exerts the strongest influence, the longitudinal the weakest. This is expected; the transverse noise directly destabilizes the transverse soliton profile, while the longitudinal noise modulates the profile longitudinally. Small longitudinal modulations tend to live for long. It is maybe unexpected that the mixed noise is in-between; being the sum of the two, one would expect its influence to be the strongest. This, actually is not the case—the longitudinal noise added to the transverse serves to dampen the influence of the transverse noise.

Thus, any amount of white noise introduced into the system makes it effectively lossy. Solitons radiate and slowly decay. However, in 2D they cannot approach a steady state, because such a dissipative steady state does not exist. The noise—distributed throughout the medium—keeps the solitons dissipative all along. Therefore, after a prolonged propagation distance, they disperse. For small noise this distance may be well beyond the experimental propagation lengths, but it is still there. In Fig. 5, this distance is considerably longer than the cell length; it naturally shortens as the level of noise is increased. For the most destructive transverse noise we found that about 0.1 % of white noise in the pre-tilt angle is enough to destroy the nematicon just beyond the reported experimental length of about 1.5 mm, which is about 20 L_D. An unreasonably high level of noise may destroy solitons upon very short propagation distance. These findings will be discussed elsewhere.

The length scale over which noise destroys real nematicons may vary, depending on the experimental conditions and the type of noise in the medium. We do not know, nor could we find in the literature, the experimental values on the fluctuations in the director's orientation angle. Nor do we know what proportion of scattering losses—which allow solitons to be visible from side windows and present a natural loss mechanism—could be attributed to noise. In any case, the fact that noise is unavoidable in any realistic set-up—be it experimental or numerical—opens the question of physical observability of shape-invariant fundamental solitons in highly nonlocal nonlinear media.

5 Conclusion

In summary, we have demonstrated the existence of a family of fundamental solitons in highly nonlocal nematic liquid crystals. We have depicted the dependence of these solitons on the boundary conditions applied. We have displayed their stability in propagation, utilizing two different propagation methods. We have discovered that adding any amount of white noise to the medium leads to the breathing of fundamental solitons and their eventual dispersal. Adding more noise leads to the speedier dissipation of fundamental solitons.

In the end, we should point out that similar comments can be made to almost any of the solitary waves observed in nonlocal media in nonlinear optics or other fields of physics. These comments do not invalidate their existence in general or the models which describe them in particular. All such models are developed under certain conditions and are valid in certain domains of physical reality. Nematicons are real and here to stay. We are pointing to what may happen when more subtle but still relevant aspects of that reality—such as the influence of noise—are imposed on their existence.

Acknowledgements This work has been supported by the Ministry of Education and Science of the Republic of Serbia, under the projects OI 171033, 171006 and III 45016, and by the Qatar National Research Fund project NPRP 09-462-1-074. Authors gladly acknowledge supercomputer time provided by the IT Research Computing group of the Texas A&M University at Qatar.

References

1. Y.S. Kivshar, G. Agrawal, *Optical Solitons: From Fibers to Photonic Crystals* (Academic, San Diego, 2003)
2. G.L. Lamb, *Elements of Soliton Theory* (Wiley, New York, 1980)
3. C. Sulem, P. Sulem, *The Nonlinear Schrödinger Equation: Self-Focusing and Wave Collapse* (Springer, Berlin, 2000)
4. C. Conti, M. Peccianti, G. Assanto, Phys. Rev. Lett. **91**, 073901 (2003)
5. C. Conti, M. Peccianti, G. Assanto, Phys. Rev. Lett. **92**, 113902 (2004)
6. J.F. Henninot, J.F. Blach, M. Warenghem, J. Opt. A **9**, 20 (2007)
7. X. Hutsebaut, C. Cambournac, M. Haelterman, J. Beeckman, K. Neyts, J. Opt. Soc. Am. B **22**, 1424 (2005)
8. J. Beeckman, K. Neyts, X. Hutsebaut, C. Cambournac, M. Haelterman, Opt. Express **12**, 1011 (2004)
9. M. Peccianti, C. Conti, G. Assanto, Opt. Lett. **30**, 415 (2005)
10. A.W. Snyder, D.J. Mitchell, Science **276**, 1538 (1997)
11. C. Rotschild, M. Segev, Z. Xu, Y. Kartashov, L. Torner, O. Cohen, Opt. Lett. **31**, 3312 (2006)
12. C. Rotschild, O. Cohen, O. Manela, M. Segev, Phys. Rev. Lett. **95**, 213904 (2005)
13. F. Ye, Y. Kartashov, B. Hu, L. Torner, Opt. Lett. **34**, 2658 (2009)
14. S. Skupin, O. Bang, D. Edmundson, W. Krolikowski, Phys. Rev. E **73**, 066603 (2006)
15. A.I. Yakimenko, Y.A. Zaliznyak, Y.S. Kivshar, Phys. Rev. E **71**, 065603(R) (2005)
16. A. Strinić, M. Petrović, D. Timotijević, N. Aleksić, M. Belić, Opt. Express **17**, 11698 (2009)
17. M. Petrović, A. Strinić, M. Belić, Phys. Scr. **85**, 015403 (2012)

18. N. Aleksić, M. Petrović, A. Strinić, M. Belić, Phys. Rev. A **85**, 033826 (2012)
19. A. Minzoni, N. Smyth, A. Worthy, J. Opt. Soc. Am. B **24**, 1549 (2007)
20. H. Ren, S. Ouyang, Q. Guo, W. Hu, C. Longgui, J. Opt. A **10**, 025102 (2008)
21. H. Zhang, D. Xu, L. Li, J. Opt. A **11**, 125203 (2009)
22. J. Beeckman, K. Neyts, P.J.M. Vanbrabant, R. James, F.A. Fernandez, Opt. Express **18**, 3311 (2010)
23. X. Hutsebaut, C. Cambournac, M. Haelterman, A. Adamski, K. Neyts, Opt. Commun. **233**, 211 (2004)
24. V.I. Petviashvili, Plasma Phys. **2**, 469 (1976)
25. J. Yang, I. Makasyuk, A. Bezryadina, Z. Chen, Stud. Appl. Math. **113**, 389 (2004)
26. D. Jović, S. Prvanović, R. Jovanović, M. Petrović, Opt. Lett. **32**, 1857 (2007)
27. J.P. Gordon, J. Opt. Soc. Am. **B 9**, 91 (1992)
28. I.C. Khoo, *Liquid Crystals* (Wiley, New York, 2007)

Conservation Laws of a Family of Reaction-Diffusion-Convection Equations

M.S. Bruzón, M.L. Gandarias, and R. de la Rosa

Abstract Ibragimov introduced the concept of nonlinear self-adjoint equations. This definition generalizes the concept of self-adjoint and quasi-self-adjoint equations. Gandarias defined the concept of weak self-adjoint. In this paper, we found a class of nonlinear self-adjoint nonlinear reaction-diffusion-convection equations which are neither self-adjoint nor quasi-self-adjoint nor weak self-adjoint. From a general theorem on conservation laws proved by Ibragimov we obtain conservation laws for these equations.

1 Introduction

Many interesting chemical, biological and physical phenomena such as pattern formation, morphogenesis, animal coats and skin pigmentation, nerve impulse propagation in nerve fibres, wall propagation in liquid crystals, nucleation kinetics and neutron action in the reactor, are strongly related with the study of nonlinear reaction-diffusion-convection equations.

We consider the class of nonlinear reaction-diffusion-convection equations

$$u_t = [A(u)u_x]_x + B(u)u_x + C(u), \tag{1}$$

where $u = u(x,t)$ is an unknown function, $A = \frac{dK(u)}{du}$ and $K(u) \neq 0$, $B(u) \neq 0$ and $C(u)$ is an arbitrary differentiable function. Equation (1) is a generalization of

M.S. Bruzón (✉) · M.L. Gandarias
Departamento de Matemáticas, Universidad de Cádiz, Puerto Real, Cádiz, 11510, Spain
e-mail: m.bruzon@uca.es

R. de la Rosa
Universidad de Cádiz, Cádiz, Spain
e-mail: fae_dasilva@hotmail.com

R. Carretero-González et al. (eds.), *Localized Excitations in Nonlinear Complex Systems*, Nonlinear Systems and Complexity 7, DOI 10.1007/978-3-319-02057-0_21,

many nonlinear evolution equations of second order. Lie symmetries of equation (1), where $A(u)$, $B(u)$, $C(u)$ are arbitrary smooth functions, were completely obtained by Cherniha and Serov [4–6]. Dorodnitsyn and Svirshchevskii [7] proved that Eq. (1) admits nontrivial conservation laws when $B = 0$, $A = k$ and $C = k_1 u + k_2$, and when $A = \frac{dK(u)}{du}$ and $C = k_1 K(u) + k_2 u + k_3$, where $K(u)$ is an arbitrary function, k, k_1 and k_2 are arbitrary constants. Cherniha and King generalized these results to higher dimensions by inspection and they obtained a new conservation law when $A = A(u)$ and $C = k_1 K(u) + k_2 u + k_3$ in which $K(u)$ is the Kirchhoff variable $K(u)$ [3].

The classical theory of Lie point symmetries for differential equations describes the groups of infinitesimal transformations in a space of dependent and independent variables that leave the manifold associated with the equation unchanged [9, 14, 15]. The methods of point transformations are a powerful tool for find exact solutions for nonlinear partial differential equations (PDE's) and for construct conservation laws [1].

The idea of a conservation law, or more particularly of a conserved quantity, has its origin in mechanics and physics. Since a large number of physical theories, including some of the "laws of nature", are usually expressed as systems of nonlinear differential equations, it follows that conservation laws are useful in both general theory and the analysis of concrete systems [16].

The concepts of self-adjoint, quasi-self-adjoint equations and nonlinear self-adjoints were introduced by Ibragimov in [9, 10, 13]. In [2] self-adjoint classes of a family of Benjamin-Bona-Mahony-Burgers equations were determined and some conservation laws were established. In [8] Gandarias generalized the concepts of self-adjoint and quasi-self-adjoint equations defining weak self-adjoint equations. We are interested in finding equations that satisfy the property of being weak self-adjoint to enable us to find conservation laws using symmetries of the differential equation.

In [11] Ibragimov introduced a new theorem on conservation laws. The theorem is valid for any system of differential equations for which the number of equations is equal to the number of dependent variables. The new theorem does not require existence of a Lagrangian and this theorem is based on a concept of an adjoint equation for non-linear equations.

The aim of this work is to determine, for Eq. (1), the subclasses of equations which are self-adjoint, quasi-self-adjoint, weak self-adjoint and nonlinear self-adjoint. We also determine, by using the notation and techniques [10, 11], some nontrivial conservation laws for Eq. (1). The paper is organized as follows. In Sect. 2 we obtain the adjoint equation to Eq. (1). In Sect. 3 we give the Lie symmetries of (1) equation obtained by Cherniha and Serov [4]. In Sect. 4 we determine the subclasses of equations of (1) which are self-adjoint, quasi-self-adjoint, weak self-adjoint and nonlinear self-adjoint. In Sect. 5 we obtain nontrivial conservation laws for Eq. (1). Finally, in Sect. 6 we give conclusions.

2 Formal Lagrangian and Adjoint Equation

The new theorem given by Ibragimov is based on a concept of an adjoint equation for nonlinear equations. In order to obtain the adjoint equation we use the following definition given in [11].

Definition 1. Consider an sth-order partial differential equation

$$F(x, u, u_{(1)}, \ldots, u_{(s)}) = 0 \tag{2}$$

with independent variables $x = (x^1, \ldots, x^n)$ and a dependent variable u, where $u_{(1)} = \{u_i\}$, $u_{(2)} = \{u_{ij}\}, \ldots$ denote the sets of the partial derivatives of the first, second, etc. orders, $u_i = \partial u/\partial x^i$, $u_{ij} = \partial^2 u/\partial x^i \partial x^j$. The formal Lagrangian is defined as

$$\mathcal{L} = v\, F\left(x, u, u_{(1)}, \ldots, u_{(s)}\right), \tag{3}$$

where $v = v(x,t)$ is a new dependent variable. The adjoint equation to (2) is

$$F^*(x, u, v, u_{(1)}, v_{(1)}, \ldots, u_{(s)}, v_{(s)}) = 0, \tag{4}$$

with

$$F^*(x, u, v, u_{(1)}, v_{(1)}, \ldots, u_{(s)}, v_{(s)}) = \frac{\delta(v\, F)}{\delta u},$$

where

$$\frac{\delta}{\delta u} = \frac{\partial}{\partial u} + \sum_{s=1}^{\infty}(-1)^s D_{i_1} \cdots D_{i_s} \frac{\partial}{\partial u_{i_1 \cdots i_s}}$$

denotes the variational derivatives (the Euler–Lagrange operator). Here

$$D_i = \frac{\partial}{\partial x^i} + u_i \frac{\partial}{\partial u} + u_{ij} \frac{\partial}{\partial u_j} + \cdots$$

are the total differentiations.

Given the nonlinear reaction-diffusion-convection equations (1), by applying Definition (1), the adjoint equation to Eq. (1) is defined by

$$F^* \equiv -v\, C_u + v_x\, B - v_{x\,x}\, A - v_t. \tag{5}$$

3 Lie Symmetries

Lie's symmetries admitted by a PDE are useful for finding invariant solutions. These solutions are obtained by using group invariants to reduce the number of independent variables. The fundamental basis of the technique is that, when a differential equation is invariant under a Lie group of transformations, a reduction transformation exists. The machinery of the Lie group theory provides a systematic method to search for these special group invariant solutions. To apply Lie classical method to Eq. (1) we consider the one-parameter Lie group of infinitesimal transformations in (x, t, u) given by

$$\hat{x} = x + \epsilon\xi(x,t,u) + O(\epsilon^2),$$
$$\hat{t} = t + \epsilon\tau(x,t,u) + O(\epsilon^2),$$
$$\hat{u} = u + \epsilon\eta(x,t,u) + O(\epsilon^2).$$

Therefore, the infinitesimal generator is given by:

$$X = \xi(x,t,u)\frac{\partial}{\partial x} + \tau(x,t,u)\frac{\partial}{\partial t} + \eta(x,t,u)\frac{\partial}{\partial u}.$$

From the previous equation we define the characteristic of the group

$$Q = \eta - \xi u_x - \tau u_t.$$

We say that the surface $u = u(x,t)$ is invariant if:

$$Q = 0 \quad \text{when} \quad u = u(x,t).$$

The previous equation is called the invariant *surface condition*. Most of the techniques for finding exact solutions of PDEs are based in the use of this condition.

Differentiating $\hat{u}$ with respect to x and respect to t we obtain:

$$\hat{u}_{\hat{x}} = u_x + \epsilon\eta^x(x,t,u,u_x,u_t) + O(\epsilon^2),$$
$$\hat{u}_{\hat{t}} = u_t + \epsilon\eta^t(x,t,u,u_x,u_t) + O(\epsilon^2),$$

where using the definition of $\hat{u}_{\hat{x}}$, $\hat{u}_{\hat{t}}$ and the definition of characteristic of the group we obtain:

$$\eta^x = D_x Q + \xi u_{xx} + \tau u_{xt},$$
$$\eta^t = D_t Q + \xi u_{xt} + \tau u_{tt}.$$

Table 1 The forms of the functions A, B and C when Eq. (1) admits non-trivial Lie symmetry [6, 7]

N	$A(u)$	$B(u)$	$C(u)$
1.	1	$\lambda_1 u^m$	$\lambda_2 u^{2m+1}$
2.	1	$\lambda_1 u$	$\lambda_2 u, (\lambda_2 \neq 0)$
3.	1	$\lambda_1 u$	0
4.	1	$\lambda_1 \ln u$	$\lambda_2 u$
5.	1	$\lambda_1 \ln u$	$u(\lambda_2 + \lambda_3 \ln u), (\lambda_3 \neq 0)$
6.	1	$\lambda_1 \ln u$	$u(\lambda_2 + \lambda_3 \ln u + \frac{\lambda_1^2}{4} \ln^2 u)$
7.	1	$\lambda_1 \exp u$	$\lambda_2 \exp(2u)$
8.	u^k	$\lambda_1 u^m, (m \neq 0)$	$\lambda_2 u^{2m-k+1}$
9.	u^k	$\alpha_1 u^k$	$\lambda_2 u + \lambda_3 u^{1+k}$
10.	$u^k, (k \neq -\frac{4}{3})$	$\lambda_1 u^{\frac{k}{2}} + (3k+4)\alpha_1 u^k$	$\lambda_2 u^{1+k} + 2\alpha_1 \lambda_1 u^{1+\frac{k}{2}} + \lambda_4 u$
11.	$u^k, (k \neq -\frac{4}{3})$	$\alpha_1 (3k+4) u^k$	$\lambda_4 u^{k+1}$
12.	$u^k, (k \neq -\frac{4}{3})$	$\alpha_1 (3k+4) u^k$	$\lambda_4 u + \lambda_3 u^{1+k}$
13.	u^k	$\lambda_1 \ln u$	$\lambda_2 u^{1-k}$
14.	$\exp(u)$	$\alpha_1 \exp(u)$	$\lambda_2 + \lambda_3 \exp(u), (\lambda_2 \neq 0)$
15.	$\exp(u)$	$\lambda_1 \exp(\frac{u}{2}) + 3\alpha_1 \exp(u)$	$\lambda_2 + 2\alpha_1^2 \exp(u) + 2\alpha_1 \lambda_1 \exp(\frac{u}{2})$
16.	$\exp(u)$	$3\alpha_1 \exp(u)$	$2\alpha_1^2 \exp(u)$
17.	$\exp(u)$	$3\alpha_1 \exp(u)$	$\lambda_2 + 2\alpha_1^2 \exp(u)$
18.	$\exp(ku)$	$\lambda_1 u$	$\lambda_2 \exp(-ku)$
19.	$\exp(ku)$	$\lambda_1 \exp(mu)$	$\lambda_2 \exp((2m-k)u)$

These transformation are prolonged to higher-order derivatives recursively. Using the definition of point symmetries of an nth order PDE:

$$\Delta(x, t, u, u_x, u_t, \ldots) = 0.$$

We obtain the linearized symmetry condition

$$X\Delta = 0 \quad \text{when} \quad \Delta = 0$$

and the explicit expressions obtained when we differentiate recursively η_x and η_y allows us to obtain a linear system of PDEs for ϵ, τ and η, called *determining equations*.

Next we proceed to give a list of non-trivial Lie symmetries of equation (1) for certain values of A, B and C obtained in [5, 6].

In Tables 1 and 2, the following appointment are introduced: $\lambda_1 \neq 0$, λ_2, λ_3, $\alpha_1 \neq 0$, $n \neq 0$, $k \neq 0$, m are arbitrary constants, $X_t = \frac{\partial}{\partial t}$, $X_x = \frac{\partial}{\partial x}$, $X_u = \frac{\partial}{\partial u}$, $\lambda_4 = 2\alpha_1^2(2-k)$, and:

Table 2 Lie symmetries of equation (1)

N	Maximal algebra of invariance
1.	X_t, X_x, X_1
2.	X_t, X_x, X_2
3.	X_t, X_x, X_3, X_4, X_5
4.	X_t, X_x, X_6
5.	X_t, X_x, X_7
6.	X_t, X_x, X_8
7.	X_t, X_x, X_9
8.	X_t, X_x, X_{10}
9.	X_t, X_x, X_{11}
10.	X_t, X_x, X_{12}
11.	X_t, X_x, X_{13}, X_{12}
12.	X_t, X_x, X_{11}, X_{12}
13.	X_t, X_x, X_{14}
14.	X_t, X_x, X_{15}
15.	X_t, X_x, X_{16}
16.	X_t, X_x, X_{17}, X_{16}
17.	X_t, X_x, X_{15}, X_{16}
18.	X_t, X_x, X_{18}
19.	X_t, X_x, X_{19}

$$
\begin{aligned}
X_1 &= 2mtX_t + mxX_x - uX_u,\\
X_2 &= \exp(\lambda_2 t)\,(X_x - \frac{\lambda_2}{\lambda_1}X_u),\\
X_3 &= tX_x - \frac{1}{\lambda_1}X_u,\\
X_4 &= 2tX_t + xX_x - uX_u,\\
X_5 &= t^2X_t + txX_x - (\frac{x}{\lambda_1} + tu)X_u,\\
X_6 &= tX_x - \frac{1}{\lambda_1}uX_u,\\
X_7 &= \exp(\lambda_3 t)\,(X_x - \frac{\lambda_3}{\lambda_1}uX_u),\\
X_8 &= \exp[(\frac{\lambda_1^2}{4} + \lambda_3)t - \frac{\lambda_1}{2}x]\,uX_u,\\
X_9 &= 2tX_t + xX_x - X_u,\\
X_{10} &= (2m-k)tX_t + (m-k)xX_x - uX_u,\\
X_{11} &= \exp(-\lambda_3 kt)\,(X_t + \lambda_3 uX_u),\\
X_{12} &= \exp(-\alpha_1 kx)\,(X_x - 2\alpha_1 uX_u),\\
X_{13} &= ktX_t - uX_u,
\end{aligned}
\tag{6}
$$

$$
\begin{aligned}
X_{14} &= ktX_t + (kx - \lambda_1 t)X_x + uX_u, \\
X_{15} &= \exp(-\lambda_2 t)\,(X_t + \lambda_2 X_u), \\
X_{16} &= \exp(-\alpha_1 x)\,(X_x - 2\alpha_1 X_u), \\
X_{17} &= tX_t - X_u, \\
X_{18} &= ktX_t + (kx - \lambda_1 t)X_x + X_u \\
X_{19} &= (2m - k)tX_t + (m - k)xX_x - X_u,
\end{aligned}
\tag{7}
$$

are operators of Lie symmetry.

In [6] the authors found additional local substitutions and proved that there are only nine nonlinear equation of the form Eq. (1) with non-zero B (see Table 2), which admit nontrivial Lie symmetry and cannot be reduced one to another.

4 Nonlinear Self-Adjoint Equations

We use the following definition given in [13].

Definition 2. Equation (2) is said to be **nonlinear self-adjoint** if the equation obtained from the adjoint equation (4) by the substitution

$$v = h(x, t, u), \tag{8}$$

such that $h(x, t, u) \neq 0$,

$$F^*(x, u, u, u_{(1)}, u_{(1)}, \ldots, u_{(s)}, u_{(s)}) = 0$$

is identical with the original Eq. (2), i.e.

$$F^* \,|_{v=h} = \lambda F. \tag{9}$$

If $h = u$ or $h = h(u)$ and $h'(u) \neq 0$, Eq. (2) is said **self-adjoint** or **quasi-self-adjoint**, respectively, [13]. If $h = h(x, u)$ such that $h_u \neq 0$ and $h_x \neq 0$ is said **weak self-adjoint**, [8].

Given the nonlinear reaction-diffusion-convection equations (1) we apply Definition 2. Taking into account the expression (5) and using (8) and its derivatives we rewrite Eq. (9) as

$$
\begin{aligned}
&(\lambda + h_u)u_t + (A\,\lambda - h_u\,A)u_{xx} + (A_u\,\lambda - h_{uu}\,A)(u_x)^2 + (B\,\lambda + h_u\,B - 2\,h_{ux}\,A)u_x \\
&+C\,\lambda - h\,C_u + h_x\,B - h_{xx}\,A - h_t = 0.
\end{aligned}
\tag{10}
$$

Equation (10) should be satisfied identically in all variables $u_t, u_{xx}, u_x, \ldots$. Equating to zero the coefficients of the derivatives of u we obtain the system

$$\begin{aligned} -\lambda - h_u &= 0, \\ A\lambda - h_u A &= 0, \\ A_u\lambda - h_{uu} A &= 0, \\ B\lambda + h_u B - 2h_{ux} A &= 0, \\ C\lambda - hC_u + h_x B - h_{xx} A - h_t &= 0. \end{aligned} \tag{11}$$

Solving system (11), $\lambda = h_u = 0$. For $A, B \neq 0$ and $h = a(x,t)$ we obtain

$$a_{xx}A - a_x B + aC_u + a_t = 0. \tag{12}$$

We found the following solutions for (12) with $a \neq 0$:

- If $C = ku$, A and B are arbitrary functions and $a = a(t)$, we obtain

$$a = k_1 \exp(-kt), \tag{13}$$

  with k_1 arbitrary constant.
- If $A = B - C_u + k$ and $a = a(x,t)$, from Eq. (12)

$$(a_{xx} - a)C_u - (a_{xx} - a_x)B - (ka_{xx} + a_t) = 0. \tag{14}$$

  Then, from (14)

$$a = k_1 \exp(x - kt), \tag{15}$$

  with k_1 arbitrary constant.
- If $C_u = B = A$ and $a = a(x)$,

$$a = \exp(\frac{x}{2})\left(k_1 \sin\left(\frac{\sqrt{3}x}{2}\right) + k_2 \cos\left(\frac{\sqrt{3}x}{2}\right)\right). \tag{16}$$

Consequently, we deduce that

Proposition 1. *Equation (1) with $A, B \neq 0$ is nonlinear self-adjoint in cases given in Table 3. We observe that Eq. (1) is not self-adjoint nor quasi-self-adjoint nor weak self-adjoint.*

Table 3 Nonlinear self-adjoint equations (1) with $A, B \neq 0$

Case_i	$A(u)$	$B(u)$	$C(u)$	h
1.	Arbitrary	Arbitrary	k	Constant
2.	Arbitrary	Arbitrary	$ku, k \neq 0$	$k_1 \exp(-kt)$
3.	$B - C_u + k$	Arbitrary	Arbitrary	$k_1 \exp(x - kt)$
4.	C_u	C_u	Arbitrary	$\exp(\frac{x}{2})\left(k_1 \sin\left(\frac{\sqrt{3}x}{2}\right) + k_2 \cos\left(\frac{\sqrt{3}x}{2}\right)\right)$

5 Conservation Laws for a Subclass of Nonlinear Self-Adjoint Equations

Conservation laws appear in many of physical, chemical and mechanical processes, such laws enable us solve problems in which certain physical properties do not change in the course of time within an isolated physical system. The construction of conservation laws is one of the most important applications of symmetries to physical problems.

Given a PDE a conservation law is a relation of the form

$$\nabla \cdot \mathbf{C} = D_t(C^1) + D_x(C^2) = 0 \tag{17}$$

where $\mathbf{C} = (C^1, C^2)$ represents the conserved flux and density, respectively, and D_x, D_t denote the total derivative operators with respect to x and t respectively.

In order to construct conservation laws we use the following theorem on conservation laws proved in [11].

Theorem 1. *Any Lie point, Lie-Bäcklund or non-local symmetry*

$$X = \xi^i(x, u, u_{(1)}, \ldots)\frac{\partial}{\partial x^i} + \eta(x, u, u_{(1)}, \ldots)\frac{\partial}{\partial u} \tag{18}$$

of equation (2) *provides a conservation law* $D_i(C^i) = 0$ *for the simultaneous system* (2), (4). *The conserved vector is given by*

$$C^i = \xi^i \mathcal{L} + W\left[\frac{\partial \mathcal{L}}{\partial u_i} - D_k\left(\frac{\partial \mathcal{L}}{\partial u_{ij}}\right)\right] + D_k(W)\frac{\partial \mathcal{L}}{\partial u_{ij}}, \tag{19}$$

where W *is*

$$W = \eta - \xi^j u_j$$

and $\mathcal{L}$ *is defined in (3).*

Case 1. Equation (1) admits symmetries when $A = 1$, $B = \lambda_1 u$ and $k = 0$ (Tables 1–2.3). In this case Eq. (1) is the Burgers equation. The conservation laws for this equation have been obtained by various authors.

Case 2.1. Equation (1) admits symmetries when $A = 1$, $B = \lambda_1 u$ and $C = \lambda_2 u$, $\lambda_1, \lambda_2 \neq 0$ (Tables 1–2.2), and h is given in (13).

- From generator $X_2 = \exp(\lambda_2 t)\frac{\partial}{\partial x} - \exp(\lambda_2 t)\frac{\lambda_2}{\lambda_1}\frac{\partial}{\partial u}$, the normal form for this group is

$$W = -\exp(\lambda_2 t)u_x - \frac{\lambda_2 \exp(\lambda_2 t)}{\lambda_1}.$$

The vector components are

$$\begin{aligned} C^1 &= -\exp(\lambda_2 t)u_x v - \frac{\lambda_2 \exp(\lambda_2 t)v}{\lambda_1}, \\ C^2 &= -\exp(\lambda_2 t)u_x v_x - \frac{\lambda_2 \exp(\lambda_2 t)v_x}{\lambda_1} + \exp(\lambda_2 t)u_t v. \end{aligned} \tag{20}$$

Setting $v = h = k_1 \exp(\lambda_2 t)$ in (20)

$$\begin{aligned} C^1 &= -k_1 u_x - \frac{\lambda_2 k_1}{\lambda_1}, \\ C^2 &= k_1 u_t. \end{aligned} \tag{21}$$

Transfering the terms $D_x(\ldots)$ from C^1 to C^2 and simplifying, it is obtained the conserved vector $C = (C^1, C^2)$ with components given by

$$\begin{aligned} C^1 &= -\frac{\lambda_2 k_1}{\lambda_1}, \\ C^2 &= 0. \end{aligned}$$

We obtain trivial conservation laws.

Case 2.2. Equation (1) admits symmetries when $A = 1$, $B = \lambda_1 \ln(u)$ and $C = \lambda_2 u$ (Tables 1–2.4) and h is given in (13): X_t, X_x and $X_6 = t\partial_x - \frac{1}{\lambda_1}u\partial_u$.

- From generator $X_6 = t\partial_x - \frac{1}{\lambda_1}u\partial_u$, the normal form for this group is

$$W = -tu_x - \frac{u}{\lambda_1}$$

The vector components are

$$
\begin{aligned}
C^1 &= -tu_x v - \frac{uv}{\lambda_1},\\
C^2 &= -tu_x v_x - \frac{uv_x}{\lambda_1} + \frac{u_x v}{\lambda_1} + tu_t v + u\log uv - \lambda_2 tuv.
\end{aligned} \tag{22}
$$

Setting $v = h = k_1\exp(-\lambda_2 t)$

$$
\begin{aligned}
C^1 &= -k_1 t\exp(-\lambda_2 t)u_x - \frac{k_1\exp(-\lambda_2 t)u}{\lambda_1},\\
C^2 &= \frac{k_1\exp(-\lambda_2 t)u_x}{\lambda_1} + k_1 t\exp(-\lambda_2 t)u_t + k_1\exp(-\lambda_2 t)u\log u\\
&\quad -\lambda_2 k_1 t\exp(-\lambda_2 t)u.
\end{aligned} \tag{23}
$$

Transfering the terms $D_x(\ldots)$ from C^1 to C^2 and simplifying

$$
\begin{aligned}
C^1 &= -\frac{k_1\exp(-\lambda_2 t)u}{\lambda_1},\\
C^2 &= \tfrac{k_1\exp(-\lambda_2 t)u_x}{\lambda_1} + k_1\exp(-\lambda_2 t)u\log u - k_1\exp(-\lambda_2 t)u.
\end{aligned}
$$

- From generator $X_t + X_x$ proceeding as before we obtain the conservation law

$$D_t(C^1) + D_x(C^2) = 0,$$

where

$$
\begin{aligned}
C^1 &= -c_2\exp(-\lambda_2 t)u,\\
C^2 &= c_2\exp(-\lambda_2 t)u_x + \lambda_1 c_2\exp(-\lambda_2 t)u\log u - \lambda_1 c_2\exp(-\lambda_2 t)u.
\end{aligned}
$$

Case 3.1. Equation (1) admits symmetries when $A = 1$, $B = \lambda_1\ln(u)$ and $C = \lambda_1 u\ln(u)$ (Tables 1–2.5): X_t, X_x and X_7. In this case $h = k_1 e^{x-(\lambda_1+1)t}$.

- From generator $X = X_x + X_t = c_1\partial_x + c_2\partial_t$ we obtain the conservation law

$$D_t(C^1) + D_x(C^2) = 0,$$

where

$$
\begin{aligned}
C^1 &= -\left(\lambda_1 c_2 + c_2 - c_1\right)u\exp(x - \lambda_1 t - t),\\
C^2 &= \left(\lambda_1 c_2 + c_2 - c_1\right)\left(u_x + \lambda_1 u\log u - \lambda_1 u - u\right)\exp(x - \lambda_1 t - t).
\end{aligned}
$$

- From generator X_7 we obtain trivial conservation law.

Case 3.2. Equation (1) admits symmetries when $A = \exp(mu)$, $B = a_2 \exp(mu)$ and $C = \frac{a_1-1}{m} \exp(mu)$ (Tables 1–2.19): X_t, X_x and $X_{19} = mt\partial_t - \partial_u$ with $k = m$. In this case $h = k_1 \exp(x)$.

- From generator $X = X_x + X_t = c_1\partial_x + c_2\partial_t$ we obtain the conservation law

$$D_t(C^1) + D_x(C^2) = 0,$$

where

$$C^1 = u\exp(x),$$

$$C^2 = -\tfrac{1}{m}\,(mu_x + \lambda_1 - 1)\exp(x + mu).$$

- From generator $X = mt\partial_t - \partial_u$ we obtain trivial conservation law

$$D_t(C^1) + D_x(C^2) = 0,$$

where

$$C^1 = -\exp(x),$$

$$C^2 = 0.$$

Case 4. In order to apply Theorem 1 to the nonlinear self-adjoint equation (1) with $C_u = B = A$ when $h(x) = \exp(\frac{x}{2})\left(k_1 \sin\left(\frac{\sqrt{3}x}{2}\right) + k_2 \cos\left(\frac{\sqrt{3}x}{2}\right)\right)$, i.e.

$$u_t = C + (u_x)^2 C_{uu} + u_{xx}C_u + u_x C_u \tag{24}$$

we consider the generators obtained in Sect. 3 which are admitted by Eq. (24).

Case 4.1. For $A = B = 1$ and $C = u$ (Tables 1–2.1), with $m = 0$ and $\lambda_1 = \lambda_2 = 1$. In this case Eq. (1) corresponds to the linear equation, which is locally-equivalent to the classical heat equation.

Case 4.2. For $A = B = u^m$ and $C = \frac{1}{m+1}u^{m+1}$ (Tables 1–2.8) we obtain the following conservation laws:

- From generator $X_t + X_x = c_2\dfrac{\partial}{\partial t} + c_1\dfrac{\partial}{\partial x}$ the normal form for this group is

$$W = -c_1u_x - c_2u_t.$$

The vector components are

$$C^1 = -c_2 u_{xx} v - c_2 u_x v - c_1 u_x v - c_2 u v,$$
$$C^2 = -c_1 u_x v_x - c_2 u_t v_x + c_2 u_{tx} v + c_2 u_t v + c_1 u_t v - c_1 u v. \quad (25)$$

Setting $v = h = \exp(\frac{x}{2}) \left(k_1 \sin\left(\frac{\sqrt{3}x}{2}\right) + k_2 \cos\left(\frac{\sqrt{3}x}{2}\right)\right)$ in (25)

$$C^1 = -e^{\frac{x}{2}} \left(k_1 \sin\left(\frac{\sqrt{3}x}{2}\right) + c_2 \cos\left(\frac{\sqrt{3}x}{2}\right)\right) (c_2 u_{xx} + (c_2 + c_1) u_x + c_2 u),$$

$$C^2 = \frac{1}{2} H_1 e^{\frac{x}{2}} \sin\left(\frac{\sqrt{3}x}{2}\right) + \frac{1}{2} H_2 e^{\frac{x}{2}} \cos\left(\frac{\sqrt{3}x}{2}\right), \quad (26)$$

where

$$H_1 = c_1 \left(\sqrt{3}k_2 - k_1\right) u_x + 2k_1 c_2 u_{tx} + \left(\sqrt{3}c_2 k_2 + (c_2 + 2c_1)\right) k_1 u_t - 2c_1 k_1 u,$$

and

$$H_2 = \left(-c_1 k_2 - \sqrt{3}c_1 k_1\right) u_x + 2c_2 k_2 u_{tx} + \left((c_2 + 2c_1)k_2 - \sqrt{3}k_1 c_2\right) u_t - 2c_1 k_2 u.$$

We simplify the conserved vector by transferring the terms of the form $D_x(\ldots)$ from C^1 to C^2 and obtain

$$D_t(C^1) + D_x(C^2) = 0$$

where

$$C^1 = \left(\left(k_1 - \sqrt{3}k_2\right) \sin\left(\frac{\sqrt{3}x}{2}\right) + \left(k_2 + \sqrt{3}k_1\right) \cos\left(\frac{\sqrt{3}x}{2}\right)\right) \frac{c_1}{2} u \exp(\tfrac{x}{2}),$$

$$C^2 = \left(\left(\sqrt{3}k_2 - k_1\right) u_x - 2k_1 u\right) \frac{1}{2} \exp(\tfrac{x}{2}) \sin\left(\frac{\sqrt{3}x}{2}\right)$$
$$+ \left(\left(-k_2 - \sqrt{3}k_1\right) u_x - 2k_2 u\right) \frac{c_1}{2} \exp(\tfrac{x}{2}) \cos\left(\frac{\sqrt{3}x}{2}\right).$$

- From generator $X_{10} = mt\partial_t - u\partial_u$, with $k = m$,

the normal form is

$$W = -mtu_t - u.$$

The vector components are

$$\begin{aligned} C^1 &= -mtu^m u_{xx} v - m^2 t u^{m-1} (u_x)^2 v - mtu^m u_x v - \frac{mtu^{m+1} v}{m+1} - uv, \\ C^2 &= -mtu^m u_t v_x - u^{m+1} v_x + m^2 t u^{m-1} u_t u_x v + mu^m u_x v + u^m u_x v \\ &\quad + mtu^m u_{tx} v + mtu^m u_t v + u^{m+1} v. \end{aligned} \tag{27}$$

Setting $v = h = \exp(\frac{x}{2})\left(k_1 \sin\left(\frac{\sqrt{3}x}{2}\right) + k_2 \cos\left(\frac{\sqrt{3}x}{2}\right)\right)$ in (27) and proceeding as before we obtain the conservation law

$$D_t(C^1) + D_x(C^2) = 0,$$

where

$$\begin{aligned} C^1 &= -u \exp(\tfrac{x}{2}) \left(k_1 \sin\left(\tfrac{\sqrt{3}x}{2}\right) + k_2 \cos\left(\tfrac{\sqrt{3}x}{2}\right)\right), \\ C^2 &= \tfrac{1}{2m+2} \left((2k_1 m + 2k_1) u^m u_x + \left(\sqrt{3}k_2 + k_1\right) u^{m+1}\right) \exp(\tfrac{x}{2}) \sin\left(\tfrac{\sqrt{3}x}{2}\right) \\ &\quad + \tfrac{1}{2m+2} \left((2k_2 m + 2k_2) u^m u_x + \left(k_2 - \sqrt{3}k_1\right) u^{m+1}\right) \exp(\tfrac{x}{2}) \cos\left(\tfrac{\sqrt{3}x}{2}\right). \end{aligned}$$

We point out that in physical systems, many conservation laws that arise can usually be identified with a physical quantity, like energy or linear momentum, being conserved. Finally, we remark that the search for conservation laws is also useful to determine potential symmetries. These symmetries will allow us to find new solutions for the class of nonlinear reaction-diffusion-convection equations (1).

6 Conclusions

The concepts of self-adjoint, quasi-self-adjoint and nonlinear self-adjoint equations were introduced by NH Ibragimov in [12, 13]. In [8] one of the present authors has generalized the concept of self-adjoint and quasi-self-adjoint equations by introducing the definition of weak self-adjoint equations. In this paper we prove that Eq. (1) is not self-adjoint neither quasi self-adjoint nor weak-self-adjoint. However, Eq. (1) is nonlinear self-adjoint. By using the property of nonlinear self-adjointness of a quasi-linear parabolic equation and the general theorem of conservation laws [10], we have constructed conservation laws for this equation associated with its Lie symmetries obtained in [4–6].

Acknowledgements The authors acknowledge the financial support from Junta de Andalucía group FQM–201 and from project MTM2009-11875. We warmly thank the referees for reading carefully the manuscript and for giving their suggestions.

References

1. M.S. Bruzón, M.L. Gandarias, Symmetry reductions and traveling wave solutions for the Krichever–Novikov equation. Math. Meth. Appl. Sci. (2012). doi:10.1002/mma.1578
2. M.S. Bruzón, M.L. Gandarias, Conservation laws for a family of Benjamin-Bona-Mahony-Burgers equations, in *Proceedings 4th IEEE International Conference on Nonlinear Science and Complexity*, Budapest, 2012
3. R. Cherniha, J.R. King, Lie symmetries and conservation laws of nonlinear multidimensional reaction-diffusion systems with variable diffusivities. IMA J. Appl. Math. **71**, 391–408 (2006)
4. R. Cherniha, M. Serov, Lie and Non-Lie symmetries of nonlinear diffusion equations with convection term. Symmetry Nonlinear Math. Phys. **2**, 444–449 (1997)
5. R. Cherniha, M. Serov, Symmetries, ansätze and exact solutions of nonlinear second-order evolution equations with convection term. Eur. J. Appl. Math. **9**, 527–542 (1998)
6. R. Cherniha, M. Serov, Symmetries, ansätze and exact solutions of nonlinear second-order evolution equations with convection term, II. Eur. J. Appl. Math. **17**, 597–605 (2006)
7. V.A. Dorodnitsyn, S.R. Svirshchevskii, in *Conservation Laws: Lie Group Analysis of Differential Equations*, vol. 1, ed. by N.H. Ibragimov (CRC Press, Boca Raton, 1994), pp. 137–139
8. M.L. Gandarias, Weak self-adjoint differential equations. J. Phys. A Math. Theor. **44**, 262001–262007 (2011)
9. N.H. Ibragimov, *Transformation Groups Applied to Mathematical Physics* (Reidel, Dordrecht, 1985)
10. N.H. Ibragimov, The answer to the question put to me by L.V. Ovsyannikov 33 years ago. Arch. ALGA **3**, 53–80 (2006)
11. N.H. Ibragimov, A new conservation theorem. J. Math. Anal. Appl. **333**, 311–328 (2007)
12. N.H. Ibragimov, Quasi self-adjoint differential equations. Arch. ALGA **4**, 55–60 (2007)
13. N.H. Ibragimov, Nonlinear self-adjointness and conservation laws. J. Phys. A Math. Theor. **44**, 432002–432010 (2011)
14. P. Olver, *Applications of Lie Groups to Differential Equations* (Springer, New York, 1993)
15. L.V. Ovsyannikov, *Group Analysis of Differential Equations* (Academic, New York, 1982)
16. A.M. Vinogradov, Local symmetries and conservation laws. Acta Appl. Math. **2**, 21–78 (1984)

Effects Induced in Complex Biological Systems by High Density Green Photons

Sorin Comorosan, Silviu Polosan, Marian Apostol, Irinel Popescu, Ileana Farcasanu, Liliana Paslaru, and Elena Ionescu

Abstract It is known that light interaction with matter may generate an optical force, which may produce modifications at the physical and chemical level. The basic technique of the field uses strongly focused laser beams that trap small objects and manipulate local structures. In our work we use irradiation of complex biological molecules with high density green photons, which may induce electric dipoles by polarization effects. The resulting dipolar interaction may lead to organized structures like molecular aggregates and microparticles. We present experimental evidence of such an optical manipulation on long alkanes chains and two specific enzymes. A preliminary physical model is suggested for acounting of these specific interaction forces.

S. Comorosan (✉)
Interdisciplinary Research Group, Romanian Academy, Bucharest, Romania
e-mail: scomo_acadro@yahoo.com

S. Polosan
National Institute of Material Physics, Magurele, Romania
e-mail: spol68@yahoo.com

M. Apostol
Department of Theoretical Physics, Institute of Nuclear Physics, Magurele, Romania,
e-mail: apoma@theory.nipne.ro

I. Popescu · E. Ionescu
Department of Basic Research, Fundeni Institute, Bucharest, Romania
e-mail: irinel.popescu@icfundeni.ro; s_valentin2003@yahoo.com

I. Farcasanu
Department of Chemistry, University of Bucharest, Bucharest, Romania
e-mail: ileanafarcasanu@yahoo.com

L. Paslaru
Department of Biochemistry, University of Medicine, Bucharest, Romania
e-mail: lilianaliviapaslaru@yahoo.fr

R. Carretero-González et al. (eds.), *Localized Excitations in Nonlinear Complex Systems*, Nonlinear Systems and Complexity 7, DOI 10.1007/978-3-319-02057-0_22,

1 Introduction

This work deals with a specific phenomenology, occurring in a highlighted domain of physical reality, termed *mesoscopia*. Mesoscopic systems encompass length scales from nanometers to hundreds of microns, force scales from femto-to-nanonewtons and time scales from microseconds up. This is the realm of basic biological processes, cellular respiration, signaling and metabolism. In physics this is the puzzling province of the interface between classical and quantum mechanics, i.e. the macroscopic quantum behavior, rendered famous by the old Wigner [1] *impossible* conjecture.

Recently, great progress in optical manipulation [2] offers new avenues for basic and applied research in this field. The basic technique, known as optical tweezer, was introduced by Ashkin [3]. An optical tweezer represents a strongly focused beam of light that may trap small objects, smaller or much larger than the light wavelength. The mechanism is rather straightforward: small objects develop electric dipole moments generated by the light electric field, which means to draw up intensity gradients. These optical forces may then compete with the radiation pressure generated by the absorbed momentum from the beam's photons. Stable trapping, as well as the manipulation of the local structures, depend on a balance with the axial gradient force.

The scientific relevance of tweezers optical manipulations for basic research in the biological and physical sciences is far reaching. For biology, fluctuation analysis of single enzyme-kinetics [4] or single-molecule studies of DNA mechanics [5] are a few among many significant contributions. For the physical sciences, the possibility to organize minute matter structures achieved the first direct measurements of macromolecular interactions in solution [6]. The study of magnetic flux lines that invade superconductors [7] represents a technical feat.

Our technology implies the use of collimated high-density-green-photons (HDGP) beams, instead of the tweezers focused laser beams. Irradiations of complex biological structures with HDGP may induce electric dipoles by polarization effects. The induced dipolor interaction may lead to organized material structures, like molecular aggregates and micro-particles, indicating an optically organized matter.

The optical manipulation with collimated HDGP-beams is particularly suited for complex biological structures (macromolecules and cells). It covers a larger area than focused tweezer and is capable of organizing mesoscopic matter into a 3-D new molecular architecture.

The optical manipulation of matter raised a series of very interesting problems at the very core of physics fundamental principles, particularly since the statistical many-body theory is not applicable to mesoscopia. A reported work on hydrodynamic fluctuations [8] may be interpreted as transient violation of thermodynamics second law.

Recently a new interest is manifested in the problem of macroscopic quantum behavior. The discussion resumed in the 1970s with the remarkable paper of Fröhlich [9] on long-range coherence in enxyme kinetics. An elegant mathematical proof reported by Rosen [10] demonstrated that a chemical reaction of the (substrate $\rightarrow$ product) type may be interpreted as a microphysical system capable to *detect* perturbations induced in the system. Using these conjectures Comorosan group designed the first experimental set-up in which enzyme substrates were optical manipulated by green photons beams ($\lambda = 515 - 525$ nm) resulting in clear-cut modifications of the enzyme kinetics [11, 12]. The relevance of this work for the problem of macroscopic quantum behavior appeared when large biological macromolecules (like bovine serum albumin) or cellular cultures were subjected to irradiations with HDGP-beams. The optical manipulation of these extended biological structures resulted in generation of a new local macromolecular architecture, metastable, that clearly protected them toward a subsequent aggression (like UV-irradiations) [13, 14]. In our context this is understood as the experimental circumstances when large number of molecules can collectively cooperate as a quantum state which (by its metastability) remains disentangled from the environment.

2 Experimental Evidence

We designed two experimental set-ups to detect possible optical manipulation of biological material under HDGP-beams.

- Alkanes mixture
  Alkanes are carbon chains, which represent the backbone of biological macromolecules. Under thermal degradation they generate free radicals that may be rigorously measured. Accordingly they are suited for the study of oxido-reduction reactions, a significant class of cellular process. For this study we used long C-chains, > 20–C atoms, previously selected by liquid chromatography. We show that under the HDGP-beams manipulation, the newly structured alkane chains generate, under thermal degradation, a significant lower concentration of free radicals.
- Enzyme structures
  In this study we used two biological significant macromolecules: superoxide dismutase (SOD) enzyme, and catalase (CAT) enzyme. Both enzymes, known as *free radicals scavengers*, are implied in the basic processes of cellular respiration. We revealed that under HDGP-beams manipulations the new *optical matter*, induced in the macromolecular structures of these enzymes, protect their function against a subsequent ultraviolet (UV) aggregation.

2.1 *Mass Spectroscopy*

This is a powerful analytical technique used to identify unknown compounds, to quantify known compounds and/or to elucidate the structure and chemical properties of molecules.

In order to analyze complex mixtures, a separation technique like liquid chromatography is coupled with the mass spectroscopy. In this way, the chromatograph is connected directly to the mass spectrometer and the mass spectra are acquired while the compounds of the mixture are eluted.

2.1.1 The Experimental Set-Up

A defined volume of mineral oil for each type of measurement is placed in two dishes (for control and probe) and kept for 30 min into an electrical oven at 165 °C. The probe is exposed to green light irradiation through a small hole in the ceiling of the oven. After 30 min the dishes were taken out for the respective determinations. The geometry and the proceedings of the experiments are rigorously arranged to be similar for the control and the probe in all the experimental set-ups. For GL (green light) irradiation powerful light-emitted diodes (LEDs) of 1,000 lumens, mounted on ventilated copper radiators, were used. In these setups, monochromatic light of $\lambda = 514$ nm was obtained, with intensity up to 4×10^5 Lx, as measured by a digital Luxmeter LX-1102, Lutron. For each experiment the GL-irradiation was arranged to deliver a standard flux of 10^4 Lx.

2.1.2 Acquisition of Mass Spectra

A Shimadzu's LCMS 2010EV mass spectrometer, with electrospray ionization technique, one step quadrupole mass analyzer, data acquisition in the range of m/z 200–1,000 and the scan speed of 1,000 amu/s was used. The mobile phase is acetonitrile 99.98 % purity, with a debit of 0.4 ml/min. The LCMS was operated in the positive acquisition mode with the nebulizing gas flow set at 1.5 l/min. The sample was obtained by mixing 10 μl mineral oil with 1 ml acetonitrile 99.98 % purity, further ultrasonicated for 20 min into an ultrasonic bath. A sample volume of 20 μl was injected into the LCMS.

2.1.3 The Mass Spectra

The mass spectra of macromolecules allow the analysis of high molecular weight, using mass-to-charge ratio (m/z). These spectra correspond to a statistical distribution of consecutive peaks, characteristic of multiple charged molecular ions, obtained through protonation or deprotonation. An ion separated from the central

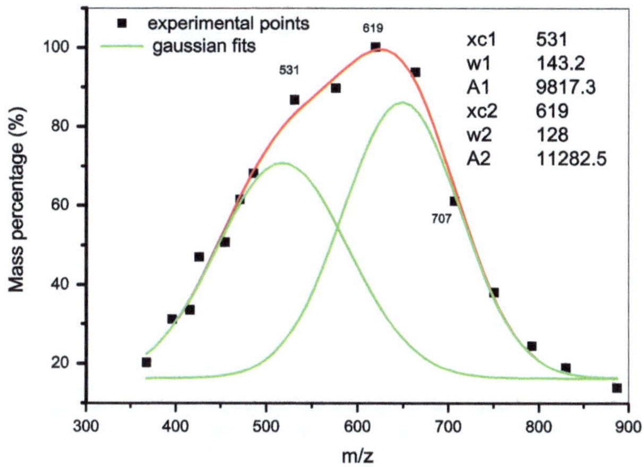

Fig. 1 Mass spectroscopy of multiply charged ions: mass-to-charge-ratio (*dot points*) of thermal degraded mineral oil (165 °C) (normalized curves of a Gaussian fit)

one by $(j - 1)$ peaks, in increasing order of mass-to-charge ratio, has a ratio of m_2 and a charge number $(z_1 - j)$ so that:

$$m_2(z_1 - j) = M(z_1 - j)m_p$$

where M = molecular mass, m_p = proton mass. These parameters may be used to uncover possible changes, in the distribution of different species, within the mineral oil mixture. In our experimental set-up the control mass-to-charge distribution displays two Gaussian fits parameters (xc = center; w = half width; A = area) with a first statistical distribution centered at m/z = 525. The thermal degraded mineral oil exhibits two statistical distributions, each one with characteristic peaks. The first distribution appears centered at m/z = 531 and the second one at m/z = 619 (Fig. 1). The difference between the central peak (619), and the $j = 3$, situated at 707, is a multiple of carbonyl mass (88). The thermal degraded mineral oil, under HDGP irradiation, (Fig. 2) exhibits a different behaviour. The mass spectroscopy pattern reveals a single statistical distribution, centered at m/z = 525, close to the first control value. The difference between the control peak at 525, and the one with $j = 2$, at 569, is again a carbonyl mass value (44). The HDGP irradiation of the probe induces *a single species* of alkanes, which undergoes the oxidative process.

This is a significant result. It reveals that the thermal degraded mixture of alkane chains undergoes, upon HDGP-manipulation, a clear rearrangement of its components. The newly formed blend displays an unexpected resistance to carbonyl generation. Hence the antioxidant effect is evident. We may call this minute fabric created under GL-irradiation, within the alkane mixtures, *optical matter*. In our context the reported antioxidant effects may put the class of chemical antioxidant reactions in a new perspective.

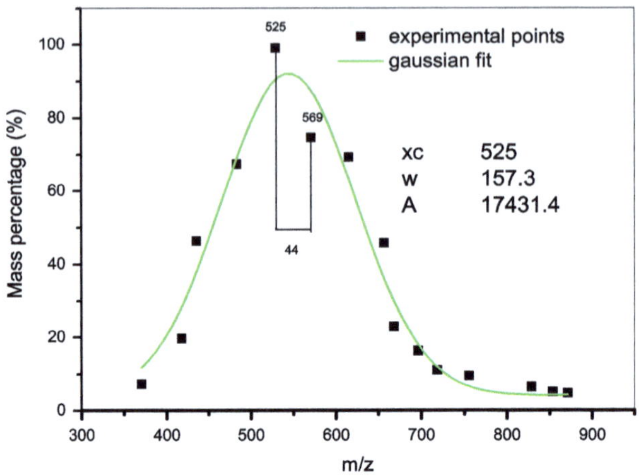

Fig. 2 Mass spectroscopy of multiply charged ions: mass-to-charge-ratio (*dot points*) of thermal degraded mineral oil (165 °C) under HDGP (normalized curves of a Gaussian fit)

2.2 *Enzyme Reactions*

2.2.1 Cells and Culture Conditions

In this study we used a *Saccharomyces cerevisiae* strain BY4741 (MATa; his3Δ1; leu2Δ0; met15Δ0; ura3Δ0) Strains were grown in standard YPD or SD (synthetic dextrose) medium supplemented with amino acids. Overnight pre-cultures were inoculated in fresh media at a density of 10^6 cells/ml, then grown with shaking (200 rpm, 28 °C) to a mid-logarithmic phase (5×10^6 cells/ml) in SD medium. Cells were washed with 0.1 M MES/Tris (2-N-morpholino ethanesulfonic acid/2-amino-2-hydroxymethyl-propane-1,3-diol), pH 6.5 and resuspended in SD medium to 2×10^8 cells/ml.

2.2.2 Enzyme Assay

For enzymatic assays, cells were washed with 0.1 M MES/Tris, pH 6.5, and concentrated to 1×10^9 cells/ml. Aliquots (100 μl) were prepared in 1.5 ml tubes, and cell lysis was achieved with acid-washed glass beads (Sigma) by vortex mixing (3 min, 4 °C). The cellular debris was removed by two centrifugation steps (10,000 rpm, 2 min, 4 °C). The supernatant was collected after each spin, and the resultant crude lysate assayed. SOD activity was determined using the SOD Assay Kit-WST (Sigma) with Dojindo's highly water-soluble tetrazolium salt WST1 (2-(4-iodophenyl)-3-(4-nitrophenyl)-5-(2.4- disulfophenyl)-2H-tetrazolium, monosodium salt). WST1 produces a water-soluble formazan dye upon reduction with a superoxide anion; the rate of reduction with O_2 is related linearly to

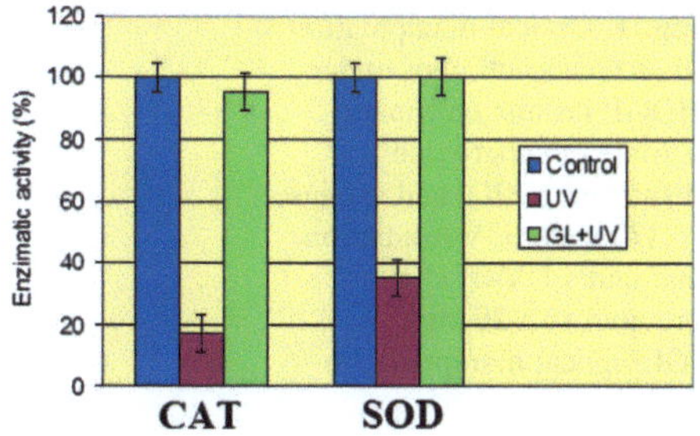

Fig. 3 Biochemical cascade linking SOD and CAT scavenge of superoxide O_2^-

the xanthine oxidase (XO) activity and is inhibited by SOD. The IC_{50} (50 % activity inhibition) was determined spectrophotometrically by the decrease in colour development at 450 nm. Catalase activity was determined using the Sigma Catalase Assay Kit. The hydrogen peroxide substrate remaining after the action of catalase is measured. A substituted phenol (3.5-dichloro-2-hydroxybenzene-sulfonic acid) couples oxidatively to 4- antipyrine in the presence of hydrogen peroxide and horseradish peroxidase (HPR) and is determined spectrophotometrically at 520 nm. One unit of catalase splits 1.0 μmol of hydrogen peroxide to oxygen and water per minute at pH 7.0, 25 °C at a substrate concentration of 50 mM hydrogen peroxide. The percent activity of the enzymes was calculated with respect to the control samples. In the optical manipulations of enzyme structures, the crude enzymic extract was GL-irradiated for 20 min previous to the denaturation with UV. GL-irradiation as described in the preceding section.

2.2.3 Enzymes Mechanism

Superoxide dismutase (SOD) is mithocondrially located, MnSod (encoded by SOD2 gene) and cytoplasmically located, Cu/ZnSod (encoded by SOD1 gene). Catalase (CAT) is peroxisome located, A (encoded by CAT1 gene) and cytosolic located, T (encoded by CTT1 gene). SOD is involved in a dismutation reaction in which a single reactant is converted into two different products. The oxidized SOD_{ox} reacts with the O_2-superoxide to form O_2 and the reduced enzyme SOD_{red}. This in turn reacts with a second superoxide and two protons to generate hydrogen peroxide and regenerate SOD_{ox}. A coupling with CAT achieves a second dismutation, converting hydrogen peroxide into H_2O and O_2 (Fig. 3). Under UV- irradiation both enzymes are drastically inhibited (Fig. 4). The UV inactivation on SOD and CAT enzymes may be interpreted through a physical mechanism, by modification of protein quaternary structure. A quaternary structure is a spatial arrangement of the polypeptide chains, called *subunits*, in a woven material, globularly packed and stabilized by cross-links. UV-irradiation generates numerous free radicals, which disrupt the cross-links, breaking the non-covalent bonds, through a powerful oxidation. The result is a fully reduced, randomly coiled (unfolded) polypeptide chain, devoid of enzymic activity. This type of transition, from a native form of an enzyme into a

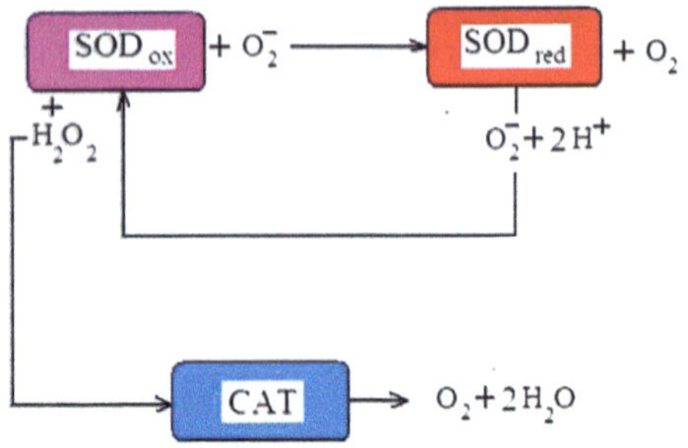

Fig. 4 Optical manipulation of enzymes activities under HDGP-beams: enzymatic activity of superoxide dismutase (*SOD*) and catalase (*CAT*) under UV-irradiation and under UV-irradiation previous to a 20 min (GL-optical manipulation)

denatured one, is well known in biochemistry. As seen from Fig. 4 under a previous optical manipulation of enzymes structure with a 20 min. HDGP-irradiation, this denaturation is almost completely inhibited. This is a significant experimental effect, suggesting generation of new localized biological optical matter.

3 Discusion and Conclusion: A Physical Model

Irradiation of complex biological structures with HDGP may induce locally electric dipoles by polarization effects. Within the enzymes macromolecules particularly the light atoms (H) and the π-bonds may be affected. The induced dipoles interact by optical forces (electromagnetic field) with each other (vey likely in a head-tail pattern), leading to chain/string structures. This particular arrangement of patterns may create string-string interactions which may in turn join together the dipole chains. As a result, the dispersed enzyme molecules from the cellular medium aggregate into clusters of macromolecules (Fig. 5). Dipolar interaction results in the lowering energy of the biological system as compared with individual molecules, generating more stable and less reactive matter. In the biological optical matter, a series of known light interaction effects, like UV denaturation, disruption of internal chemical bonds, generation of free radical, are partially inhibited.

The green light with its wavelength $\simeq$514 nm is not able to produce photo-ionization in typical biological matter (like other spectral components, except ultraviolet light). However, it is likely that biological matter responds, especially to the green light, which is poorly absorbed, by induced electric dipoles. As it is well known, these dipoles interact with the external electromagnetic field and, by their own polarization field, they interact with each other. This interaction is known as the optical force, and it may lead to organized material structures of polarizable molecules, molecular aggregates, micro-particles, etc. [15]. The optical force is extensively used in manipulating atoms by lasers, atom trapping, tweezers, etc, with relevant practical applications [16, 17].

An external field **E** interacting with matter may induce two *point* dipoles separated by distace **R**; the field act upon these dipoles (with an external force), but, at the same time, the dipoles interact with each other. Leaving aside the external force we can estimate the mean interaction force (Lorentz force) as

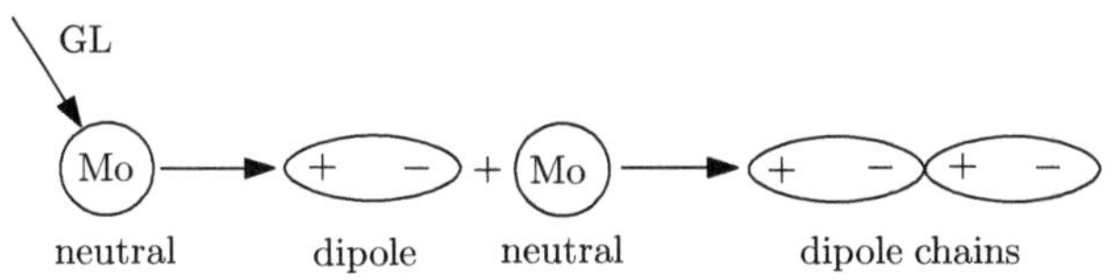

Fig. 5 A suggested representation of dipole chains/strings interactions. *GL* green light, *Mo* neutral molecules

$$\mathbf{F} = \frac{v_1 v_2}{2\pi T} \int d\omega \alpha_1^*(\omega)\alpha_2(\omega) E_i^*(\omega; \mathbf{R}) E_j(\omega; \mathbf{R} = 0) \left(\frac{\omega^2}{c^2}\delta_{ij} + \frac{\partial^2}{\partial R_i \partial R_j} \right) \frac{\partial}{\partial \mathbf{R}} \frac{e^{i\frac{\omega}{c}R}}{R} \tag{1}$$

where $v_{1,2}$ are the volumes of the two dipoles, $\alpha_{1,2}$ are their polarizabilities, T is a long averaging time and i, j denote the cartesian components of the field. For identical particles and a plane wave ($v_1 = v_2 = v$, $\alpha_1 = \alpha_2 = \alpha$) the force becomes

$$\mathbf{F} = \frac{1}{4} v^2 \alpha^2 e^{i\mathbf{k}\cdot\mathbf{R}} \left\{ \lambda^2 E^2 + (\mathbf{E}\cdot\nabla)^2 \right\} \frac{\partial}{\partial \mathbf{R}} \frac{e^{-i\lambda R}}{R} + c.c. \tag{2}$$

where $\mathbf{k}$ is the wavevector and $\lambda = \omega/c = k$. This force exhibits oscilaltions, delineating regions of repulsion and regions of attraction. We suggest that under the action of such a force the matter organizes itself into quasi-bound structures with specific properties. A preliminary investigation shows that the induced dipoles bind in chains oriented along the direction of the electric field. In the bound state the macromolecular chains (filaments) have a different electronic structure, which makes them more resistent to the oxidant action of the ultraviolet radiation. The matter so organized by optical means may exhibit a different electronic and optical structure and activity, as observed experimentally. This significant result may open new directions for the basic and practical research at the border between biology and physics. All these properties are subjected to further investigation.

This work was presented at the International Conference Localized Excitations in Nonlinear Complex Systems (LENCOS'12), Sevilla, Spain.

Acknowledgements The authors thank to Anamaria Boros, professor of Mathematics at the High School *Radu Petrescu* (Prundu Bargaului), for her contribution to edit this work.

References

1. E. Wigner, *The Probability of a Self-Reproducing Unit. Symmetries and Reflections* (Indiana University Press, Bloomington, 1967)
2. D.G. Grien, Nature **424**, 810–816 (2003)
3. A. Ashkin, J.M. Dziedzic, J.E. Bjorkholm, S. Chu, Opt. Lett. **11**, 288–290 (1986)

4. K. Svoboda, P.P. Mitra, S.M. Block, Proc. Nat. Acad. Sci. **91**, 11782–11786 (1994)
5. C. Bustamante, S.B. Smith, J. Liphardt, D. Smith, Curr. Opin. Struct. Biol. **10**, 279–285 (2000)
6. J.C. Crocker, D.G. Grier, Phys. Rev. Lett. **73**, 352–355 (1994)
7. K. Mangold, P. Leidered, C. Bechinger, Phys. Rev. Lett. **90**, 158302 (2003)
8. G.M. Wang, E.M. Sevick, E. Mittag, D.J. Searles, D.J. Evans, Phys. Rev. Lett. **89**, 050601 (2002)
9. H. Fröhlich, Nature **228**, 228–234 (1970)
10. R. Rosen, Bull. Math. Biophys. **25**, 183–187 (1963)
11. S. Comorosan, Nature **227**, 64–65 (1970)
12. S. Comorosan, S. Vieru, D. Sandru, Int. J. Radiat. Biol. **17**, 105–115 (1970)
13. S. Comorosan, S. Polosan, I. Popescu, E. Ionescu, R. Mitrica, L. Cristache, A.E. State, Eur. Biophys. J. **39**, 1483 (2010)
14. S. Comorosan, S. Polosan, S. Jipa, I. Popescu, G. Marton, E. Ionescu, L. Cristache, D. Badila, R. Mitrica, J. Photochem. Photobiol. B **102**, 39–44 (2011)
15. M.M. Burns, J.-M. Fournier, J.E. Golovchenko, Science **249**, 749–754 (1990)
16. M. Nieto-Vesperinas, J.J. Sáenz, R. Gómez-Medina, L. Chantada, Opt. Express **18** 11428–11443 (2010)
17. J. Chen, J. Ng, Z. Lin, C.T. Chan, Nat. Photon. **5**, 531–534 (2011)

Index

R. Carretero-González et al. (eds.), *Localized Excitations in Nonlinear Complex Systems*, Nonlinear Systems and Complexity 7, DOI 10.1007/978-3-319-02057-0,

W

MIX
Papier aus verantwortungsvollen Quellen
Paper from responsible sources
FSC® C105338

If you have any concerns about our products,
you can contact us on
ProductSafety@springernature.com

In case Publisher is established outside the EU,
the EU authorized representative is:
Springer Nature Customer Service Center GmbH
Europaplatz 3, 69115 Heidelberg, Germany

Printed by Libri Plureos GmbH
in Hamburg, Germany